S0-BSG-084

Matrix Analysis for Statistics

WILEY SERIES IN PROBABILITY AND STATISTICS

Established by WALTER A. SHEWHART and SAMUEL S. WILKS

A complete list of the titles in this series appears at the end of this volume.

Matrix Analysis for Statistics

JAMES R. SCHOTT

A Wiley-Interscience Publication
JOHN WILEY & SONS, INC.
New York • Chichester • Brisbane • Toronto • Singapore • Weinheim

This text is printed on acid-free paper.

Library Of Congress Cataloging In Publication Data:

Schott, James R., 1955-
Matrix Analysis For Statistics / James R. Schott.
p. cm. – (Wiley Series In Probability And Statistics. Applied Probability And Statistics)
"A Wiley-Interscience Publication."
Includes bibliographical references and index.
ISBN 0-471-15409-1 (cloth : alk. paper)
1. Matrices. 2. Mathematical Statistics. I. Title. II. Series.
QA188.S24 1996
512.9'434–dc20 96-12133

Printed in the United States of America

10 9 8 7 6 5 4 3 2 1

To
Susan, Adam, and Sarah

Contents

Preface

As the field of statistics has developed over the years, the role of matrix methods has evolved from a tool through which statistical problems could be more conveniently expressed to an absolutely essential part in the development, understanding, and use of the more complicated statistical analyses that have appeared in recent years. As such, a background in matrix analysis has become a vital part of a graduate education in statistics. Too often, the statistics graduate student gets his or her matrix background in bits and pieces through various courses on topics such as regression analysis, multivariate analysis, linear models, stochastic processes, and so on. An alternative to this fragmented approach is an entire course devoted to matrix methods useful in statistics. This text has been written with such a course in mind. It also could be used as a text for an advanced undergraduate course with an unusually bright group of students and should prove to be useful as a reference for both applied and research statisticians.

Students beginning a graduate program in statistics often have their previous degrees in other fields, such as mathematics, and so initially their statistical backgrounds may not be all that extensive. With this in mind, I have tried to make the statistical topics presented as examples in this text as self-contained as possible. This has been accomplished by including a section in the first chapter that covers some basic statistical concepts and by having most of the statistical examples deal with applications that are fairly simple to understand; for instance, many of these examples involve least squares regression or applications that utilize the simple concepts of mean vectors and covariance matrices. Thus, an introductory statistics course should provide the reader of this text with a sufficient background in statistics. An additional prerequisite is an undergraduate course in matrices or linear algebra, while a calculus background is necessary for some portions of the book, most notably Chapter 8.

By selectively omitting some sections, all nine chapters of this book can be covered in a one-semester course. For instance, in a course targeted at students who end their educational careers with the master's degree, I typically omit Sections 2.10, 3.5 3.7, 4.8, 5.4–5.7, and 8.6, along with a few other sections.

Anyone writing a book on a subject for which other texts have already been written stands to benefit from these earlier works, and that certainly has been the case here. The texts by Basilevsky (1983), Graybill (1983), Healy (1986), and Searle (1982), all books on matrices for statistics, have helped me, in varying degrees, to formulate my ideas on matrices. Graybill's book has been particularly influential, since this is the book that I referred to extensively, first as a graduate student and then in the early stages of my research career. Other texts that have proven to be quite helpful are Horn and Johnson (1985, 1991), Magnus and Neudecker (1988), particularly in the writing of Chapter 8, and Magnus (1988).

I wish to thank several anonymous reviewers who offered many very helpful suggestions and Mark Johnson for his support and encouragement throughout this project. I am also grateful to the numerous students who have alerted me to various mistakes and typos in earlier versions of this book. In spite of their help and my diligent efforts at proofreading, undoubtedly some mistakes remain, and I would appreciate being informed of any that are spotted.

JIM SCHOTT

Orlando, Florida

CHAPTER ONE

A Review of Elementary Matrix Algebra

1. INTRODUCTION

In this chapter we review some of the basic operations and fundamental properties involved in matrix algebra. In most cases properties will be stated without proof, but in some cases, when instructive, proofs will be presented. We end the chapter with a brief discussion of random variables and random vectors, expected values of random variables, and some important distributions encountered elsewhere in the book.

2. DEFINITIONS AND NOTATION

Except when stated otherwise, a scalar such as α will represent a real number. A matrix A of size $m \times n$, is the $m \times n$ rectangular array of scalars given by

$$A = \begin{bmatrix} a_{11} & a_{12} & \cdots & a_{1n} \\ a_{21} & a_{22} & \cdots & a_{2n} \\ \vdots & \vdots & & \vdots \\ a_{m1} & a_{m2} & \cdots & a_{mn} \end{bmatrix},$$

and sometimes simply identified as $A = (a_{ij})$. Sometimes it also will be convenient to refer to the (i,j)th element of A, as $(A)_{ij}$; that is, $a_{ij} = (A)_{ij}$. If $m = n$, then A is called a square matrix of order m. An $m \times 1$ matrix

$$\boldsymbol{a} = \begin{bmatrix} a_1 \\ a_2 \\ \vdots \\ a_m \end{bmatrix}$$

is called a column vector or simply a vector. The element a_i is referred to as the ith component of $\boldsymbol{a}$. A $1 \times n$ matrix is called a row vector. The ith row and jth column of the matrix A sometimes will be referred to by $(A)_{i\cdot}$ and $(A)_{\cdot j}$, respectively. We will usually use capital letters to represent matrices and lowercase bold letters for vectors.

The diagonal element of the $m \times m$ matrix A are $a_{11}, a_{22}, \dots, a_{mm}$. If all other elements of A are equal to 0, A is called a diagonal matrix and can be identified as $A = \text{diag}(a_{11}, \dots, a_{mm})$. If, in addition, $a_{ii} = 1$ for $i = 1, \dots, m$ so that $A = \text{diag}(1, \dots, 1)$, then the matrix A is called the identity matrix of order m and will be written as $A = \mathrm{I}_m$ or simply $A = \mathrm{I}$ if the order is obvious. If $A = \text{diag}(a_1, \dots, a_m)$ and b is a scalar, then we will use A^b to denote the diagonal matrix $\text{diag}(a_1^b, \dots, a_m^b)$. For any $m \times m$ matrix A, D_A will denote the diagonal matrix with diagonal elements equal to the diagonal elements of A and, for any $m \times 1$ vector $\boldsymbol{a}$, $D_{\boldsymbol{a}}$ denotes the diagonal matrix with diagonal elements equal to the components of $\boldsymbol{a}$; that is, $D_A = \text{diag}(a_{11}, \dots, a_{mm})$ and $D_{\boldsymbol{a}} = \text{diag}(a_1, \dots, a_m)$.

A triangular matrix is a square matrix that is either an upper triangular matrix or a lower triangular matrix. An upper triangular matrix is one which has all of its elements below the diagonal equal to 0, while a lower triangular matrix has all of its elements above the diagonal equal to 0.

The ith column of the $m \times m$ identity matrix will be denoted by $\boldsymbol{e}_i$; that is, $\boldsymbol{e}_i$ is the $m \times 1$ vector which has its ith component equal to 1 and all of its other components equal to 0. When the value of m is not obvious, we will make it more explicit by writing $\boldsymbol{e}_i$ as $\boldsymbol{e}_{i,m}$. The $m \times m$ matrix whose only nonzero element is a 1 in the (i, j)th position will be identified as E_{ij}.

The scalar zero is written 0, while a vector of zeros, called a null vector, will be denoted by $\mathbf{0}$, and a matrix of zeros, called a null matrix, will be denoted by (0). The $m \times 1$ vector having each component equal to 1 will be denoted $\mathbf{1}_m$ or simply $\mathbf{1}$ when the size of the vector is obvious.

3. MATRIX ADDITION AND MULTIPLICATION

The sum of two matrices A and B is defined if they have the same number of rows and the same number of columns; in this case

$$A + B = (a_{ij} + b_{ij})$$

The product of a scalar α and a matrix A is

$$\alpha A = A\alpha = (\alpha a_{ij})$$

The premultiplication of the matrix B by the matrix A is defined only if the number of columns of A equals the number of rows of B. Thus, if A is $m \times p$ and B is $p \times n$, then $C = AB$ will be the $m \times n$ matrix, which has its (i, j)th element, c_{ij}, given by

$$c_{ij} = (A)_{i\cdot}(B)_{\cdot j} = \sum_{k=1}^{p} a_{ik}b_{kj}$$

There is a similar definition for BA, the postmultiplication of B by A. When both products are defined, we will not have, in general, $AB = BA$. If the matrix A is square, then the product AA, or simply A^2, is defined. In this case, if we have $A^2 = A$, then A is said to be an idempotent matrix.

The following basic properties of matrix addition and multiplication are easy to verify.

Theorem 1.1. Let α and β be scalars and A, B, and C be matrices. Then, when the operations involved are defined, the following properties hold.

(a) $A + B = B + A$.
(b) $(A + B) + C = A + (B + C)$.
(c) $\alpha(A + B) = \alpha A + \alpha B$.
(d) $(\alpha + \beta)A = \alpha A + \beta A$.
(e) $A - A = A + (-A) = (0)$.
(f) $A(B + C) = AB + AC$.
(g) $(A + B)C = AC + BC$.
(h) $(AB)C = A(BC)$.

4. THE TRANSPOSE

The transpose of an $m \times n$ matrix A is the $n \times m$ matrix A' obtained by interchanging the rows and columns of A. Thus, the (i,j)th element of A' is a_{ji}. If A is $m \times p$ and B is $p \times n$, then the (i,j)th element of $(AB)'$ can be expressed as

$$((AB)')_{ij} = (AB)_{ji} = (A)_{j\cdot}(B)_{\cdot i} = \sum_{k=1}^{p} a_{jk}b_{ki} = (B')_{i\cdot}(A')_{\cdot j} = (B'A')_{ij}$$

Thus, evidently $(AB)' = B'A'$. This along with some other results involving the transpose are summarized below.

Theorem 1.2. Let α and β be scalars and A and B be matrices. Then, when defined, the following hold.

(a) $(\alpha A)' = \alpha A'$.
(b) $(A')' = A$.

(c) $(\alpha A + \beta B)' = \alpha A' + \beta B'$.
(d) $(AB)' = B'A'$.

If A is $m \times m$, that is, A is a square matrix, then A' is also $m \times m$. In this case, if $A = A'$, then A is called a symmetric matrix, while A is called a skew-symmetric matrix if $A = -A'$.

The transpose of a column vector is a row vector, and in some situtations we may write a matrix as a column vector times a row vector. For instance, the matrix E_{ij} defined in the previous section can be expressed as $E_{ij} = \boldsymbol{e}_i \boldsymbol{e}_j'$. More generally, $\boldsymbol{e}_{i,m} \boldsymbol{e}_{j,n}'$ yields an $m \times n$ matrix having 1, as its only nonzero element, in the (i,j)th position, and if A is an $m \times n$ matrix then

$$A = \sum_{i=1}^{m} \sum_{j=1}^{n} a_{ij} \boldsymbol{e}_{i,m} \boldsymbol{e}_{j,n}'$$

5. THE TRACE

The trace is a function that is defined only on square matrices. If A is an $m \times m$ matrix, then the trace of A, denoted by $\text{tr}(A)$, is defined to be the sum of the diagonal elements of A; that is,

$$\text{tr}(A) = \sum_{i=1}^{m} a_{ii}$$

Now if A is $m \times n$ and B is $n \times m$, then AB is $m \times m$ and

$$\begin{aligned} \text{tr}(AB) &= \sum_{i=1}^{m} (AB)_{ii} = \sum_{i=1}^{m} (A)_{i\cdot}(B)_{\cdot i} = \sum_{i=1}^{m} \sum_{j=1}^{n} a_{ij} b_{ji} = \sum_{j=1}^{n} \sum_{i=1}^{m} b_{ji} a_{ij} \\ &= \sum_{j=1}^{n} (B)_{j\cdot}(A)_{\cdot j} = \sum_{j=1}^{n} (BA)_{jj} = \text{tr}(BA) \end{aligned}$$

This property of the trace, along with some others, is summarized in the following theorem.

Theorem 1.3. Let α be a scalar and A and B be matrices. Then, when the appropriate operations are defined, we have

(a) $\text{tr}(A') = \text{tr}(A)$,
(b) $\text{tr}(\alpha A) = \alpha \, \text{tr}(A)$,

(c) $\text{tr}(A + B) = \text{tr}(A) + \text{tr}(B)$,
(d) $\text{tr}(AB) = \text{tr}(BA)$,
(e) $\text{tr}(A'A) = 0$ if and only if $A = (0)$.

6. THE DETERMINANT

The determinant is another function defined on square matrices. If A is an $m\times m$ matrix, then its determinant, denoted by $|A|$, is given by

$$|A| = \sum (-1)^{f(i_1,\ldots,i_m)} a_{1i_1}a_{2i_2}\cdots a_{mi_m}$$

$$= \sum (-1)^{f(i_1,\ldots,i_m)} a_{i_1 1}a_{i_2 2}\cdots a_{i_m m},$$

where the summation is taken over all permutations, $(i_1,\ldots,i_m)$ of the set of integers $(1,\ldots,m)$, and the function $f(i_1,\ldots,i_m)$ equals the number of transpositions necessary to change $(i_1,\ldots,i_m)$ to $(1,\ldots,m)$. A transposition is the interchange of two of the integers. Although f is not unique, it is uniquely even or odd, so that $|A|$ is uniquely defined. Note that the determinant produces all products of m terms of the elements of the matrix A such that exactly one element is selected from each row and each column of A.

An alternative expression for $|A|$ can be given in terms of the cofactors of A. The minor of the element a_{ij}, denoted by m_{ij}, is the determinant of the $(m-1)\times(m-1)$ matrix obtained after removing the ith row and jth column from A. The corresponding cofactor of a_{ij}, denoted by A_{ij}, is then given as $A_{ij} = (-1)^{i+j}m_{ij}$. For any $i = 1,\ldots,m$, the determinant of A can be obtained by expanding along the ith row,

$$|A| = \sum_{j=1}^{m} a_{ij}A_{ij}, \tag{1.1}$$

or expanding along the jth column,

$$|A| = \sum_{i=1}^{m} a_{ij}A_{ij} \tag{1.2}$$

On the other hand, if the cofactors of a row or column are matched with the elements from a different row or column, the expansion reduces to 0; that is, if $k \neq i$, then

$$\sum_{j=1}^{m} a_{ij}A_{kj} = \sum_{j=1}^{m} a_{ji}A_{jk} = 0 \tag{1.3}$$

Example 1.1. We will find the determinant of the 5×5 matrix given by

$$A = \begin{bmatrix} 2 & 1 & 2 & 1 & 1 \\ 0 & 0 & 3 & 0 & 0 \\ 0 & 0 & 2 & 2 & 0 \\ 0 & 0 & 1 & 1 & 1 \\ 0 & 1 & 2 & 2 & 1 \end{bmatrix}$$

Using the cofactor expansion formula on the first column of A, we obtain

$$|A| = 2 \begin{vmatrix} 0 & 3 & 0 & 0 \\ 0 & 2 & 2 & 0 \\ 0 & 1 & 1 & 1 \\ 1 & 2 & 2 & 1 \end{vmatrix},$$

and then using the same expansion formula on the first column of this 4×4 matrix, we get

$$|A| = 2 \cdot (-1) \cdot \begin{vmatrix} 3 & 0 & 0 \\ 2 & 2 & 0 \\ 1 & 1 & 1 \end{vmatrix}$$

Since the determinant of the 3×3 matrix above is 6 we have

$$|A| = 2 \cdot (-1) \cdot 6 = -12$$

The following properties of the determinant are fairly straightforward to verify using the definition of a determinant or the expansion formulas given in (1.1) and (1.2).

Theorem 1.4. If α is a scalar and A is an $m \times m$ matrix, then the following properties hold.

(a) $|A'| = |A|$.
(b) $|\alpha A| = \alpha^m |A|$.
(c) If A is a diagonal matrix, then $|A| = a_{11} \cdots a_{mm} = \prod_{i=1}^m a_{ii}$.
(d) If all elements of a row (or column) of A are zero, $|A| = 0$.
(e) If two rows (or columns) of A are proportional to one another, $|A| = 0$.
(f) The interchange of two rows (or columns) of A changes the sign of $|A|$.
(g) If all the elements of a row (or column) of A are multiplied by α, then the determinant is multiplied by α.

(h) The determinant of A is unchanged when a multiple of one row (or column) is added to another row (or column).

Consider the $m \times m$ matrix C whose columns are given by the vectors $\boldsymbol{c}_1, \dots, \boldsymbol{c}_m$; that is, we can write $C = (\boldsymbol{c}_1, \dots, \boldsymbol{c}_m)$. Suppose that, for some vector $\boldsymbol{b} = (b_1, \dots, b_m)'$ and matrix $A = (\boldsymbol{a}_1, \dots, \boldsymbol{a}_m)$, we have

$$\boldsymbol{c}_1 = A\boldsymbol{b} = \sum_{i=1}^{m} b_i \boldsymbol{a}_i$$

Then, if we find the determinant of C by expanding along the first column of C, we get

$$\begin{aligned} |C| &= \sum_{j=1}^{m} c_{j1} C_{j1} = \sum_{j=1}^{m} \left(\sum_{i=1}^{m} b_i a_{ji} \right) C_{j1} \\ &= \sum_{i=1}^{m} b_i \left(\sum_{j=1}^{m} a_{ji} C_{j1} \right) = \sum_{i=1}^{m} b_i |(\boldsymbol{a}_i, \boldsymbol{c}_2, \dots, \boldsymbol{c}_m)|, \end{aligned}$$

so that the determinant of C is a linear combination of m determinants. If B is an $m \times m$ matrix and we now define $C = AB$, then by applying the derivation above on each column of C we find that

$$\begin{aligned} |C| &= \left| \left(\sum_{i_1=1}^{m} b_{i_1 1} \boldsymbol{a}_{i_1}, \dots, \sum_{i_m=1}^{m} b_{i_m m} \boldsymbol{a}_{i_m} \right) \right| \\ &= \sum_{i_1=1}^{m} \cdots \sum_{i_m=1}^{m} b_{i_1 1} \cdots b_{i_m m} |(\boldsymbol{a}_{i_1}, \dots, \boldsymbol{a}_{i_m})| \\ &= \sum b_{i_1 1} \cdots b_{i_m m} |(\boldsymbol{a}_{i_1}, \dots, \boldsymbol{a}_{i_m})|, \end{aligned}$$

where this final sum is only over all permutations of $(1, \dots, m)$, since Theorem 1.4(e) implies that

$$|(\boldsymbol{a}_{i_1}, \dots, \boldsymbol{a}_{i_m})| = 0$$

if $i_j = i_k$ for any $j \neq k$. Finally, reordering the columns in $|(\boldsymbol{a}_{i_1}, \dots, \boldsymbol{a}_{i_m})|$ and using Theorem 1.4(f), we have

$$|C| = \sum b_{i_1 1} \cdots b_{i_m m}(-1)^{f(i_1,\ldots,i_m)}|(\boldsymbol{a}_1,\ldots,\boldsymbol{a}_m)| = |B||A|.$$

This very useful result is summarized below.

Theorem 1.5. If both A and B are square matrices of the same order, then

$$|AB| = |A|\ |B|$$

7. THE INVERSE

An $m \times m$ matrix A for which $|A| \neq 0$ is said to be a nonsingular matrix. In this case, there exists a nonsingular matrix denoted by A^{-1} and called the inverse of A, such that

$$AA^{-1} = A^{-1}A = \mathrm{I}_m \tag{1.4}$$

This inverse is unique since, if B is another $m \times m$ matrix satisfying the inverse formula (1.4) for A, then $BA = \mathrm{I}_m$, and so

$$B = B\mathrm{I}_m = BAA^{-1} = \mathrm{I}_m A^{-1} = A^{-1}$$

The following basic properties of the matrix inverse can be easily verified by utilizing (1.4).

Theorem 1.6. If α is a nonzero scalar, and A and B are nonsingular $m \times m$ matrices, then

(a) $(\alpha A)^{-1} = \alpha^{-1}A^{-1}$,
(b) $(A')^{-1} = (A^{-1})'$,
(c) $(A^{-1})^{-1} = A$,
(d) $|A^{-1}| = |A|^{-1}$,
(e) if $A = \mathrm{diag}(a_{11},\ldots,a_{mm})$, then $A^{-1} = \mathrm{diag}(a_{11}^{-1},\ldots,a_{mm}^{-1})$,
(f) if $A = A'$, then $A^{-1} = (A^{-1})'$,
(g) $(AB)^{-1} = B^{-1}A^{-1}$.

As with the determinant of A, the inverse of A can be expressed in terms of the cofactors of A. Let $A_\#$, called the adjoint of A, be the transpose of the matrix of cofactors of A; that is, the (i,j)th element of $A_\#$ is A_{ji}, the cofactor of a_{ji}. Then

$$AA_{\#} = A_{\#}A = \text{diag}(|A|, \ldots, |A|) = |A|\text{I}_m,$$

since $(A)_{i\cdot}(A_{\#})_{\cdot i} = (A_{\#})_{i\cdot}(A)_{\cdot i} = |A|$ follows from (1.1) and (1.2), and $(A)_{i\cdot}(A_{\#})_{\cdot j} = (A_{\#})_{i\cdot}(A)_{\cdot j} = 0$, for $i \neq j$ follows from (1.3). The equation above then yields the relationship

$$A^{-1} = |A|^{-1}A_{\#}$$

The relationship between the inverse of a matrix product and the product of the inverses, given in Theorem 1.6(g), is a very useful property. Unfortunately, such a nice relationship does not exist between the inverse of a sum and the sum of the inverses. We do, however, have the following result, which is sometimes useful.

Theorem 1.7. Suppose A and B are nonsingular matrices, with A being $m \times m$ and B being $n \times n$. For any $m \times n$ matrix C and any $n \times m$ matrix D, it follows that if $A + CBD$ is nonsingular then

$$(A + CBD)^{-1} = A^{-1} - A^{-1}C(B^{-1} + DA^{-1}C)^{-1}DA^{-1}$$

Proof. The proof simply involves verifying that $(A + CBD)(A + CBD)^{-1} = \text{I}_m$ for $(A + CBD)^{-1}$ given above. We have

$$\begin{aligned}
&(A + CBD)\{A^{-1} - A^{-1}C(B^{-1} + DA^{-1}C)^{-1}DA^{-1}\} \\
&\quad = \text{I}_m - C(B^{-1} + DA^{-1}C)^{-1}DA^{-1} + CBDA^{-1} \\
&\qquad - CBDA^{-1}C(B^{-1} + DA^{-1}C)^{-1}DA^{-1} \\
&\quad = \text{I}_m - C\left\{(B^{-1} + DA^{-1}C)^{-1} - B + BDA^{-1}C(B^{-1} + DA^{-1}C)^{-1}\right\}DA^{-1} \\
&\quad = \text{I}_m - C\left\{B(B^{-1} + DA^{-1}C)(B^{-1} + DA^{-1}C)^{-1} - B\right\}DA^{-1} \\
&\quad = \text{I}_m - C\{B - B\}DA^{-1} = \text{I}_m
\end{aligned}$$

□

If $m = n$ and C and D are identity matrices, then we obtain the following special case of Theorem 1.7.

Corollary 1.7.1. Suppose that A, B and $A + B$ are all $m \times m$ nonsingular matrices. Then

$$(A + B)^{-1} = A^{-1} - A^{-1}(B^{-1} + A^{-1})^{-1}A^{-1}$$

We obtain another special case of Theorem 1.7 when $n = 1$.

Corollary 1.7.2. Let A be an $m\times m$ nonsingular matrix. If $\boldsymbol{c}$ and $\boldsymbol{d}$ are both $m\times 1$ vectors and $A+\boldsymbol{cd}'$ is nonsingular, then

$$(A+\boldsymbol{cd}')^{-1}=A^{-1}-A^{-1}\boldsymbol{cd}'A^{-1}/(1+\boldsymbol{d}'A^{-1}\boldsymbol{c})$$

Example 1.2. Theorem 1.7 can be particularly useful when m is larger than n and the inverse of A is fairly easy to compute. For instance, suppose we have $A = \mathrm{I}_5$,

$$B=\begin{bmatrix} 1 & 1 \\ 1 & 2 \end{bmatrix},\qquad C=\begin{bmatrix} 1 & 0 \\ 2 & 1 \\ -1 & 1 \\ 0 & 2 \\ 1 & 1 \end{bmatrix},\qquad D'=\begin{bmatrix} 1 & -1 \\ -1 & 2 \\ 0 & 1 \\ 1 & 0 \\ -1 & 1 \end{bmatrix},$$

from which we obtain

$$G=A+CBD=\begin{bmatrix} 1 & 1 & 1 & 1 & 0 \\ -1 & 6 & 4 & 3 & 1 \\ -1 & 2 & 2 & 0 & 1 \\ -2 & 6 & 4 & 3 & 2 \\ -1 & 4 & 3 & 2 & 2 \end{bmatrix}$$

It is somewhat tedious to compute the inverse of this 5×5 matrix directly. However, the calculations in Theorem 1.7 are fairly straightforward. Clearly, $A^{-1} = \mathrm{I}_5$ and

$$B^{-1}=\begin{bmatrix} 2 & -1 \\ -1 & 1 \end{bmatrix},$$

so that

$$(B^{-1}+DA^{-1}C)=\begin{bmatrix} 2 & -1 \\ -1 & 1 \end{bmatrix}+\begin{bmatrix} -2 & 0 \\ 3 & 4 \end{bmatrix}=\begin{bmatrix} 0 & -1 \\ 2 & 5 \end{bmatrix},$$

and

$$(B^{-1}+DA^{-1}C)^{-1}=\begin{bmatrix} 2.5 & 0.5 \\ -1 & 0 \end{bmatrix}$$

Thus, we find that

$$G^{-1} = \mathrm{I}_5 - C(B^{-1} + DA^{-1}C)^{-1}D$$

$$= \begin{bmatrix} -1 & 1.5 & -0.5 & -2.5 & 2 \\ -3 & 3 & -1 & -4 & 3 \\ 3 & -2.5 & 1.5 & 3.5 & -3 \\ 2 & -2 & 0 & 3 & -2 \\ -1 & 0.5 & -0.5 & -1.5 & 2 \end{bmatrix}$$

8. PARTITIONED MATRICES

Occasionally we will find it useful to partition a given matrix into submatrices. For instance, suppose A is $m\times n$ and the positive integers m_1, m_2, n_1, n_2 are such that $m = m_1 + m_2$ and $n = n_1 + n_2$. Then one way of writing A as a partitioned matrix is

$$A = \begin{bmatrix} A_{11} & A_{12} \\ A_{21} & A_{22} \end{bmatrix},$$

where A_{11} is $m_1 \times n_1$, A_{12} is $m_1 \times n_2$, A_{21} is $m_2 \times n_1$, and A_{22} is $m_2 \times n_2$. That is, A_{11} is the matrix consisting of the first m_1 rows and n_1 columns of A, A_{12} is the matrix consisting of the first m_1 rows and last n_2 columns of A, and so on. Matrix operations can be expressed in terms of the submatrices of the partitioned matrix. For example, suppose B is an $n \times p$ matrix partitioned as

$$B = \begin{bmatrix} B_{11} & B_{12} \\ B_{21} & B_{22} \end{bmatrix},$$

where B_{11} is $n_1 \times p_1$, B_{12} is $n_1 \times p_2$, B_{21} is $n_2 \times p_1$, B_{22} is $n_2 \times p_2$, and $p_1 + p_2 = p$. Then the premultiplication of B by A can be expressed in partitioned form as

$$AB = \begin{bmatrix} A_{11}B_{11} + A_{12}B_{21} & A_{11}B_{12} + A_{12}B_{22} \\ A_{21}B_{11} + A_{22}B_{21} & A_{21}B_{12} + A_{22}B_{22} \end{bmatrix}$$

Matrices can be partitioned into submatrices in other ways besides the 2×2 partitioning given above. For instance, we could partition only the columns of A, yielding the expression

$$A = [A_1 \quad A_2],$$

where A_1 is $m \times n_1$ and A_2 is $m \times n_2$. A more general situation is one in which the rows of A are partitioned into r groups and the columns of A are partitioned into c groups so that A can be written as

$$A = \begin{bmatrix} A_{11} & A_{12} & \cdots & A_{1c} \\ A_{21} & A_{22} & \cdots & A_{2c} \\ \vdots & \vdots & & \vdots \\ A_{r1} & A_{r2} & \cdots & A_{rc} \end{bmatrix},$$

where the submatrix A_{ij} is $m_i \times n_j$ and the integers $m_1, \dots, m_r$ and $n_1, \dots, n_c$ are such that

$$\sum_{i=1}^{r} m_i = m \quad \text{and} \quad \sum_{j=1}^{c} n_j = n$$

The matrix A above is said to be in block diagonal form if $r = c$, A_{ii} is a square matrix for each i, and A_{ij} is a null matrix for all i and j for which $i \neq j$. In this case we will write $A = \text{diag}(A_{11}, \dots, A_{rr})$; that is,

$$\text{diag}(A_{11}, \dots, A_{rr}) = \begin{bmatrix} A_{11} & (0) & \cdots & (0) \\ (0) & A_{22} & \cdots & (0) \\ \vdots & \vdots & & \vdots \\ (0) & (0) & \cdots & A_{rr} \end{bmatrix}$$

Example 1.3. Suppose we wish to compute the transpose product AA', where the 5×5 matrix A is given by

$$A = \begin{bmatrix} 1 & 0 & 0 & 1 & 1 \\ 0 & 1 & 0 & 1 & 1 \\ 0 & 0 & 1 & 1 & 1 \\ -1 & -1 & -1 & 2 & 0 \\ -1 & -1 & -1 & 0 & 2 \end{bmatrix}$$

The computation can be simplified by observing that A may be written

$$A = \begin{bmatrix} I_3 & \mathbf{1}_3\mathbf{1}_2' \\ -\mathbf{1}_2\mathbf{1}_3' & 2I_2 \end{bmatrix}$$

As a result, we have

$$AA' = \begin{bmatrix} I_3 & \mathbf{1}_3\mathbf{1}_2' \\ -\mathbf{1}_2\mathbf{1}_3' & 2I_2 \end{bmatrix} \begin{bmatrix} I_3 & -\mathbf{1}_3\mathbf{1}_2' \\ \mathbf{1}_2\mathbf{1}_3' & 2I_2 \end{bmatrix} = \begin{bmatrix} I_3 + \mathbf{1}_3\mathbf{1}_2'\mathbf{1}_2\mathbf{1}_3' & -\mathbf{1}_3\mathbf{1}_2' + 2\mathbf{1}_3\mathbf{1}_2' \\ -\mathbf{1}_2\mathbf{1}_3' + 2\mathbf{1}_2\mathbf{1}_3' & \mathbf{1}_2\mathbf{1}_3'\mathbf{1}_3\mathbf{1}_2' + 4I_2 \end{bmatrix}$$

$$= \begin{bmatrix} I_3 + 2\mathbf{1}_3\mathbf{1}_3' & \mathbf{1}_3\mathbf{1}_2' \\ \mathbf{1}_2\mathbf{1}_3' & 3\mathbf{1}_2\mathbf{1}_2' + 4I_2 \end{bmatrix} = \begin{bmatrix} 3 & 2 & 2 & 1 & 1 \\ 2 & 3 & 2 & 1 & 1 \\ 2 & 2 & 3 & 1 & 1 \\ 1 & 1 & 1 & 7 & 3 \\ 1 & 1 & 1 & 3 & 7 \end{bmatrix}$$

9. THE RANK OF A MATRIX

Our initial definition of the rank of an $m \times n$ matrix A is given in terms of submatrices. In general, any matrix formed by deleting rows or columns of A is called a submatrix of A. The determinant of an $r \times r$ submatrix of A is called a minor of order r. For instance, for an $m \times m$ matrix A, we have previously defined what we called the minor of a_{ij}; this is, an example of a minor of order $m - 1$. Now the rank of a nonnull $m \times n$ matrix A is r, written $\text{rank}(A) = r$, if at least one of its minors of order r is nonzero while all minors of order $r + 1$ (if there are any) are zero. If A is a null matrix, then $\text{rank}(A) = 0$.

The rank of a matrix A is unchanged by any of the following operations, called elementary transformations.

(a) The interchange of two rows (or columns) of A.
(b) The multiplication of a row (or column) of A by a nonzero scalar.
(c) The addition of a scalar multiple of a row (or column) of A to another row (or column) of A.

Any elementary transformation of A can be expressed as the multiplication of A by a matrix referred to as an elementary transformation matrix. An elementary transformation of the rows of A will be given by the premultiplication of A by an elementary transformation matrix, while an elementary transformation of the columns corresponds to a postmultiplication. Elementary transformation matrices are nonsingular and any nonsingular matrix can be expressed as the product of elementary transformation matrices. Consequently, we have the following very useful result.

Theorem 1.8. Let A be an $m \times n$ matrix, B an $m \times m$ matrix, and C an $n \times n$ matrix. Then if B and C are nonsingular matrices, it follows that

$$\text{rank}(BAC) = \text{rank}(BA) = \text{rank}(AC) = \text{rank}(A)$$

By using elementary transformation matrices, any matrix A can be transformed to another matrix of simpler form having the same rank as A.

Theorem 1.9. If A is an $m \times n$ matrix of rank $r > 0$, then there exist nonsingular $m \times m$ and $n \times n$ matrices B and C, such that $H = BAC$ and $A = B^{-1}HC^{-1}$, where H is given by

(a) I_r if $r = m = n$, (b) $[\mathrm{I}_r \quad (0)]$ if $r = m < n$,

(c) $\begin{bmatrix} \mathrm{I}_r \\ (0) \end{bmatrix}$ if $r = n < m$, (d) $\begin{bmatrix} \mathrm{I}_r & (0) \\ (0) & (0) \end{bmatrix}$ if $r < m, r < n$

The following is an immediate consequence of Theorem 1.9.

Corollary 1.9.1. Let A be an $m \times n$ matrix with $\mathrm{rank}(A) = r > 0$. Then there exist an $m \times r$ matrix F and an $r \times n$ matrix G such that $\mathrm{rank}(F) = \mathrm{rank}(G) = r$ and $A = FG$.

10. ORTHOGONAL MATRICES

An $m \times 1$ vector $\boldsymbol{p}$ is said to be a normalized vector or a unit vector if $\boldsymbol{p}'\boldsymbol{p} = 1$. The $m \times 1$ vectors, $\boldsymbol{p}_1, \ldots, \boldsymbol{p}_n$, where $n \leq m$, are said to be orthogonal if $\boldsymbol{p}_i'\boldsymbol{p}_j = 0$ for all $i \neq j$. If in addition, each $\boldsymbol{p}_i$ is a normalized vector then the vectors are said to be orthonormal. An $m \times m$ matrix P whose columns form an orthonormal set of vectors is called an orthogonal matrix. It immediately follows that

$$P'P = \mathrm{I}$$

Taking the determinant of both sides, we see that

$$|P'P| = |P'|\,|P| = |P|^2 = |\mathrm{I}| = 1$$

Thus, $|P| = +1$ or -1, so that P is nonsingular, $P^{-1} = P'$, and $PP' = \mathrm{I}$ in addition to $P'P = \mathrm{I}$; that is, the rows of P also form an orthonormal set of $m \times 1$ vectors. Some basic properties of orthogonal matrices are summarized in the following theorem.

Theorem 1.10. Let P and Q be $m \times m$ orthogonal matrices and A be any $m \times m$ matrix. Then

(a) $|P| = \pm 1$,
(b) $|P'AP| = |A|$,
(c) PQ is an orthogonal matrix.

An $m \times m$ matrix P is called a permutation matrix if each row and each column of P has a single element 1, while all the remaining elements are zeros. As a result the columns of P will be $\boldsymbol{e}_1, \ldots, \boldsymbol{e}_m$, the columns of I_m, in some order. Note then that the (h, h)th element of $P'P$ will be $\boldsymbol{e}_i'\boldsymbol{e}_i = 1$ for some i, and the (h, l)th element of $P'P$ will be $\boldsymbol{e}_i'\boldsymbol{e}_j = 0$ for some $i \neq j$ if $h \neq l$; that is, a permutation matrix is a special orthogonal matrix. Since there are $m!$ ways of permuting the columns of I_m, there are $m!$ different permutation matrices of order m. If A is also $m \times m$, then PA creates an $m \times m$ matrix by permuting the rows of A, and AP produces a matrix by permuting the columns of A.

11. QUADRATIC FORMS

Let $\boldsymbol{x}$ be an $m \times 1$ vector, $\boldsymbol{y}$ an $n \times 1$ vector, and A an $m \times n$ matrix. Then the function of $\boldsymbol{x}$ and $\boldsymbol{y}$ given by

$$\boldsymbol{x}'A\boldsymbol{y} = \sum_{i=1}^{m} \sum_{j=1}^{n} x_i y_j a_{ij}$$

is sometimes called a bilinear form in $\boldsymbol{x}$ and $\boldsymbol{y}$. We will be most interested in the special case in which $m = n$ so that A is $m \times m$ and $\boldsymbol{x} = \boldsymbol{y}$. In this case, the function above reduces to the function of $\boldsymbol{x}$,

$$f(\boldsymbol{x}) = \boldsymbol{x}'A\boldsymbol{x} = \sum_{i=1}^{m} \sum_{j=1}^{m} x_i x_j a_{ij},$$

which is called a quadratic form in $\boldsymbol{x}$; A is referred to as the matrix of the quadratic form. We will always assume that A is a symmetric matrix since, if it is not, A may be replaced by $B = \frac{1}{2}(A + A')$, which is symmetric, without altering $f(\boldsymbol{x})$; that is,

$$\boldsymbol{x}'B\boldsymbol{x} = \frac{1}{2}\,\boldsymbol{x}'(A + A')\boldsymbol{x} = \frac{1}{2}\,(\boldsymbol{x}'A\boldsymbol{x} + \boldsymbol{x}'A'\boldsymbol{x}) = \frac{1}{2}\,(\boldsymbol{x}'A\boldsymbol{x} + \boldsymbol{x}'A\boldsymbol{x}) = \boldsymbol{x}'A\boldsymbol{x}$$

since $\boldsymbol{x}'A'\boldsymbol{x} = (\boldsymbol{x}'A'\boldsymbol{x})' = \boldsymbol{x}'A\boldsymbol{x}$.

Every symmetric matrix A and its associated quadratic form is classified into one of the following five categories.

(a) If $\boldsymbol{x}'A\boldsymbol{x} > 0$ for all $\boldsymbol{x} \neq \boldsymbol{0}$, then A is positive definite.

(b) If $\boldsymbol{x}'A\boldsymbol{x} \geq 0$ for all $\boldsymbol{x} \neq \boldsymbol{0}$ and $\boldsymbol{x}'A\boldsymbol{x} = 0$ for some $\boldsymbol{x} \neq \boldsymbol{0}$, then A is positive semidefinite.

(c) If $\boldsymbol{x}'A\boldsymbol{x} < 0$ for all $\boldsymbol{x} \neq \boldsymbol{0}$, then A is negative definite.

(d) If $\boldsymbol{x}'A\boldsymbol{x} \leq 0$ for all $\boldsymbol{x} \neq \boldsymbol{0}$ and $\boldsymbol{x}'A\boldsymbol{x} = 0$ for some $\boldsymbol{x} \neq \boldsymbol{0}$, then A is negative semidefinite.

(e) If $\boldsymbol{x}'A\boldsymbol{x} > 0$ for some $\boldsymbol{x}$ and $\boldsymbol{x}'A\boldsymbol{x} < 0$ for some $\boldsymbol{x}$, then A is indefinite.

Note that the null matrix is actually both positive semidefinite and negative semidefinite.

Positive definite and negative definite matrices are nonsingular, whereas positive semidefinite and negative semidefinite matrices are singular. Sometimes the term nonnegative definite will be used to refer to a symmetric matrix that is either positive definite or positive semidefinite. An $m \times m$ matrix B is called a square root of the nonnegative definite $m \times m$ matrix A if $A = BB'$. Sometimes we will denote such a matrix B as $A^{1/2}$. If B is also symmetric, so that $A = B^2$, then B is called the symmetric square root of A.

Quadratic forms play a prominent role in inferential statistics. In Chapter 9, we will develop some of the most important results involving quadratic forms that are of particular interest in statistics.

12. COMPLEX MATRICES

Throughout this entire text we will be dealing with the analysis of vectors and matrices composed of real numbers or variables. However, there are occasions in which an analysis of a real matrix, such as the decomposition of a matrix in the form of a product of other matrices, leads to matrices that contain complex numbers. For this reason, we will briefly summarize in this section some of the basic notation and terminology regarding complex numbers.

Any complex number c can be written in the form

$$c = a + ib,$$

where a and b are real numbers and i represents the imaginary number $\sqrt{-1}$. The real number a is called the real part of c, while b is referred to as the imaginary part of c. Thus, the number c is a real number only if b is 0. If we have two complex numbers, $c_1 = a_1 + ib_1$ and $c_2 = a_2 + ib_2$, then their sum is given by

$$c_1 + c_2 = (a_1 + a_2) + i(b_1 + b_2),$$

while their product is given by

$$c_1c_2 = a_1a_2 - b_1b_2 + i(a_1b_2 + a_2b_1)$$

Corresponding to each complex number $c = a + ib$ is another complex number denoted by $\overline{c}$ and called the complex conjugate of c. The complex conjugate of c is given by $\overline{c} = a - ib$ and satisfies $c\overline{c} = a^2 + b^2$, so that the product of a complex number and its conjugate results in a real number.

A complex number can be represented geometrically by a point in the complex plane, where one of the axes is the real axis and the other axis is the complex or imaginary axis. Thus, the complex number $c = a + ib$, would be represented by the point (a, b) in this complex plane. Alternatively, we can use the polar coordinates (r, θ), where r is the length of the line from the origin to the point (a, b) and θ is the angle between this line and the positive half of the real axis. The relationship between a and b and r and θ is then given by

$$a = r\cos\theta, \qquad b = r\sin\theta$$

Writing c in terms of the polar coordinates, we have

$$c = r\cos\theta + ir\sin\theta,$$

or, after using Euler's formula, simply $c = re^{i\theta}$. The absolute value, also sometimes called the modulus, of the complex number c is defined to be r. This is, of course, always a nonnegative real number, and since $a^2 + b^2 = r^2$ we have

$$|c| = |a + ib| = \sqrt{a^2 + b^2}$$

We also find that

$$\begin{aligned} |c_1c_2| &= \sqrt{(a_1a_2 - b_1b_2)^2 + (a_1b_2 + a_2b_1)^2} \\ &= \sqrt{(a_1^2 + b_1^2)(a_2^2 + b_2^2)} = |c_1|\,|c_2| \end{aligned}$$

Using the identity above repeatedly, we also see that for any complex number c and any positive integer n, $|c^n| = |c|^n$.

A useful identity relating a complex number c and its conjugate to the absolute value of c is

$$c\overline{c} = |c|^2$$

Applying this to the sum of two complex numbers $c_1 + c_2$ and noting that $c_1\overline{c}_2 + \overline{c}_1c_2 \leq 2|c_1|\,|c_2|$, we get

$$\begin{aligned}|c_1 + c_2|^2 &= (c_1 + c_2)\overline{(c_1 + c_2)} = (c_1 + c_2)(\bar{c}_1 + \bar{c}_2)\\ &= c_1\bar{c}_1 + c_1\bar{c}_2 + c_2\bar{c}_1 + c_2\bar{c}_2\\ &\leq |c_1|^2 + 2|c_1|\,|c_2| + |c_2|^2\\ &= (|c_1| + |c_2|)^2\end{aligned}$$

From this we get the important inequality $|c_1 + c_2| \leq |c_1| + |c_2|$, known as the triangle inequality.

A complex matrix is simply a matrix whose elements are complex numbers. As a result, a complex matrix can be written as the sum of a real matrix and an imaginary matrix; that is, if C is an $m \times n$ complex matrix then it can be expressed as

$$C = A + iB,$$

where both A and B are $m \times n$ real matrices. The complex conjugate of C, denoted $\overline{C}$, is simply the matrix containing the complex conjugates of the elements of C; that is,

$$\overline{C} = A - iB$$

The conjugate transpose of C is $C* = \overline{C}'$. If the complex matrix C is square and $C^* = C$, so that $c_{ij} = \bar{c}_{ji}$, then C is said to be Hermitian. Note that if C is Hermitian and C is a real matrix, then C is symmetric. The $m \times m$ matrix C is said to be unitary if $C^*C = \mathrm{I}_m$. This is the generalization of the concept of orthogonal matrices to complex matrices since if C is real then $C^* = C'$.

13. RANDOM VECTORS AND SOME RELATED STATISTICAL CONCEPTS

In this section, we review some of the basic definitions and results in distribution theory which will be needed later in this text. A more comprehensive treatment of this subject can be found in books on statistical theory such as Casella and Berger (1990) or Lindgren (1993). To be consistent with our notation, which uses a capital letter to denote a matrix, a bold lowercase letter for a vector, and a lowercase letter for a scalar, we will use a lowercase letter instead of the more conventional capital letter to denote a scalar random variable.

A random variable x is said to be discrete if its collection of possible values, R_x, is a countable set. In this case, x has a probability function $p_x(t)$ satisfying $p_x(t) = P(x = t)$, for $t \in R_x$, and $p_x(t) = 0$, for $t \notin R_x$. A continuous random variable x, on the other hand, has for its range, R_x, an uncountably infinite set.

Associated with each continuous random variable x is a density function $f_x(t)$ satisfying $f_x(t) > 0$, for $t \in R_x$ and $f_x(t) = 0$, for $t \notin R_x$. Probabilities for x are obtained by integration; if $\mathcal{B}$ is a subset of the real line, then

$$P(x \in \mathcal{B}) = \int_{\mathcal{B}} f_x(t)\,\mathrm{d}t$$

For both discrete and continuous x, we have $P(x \in R_x) = 1$.

The expected value of a real-valued function of $x, g(x)$, gives the average observed value of $g(x)$. This expectation, denoted $\mathrm{E}[g(x)]$, is given by

$$\mathrm{E}[g(x)] = \sum_{t \in R_x} g(t)p_x(t),$$

if x is discrete and

$$\mathrm{E}[g(x)] = \int_{-\infty}^{\infty} g(t)f_x(t)\,\mathrm{d}t,$$

if x is continuous. Properties of the expectation operator follow directly from properties of sums and integrals. For instance, if x and y are random variables and α and β are constants, then the expectation operator satisfies the properties

$$\mathrm{E}(\alpha) = \alpha,$$

and

$$\mathrm{E}[\alpha g_1(x) + \beta g_2(y)] = \alpha\mathrm{E}[g_1(x)] + \beta\mathrm{E}[g_2(y)],$$

where g_1 and g_2 are any real-valued functions. The set of expected values of a random variable x given by $\mathrm{E}(x^k), k = 1, 2, \ldots,$ are known as the moments of x. These are important for both descriptive and theoretical purposes. The first few moments can be used to describe certain features of the distribution of x. For instance, the first moment or mean of x, $\mu_x = \mathrm{E}(x)$, locates a central value of the distribution. The variance of x, denoted σ_x^2 or $\mathrm{var}(x)$, is defined as

$$\sigma_x^2 = \mathrm{var}(x) = \mathrm{E}[(x - \mu_x)^2] = \mathrm{E}(x^2) - \mu_x^2,$$

so that it is a function of the first and second moments of x. The variance gives a measure of the dispersion of the observed values of x about the central value

μ_x. Using properties of expectation, it is easily verified that

$$\text{var}(\alpha + \beta x) = \beta^2 \text{var}(x)$$

All of the moments of a random variable x are imbedded in a function called the moment generating function of x. This function is defined as a particular expectation; specifically, the moment generating function of x, $m_x(t)$, is given by

$$m_x(t) = \text{E}(e^{tx}),$$

provided this expectation exists for values of t in a neighborhood of 0. Otherwise, the moment generating function does not exist. If the moment generating function of x does exist, then we can obtain any moment from it since

$$\left. \frac{\text{d}^k}{\text{d}t^k} m_x(t) \right|_{t=0} = \text{E}(x^k)$$

More importantly, the moment generating function characterizes the distribution of x in that no two different distributions have the same moment generating function.

We now focus on two particular families of distributions that we will encounter later in this text. A random variable x is said to have a normal distribution with mean μ and variance σ^2, indicated by $x \sim \text{N}(\mu, \sigma^2)$, if the density of x is given by

$$f_x(t) = \frac{1}{\sqrt{2\pi}\,\sigma} e^{-(t-\mu)^2/2\sigma^2}, \qquad -\infty < t < \infty$$

The corresponding moment generating function is

$$m_x(t) = e^{\mu t + \sigma^2 t^2/2}$$

A special member of this family of normal distributions is the standard normal distribution N(0, 1). The importance of this distribution follows from the fact that if $x \sim \text{N}(\mu, \sigma^2)$, then the standardizing transformation $z = (x - \mu)/\sigma$ yields a random variable z which has the standard normal distribution. By differentiating the moment generating function of $z \sim \text{N}(0, 1)$, it is easy to verify that the first six moments of z, which we will need in a later chapter, are 0, 1, 0, 3, 0, and 15, respectively.

If r is a positive integer, then a random variable v has a chi-squared distribution with r degrees of freedom, written $v \sim \chi_r^2$, if its density function is

$$f_v(t) = \frac{t^{(r/2)-1} e^{-t/2}}{2^{r/2}\Gamma(r/2)}, \qquad t > 0,$$

where $\Gamma(r/2)$ is the gamma function evaluated at $r/2$. The moment generating function of v is given by $m_v(t) = (1 - 2t)^{-r/2}$, for $t < \frac{1}{2}$. The importance of the chi-squared distribution arises from its connection to the normal distribution. If $z \sim N(0, 1)$, then $z^2 \sim \chi_1^2$. Further, if $z_1, \ldots, z_r$ are independent random variables with $z_i \sim N(0, 1)$ for $i = 1, \ldots, r$, then

$$\sum_{i=1}^{r} z_i^2 \sim \chi_r^2 \tag{1.5}$$

The chi-squared distribution mentioned above is sometimes referred to as a central chi-squared distribution since it is actually a special case of a more general family of distributions known as the noncentral chi-squared distributions. These noncentral chi-squared distributions are also related to the normal distribution. If $x_1, \ldots, x_r$ are independent random variables with $x_i \sim N(\mu_i, 1)$, then

$$\sum_{i=1}^{r} x_i^2 \sim \chi_r^2(\lambda), \tag{1.6}$$

where $\chi_r^2(\lambda)$ denotes the noncentral chi-squared distribution with r degrees of freedom and noncentrality parameter

$$\lambda = \frac{1}{2} \sum_{i=1}^{r} \mu_i^2;$$

that is, the noncentral chi-squared density, which we will not give here, depends not only on the parameter r but also on the parameter λ. Since (1.6) reduces to (1.5) when $\mu_i = 0$ for all i, we see that the distribution $\chi_r^2(\lambda)$ corresponds to χ_r^2 when $\lambda = 0$.

A distribution related to the chi-squared distribution is the F distribution with r_1 and r_2 degrees of freedom, denoted by F_{r_1,r_2}. If $y \sim F_{r_1,r_2}$, then the density function of y is

$$f_y(t) = \frac{\Gamma\{(r_1 + r_2)/2\}}{\Gamma(r_1/2)\Gamma(r_2/2)} \left(\frac{r_1}{r_2}\right)^{r_1/2} t^{(r_1-2)/2} \left(1 + \frac{r_1}{r_2} t\right)^{-(r_1+r_2)/2}, \qquad t > 0$$

The importance of this distribution arises from the fact that if v_1 and v_2 are independent random variables with $v_1 \sim \chi^2_{r_1}$ and $v_2 \sim \chi^2_{r_2}$, then the ratio

$$t = \frac{v_1/r_1}{v_2/r_2}$$

has the F distribution with r_1 and r_2 degrees of freedom.

The concept of a random variable can be extended to that of a random vector. A sequence of related random variables $x_1, \dots, x_m$ is modeled by a joint or multivariate probability function, $p_{\boldsymbol{x}}(\boldsymbol{t})$ if all of the random variables are discrete, and a multivariate density function $f_{\boldsymbol{x}}(\boldsymbol{t})$, if all of the random variables are continuous, where $\boldsymbol{x} = (x_1, \dots, x_m)'$ and $\boldsymbol{t} = (t_1, \dots, t_m)'$. For instance, if they are continuous and $\mathcal{B}$ is a region in R^m, then the probability that $\boldsymbol{x}$ falls in $\mathcal{B}$ is

$$P(\boldsymbol{x} \in \mathcal{B}) = \int \cdots \int_{\mathcal{B}} f_{\boldsymbol{x}}(\boldsymbol{t})\, \mathrm{d}t_1 \cdots \mathrm{d}t_m,$$

while the expected value of the real-valued function $g(\boldsymbol{x})$ of $\boldsymbol{x}$ is given by

$$\mathrm{E}[g(\boldsymbol{x})] = \int_{-\infty}^{\infty} \cdots \int_{-\infty}^{\infty} g(\boldsymbol{x}) f_{\boldsymbol{x}}(\boldsymbol{t}) \mathrm{d}t_1 \cdots \mathrm{d}t_m$$

The random variables $x_1, \dots, x_m$ are said to be independent, a concept we have already referred to, if and only if the joint probability function or density function factors into the product of the marginal probability or density functions; that is, in the continuous case, $x_1, \dots, x_m$ are independent if and only if

$$f_{\boldsymbol{x}}(\boldsymbol{t}) = f_{x_1}(t_1) \cdots f_{x_m}(t_m),$$

for all $\boldsymbol{t}$.

The mean vector of $\boldsymbol{x}$, denoted by $\boldsymbol{\mu}$, is the vector of expected values of the x_is; that is,

$$\boldsymbol{\mu} = (\mu_1, \dots, \mu_m)' = \mathrm{E}(\boldsymbol{x}) = [\mathrm{E}(x_1), \dots, \mathrm{E}(x_m)]'$$

A measure of the linear relationship between x_i and x_j is given by the covariance of x_i and x_j, which is denoted $\mathrm{cov}(x_i, x_j)$ or σ_{ij} and is defined by

$$\sigma_{ij} = \mathrm{cov}(x_i, x_j) = \mathrm{E}[(x_i - \mu_i)(x_j - \mu_j))] = \mathrm{E}(x_i x_j) - \mu_i \mu_j \tag{1.7}$$

When $i = j$, this covariance reduces to the variance of x_i; that is, $\sigma_{ii} = \sigma_i^2 = \text{var}(x_i)$. When $i \neq j$ and x_i and x_j are independent, then $\text{cov}(x_i, x_j) = 0$ since $\text{E}(x_i x_j) = \mu_i \mu_j$. If α_1, α_2, β_1, and β_2 are scalars, then

$$\text{cov}(\alpha_1 + \beta_1 x_i, \alpha_2 + \beta_2 x_j) = \beta_1 \beta_2 \,\text{cov}\,(x_i, x_j)$$

The matrix Ω, which has σ_{ij} as its (i,j)th element, is called the variance-covariance matrix, or simply the covariance matrix, of $\boldsymbol{x}$. This matrix will be also denoted sometimes by $\text{var}(\boldsymbol{x})$ or $\text{cov}(\boldsymbol{x},\boldsymbol{x})$. Clearly, $\sigma_{ij} = \sigma_{ji}$ so that Ω is a symmetric matrix. Using (1.7) we obtain the matrix formulation for Ω,

$$\Omega = \text{var}(\boldsymbol{x}) = \text{E}[(\boldsymbol{x} - \boldsymbol{\mu})(\boldsymbol{x} - \boldsymbol{\mu})'] = \text{E}(\boldsymbol{x}\boldsymbol{x}') - \boldsymbol{\mu}\boldsymbol{\mu}'$$

If $\boldsymbol{\alpha}$ is an $m \times 1$ vector of constants and we define the random variable $y = \boldsymbol{\alpha}'\boldsymbol{x}$, then

$$\text{E}(y) = \text{E}(\boldsymbol{\alpha}'\boldsymbol{x}) = \text{E}\left(\sum_{i=1}^{m} \alpha_i x_i\right) = \sum_{i=1}^{m} \alpha_i \text{E}(x_i) = \sum_{i=1}^{m} \alpha_i \mu_i = \boldsymbol{\alpha}'\boldsymbol{\mu}$$

If, in addition, $\boldsymbol{\beta}$ is another $m \times 1$ vector of constants and $w = \boldsymbol{\beta}'\boldsymbol{x}$, then

$$\begin{aligned}\text{cov}(y, w) &= \text{cov}(\boldsymbol{\alpha}'\boldsymbol{x}, \boldsymbol{\beta}'\boldsymbol{x}) = \text{cov}\left(\sum_{i=1}^{m} \alpha_i x_i, \sum_{j=1}^{m} \beta_j x_j\right) \\ &= \sum_{i=1}^{m}\sum_{j=1}^{m} \alpha_i \beta_j \text{cov}(x_i, x_j) = \sum_{i=1}^{m}\sum_{j=1}^{m} \alpha_i \beta_j \sigma_{ij} = \boldsymbol{\alpha}'\Omega\boldsymbol{\beta}\end{aligned}$$

In particular, $\text{var}(y) = \text{cov}(y, y) = \boldsymbol{\alpha}'\Omega\boldsymbol{\alpha}$. Since this holds for any choice of $\boldsymbol{\alpha}$ and since the variance is always nonnegative, Ω must be a nonnegative definite matrix. More generally, if A is a $p \times m$ matrix of constants and $\boldsymbol{y} = A\boldsymbol{x}$, then

$$\text{E}(\boldsymbol{y}) = \text{E}(A\boldsymbol{x}) = A\text{E}(\boldsymbol{x}) = A\boldsymbol{\mu}, \tag{1.8}$$

$$\begin{aligned}\text{var}(\boldsymbol{y}) &= \text{E}[\{\boldsymbol{y} - \text{E}(\boldsymbol{y})\}\{\boldsymbol{y} - \text{E}(\boldsymbol{y})\}'] = \text{E}[(A\boldsymbol{x} - A\boldsymbol{\mu})(A\boldsymbol{x} - A\boldsymbol{\mu})'] \\ &= \text{E}[A(\boldsymbol{x} - \boldsymbol{\mu})(\boldsymbol{x} - \boldsymbol{\mu})'A'] = A\{\text{E}[(\boldsymbol{x} - \boldsymbol{\mu})(\boldsymbol{x} - \boldsymbol{\mu})']\}A' = A\Omega A' \end{aligned} \tag{1.9}$$

Thus, the mean vector and covariance matrix of the transformed vector, $A\boldsymbol{x}$, is $A\boldsymbol{\mu}$ and $A\Omega A'$. If $\boldsymbol{v}$ and $\boldsymbol{w}$ are random vectors, then the matrix of covariances between components of $\boldsymbol{v}$ and components of $\boldsymbol{w}$ is given by

$$\text{cov}(\boldsymbol{v}, \boldsymbol{w}) = \text{E}(\boldsymbol{v}\boldsymbol{w}') - \text{E}(\boldsymbol{v})\text{E}(\boldsymbol{w})'$$

In particular, if $\boldsymbol{v} = A\boldsymbol{x}$ and $\boldsymbol{w} = B\boldsymbol{x}$, then

$$\text{cov}(\boldsymbol{v}, \boldsymbol{w}) = A\,\text{cov}(\boldsymbol{x}, \boldsymbol{x})B' = A\,\text{var}(\boldsymbol{x})B' = A\Omega B'$$

A measure of the linear relationship between x_i and x_j that is unaffected by the measurement scales of x_i and x_j is called the correlation coefficient ρ_{ij}, defined by

$$\rho_{ij} = \frac{\text{cov}(x_i, x_j)}{\sqrt{\text{var}(x_i)\text{var}(x_j)}} = \frac{\sigma_{ij}}{\sqrt{\sigma_{ii}\sigma_{jj}}}$$

When $i = j$, $\rho_{ij} = 1$. The correlation matrix P, which has ρ_{ij} as its (i, j)th element, can be expressed in terms of the corresponding covariance matrix Ω and the diagonal matrix $D_\Omega^{-1/2} = \text{diag}(\sigma_{11}^{-1/2}, \ldots, \sigma_{mm}^{-1/2})$; specifically,

$$P = D_\Omega^{-1/2}\Omega D_\Omega^{-1/2} \tag{1.10}$$

For any $m \times 1$ vector $\boldsymbol{\alpha}$, we have

$$\boldsymbol{\alpha}'P\boldsymbol{\alpha} = \boldsymbol{\alpha}'D_\Omega^{-1/2}\Omega D_\Omega^{-1/2}\boldsymbol{\alpha} = \boldsymbol{\beta}'\Omega\boldsymbol{\beta},$$

where $\boldsymbol{\beta} = D_\Omega^{-1/2}\boldsymbol{\alpha}$, and so P must be nonnegative definite because Ω is. In particular, if $\boldsymbol{e}_i$ is the ith column of the $m \times m$ identity matrix, then

$$\begin{aligned}(\boldsymbol{e}_i + \boldsymbol{e}_j)'P(\boldsymbol{e}_i + \boldsymbol{e}_j) &= (P)_{ii} + (P)_{ij} + (P)_{ji} + (P)_{jj} \\ &= 2(1 + \rho_{ij}) \geq 0,\end{aligned}$$

and

$$\begin{aligned}(\boldsymbol{e}_i - \boldsymbol{e}_j)'P(\boldsymbol{e}_i - \boldsymbol{e}_j) &= (P)_{ii} - (P)_{ij} - (P)_{ji} + (P)_{jj} \\ &= 2(1 - \rho_{ij}) \geq 0,\end{aligned}$$

from which we obtain the inequality, $-1 \leq \rho_{ij} \leq 1$.

Typically, means, variances, and covariances are unknown and so they must be estimated from a sample. Suppose $x_1, \ldots, x_n$ represents a random sample of a random variable x that has some distribution with mean μ and variance σ^2. These quantities can be estimated by the sample mean and sample variance given by

$$\bar{x} = \frac{1}{n} \sum_{i=1}^{n} x_i, \qquad s^2 = \frac{1}{n-1} \sum_{i=1}^{n} (x_i - \bar{x})^2$$

In the multivariate setting we have analogous estimators for $\boldsymbol{\mu}$ and Ω; if $\boldsymbol{x}_1, \ldots, \boldsymbol{x}_n$ is a random sample of an $m \times 1$ random vector $\boldsymbol{x}$ having mean vector $\boldsymbol{\mu}$ and covariance matrix Ω, then the sample mean vector and sample covariance matrix are given by

$$\bar{\boldsymbol{x}} = \frac{1}{n} \sum_{i=1}^{n} \boldsymbol{x}_i, \qquad S = \frac{1}{n-1} \sum_{i=1}^{n} (\boldsymbol{x}_i - \bar{\boldsymbol{x}})(\boldsymbol{x}_i - \bar{\boldsymbol{x}})'$$

The sample covariance matrix can be then used in (1.10) to obtain an estimator of the correlation matrix, P; that is, if we define the diagonal matrix $D_S^{-1/2} = \text{diag}(s_{11}^{-1/2}, \ldots, s_{mm}^{-1/2})$, the correlation matrix can be estimated by the sample correlation matrix defined as

$$R = D_S^{-1/2} S D_S^{-1/2}$$

The one particular joint distribution that we will consider is the multivariate normal distribution. This distribution can be defined in terms of independent standard normal random variables. Let $z_1, \ldots, z_m$ be independently distributed as $\text{N}(0, 1)$ and put $\boldsymbol{z} = (z_1, \ldots, z_m)'$. The density function of $\boldsymbol{z}$ is then given by

$$f(\boldsymbol{z}) = \prod_{i=1}^{m} \frac{1}{\sqrt{2\pi}} \exp\left(-\frac{1}{2} z_i^2\right) = \frac{1}{(2\pi)^{m/2}} \exp\left(-\frac{1}{2} \boldsymbol{z}'\boldsymbol{z}\right).$$

Since $\text{E}(\boldsymbol{z}) = \boldsymbol{0}$ and $\text{var}(\boldsymbol{z}) = \text{I}_m$, this particular m-dimensional multivariate normal distribution is denoted as $\text{N}_m(\boldsymbol{0}, \text{I}_m)$. If $\boldsymbol{\mu}$ is an $m \times 1$ vector of constants and T is an $m \times m$ nonsingular matrix, then $\boldsymbol{x} = \boldsymbol{\mu} + T\boldsymbol{z}$ is said to have the m-dimensional multivariate normal distribution with mean vector $\boldsymbol{\mu}$ and covariance matrix $\Omega = TT'$. This is indicated by $\boldsymbol{x} \sim \text{N}_m(\boldsymbol{\mu}, \Omega)$. For instance, if $m = 2$, the vector $\boldsymbol{x} = (x_1, x_2)'$ has a bivariate normal distribution and its density, induced by the transformation $\boldsymbol{x} = \boldsymbol{\mu} + T\boldsymbol{z}$, can be shown to be

$$f(\boldsymbol{x}) = \frac{1}{2\pi\sqrt{\sigma_{11}\sigma_{22}(1-\rho^2)}} \exp\left(-\frac{1}{2(1-\rho^2)}\left\{\frac{(x_1-\mu_1)^2}{\sigma_{11}} - 2\rho\left(\frac{x_1-\mu_1}{\sqrt{\sigma_{11}}}\right)\left(\frac{x_2-\mu_2}{\sqrt{\sigma_{22}}}\right) + \frac{(x_2-\mu_2)^2}{\sigma_{22}}\right\}\right), \tag{1.11}$$

for all $\boldsymbol{x} \in R^2$, where $\rho = \rho_{12}$ is the correlation coefficient. When $\rho = 0$, this density factors into the product of the marginal densities, so x_1 and x_2 are independent if and only if $\rho = 0$. The rather cumbersome looking density function given in (1.11) can be more conveniently expressed by utilizing matrix notation. It is straightforward to verify that this density is identical to

$$f(\boldsymbol{x}) = \frac{1}{2\pi|\Omega|^{1/2}} \exp\left\{-\frac{1}{2}(\boldsymbol{x}-\boldsymbol{\mu})'\Omega^{-1}(\boldsymbol{x}-\boldsymbol{\mu})\right\} \tag{1.12}$$

The density function of an m-variate normal random vector is very similar to the function given in (1.12). If $\boldsymbol{x} \sim \mathrm{N}_m(\boldsymbol{\mu}, \Omega)$, then its density is

$$f(\boldsymbol{x}) = \frac{1}{(2\pi)^{m/2}|\Omega|^{1/2}} \exp\left\{-\frac{1}{2}(\boldsymbol{x}-\boldsymbol{\mu})'\Omega^{-1}(\boldsymbol{x}-\boldsymbol{\mu})\right\}, \tag{1.13}$$

for all $\boldsymbol{x} \in R^m$.

If Ω is positive semidefinite, then $\boldsymbol{x} \sim \mathrm{N}_m(\boldsymbol{\mu}, \Omega)$ is said to have a singular normal distribution. In this case Ω^{-1} does not exist, and so the multivariate normal density cannot be written in the form given in (1.13). However, the random vector $\boldsymbol{x}$ can still be expressed in terms of independent standard normal random variables. Suppose that $\mathrm{rank}(\Omega) = r$ and U is an $m \times r$ matrix satisfying $UU' = \Omega$. Then $\boldsymbol{x} \sim \mathrm{N}_m(\boldsymbol{\mu}, \Omega)$ if $\boldsymbol{x}$ is distributed the same as $\boldsymbol{\mu} + U\boldsymbol{z}$, where now $\boldsymbol{z} \sim \mathrm{N}_r(\boldsymbol{0}, \mathrm{I}_r)$.

An important property of the multivariate normal distribution is that a linear transformation of a multivariate normal vector yields a multivariate normal vector; that is, if $\boldsymbol{x} \sim \mathrm{N}_m(\boldsymbol{\mu}, \Omega)$ and A is a $p \times m$ matrix of constants, then $\boldsymbol{y} = A\boldsymbol{x}$ has a p-variate normal distribution. In particular, from (1.8) and (1.9) we know that $\boldsymbol{y} \sim \mathrm{N}_p(A\boldsymbol{\mu}, A\Omega A')$.

One of the most widely used procedures in statistics is regression analysis. We will briefly describe this analysis here and subsequently use regression analysis to illustrate some of the matrix methods developed in this text. Some good references on regression are Neter, Wasserman, and Kutner (1985) and Sen and Srivastava (1990). In the typical regression problem, one wishes to study the relationship between some response variable, say y, and k explanatory variables $x_1, \ldots, x_k$. For instance, y might be the yield of some product of a manufacturing process, while the explanatory variables are conditions affecting the production process, such as temperature, humidity, pressure, and so on. A model relating the x_js to y is given by

$$y = \beta_0 + \beta_1 x_1 + \cdots + \beta_k x_k + \epsilon, \tag{1.14}$$

where $\beta_0, \dots, \beta_k$ are unknown parameters and ϵ is a random error, that is, a random variable, with $E(\epsilon) = 0$. In what is known as ordinary least squares regression, we also have the errors being independent random variables with common variance σ^2; that is, if ϵ_i and ϵ_j are random errors associated with the responses y_i and y_j, then $\text{var}(\epsilon_i) = \text{var}(\epsilon_j) = \sigma^2$ and $\text{cov}(\epsilon_i, \epsilon_j) = 0$. The model given in (1.14) is an example of a linear model since it is a linear function of the parameters. It need not be linear in the x_js so that, for instance, we might have $x_2 = x_1^2$. Since the parameters are unknown, they must be estimated and this will be possible if we have some observed values of y and the corresponding x_js. Thus, for the ith observation suppose that the explanatory variables are set to the values $x_{i1}, \dots, x_{ik}$ yielding the response y_i, and this is done for $i = 1, \dots, N$, where $N > k + 1$. If model (1.14) holds, then we should have, approximately,

$$y_i = \beta_0 + \beta_1 x_{i1} + \cdots + \beta_k x_{ik}$$

for each i. This can be written as the matrix equation

$$\boldsymbol{y} = X\boldsymbol{\beta}$$

if we define

$$\boldsymbol{y} = \begin{bmatrix} y_1 \\ y_2 \\ \vdots \\ y_N \end{bmatrix}, \qquad \boldsymbol{\beta} = \begin{bmatrix} \beta_0 \\ \beta_1 \\ \vdots \\ \beta_k \end{bmatrix}, \qquad X = \begin{bmatrix} 1 & x_{11} & \dots & x_{1k} \\ 1 & x_{21} & \dots & x_{2k} \\ \vdots & \vdots & & \vdots \\ 1 & x_{N1} & \dots & x_{Nk} \end{bmatrix}$$

One method of estimating the β_js, which we will discuss from time to time in this text, is called the method of least squares. If $\hat{\boldsymbol{\beta}} = (\hat{\beta}_1, \dots, \hat{\beta}_k)'$ is an estimate of the parameter vector $\boldsymbol{\beta}$, then $\hat{\boldsymbol{y}} = X\hat{\boldsymbol{\beta}}$ is the vector of fitted values, while $\hat{\boldsymbol{y}} - \boldsymbol{y}$ gives the vector of errors or deviations of the actual responses from the corresponding fitted values, and

$$f(\hat{\boldsymbol{\beta}}) = (\boldsymbol{y} - X\hat{\boldsymbol{\beta}})'(\boldsymbol{y} - X\hat{\boldsymbol{\beta}})$$

gives the sum of squares of these errors. The method of least squares selects as $\hat{\boldsymbol{\beta}}$ any vector that minimizes the function $f(\hat{\boldsymbol{\beta}})$. We will see later that any such vector satisfies the system of linear equations, sometimes referred to as the normal equations,

$$X'X\hat{\boldsymbol{\beta}} = X'\boldsymbol{y}$$

If X has full column rank, that is, $\text{rank}(X) = k + 1$, then $(X'X)^{-1}$ exists and so the least squares estimator of $\boldsymbol{\beta}$ is unique and is given by

$$\hat{\boldsymbol{\beta}} = (X'X)^{-1}X'\boldsymbol{y}$$

PROBLEMS

1. Prove Theorem 1.3(e); that is, if A is an $m \times n$ matrix, show that $\text{tr}(A'A) = 0$ if and only if $A = (0)$.

2. Show that if $\boldsymbol{x}$ and $\boldsymbol{y}$ are $m \times 1$ vectors, $\text{tr}(\boldsymbol{x}\boldsymbol{y}') = \boldsymbol{x}'\boldsymbol{y}$. Show that if A and B are $m \times m$ matrices and B is nonsingular, $\text{tr}(BAB^{-1}) = \text{tr}(A)$.

3. Prove Theorem 1.4.

4. Show that any square matrix can be written as the sum of a symmetric matrix and a skew-symmetric matrix.

5. Define the $m \times m$ matrices, A, B, and C as

$$A = \begin{bmatrix} b_{11} + c_{11} & b_{12} + c_{12} & \cdots & b_{1m} + c_{1m} \\ a_{21} & a_{22} & \cdots & a_{2m} \\ \vdots & \vdots & & \vdots \\ a_{m1} & a_{m2} & \cdots & a_{mm} \end{bmatrix}$$

$$B = \begin{bmatrix} b_{11} & b_{12} & \cdots & b_{1m} \\ a_{21} & a_{22} & \cdots & a_{2m} \\ \vdots & \vdots & & \vdots \\ a_{m1} & a_{m2} & \cdots & a_{mm} \end{bmatrix}, \qquad C = \begin{bmatrix} c_{11} & c_{12} & \cdots & c_{1m} \\ a_{21} & a_{22} & \cdots & a_{2m} \\ \vdots & \vdots & & \vdots \\ a_{m1} & a_{m2} & \cdots & a_{mm} \end{bmatrix}$$

   Prove that $|A| = |B| + |C|$.

6. Verify the results of Theorem 1.6.

7. Consider the 4×4 matrix

$$A = \begin{bmatrix} 1 & 2 & 1 & 1 \\ 0 & 1 & 2 & 0 \\ 1 & 2 & 2 & 1 \\ 0 & -1 & 1 & 2 \end{bmatrix}$$

Find the determinant of A by using the cofactor expansion formula on the first column of A.

8. Using the matrix A from the previous problem, verify equation (1.3) when $i = 1$ and $k = 2$.

9. Let λ be a variable and consider the determinant of $A - \lambda \mathrm{I}_m$, where A is an $m \times m$ matrix, as a function of λ. What type of function of λ is this?

10. Find the adjoint matrix of the matrix A given in Problem 7. Use this to obtain the inverse of A.

11. Using elementary transformations, determine matrices B and C so that $BAC = \mathrm{I}_4$ for the matrix A given in Problem 7. Use B and C to compute the inverse of A; that is, take the inverse of both sides of the equation $BAC = \mathrm{I}_4$ and then solve for A^{-1}.

12. Show that the determinant of a triangular matrix is the product of its diagonal elements. In addition, show that the inverse of a lower triangular matrix is a lower triangular matrix.

13. Let $\boldsymbol{a}$ and $\boldsymbol{b}$ be $m \times 1$ vectors and D be an $m \times m$ diagonal matrix. Use Corollary 1.7.2 to find an expression for the inverse of $D + \alpha \boldsymbol{ab}'$, where α is a scalar.

14. Consider the $m \times m$ partitioned matrix

$$A = \begin{bmatrix} A_{11} & (0) \\ A_{21} & A_{22} \end{bmatrix},$$

where the $m_1 \times m_1$ matrix A_{11} and the $m_2 \times m_2$ matrix A_{22} are nonsingular. Obtain an expression for A^{-1} in terms of A_{11}, A_{22}, and A_{21}.

15. Let

$$A = \begin{bmatrix} A_{11} & A_{12} \\ A'_{12} & A_{22} \end{bmatrix},$$

where A_{11} is $m_1 \times m_1$, A_{22} is $m_2 \times m_2$, and A_{12} is $m_1 \times m_2$. Show that if A is positive definite then A_{11} and A_{22} are also positive definite.

16. Find the rank of the 4×4 matrix

$$A = \begin{bmatrix} 2 & 0 & 1 & -1 \\ 1 & -1 & 1 & -1 \\ 1 & -1 & 2 & 0 \\ 2 & 0 & 0 & -2 \end{bmatrix}$$

17. Use elementary transformations to transform the matrix A given in problem 16 to a matrix H having the form given in Theorem 1.9. Consequently, determine matrices B and C so $BAC = H$.

18. Prove parts (b) and (c) of Theorem 1.10.

19. List all permutation matrices of order 3.

20. Consider the 3×3 matrix

$$P = \frac{1}{\sqrt{6}} \begin{bmatrix} \sqrt{2} & \sqrt{2} & \sqrt{2} \\ \sqrt{3} & -\sqrt{3} & 0 \\ p_{31} & p_{32} & p_{33} \end{bmatrix}$$

Find values for p_{31}, p_{32}, and p_{33} so that P is an orthogonal matrix. Is your solution unique?

21. Suppose the $m \times m$ orthogonal matrix P is partitioned as $P = [P_1 \quad P_2]$, where P_1 is $m \times m_1$, P_2 is $m \times m_2$, and $m_1 + m_2 = m$. Show that $P_1'P_1 = I_{m_1}$, $P_2'P_2 = I_{m_2}$, and $P_1P_1' + P_2P_2' = I_m$.

22. Let A be an $m \times m$ matrix and suppose there exists a real $n \times m$ matrix T such that $T'T = A$. Show that A must be nonnegative definite.

23. Prove that a nonnegative definite matrix must have nonnegative diagonal elements; that is, show that if a symmetric matrix has any negative diagonal elements then it is not nonnegative definite. Show that the converse is not true; that is, find a symmetric matrix that has nonnegative diagonal elements but is not nonnegative definite.

24. Let A be an $m \times m$ nonnegative definite matrix, while B is an $n \times m$ matrix. Show that BAB' is a nonnegative definite matrix.

25. Use the standard normal moment-generating function, $m_z(t) = e^{t^2/2}$ to show that the first six moments of the standard normal distribution are 0, 1, 0, 3, 0, and 15.

26. Use properties of expectation to show that for random variables x_1 and x_2, and scalars α_1, α_2, β_1, and β_2

$$\text{cov}(\alpha_1 + \beta_1 x_1, \alpha_2 + \beta_2 x_2) = \beta_1\beta_2 \,\text{cov}(x_1, x_2).$$

27. Suppose $\boldsymbol{x} \sim N_3(\boldsymbol{\mu}, \Omega)$, where

$$\boldsymbol{\mu} = \begin{bmatrix} 1 \\ 2 \\ 3 \end{bmatrix}, \qquad \Omega = \begin{bmatrix} 2 & 1 & -1 \\ 1 & 2 & 1 \\ -1 & 1 & 3 \end{bmatrix},$$

and let the 3×3 matrix A and 2×3 matrix B be given by

$$A = \begin{bmatrix} 2 & 2 & 1 \\ 1 & 0 & -1 \\ 0 & 1 & -1 \end{bmatrix}, \qquad B = \begin{bmatrix} 1 & 1 & 1 \\ -1 & 1 & 0 \end{bmatrix}$$

(a) Find the correlation matrix of $\boldsymbol{x}$.
(b) Determine the distribution of $u = \mathbf{1}'\boldsymbol{x}$.
(c) Determine the distribution of $\boldsymbol{v} = A\boldsymbol{x}$.
(d) Determine the distribution of

$$\boldsymbol{w} = \begin{bmatrix} A\boldsymbol{x} \\ B\boldsymbol{x} \end{bmatrix}$$

(e) Which, if any, of the distributions obtained in (b), (c), and (d) are singular distributions?

28. Suppose $\boldsymbol{x}$ is an $m \times 1$ random vector with mean vector $\boldsymbol{\mu}$ and covariance matrix Ω. If A is an $n \times m$ matrix of constants and $\boldsymbol{c}$ is an $m \times 1$ vector of constants, give expressions for
(a) $E[A(\boldsymbol{x} + \boldsymbol{c})]$,
(b) $\text{var}[A(\boldsymbol{x} + \boldsymbol{c})]$.

CHAPTER TWO

Vector Spaces

1. INTRODUCTION

In statistics, observations typically take the form of vectors of values of different variables; for example, for each subject in a sample, one might record height, weight, age, and so on. In estimation and hypotheses testing situations, we are usually interested in inferences regarding a vector of parameters. As a result, the topic of this chapter, vector spaces, has important applications in statistics. In addition, the concept of linearly independent and dependent vectors, which we discuss in Section 3, is very useful in the understanding and determination of the rank of a matrix.

2. DEFINITIONS

A vector space is a collection of vectors that satisfies some special properties; in particular, the collection is closed under the addition of vectors and under the multiplication of a vector by a scalar.

Definition 2.1. Let S be a collection of $m \times 1$ vectors satisfying the following.

(a) If $\boldsymbol{x}_1 \in S$ and $\boldsymbol{x}_2 \in S$, then $\boldsymbol{x}_1 + \boldsymbol{x}_2 \in S$.
(b) If $\boldsymbol{x} \in S$ and α is any real scalar, then $\alpha\boldsymbol{x} \in S$.

Then S is called a vector space in m-dimensional space. If S is a subset of T, which is another vector space in m-dimensional space, then S is called a vector subspace of T. This will be indicated by writing $S \subseteq T$.

The choice of $\alpha = 0$ in Definition 2.1(b) implies that the null vector $\mathbf{0} \in S$; that is, every vector space must contain the null vector. In fact, the set $S = \{\mathbf{0}\}$, consisting of the null vector only, is itself a vector space. Note also that the two conditions (a) and (b) are equivalent to the one condition that says if $\boldsymbol{x}_1 \in S$,

$\boldsymbol{x}_2 \in S$, and α_1 and α_2 are any real scalars, then $(\alpha_1\boldsymbol{x}_1 + \alpha_2\boldsymbol{x}_2) \in S$. This can be easily generalized to more than two, say n, vectors; that is, if $\alpha_1, \dots, \alpha_n$ are real scalars and $\boldsymbol{x}_1, \dots, \boldsymbol{x}_n$ vectors such that $\boldsymbol{x}_i \in S$, for all i, then for S to be a vector space we must have

$$\sum_{i=1}^{n} \alpha_i \boldsymbol{x}_i \in S \tag{2.1}$$

The left-hand side of (2.1) is called a linear combination of the vectors $\boldsymbol{x}_1, \dots, \boldsymbol{x}_n$. Since a vector space is closed under the formation of linear combinations, vector spaces are sometimes also referred to as linear spaces.

Example 2.1. Consider the sets of vectors given by

$$S_1 = \{(a, 0, a)': -\infty < a < \infty\},$$
$$S_2 = \{(a, b, a + b)': -\infty < a < \infty, -\infty < b < \infty\},$$
$$S_3 = \{(a, a, a)': a \geq 0\}.$$

Let $\boldsymbol{x}_1 = (a_1, 0, a_1)'$ and $\boldsymbol{x}_2 = (a_2, 0, a_2)'$, where a_1 and a_2 are arbitrary scalars. Then $\boldsymbol{x}_1 \in S_1$, $\boldsymbol{x}_2 \in S_1$, and

$$\alpha_1\boldsymbol{x}_1 + \alpha_2\boldsymbol{x}_2 = (\alpha_1 a_1 + \alpha_2 a_2, 0, \alpha_1 a_1 + \alpha_2 a_2)' \in S_1,$$

so that S_1 is a vector space. By a similar argument, we find that S_2 is also a vector space. Further, S_1 consists of all the vectors of S_2 for which $b = 0$, so S_1 is a subset of S_2, and thus S_1 is a vector subspace of S_2. On the other hand, S_3 is not a vector space since, for example, if we take $\alpha = -1$ and $\boldsymbol{x} = (1, 1, 1)'$, then $\boldsymbol{x} \in S_3$ but

$$\alpha\boldsymbol{x} = -(1, 1, 1)' \notin S_3$$

Every vector space with the exception of the vector space $\{\mathbf{0}\}$ has infinitely many vectors. However, by utilizing the process of forming linear combinations, a vector space can be associated with a finite set of vectors as long as each vector in the vector space can be expressed as some linear combination of the vectors in this set.

Definition 2.2. Let $\{\boldsymbol{x}_1, \dots, \boldsymbol{x}_n\}$ be a set of $m \times 1$ vectors in the vector space S. If each vector in S can be expressed as a linear combination of the vectors $\boldsymbol{x}_1, \dots, \boldsymbol{x}_n$, then the set $\{\boldsymbol{x}_1, \dots, \boldsymbol{x}_n\}$ is said to span or generate the vector space S, and $\{\boldsymbol{x}_1, \dots, \boldsymbol{x}_n\}$ is called a spanning set of S.

Suppose we select from the vector space S a set of vectors $\{\boldsymbol{x}_1, \dots, \boldsymbol{x}_n\}$. In general, we cannot be assured that every $\boldsymbol{x} \in S$ is a linear combination of

$\boldsymbol{x}_1, \ldots, \boldsymbol{x}_n$, and so it is possible that the set $\{\boldsymbol{x}_1, \ldots, \boldsymbol{x}_n\}$ does not span S. This set must, however, span a vector space which is a subspace of S.

Theorem 2.1. Let $\{\boldsymbol{x}_1, \ldots, \boldsymbol{x}_n\}$ be a set of $m \times 1$ vectors in the vector space S, and let W be the set of all possible linear combinations of these vectors; that is,

$$W = \left\{ \boldsymbol{x}\colon \boldsymbol{x} = \sum_{i=1}^{n} \alpha_i \boldsymbol{x}_i, -\infty < \alpha_i < \infty \quad \text{for all } i \right\}$$

Then W is a vector subspace of S.

Proof. Clearly, W is a subset of S since the vectors $\boldsymbol{x}_1, \ldots, \boldsymbol{x}_n$ are in S, and S is closed under the formation of linear combinations. To prove that W is a subspace of S, we must show that, for arbitrary vectors $\boldsymbol{u}$ and $\boldsymbol{v}$ in W and scalars a and b, $a\boldsymbol{u} + b\boldsymbol{v}$ is in W. Since $\boldsymbol{u}$ and $\boldsymbol{v}$ are in W, by the definition of W, there must exist scalars $c_1, \ldots, c_n$ and $d_1, \ldots, d_n$ such that

$$\boldsymbol{u} = \sum_{i=1}^{n} c_i \boldsymbol{x}_i, \qquad \boldsymbol{v} = \sum_{i=1}^{n} d_i \boldsymbol{x}_i$$

It then follows that

$$a\boldsymbol{u} + b\boldsymbol{v} = a\left(\sum_{i=1}^{n} c_i \boldsymbol{x}_i \right) + b\left(\sum_{i=1}^{n} d_i \boldsymbol{x}_i \right) = \sum_{i=1}^{n} (ac_i + bd_i)\boldsymbol{x}_i,$$

so that $a\boldsymbol{u} + b\boldsymbol{v}$ is a linear combination of $\boldsymbol{x}_1, \ldots, \boldsymbol{x}_n$ and thus $a\boldsymbol{u} + b\boldsymbol{v} \in W$. □

The notions of the size or length of a vector, or the distance between two vectors are important concepts when dealing with vector spaces. Although we are most familiar with the standard Euclidean formulas for length and distance, there are a variety of ways of defining length and distance. These measures of length and distance sometimes utilize a product of vectors called an inner product.

Definition 2.3. Let S be a vector space. A function, $\langle \boldsymbol{x}, \boldsymbol{y} \rangle$, defined for all $\boldsymbol{x} \in S$ and $\boldsymbol{y} \in S$, is an inner product if for any $\boldsymbol{x}$, $\boldsymbol{y}$, and $\boldsymbol{z}$ in S, and any scalar c,

(a) $\langle \boldsymbol{x}, \boldsymbol{x} \rangle \geq 0$ and $\langle \boldsymbol{x}, \boldsymbol{x} \rangle = 0$ if and only if $\boldsymbol{x} = \boldsymbol{0}$,
(b) $\langle \boldsymbol{x}, \boldsymbol{y} \rangle = \langle \boldsymbol{y}, \boldsymbol{x} \rangle$,
(c) $\langle \boldsymbol{x} + \boldsymbol{y}, \boldsymbol{z} \rangle = \langle \boldsymbol{x}, \boldsymbol{z} \rangle + \langle \boldsymbol{y}, \boldsymbol{z} \rangle$,
(d) $\langle c\boldsymbol{x}, \boldsymbol{y} \rangle = c\langle \boldsymbol{x}, \boldsymbol{y} \rangle$.

A useful result regarding inner products is given by the Cauchy–Schwarz inequality.

Theorem 2.2. If $\boldsymbol{x}$ and $\boldsymbol{y}$ are in the vector space S and $\langle \boldsymbol{x},\boldsymbol{y}\rangle$ is an inner product defined on S, then

$$\langle \boldsymbol{x},\boldsymbol{y}\rangle^2 \leq \langle \boldsymbol{x},\boldsymbol{x}\rangle\langle \boldsymbol{y},\boldsymbol{y}\rangle$$

Proof. The result is trivial if $\boldsymbol{x} = \boldsymbol{0}$ since it is easily shown that, in this case, $\langle \boldsymbol{x},\boldsymbol{y}\rangle = \langle \boldsymbol{x},\boldsymbol{x}\rangle = 0$. Suppose that $\boldsymbol{x} \neq \boldsymbol{0}$, and let $a = \langle \boldsymbol{x},\boldsymbol{x}\rangle$, $b = 2\langle \boldsymbol{x},\boldsymbol{y}\rangle$, and $c = \langle \boldsymbol{y},\boldsymbol{y}\rangle$. Then using Definition 2.3, we find that for any scalar t

$$0 \leq \langle t\boldsymbol{x}+\boldsymbol{y}, t\boldsymbol{x}+\boldsymbol{y}\rangle = \langle \boldsymbol{x},\boldsymbol{x}\rangle t^2 + 2\langle \boldsymbol{x},\boldsymbol{y}\rangle t + \langle \boldsymbol{y},\boldsymbol{y}\rangle = at^2 + bt + c$$

Consequently, the polynomial $at^2 + bt + c$ either has a repeated real root or no real roots. This means that the discriminant $b^2 - 4ac$ must be nonpositive, and this leads to the inequality

$$b^2 \leq 4ac,$$

which simplifies to $\langle \boldsymbol{x},\boldsymbol{y}\rangle^2 \leq \langle \boldsymbol{x},\boldsymbol{x}\rangle\langle \boldsymbol{y},\boldsymbol{y}\rangle$. □

The most common inner product is the Euclidean inner product given by $\langle \boldsymbol{x},\boldsymbol{y}\rangle = \boldsymbol{x}'\boldsymbol{y}$. Applying the Cauchy–Schwarz inequality to this inner product, we find that

$$\left(\sum_{i=1}^m x_iy_i\right)^2 \leq \left(\sum_{i=1}^m x_i^2\right)\left(\sum_{i=1}^m y_i^2\right)$$

holds for any $m \times 1$ vectors $\boldsymbol{x}$ and $\boldsymbol{y}$.

A vector norm and a distance function provide us with the means of measuring the length of a vector and the distance between two vectors.

Definition 2.4. A function $\|\boldsymbol{x}\|$ is a vector norm on the vector space S if, for any vectors $\boldsymbol{x}$ and $\boldsymbol{y}$ in S, we have

(a) $\|\boldsymbol{x}\| \geq 0$,
(b) $\|\boldsymbol{x}\| = 0$ if and only if $\boldsymbol{x} = \boldsymbol{0}$,
(c) $\|c\boldsymbol{x}\| = |c|\,\|\boldsymbol{x}\|$ for any scalar c,
(d) $\|\boldsymbol{x}+\boldsymbol{y}\| \leq \|\boldsymbol{x}\| + \|\boldsymbol{y}\|$.

Definition 2.5. A function $d(\boldsymbol{x},\boldsymbol{y})$ is a distance function defined on the vector space S if for any vectors $\boldsymbol{x}$, $\boldsymbol{y}$, and $\boldsymbol{z}$ in S, we have

(a) $d(\boldsymbol{x},\boldsymbol{y}) \geq 0$,
(b) $d(\boldsymbol{x},\boldsymbol{y}) = 0$ if and only if $\boldsymbol{x} = \boldsymbol{y}$,
(c) $d(\boldsymbol{x},\boldsymbol{y}) = d(\boldsymbol{y},\boldsymbol{x})$,
(d) $d(\boldsymbol{x},\boldsymbol{z}) \leq d(\boldsymbol{x},\boldsymbol{y}) + d(\boldsymbol{y},\boldsymbol{z})$.

Property (d) given in the two definitions above is known as the triangle inequality because it is a generalization of the familiar relationship in two-dimensional geometry. One common way of defining a vector norm and a distance function is in terms of an inner product. The reader can verify that for any inner product, $\langle \boldsymbol{x},\boldsymbol{y} \rangle$, the functions, $\|\boldsymbol{x}\| = \langle \boldsymbol{x},\boldsymbol{x} \rangle^{1/2}$ and $d(\boldsymbol{x},\boldsymbol{y}) = \langle \boldsymbol{x}-\boldsymbol{y},\boldsymbol{x}-\boldsymbol{y} \rangle^{1/2}$ satisfy the conditions given in Definitions 2.4 and 2.5.

We will use R^m to denote the vector space consisting of all $m \times 1$ vectors with real components; that is, $R^m = \{(x_1,\dots,x_m): -\infty < x_i < \infty, i = 1,\dots,m\}$. We usually have associated with this vector space the Euclidean distance function $d_{\mathrm{I}}(\boldsymbol{x},\boldsymbol{y}) = \|\boldsymbol{x}-\boldsymbol{y}\|_2$; $\|\boldsymbol{x}\|_2$ is the Euclidean norm given by

$$\|\boldsymbol{x}\|_2 = \{\boldsymbol{x}'\boldsymbol{x}\}^{1/2} = \left\{\sum_{i=1}^{m} x_i^2\right\}^{1/2}$$

and based on the Euclidean inner product $\langle \boldsymbol{x},\boldsymbol{y} \rangle = \boldsymbol{x}'\boldsymbol{y}$. This distance formula is a generalization of the familiar formulas that we have for distance in two and three-dimensional geometry. The space with this distance function is called Euclidean m-dimensional space. Whenever this text works with the vector space R^m, the associated distance will be this Euclidean distance unless stated otherwise. There are, however, many situations in statistics in which non-Euclidean distance functions are appropriate.

Example 2.2. Suppose we wish to compute the distance between the $m \times 1$ vectors $\boldsymbol{x}$ and $\boldsymbol{\mu}$, where $\boldsymbol{x}$ is an observation from a distribution having mean vector $\boldsymbol{\mu}$ and covariance matrix Ω. If we want to take into account the effect of the covariance structure, then the Euclidean distance defined above would not be appropriate unless $\Omega = \mathrm{I}$. For example, if $m = 2$ and $\Omega = \mathrm{diag}(0.5, 2)$, then a large value of $(x_1 - \mu_1)^2$ would be more surprising than a similar value of $(x_2 - \mu_2)^2$ because the variance of the first component of $\boldsymbol{x}$ is smaller than the variance of the second component; that is, it seems reasonable in defining distance to put more weight on $(x_1 - \mu_1)^2$ than on $(x_2 - \mu_2)^2$. A more appropriate distance function is given by

$$d_\Omega(\boldsymbol{x}, \boldsymbol{\mu}) = \{(\boldsymbol{x} - \boldsymbol{\mu})'\Omega^{-1}(\boldsymbol{x} - \boldsymbol{\mu})\}^{1/2},$$

and is called the Mahalanobis distance between $\boldsymbol{x}$ and $\boldsymbol{\mu}$. This is sometimes also referred to as the distance between $\boldsymbol{x}$ and $\boldsymbol{\mu}$ in the metric of Ω and is useful in a multivariate statistical procedure known as discriminant analysis [see McLachlan (1992) or Huberty (1994)]. Note that if $\Omega = I$ this distance function reduces to the Euclidean distance function. For $\Omega = \text{diag}(0.5, 2)$, this distance function simplifies to

$$d_\Omega(\boldsymbol{x}, \boldsymbol{\mu}) = \{2(x_1 - \mu_1)^2 + 0.5(x_2 - \mu_2)^2\}^{1/2}$$

As a second illustration suppose that again $m = 2$, but now

$$\Omega = \begin{bmatrix} 1 & 0.5 \\ 0.5 & 1 \end{bmatrix}$$

Because of the positive correlation, $(x_1 - \mu_1)$ and $(x_2 - \mu_2)$ will tend to have the same sign. This is reflected in the Mahalanobis distance,

$$d_\Omega(\boldsymbol{x}, \boldsymbol{\mu}) = \left(\frac{4}{3} \{(x_1 - \mu_1)^2 + (x_2 - \mu_2)^2 - (x_1 - \mu_1)(x_2 - \mu_2)\} \right)^{1/2},$$

through the last term, which increases or decreases the distance according to whether $(x_1 - \mu_1)(x_2 - \mu_2)$ is negative or positive. In Chapter 4, we will take a closer look at the construction of this distance function.

We end this section with examples of some other commonly used vector norms. The norm $\|\boldsymbol{x}\|_1$, called the sum norm, is defined by

$$\|\boldsymbol{x}\|_1 = \sum_{i=1}^{m} |x_i|$$

Both the sum norm and the Euclidean norm $\|\boldsymbol{x}\|_2$ are members of the family of norms given by

$$\|\boldsymbol{x}\|_p = \left\{ \sum_{i=1}^{m} |x_i|^p \right\}^{1/p},$$

where $p \geq 1$. Yet another example of a vector norm, known as the infinity norm or max norm, is given by

$$\|\boldsymbol{x}\|_\infty = \max_{1 \leq i \leq m} |x_i|$$

Although we have been confining attention to real vectors, the norms defined above also serve as norms for complex vectors. However, in this case, the absolute values appearing in the expression for $\|\boldsymbol{x}\|_p$ are necessary even when p is even. In particular, the Euclidean norm, valid for complex as well as real vectors, is

$$\|\boldsymbol{x}\|_2 = \left\{ \sum_{i=1}^{m} |x_i|^2 \right\}^{1/2}$$

3. LINEAR INDEPENDENCE AND DEPENDENCE

We have seen that the formation of linear combinations of vectors is a fundamental operation of vector spaces. This operation is what establishes a link between a spanning set and its vector space. In many situations, our investigation of a vector space can be reduced simply to an investigation of a spanning set for that vector space. In this case, it will be advantageous to make the spanning set as small as possible. In order to do this, it is first necessary to understand the concepts of linear independence and linear dependence.

Definition 2.6. The set of $m \times 1$ vectors $\{\boldsymbol{x}_1, \dots, \boldsymbol{x}_n\}$ is said to be a linearly independent set if the only solution to the equation

$$\sum_{i=1}^{n} \alpha_i \boldsymbol{x}_i = \boldsymbol{0}$$

is given by $\alpha_1 = \cdots = \alpha_n = 0$. If there are other solutions, the set is called a linearly dependent set.

Example 2.3. Consider the three vectors $\boldsymbol{x}_1 = (1, 1, 1)'$, $\boldsymbol{x}_2 = (1, 0, -1)'$, and $\boldsymbol{x}_3 = (3, 2, 1)'$. To determine whether these vectors are linearly independent, we solve the system of equations $\alpha_1 \boldsymbol{x}_1 + \alpha_2 \boldsymbol{x}_2 + \alpha_3 \boldsymbol{x}_3 = \boldsymbol{0}$ or, equivalently,

$$\begin{aligned} \alpha_1 + \alpha_2 + 3\alpha_3 &= 0, \\ \alpha_1 + 2\alpha_3 &= 0, \\ \alpha_1 - \alpha_2 + \alpha_3 &= 0 \end{aligned}$$

These equations yield the constraints $\alpha_2 = 0.5\alpha_1$ and $\alpha_3 = -0.5\alpha_1$. Thus, for any scalar α, a solution will be given by $\alpha_1 = \alpha$, $\alpha_2 = 0.5\alpha$, $\alpha_3 = -0.5\alpha$, and so the vectors are linearly dependent. On the other hand, any pair of the three vectors are linearly independent; that is, $\{\boldsymbol{x}_1, \boldsymbol{x}_2\}$, $\{\boldsymbol{x}_1, \boldsymbol{x}_3\}$, and $\{\boldsymbol{x}_2, \boldsymbol{x}_3\}$ is each a linearly independent set of vectors.

The proofs of the following results are left to the reader.

Theorem 2.3. Let $\{\boldsymbol{x}_1, \ldots, \boldsymbol{x}_n\}$ be a set of $m \times 1$ vectors. Then the following statements hold.

(a) The set is linearly dependent if the null vector $\mathbf{0}$ is in the set.
(b) If this set of vectors is linearly dependent, any nonempty subset of it is also linearly independent.
(c) If this set of vectors is linearly dependent, any other set containing this set as a subset is also linearly dependent.

Note that in Definition 2.6, if $n = 1$, that is, there is only one vector in the set, then the set is linearly independent unless that vector is $\mathbf{0}$. If $n = 2$, the set is linearly independent unless one of the vectors is the null vector, or each vector is a nonzero scalar multiple of the other vector; that is, a set of two vectors is linearly dependent if and only if at least one of the vectors is a scalar multiple of the other. In general, we have the following.

Theorem 2.4. The set of $m \times 1$ vectors $\{\boldsymbol{x}_1, \ldots, \boldsymbol{x}_n\}$, where $n > 1$, is a linearly dependent set if and only if at least one vector in the set can be expressed as a linear combination of the remaining vectors.

Proof. The result is obvious if one of the vectors in the set is the null vector since then the set must be linearly dependent, and the $m \times 1$ null vector is a linear combination of any set of $m \times 1$ vectors. Now assume the set does not include the null vector. First suppose one of the vectors, say $\boldsymbol{x}_n$, can be expressed as a linear combination of the others; that is, we can find scalars $\alpha_1, \ldots, \alpha_{n-1}$ such that $\boldsymbol{x}_n = \alpha_1 \boldsymbol{x}_1 + \cdots + \alpha_{n-1}\boldsymbol{x}_{n-1}$. But this implies that

$$\sum_{i=1}^{n} \alpha_i \boldsymbol{x}_i = \mathbf{0}, \tag{2.2}$$

if we define $\alpha_n = -1$, so the vectors $\boldsymbol{x}_1, \ldots \boldsymbol{x}_n$ are linearly dependent. Conversely, suppose that the vectors $\boldsymbol{x}_1, \ldots, \boldsymbol{x}_n$ are linearly dependent so that (2.2) holds for some choice of $\alpha_1, \ldots, \alpha_n$, with at least one of the α_is, say α_n, not equal to zero. Thus, we can solve (2.2) for $\boldsymbol{x}_n$, in which case we get

$$\boldsymbol{x}_n = \sum_{i=1}^{n-1} \left(\frac{-\alpha_i}{\alpha_n} \right) \boldsymbol{x}_i,$$

so that $\boldsymbol{x}_n$ is a linear combination of $\boldsymbol{x}_1, \ldots, \boldsymbol{x}_{n-1}$. This completes the proof. □

We end this section by proving two additional results that we will need later. Note that the first of these theorems, although stated in terms of the columns of a matrix, applies as well to the rows of a matrix.

Theorem 2.5. Consider the $m \times m$ matrix X with columns $\boldsymbol{x}_1, \ldots, \boldsymbol{x}_m$. Then $|X| \neq 0$ if and only if the vectors $\boldsymbol{x}_1, \ldots, \boldsymbol{x}_m$ are linearly independent.

Proof. If $|X| = 0$, then $\text{rank}(X) = r < m$ and so it follows from Theorem 1.9 that there are nonsingular $m \times m$ matrices U and $V = [V_1 \quad V_2]$, with V_1 $m \times r$ such that

$$XU = V \begin{bmatrix} I_r & (0) \\ (0) & (0) \end{bmatrix} = [V_1 \quad (0)]$$

But then the last column of U gives coefficients for a linear combination of $\boldsymbol{x}_1, \ldots, \boldsymbol{x}_m$ which equals the null vector. Thus, if these vectors are to be linearly independent, we must have $|X| \neq 0$. Conversely, if $\boldsymbol{x}_1, \ldots, \boldsymbol{x}_m$ are linearly dependent we can find a vector $\boldsymbol{u} \neq \boldsymbol{0}$ satisfying $X\boldsymbol{u} = \boldsymbol{0}$ and then construct a nonsingular matrix U with $\boldsymbol{u}$ as its last column. In this case, $XU = [W \quad \boldsymbol{0}]$, where W is an $m \times (m-1)$ matrix and, since U is nonsingular,

$$\text{rank}(X) = \text{rank}(XU) = \text{rank}([W \quad \boldsymbol{0}]) \leq m - 1$$

Consequently, if $|X| \neq 0$, so that $\text{rank}(X) = m$, then $\boldsymbol{x}_1, \ldots, \boldsymbol{x}_m$ must be linearly independent. □

Theorem 2.6. The set $\{\boldsymbol{x}_1, \ldots, \boldsymbol{x}_n\}$ of $m \times 1$ vectors is linearly dependent if $n > m$.

Proof. Consider the subset of vectors $\{\boldsymbol{x}_1, \ldots, \boldsymbol{x}_m\}$. If this is a linearly dependent set, it follows from Theorem 2.3(c) that so is the set $\{\boldsymbol{x}_1, \ldots, \boldsymbol{x}_n\}$. Thus, the proof will be complete if we can show that when $\boldsymbol{x}_1, \ldots, \boldsymbol{x}_m$ are linearly independent, then one of the other vectors, say $\boldsymbol{x}_{m+1}$, can be expressed as a linear combination of $\boldsymbol{x}_1, \ldots, \boldsymbol{x}_m$. When $\boldsymbol{x}_1, \ldots, \boldsymbol{x}_m$ are linearly independent, it follows from the previous theorem that if we define X as the $m \times m$ matrix with $\boldsymbol{x}_1, \ldots, \boldsymbol{x}_m$ as its columns, then $|X| \neq 0$ and so X^{-1} exists. Let $\boldsymbol{\alpha} = X^{-1}\boldsymbol{x}_{m+1}$ and note that $\boldsymbol{\alpha} \neq \boldsymbol{0}$ unless $\boldsymbol{x}_{m+1} = \boldsymbol{0}$, in which case the theorem is trivially true due to Theorem 2.3(a). Thus, we have

$$\sum_{i=1}^{m} \alpha_i \boldsymbol{x}_i = X\boldsymbol{\alpha} = XX^{-1}\boldsymbol{x}_{m+1} = \boldsymbol{x}_{m+1},$$

and so the set $\{\boldsymbol{x}_1, \ldots, \boldsymbol{x}_{m+1}\}$ and hence also the set $\{\boldsymbol{x}_1, \ldots, \boldsymbol{x}_n\}$ is linearly dependent. □

4. BASES AND DIMENSION

The concept of dimension is a familiar one from geometry. For example, we recognize a line as a one-dimensional region and a plane as a two-dimensional region. In this section, we generalize this notion to any vector space. The dimension of a vector space can be determined by looking at spanning sets for that vector space. In particular, we need to be able to find the minimum number of vectors necessary for a spanning set.

Definition 2.7. Let $\{\boldsymbol{x}_1, \dots, \boldsymbol{x}_n\}$ be a set of $m \times 1$ vectors in a vector space S. This set is called a basis of S if it spans the vector space S and the vectors $\boldsymbol{x}_1, \dots, \boldsymbol{x}_n$ are linearly independent.

Every vector space, except the vector space consisting only of the null vector $\boldsymbol{0}$, has a basis. Although a basis for a vector space is not uniquely defined, the number of vectors in a basis is unique, and this is what gives us the dimension of a vector space.

Definition 2.8. If the vector space S is $\{\boldsymbol{0}\}$, then the dimension of S, denoted by $\dim(S)$, is defined to be zero. Otherwise, the dimension of the vector space S is the number of vectors in any basis for S.

Example 2.4. Consider the set of $m \times 1$ vectors $\{\boldsymbol{e}_1, \dots, \boldsymbol{e}_m\}$, where for each i, $\boldsymbol{e}_i$ is defined to be the vector whose only nonzero component is the ith component, which is one. Now, the linear combination of the $\boldsymbol{e}_i$s,

$$\sum_{i=1}^{m} \alpha_i \boldsymbol{e}_i = (\alpha_1, \dots, \alpha_m)',$$

will equal $\boldsymbol{0}$ only if $\alpha_1 = \cdots = \alpha_m = 0$, so the vectors $\boldsymbol{e}_1, \dots, \boldsymbol{e}_m$ are linearly independent. Also, if $\boldsymbol{x} = (x_1, \dots, x_m)'$ is an arbitrary vector in R^m, then

$$\boldsymbol{x} = \sum_{i=1}^{m} x_i \boldsymbol{e}_i,$$

so that $\{\boldsymbol{e}_1, \dots, \boldsymbol{e}_m\}$ spans R^m. Thus, $\{\boldsymbol{e}_1, \dots, \boldsymbol{e}_m\}$ is a basis for the m-dimensional space R^m and, in fact, any linearly independent set of m $m \times 1$ vectors will be a basis for R^m. For instance, if the $m \times 1$ vector γ_i has its first i components equal to one while the rest are all zero, then $\{\gamma_1, \dots, \gamma_m\}$ is also a basis of R^m.

Example 2.5. Consider the vector space S spanned by the vectors $\boldsymbol{x}_1 = (1,1,1)'$, $\boldsymbol{x}_2 = (1,0,-1)'$, and $\boldsymbol{x}_3 = (3,2,1)'$. We saw in Example 2.3, that $\{\boldsymbol{x}_1, \boldsymbol{x}_2, \boldsymbol{x}_3\}$ is a linearly dependent set of vectors so that this set is not a basis for S. However, the set $\{\boldsymbol{x}_1, \boldsymbol{x}_2\}$ is linearly independent and $\boldsymbol{x}_3 = 2\boldsymbol{x}_1 + \boldsymbol{x}_2$, so that $\{\boldsymbol{x}_1, \boldsymbol{x}_2\}$ and $\{\boldsymbol{x}_1, \boldsymbol{x}_2, \boldsymbol{x}_3\}$ must span the same vector space. Thus, $\{\boldsymbol{x}_1, \boldsymbol{x}_2\}$ is a basis for S and so S is a two-dimensional subspace, that is, a plane in R^3. Any pair of linearly independent vectors in S will be a basis for S; for example $\{\boldsymbol{x}_1, \boldsymbol{x}_3\}$ and $\{\boldsymbol{x}_2, \boldsymbol{x}_3\}$ are also bases for S.

Every vector $\boldsymbol{x}$ in a vector space can be expressed as a linear combination of the vectors in a spanning set. However, in general, there may be more than one linear combination that yields a particular $\boldsymbol{x}$. Our next result indicates that this is not the case when the spanning set is a basis.

Theorem 2.7. Suppose the set of $m \times 1$ vectors $\{\boldsymbol{x}_1, \dots, \boldsymbol{x}_n\}$ is a basis for the vector space S. Then any vector $\boldsymbol{x} \in S$ has a unique representation as a linear combination of the vectors $\boldsymbol{x}_1, \dots, \boldsymbol{x}_n$.

Proof. Since $\{\boldsymbol{x}_1, \dots, \boldsymbol{x}_n\}$ spans S and $\boldsymbol{x} \in S$, there must exist scalars $\alpha_1, \dots, \alpha_n$ such that

$$\boldsymbol{x} = \sum_{i=1}^{n} \alpha_i \boldsymbol{x}_i$$

Thus, we only need to prove that the representation above is unique. Suppose it is not unique so there exists another set of scalars $\beta_1, \dots, \beta_n$ for which

$$\boldsymbol{x} = \sum_{i=1}^{n} \beta_i \boldsymbol{x}_i$$

But this then implies that

$$\sum_{i=1}^{n} (\alpha_i - \beta_i)\boldsymbol{x}_i = \sum_{i=1}^{n} \alpha_i \boldsymbol{x}_i - \sum_{i=1}^{n} \beta_i \boldsymbol{x}_i = \boldsymbol{x} - \boldsymbol{x} = \boldsymbol{0}$$

Since $\{\boldsymbol{x}_1, \dots, \boldsymbol{x}_n\}$ is a basis, the vectors $\boldsymbol{x}_1, \dots, \boldsymbol{x}_n$ must be linearly independent and so the equation above is satisfied only if $\alpha_i - \beta_i = 0$ for all i. Thus we must have $\alpha_i = \beta_i$, for $i = 1, \dots, n$ and so the representation is unique. □

Some additional useful results regarding vector spaces and their bases are summarized below. The proofs are left to the reader.

Theorem 2.8

(a) Any two bases of a vector space S must have the same number of vectors.
(b) If $\{\boldsymbol{x}_1, \ldots, \boldsymbol{x}_n\}$ is a set of linearly independent vectors in a vector space S and the dimension of S is n, then $\{\boldsymbol{x}_1, \ldots, \boldsymbol{x}_n\}$ is a basis for S.
(c) If the set $\{\boldsymbol{x}_1, \ldots, \boldsymbol{x}_n\}$ spans the vector space S and the dimension of S is n, then the set $\{\boldsymbol{x}_1, \ldots, \boldsymbol{x}_n\}$ must be linearly independent and thus a basis for S.
(d) If the vector space S has dimension n and the set of linearly independent vectors $\{\boldsymbol{x}_1, \ldots \boldsymbol{x}_r\}$ is in S, where $r < n$, there are bases for S which contain this set as a subset.

5. MATRIX RANK AND LINEAR INDEPENDENCE

We have seen that we often work with a vector space through one of its spanning sets. In many situations our vector space has, as a spanning set, vectors that are either the columns or rows of some matrix. We define the following terminology appropriate for such situations.

Definition 2.9. Let X be an $m \times n$ matrix. The subspace of R^n spanned by the m row vectors of X is called the row space of X. The subspace of R^m spanned by the n column vectors of X is called the column space of X.

The column space of X is sometimes also referred to as the range of X, and we will identify it by $R(X)$; that is, $R(X)$ is the vector space given by

$$R(X) = \{\boldsymbol{y} : \boldsymbol{y} = X\boldsymbol{a}, \boldsymbol{a} \in R^n\}$$

Note that the row space of X may be written as $R(X')$.

A consequence of Theorem 2.5 is that the dimension of the column space of a square matrix, that is, the number of linearly independent column vectors, is identical to the rank of that matrix when it is nonsingular. The following result shows that this connection between the number of linearly independent columns of a matrix and the rank of that matrix always holds.

Theorem 2.9. Let X be an $m \times n$ matrix. If r is the number of linearly independent rows of X and c is the number of linearly independent columns of X, then $\text{rank}(X) = r = c$.

Proof. We will only need to prove that $\text{rank}(X) = r$ since this proof can be repeated on X' to prove that $\text{rank}(X) = c$. We will assume that the first r rows of X are linearly independent since, if they are not, elementary row transformations on X will produce such a matrix having the same rank as X. It then follows that the remaining rows of X can be expressed as linear combinations of the first r rows; that is, if X_1 is the $r \times n$ matrix consisting of the first r rows of X, there exists some $(m - r) \times r$ matrix A such that

$$X = \begin{bmatrix} X_1 \\ AX_1 \end{bmatrix} = \begin{bmatrix} \mathrm{I}_r \\ A \end{bmatrix} X_1$$

Now from Theorem 2.6 we know that there can be at most r linearly independent columns in X_1 since these are $r \times 1$ vectors. Thus, we may assume that the last $n - r$ columns of X_1 can be expressed as linear combinations of the first r columns since, if this is not the case, elementary column transformations on X_1 will produce such a matrix having the same rank as X_1. Consequently, if X_{11} is the $r \times r$ matrix with the first r columns of X_1, then there exists an $r \times (n - r)$ matrix B satisfying

$$X = \begin{bmatrix} \mathrm{I}_r \\ A \end{bmatrix} [X_{11} \quad X_{11}B] = \begin{bmatrix} \mathrm{I}_r \\ A \end{bmatrix} X_{11}[\mathrm{I}_r \quad B]$$

If we define the $m \times m$ and $n \times n$ matrices U and V by

$$U = \begin{bmatrix} \mathrm{I}_r & (0) \\ -A & \mathrm{I}_{m-r} \end{bmatrix} \quad \text{and} \quad V = \begin{bmatrix} \mathrm{I}_r & -B \\ (0) & \mathrm{I}_{n-r} \end{bmatrix},$$

then we have

$$UXV = \begin{bmatrix} X_{11} & (0) \\ (0) & (0) \end{bmatrix}$$

Since the determinant of a triangular matrix is equal to the product of its diagonal elements, we find that $|U| = |V| = 1$, so that U and V are nonsingular and thus

$$\mathrm{rank}(X) = \mathrm{rank}(UXV) = \mathrm{rank}(X_{11})$$

Finally, we must have $\mathrm{rank}(X_{11}) = r$, since if not, by Theorem 2.5 the rows of X_{11} would be linearly dependent and this would contradict the already stated linear independence of the rows of $X_1 = [X_{11} \quad X_{11}B]$. □

Example 2.6. An implication of Theorem 2.9 is that the dimension of the column space of a matrix is the same as the dimension of the row space. However, this does not mean that the two vector spaces are the same. As a simple example consider the matrix

$$X = \begin{bmatrix} 0 & 0 & 1 \\ 0 & 1 & 0 \\ 0 & 0 & 0 \end{bmatrix},$$

which has rank 2. The column space of X is the two-dimensional subspace of R^3 composed of all vectors of the form $(a, b, 0)'$, while the row space of X is the two-dimensional subspace of R^3 containing all vectors of the form $(0, a, b)'$. If X is not square then the column space and row space will be subspaces of different Euclidean spaces. For instance, if

$$X = \begin{bmatrix} 1 & 0 & 0 & 1 \\ 0 & 1 & 0 & 1 \\ 0 & 0 & 1 & 1 \end{bmatrix},$$

the column space is R^3, while the row space is the three-dimensional subspace of R^4 consisting of all vectors of the form $(a, b, c, a + b + c)'$.

The formulation of matrix rank in terms of the number of linearly independent rows or columns of the matrix is often easier to work with than our original definition in terms of submatrices. This is evidenced in the proof of the following basic results regarding the rank of a matrix.

Theorem 2.10. Let A be an $m \times n$ matrix. Then the following hold.

(a) If B is an $n \times p$ matrix, $\text{rank}(AB) \leq \min\{\text{rank}(A), \text{rank}(B)\}$.
(b) If B is an $m \times n$ matrix, $\text{rank}(A + B) \leq \text{rank}(A) + \text{rank}(B)$.
(c) $\text{rank}(A) = \text{rank}(A') = \text{rank}(AA') = \text{rank}(A'A)$.

Proof. Note that we can write

$$(AB)_{\cdot i} = \sum_{j=1}^{n} b_{ji}(A)_{\cdot j};$$

that is, each column of AB can be expressed as a linear combination of the columns of A, and so the number of linearly independent columns in AB can be no more than the number of linearly independent columns in A. Thus $\text{rank}(AB) \leq \text{rank}(A)$. Similarly, each row of AB can be expressed as a linear combination of the rows of B from which we get $\text{rank}(AB) \leq \text{rank}(B)$, and so property (a) is proven. To prove (b), note that by using partitioned matrices we can write

$$A + B = [A \quad B]\begin{bmatrix} \text{I}_n \\ \text{I}_n \end{bmatrix}$$

So using property (a) on the right-hand side of the equation above, we find that

$$\text{rank}(A + B) \leq \text{rank}([A \quad B]) \leq \text{rank}(A) + \text{rank}(B),$$

where the final inequality follows from the fact that the number of linearly independent columns of $[A \quad B]$ cannot exceed the sum of the numbers of linearly independent columns in A and B. In proving (c), note that it follows immediately that $\text{rank}(A) = \text{rank}(A')$. It will suffice to prove that $\text{rank}(A) = \text{rank}(A'A)$ since this can then be used on A' to prove that $\text{rank}(A') = \text{rank}\{(A')'A'\} = \text{rank}(AA')$. If $\text{rank}(A) = r$, there exists a full column rank $m \times r$ matrix A_1 such that, after possibly interchanging some of the columns of A, $A = [A_1 \quad A_1C] = A_1[\text{I}_r \quad C]$, where C is an $r \times n - r$ matrix. As a result, we have

$$A'A = \begin{bmatrix} \text{I}_r \\ C' \end{bmatrix} A_1'A_1[\text{I}_r \quad C]$$

Note that

$$EA'AE' = \begin{bmatrix} A_1'A_1 & (0) \\ (0) & (0) \end{bmatrix} \quad \text{if } E = \begin{bmatrix} \text{I}_r & (0) \\ -C' & \text{I}_{n-r} \end{bmatrix},$$

and since the triangular matrix E has $|E| = 1$, E is nonsingular, so $\text{rank}(A'A) = \text{rank}(EA'AE') = \text{rank}(A_1'A_1)$. If $A_1'A_1$ is less than full rank, then by Theorem 2.5 its columns are linearly dependent so we can find an $r \times 1$ vector $\boldsymbol{x} \neq \boldsymbol{0}$ such that $A_1'A_1\boldsymbol{x} = \boldsymbol{0}$, which implies that $\boldsymbol{x}'A_1'A_1\boldsymbol{x} = (A_1\boldsymbol{x})'(A_1\boldsymbol{x}) = 0$. However, for any real vector $\boldsymbol{y}$, $\boldsymbol{y}'\boldsymbol{y} = 0$ only if $\boldsymbol{y} = \boldsymbol{0}$ and hence $A_1\boldsymbol{x} = \boldsymbol{0}$. But this contradicts $\text{rank}(A_1) = r$, and so we must have $\text{rank}(A'A) = \text{rank}(A_1'A_1) = r$. □

The next result gives some relationships between the rank of a partitioned matrix and the ranks of its submatrices. The proofs, which are straightforward, are left to the reader.

Theorem 2.11. Let A, B, and C be any matrices for which the partitioned matrices below are defined. Then

(a) $\text{rank}([A \quad B]) \geq \max\{\text{rank}(A), \text{rank}(B)\}$

(b) $\operatorname{rank}\left(\begin{bmatrix} A & (0) \\ (0) & B \end{bmatrix}\right) = \operatorname{rank}\left(\begin{bmatrix} (0) & B \\ A & (0) \end{bmatrix}\right) = \operatorname{rank}(A) + \operatorname{rank}(B)$

(c) $\operatorname{rank}\left(\begin{bmatrix} A & (0) \\ C & B \end{bmatrix}\right) = \operatorname{rank}\left(\begin{bmatrix} C & B \\ A & (0) \end{bmatrix}\right) = \operatorname{rank}\left(\begin{bmatrix} B & C \\ (0) & A \end{bmatrix}\right)$

$$= \operatorname{rank}\left(\begin{bmatrix} (0) & A \\ B & C \end{bmatrix}\right) \geq \operatorname{rank}(A) + \operatorname{rank}(B)$$

Our next result gives a useful inequality for the rank of the product of three matrices.

Theorem 2.12. Let A, B, and C be $p \times m$, $m \times n$, and $n \times q$ matrices, respectively. Then

$$\operatorname{rank}(ABC) \geq \operatorname{rank}(AB) + \operatorname{rank}(BC) - \operatorname{rank}(B)$$

Proof. It follows from Theorem 2.11(c) that

$$\operatorname{rank}\left(\begin{bmatrix} B & BC \\ AB & (0) \end{bmatrix}\right) \geq \operatorname{rank}(AB) + \operatorname{rank}(BC) \tag{2.3}$$

But, since

$$\begin{bmatrix} B & BC \\ AB & (0) \end{bmatrix} = \begin{bmatrix} \mathrm{I}_m & (0) \\ A & \mathrm{I}_p \end{bmatrix} \begin{bmatrix} B & (0) \\ (0) & -ABC \end{bmatrix} \begin{bmatrix} \mathrm{I}_n & C \\ (0) & \mathrm{I}_q \end{bmatrix},$$

where, clearly, the first and last matrices on the right-hand side are nonsingular, we must also have

$$\operatorname{rank}\left(\begin{bmatrix} B & BC \\ AB & (0) \end{bmatrix}\right) = \operatorname{rank}\left(\begin{bmatrix} B & (0) \\ (0) & -ABC \end{bmatrix}\right) = \operatorname{rank}(B) + \operatorname{rank}(ABC) \tag{2.4}$$

Combining (2.3) and (2.4) we obtain the desired result. □

A special case of Theorem 2.12 is obtained when $n = m$ and B is the $m \times m$ identity matrix. The resulting inequality gives a lower bound for the rank of a matrix product complementing the upper bound given in Theorem 2.10(a).

Corollary 2.12.1. If A is an $m \times n$ matrix and B an $n \times p$ matrix, then

$$\operatorname{rank}(AB) \geq \operatorname{rank}(A) + \operatorname{rank}(B) - n$$

6. ORTHONORMAL BASES AND PROJECTIONS

If each vector in a basis for a vector space S is orthogonal to every other vector in that basis, the basis is called an orthogonal basis. In this case, the vectors can be viewed as a set of coordinate axes for the vector space S. We will find it useful also to have each vector in our basis scaled to unit length, in which case we would have an orthonormal basis.

Suppose the set $\{\boldsymbol{x}_1,\ldots,\boldsymbol{x}_r\}$ forms a basis for the vector space S, and we wish to obtain an orthonormal basis for S. Unless $r = 1$, an orthonormal basis is not unique so there are many different orthonormal bases that we can construct. One method of obtaining an orthonormal basis from a given basis $\{\boldsymbol{x}_1,\ldots,\boldsymbol{x}_r\}$ is called Gram–Schmidt orthonormalization. First, we construct the set $\{\boldsymbol{y}_1,\ldots,\boldsymbol{y}_r\}$ of orthogonal vectors given by

$$\begin{aligned}
\boldsymbol{y}_1 &= \boldsymbol{x}_1,\\
\boldsymbol{y}_2 &= \boldsymbol{x}_2 - \frac{\boldsymbol{x}_2'\boldsymbol{y}_1}{\boldsymbol{y}_1'\boldsymbol{y}_1}\boldsymbol{y}_1,\\
&\vdots\\
\boldsymbol{y}_r &= \boldsymbol{x}_r - \frac{\boldsymbol{x}_r'\boldsymbol{y}_1}{\boldsymbol{y}_1'\boldsymbol{y}_1}\boldsymbol{y}_1 - \cdots - \frac{\boldsymbol{x}_r'\boldsymbol{y}_{r-1}}{\boldsymbol{y}_{r-1}'\boldsymbol{y}_{r-1}}\boldsymbol{y}_{r-1},
\end{aligned}\tag{2.5}$$

and then the set of orthonormal vectors $\{\boldsymbol{z}_1,\ldots,\boldsymbol{z}_r\}$, where for each i,

$$\boldsymbol{z}_i = \frac{\boldsymbol{y}_i}{(\boldsymbol{y}_i'\boldsymbol{y}_i)^{1/2}}$$

Note that the linear independence of $\boldsymbol{x}_1,\ldots,\boldsymbol{x}_r$ guarantees the linear independence of $\boldsymbol{y}_1,\ldots,\boldsymbol{y}_r$. Thus, we have the following result.

Theorem 2.13. Every r-dimensional vector space, except the zero-dimensional space $\{\boldsymbol{0}\}$, has an orthonormal basis.

If $\{\boldsymbol{z}_1,\ldots,\boldsymbol{z}_r\}$ is a basis for the vector space S and $\boldsymbol{x} \in S$, then from Theorem 2.7 we know that $\boldsymbol{x}$ can be uniquely expressed in the form $\boldsymbol{x} = \alpha_1\boldsymbol{z}_1 + \cdots + \alpha_r\boldsymbol{z}_r$. When $\{\boldsymbol{z}_1,\ldots,\boldsymbol{z}_r\}$ is an orthonormal basis, each of the scalars $\alpha_1,\ldots,\alpha_r$ has a rather simple form; premultiplication of the equation for $\boldsymbol{x}$ above by $\boldsymbol{z}_i'$ yields the identity $\alpha_i = \boldsymbol{z}_i'\boldsymbol{x}$.

Example 2.7. We will find an orthonormal basis for the three-dimensional vector space S which has as a basis $\{\boldsymbol{x}_1,\boldsymbol{x}_2,\boldsymbol{x}_3\}$, where

$$x_1 = \begin{bmatrix} 1 \\ 1 \\ 1 \\ 1 \end{bmatrix}, \qquad x_2 = \begin{bmatrix} 1 \\ -2 \\ 1 \\ -2 \end{bmatrix}, \qquad x_3 = \begin{bmatrix} 3 \\ 1 \\ 1 \\ -1 \end{bmatrix}$$

The orthogonal $\boldsymbol{y}_i$s are given by $\boldsymbol{y}_1 = (1, 1, 1, 1)'$,

$$\boldsymbol{y}_2 = \begin{bmatrix} 1 \\ -2 \\ 1 \\ -2 \end{bmatrix} - \frac{(-2)}{4} \begin{bmatrix} 1 \\ 1 \\ 1 \\ 1 \end{bmatrix} = \begin{bmatrix} 3/2 \\ -3/2 \\ 3/2 \\ -3/2 \end{bmatrix},$$

and

$$\boldsymbol{y}_3 = \begin{bmatrix} 3 \\ 1 \\ 1 \\ -1 \end{bmatrix} - \frac{(4)}{(4)} \begin{bmatrix} 1 \\ 1 \\ 1 \\ 1 \end{bmatrix} - \frac{(6)}{(9)} \begin{bmatrix} 3/2 \\ -3/2 \\ 3/2 \\ -3/2 \end{bmatrix} = \begin{bmatrix} 1 \\ 1 \\ -1 \\ -1 \end{bmatrix}$$

Normalizing these vectors yields the orthonormal basis $\{z_1, z_2, z_3\}$, where

$$z_1 = \begin{bmatrix} 1/2 \\ 1/2 \\ 1/2 \\ 1/2 \end{bmatrix}, \qquad z_2 = \begin{bmatrix} 1/2 \\ -1/2 \\ 1/2 \\ -1/2 \end{bmatrix}, \qquad z_3 = \begin{bmatrix} 1/2 \\ 1/2 \\ -1/2 \\ -1/2 \end{bmatrix}$$

Thus, for any $\boldsymbol{x} \in S$, $\boldsymbol{x} = \alpha_1 z_1 + \alpha_2 z_2 + \alpha_3 z_3$, where $\alpha_i = \boldsymbol{x}' z_i$. For instance, since $\boldsymbol{x}_3' z_1 = 2$, $\boldsymbol{x}_3' z_2 = 2$, $\boldsymbol{x}_3' z_3 = 2$, we have $\boldsymbol{x}_3 = 2z_1 + 2z_2 + 2z_3$.

Now if S is a vector subspace of R^m and $\boldsymbol{x} \in R^m$, the following indicates how the vector $\boldsymbol{x}$ can be decomposed into the sum of a vector in S and another vector.

Theorem 2.14. Let $\{z_1, \dots, z_r\}$ be an orthonormal basis for some vector subspace, S, of R^m. Then each $\boldsymbol{x} \in R^m$ can be expressed uniquely as

$$\boldsymbol{x} = \boldsymbol{u} + \boldsymbol{v},$$

where $\boldsymbol{u} \in S$ and $\boldsymbol{v}$ is a vector that is orthogonal to every vector in S.

Proof. It follows from Theorem 2.8(d) that we can find vectors $z_{r+1}, \dots, z_m$ so that the set $\{z_1, \dots, z_m\}$ is an orthonormal basis for the m-dimensional Euclidean space R^m. It also follows from Theorem 2.7 that there is a unique set of scalars $\alpha_1, \dots, \alpha_m$ such that

$$\boldsymbol{x} = \sum_{i=1}^{m} \alpha_i \boldsymbol{z}_i$$

Thus, if we let $\boldsymbol{u} = \alpha_1 \boldsymbol{z}_1 + \cdots + \alpha_r \boldsymbol{z}_r$ and $\boldsymbol{v} = \alpha_{r+1} \boldsymbol{z}_{r+1} + \cdots + \alpha_m \boldsymbol{z}_m$, we have, uniquely, $\boldsymbol{x} = \boldsymbol{u} + \boldsymbol{v}$, $\boldsymbol{u} \in S$, and $\boldsymbol{v}$ will be orthogonal to every vector in S due to the orthogonality of the vectors $\boldsymbol{z}_1, \ldots, \boldsymbol{z}_m$. □

The vector $\boldsymbol{u}$ in the theorem above is known as the orthogonal projection of $\boldsymbol{x}$ onto S. When $m = 3$, the orthogonal projection has a simple geometrical description that allows for visualization. If, for instance, $\boldsymbol{x}$ is a point in three-dimensional space and S is a two-dimensional subspace, then the orthogonal projection $\boldsymbol{u}$ of $\boldsymbol{x}$ will be the point of intersection of the plane S and the line that is perpendicular to S and passes through $\boldsymbol{x}$.

The importance of the orthogonal projection $\boldsymbol{u}$ in many applications arises out of the fact that it is the closest point in S to $\boldsymbol{x}$. That is, if $\boldsymbol{y}$ is any other point in S and $\mathrm{d_I}$ is the Euclidean distance function, then $\mathrm{d_I}(\boldsymbol{x}, \boldsymbol{u}) \leq \mathrm{d_I}(\boldsymbol{x}, \boldsymbol{y})$. This is fairly simple to verify. Since $\boldsymbol{u}$ and $\boldsymbol{y}$ are in S, it follows from the decomposition $\boldsymbol{x} = \boldsymbol{u} + \boldsymbol{v}$ that the vector $\boldsymbol{u} - \boldsymbol{y}$ is orthogonal to $\boldsymbol{v} = \boldsymbol{x} - \boldsymbol{u}$ and, hence, $(\boldsymbol{x} - \boldsymbol{u})'(\boldsymbol{u} - \boldsymbol{y}) = 0$. Consequently,

$$\begin{aligned}
\{\mathrm{d_I}(\boldsymbol{x}, \boldsymbol{y})\}^2 &= (\boldsymbol{x} - \boldsymbol{y})'(\boldsymbol{x} - \boldsymbol{y}) = \{(\boldsymbol{x} - \boldsymbol{u}) + (\boldsymbol{u} - \boldsymbol{y})\}'\{(\boldsymbol{x} - \boldsymbol{u}) + (\boldsymbol{u} - \boldsymbol{y})\} \\
&= (\boldsymbol{x} - \boldsymbol{u})'(\boldsymbol{x} - \boldsymbol{u}) + (\boldsymbol{u} - \boldsymbol{y})'(\boldsymbol{u} - \boldsymbol{y}) + 2(\boldsymbol{x} - \boldsymbol{u})'(\boldsymbol{u} - \boldsymbol{y}) \\
&= (\boldsymbol{x} - \boldsymbol{u})'(\boldsymbol{x} - \boldsymbol{u}) + (\boldsymbol{u} - \boldsymbol{y})'(\boldsymbol{u} - \boldsymbol{y}) = \{\mathrm{d_I}(\boldsymbol{x}, \boldsymbol{u})\}^2 + \{\mathrm{d_I}(\boldsymbol{u}, \boldsymbol{y})\}^2,
\end{aligned}$$

from which $\mathrm{d_I}(\boldsymbol{x}, \boldsymbol{u}) \leq \mathrm{d_I}(\boldsymbol{x}, \boldsymbol{y})$ follows since $\{\mathrm{d_I}(\boldsymbol{u}, \boldsymbol{y})\}^2 \geq 0$.

Example 2.8. Simple linear regression relates a response variable y to one explanatory variable x through the model

$$y = \beta_0 + \beta_1 x + \epsilon;$$

that is, if this model is correct, then observed ordered pairs (x, y) should be clustered about some line in the x, y plane. Suppose we have N observations, (x_i, y_i), $i = 1, \ldots, N$, and we form the $N \times 1$ vector $\boldsymbol{y} = (y_1, \ldots, y_N)'$ and the $N \times 2$ matrix

$$X = \begin{bmatrix} 1 & x_1 \\ 1 & x_2 \\ \vdots & \vdots \\ 1 & x_N \end{bmatrix} = [\mathbf{1}_N \quad \boldsymbol{x}]$$

The least squares estimator $\hat{\boldsymbol{\beta}}$ of $\boldsymbol{\beta} = (\beta_0, \beta_1)'$ minimizes the sum of squared errors given by

$$(\boldsymbol{y} - \hat{\boldsymbol{y}})'(\boldsymbol{y} - \hat{\boldsymbol{y}}) = (\boldsymbol{y} - X\hat{\boldsymbol{\beta}})'(\boldsymbol{y} - X\hat{\boldsymbol{\beta}}).$$

In Chapter 8 we will see how to find $\hat{\boldsymbol{\beta}}$ using differential methods. Here we will use the geometrical properties of projections to determine $\hat{\boldsymbol{\beta}}$. For any choice of $\hat{\boldsymbol{\beta}}$, $\hat{\boldsymbol{y}} = X\hat{\boldsymbol{\beta}}$ gives a point in the subspace of R^N spanned by the columns of X, that is, the plane spanned by the two vectors $\mathbf{1}_N$ and $\boldsymbol{x}$. Thus, the point $\hat{\boldsymbol{y}}$ that minimizes the distance from $\boldsymbol{y}$ will be given by the orthogonal projection of $\boldsymbol{y}$ onto this plane spanned by $\mathbf{1}_N$ and $\boldsymbol{x}$. This means that $\boldsymbol{y} - \hat{\boldsymbol{y}}$ must be orthogonal to both $\mathbf{1}_N$ and $\boldsymbol{x}$. This leads to the two normal equations

$$0 = (\boldsymbol{y} - \hat{\boldsymbol{y}})'\mathbf{1}_N = \boldsymbol{y}'\mathbf{1}_N - \hat{\boldsymbol{\beta}}'X'\mathbf{1}_N = \sum_{i=1}^{N} y_i - \hat{\beta}_0 N - \hat{\beta}_1 \sum_{i=1}^{N} x_i,$$

$$0 = (\boldsymbol{y} - \hat{\boldsymbol{y}})'\boldsymbol{x} = \boldsymbol{y}'\boldsymbol{x} - \hat{\boldsymbol{\beta}}'X'\boldsymbol{x} = \sum_{i=1}^{N} x_i y_i - \hat{\beta}_0 \sum_{i=1}^{N} x_i - \hat{\beta}_i \sum_{i=1}^{N} x_i^2,$$

which when solved simultaneously for $\hat{\beta}_0$ and $\hat{\beta}_1$, yields

$$\hat{\beta}_1 = \frac{\sum_{i=1}^{N} x_i y_i - N\bar{x}\bar{y}}{\sum_{i=1}^{N} x_i^2 - N\bar{x}^2}, \qquad \hat{\beta}_0 = \bar{y} - \hat{\beta}_1 \bar{x}$$

If we want to test the hypothesis that $\beta_1 = 0$, we would consider the reduced model

$$y = \beta_0 + \epsilon$$

and least squares estimation here only requires an estimate of β_0. In this case, the vector of fitted values satisfies $\hat{\boldsymbol{y}} = \hat{\beta}_0 \mathbf{1}_N$, so for any choice of $\hat{\beta}_0$, $\hat{\boldsymbol{y}}$ will be given by a point on the line passing through the origin and $\mathbf{1}_N$. Thus if $\hat{\boldsymbol{y}}$ is to minimize the sum of squared errors and hence the distance from $\boldsymbol{y}$, then it must be given by the orthogonal projection of $\boldsymbol{y}$ onto this line. Consequently, we must have

$$0 = (\boldsymbol{y} - \hat{\boldsymbol{y}})'\mathbf{1}_N = (\boldsymbol{y} - \hat{\beta}_0 \mathbf{1}_N)'\mathbf{1}_N = \sum_{i=1}^{N} y_i - \hat{\beta}_0 N,$$

or simply

$$\hat{\beta}_0 = \bar{y}$$

The vector $\boldsymbol{v}$ in Theorem 2.14 is called the component of $\boldsymbol{x}$ orthogonal to S. It is one vector belonging to what is known as the orthogonal complement of S.

Definition 2.10. Let S be a vector subspace of R^m. The orthogonal complement of S, denoted by $S^\perp$, is the collection of all vectors in R^m that are orthogonal to every vector in S; that is, $S^\perp = \{\boldsymbol{x}\colon \boldsymbol{x} \in R^m \text{ and } \boldsymbol{x}'\boldsymbol{y} = 0 \text{ for all } \boldsymbol{y} \in S\}$.

Theorem 2.15. If S is a vector subspace of R^m then its orthogonal complement $S^\perp$ is also a vector subspace of R^m.

Proof. Suppose that $\boldsymbol{x}_1 \in S^\perp$ and $\boldsymbol{x}_2 \in S^\perp$ so that $\boldsymbol{x}_1'\boldsymbol{y} = \boldsymbol{x}_2'\boldsymbol{y} = 0$ for any $\boldsymbol{y} \in S$. Consequently, for any $\boldsymbol{y} \in S$ and any scalars α_1 and α_2, we have

$$(\alpha_1\boldsymbol{x}_1 + \alpha_2\boldsymbol{x}_2)'\boldsymbol{y} = \alpha_1\boldsymbol{x}_1'\boldsymbol{y} + \alpha_2\boldsymbol{x}_2'\boldsymbol{y} = 0,$$

and so $(\alpha_1\boldsymbol{x}_1 + \alpha_2\boldsymbol{x}_2) \in S^\perp$ and thus $S^\perp$ is a vector space. □

A consequence of the following theorem is that if S is a vector subspace of R^m and the dimension of S is r, then the dimension of $S^\perp$ is $m - r$.

Theorem 2.16. Suppose $\{\boldsymbol{z}_1, \ldots, \boldsymbol{z}_m\}$ is an orthonormal basis for R^m and $\{\boldsymbol{z}_1, \ldots, \boldsymbol{z}_r\}$ is an orthonormal basis for the vector subspace S. Then $\{\boldsymbol{z}_{r+1}, \ldots, \boldsymbol{z}_m\}$ is an orthonormal basis for $S^\perp$.

Proof. Let T be the vector space spanned by $\{\boldsymbol{z}_{r+1}, \ldots, \boldsymbol{z}_m\}$. We must show that this vector space is the same as $S^\perp$. If $\boldsymbol{x} \in T$ and $\boldsymbol{y} \in S$, then there exist scalars $\alpha_1, \ldots, \alpha_m$ such that $\boldsymbol{y} = \alpha_1\boldsymbol{z}_1 + \cdots + \alpha_r\boldsymbol{z}_r$ and $\boldsymbol{x} = \alpha_{r+1}\boldsymbol{z}_{r+1} + \cdots + \alpha_m\boldsymbol{z}_m$. Due to the orthogonality of the $\boldsymbol{z}_i$s, $\boldsymbol{x}'\boldsymbol{y} = 0$, so $\boldsymbol{x} \in S^\perp$ and thus $T \subseteq S^\perp$. Conversely, suppose that $\boldsymbol{x} \in S^\perp$. Since $\boldsymbol{x}$ is also in R^m, there exist scalars $\alpha_1, \ldots, \alpha_m$ such that $\boldsymbol{x} = \alpha_1\boldsymbol{z}_1 + \cdots + \alpha_m\boldsymbol{z}_m$. Now if we let $\boldsymbol{y} = \alpha_1\boldsymbol{z}_1 + \cdots + \alpha_r\boldsymbol{z}_r$, then $\boldsymbol{y} \in S$, and since $\boldsymbol{x} \in S^\perp$ we must have $\boldsymbol{x}'\boldsymbol{y} = \alpha_1^2 + \cdots + \alpha_r^2 = 0$. But this can only happen if $\alpha_1 = \cdots = \alpha_r = 0$, in which case $\boldsymbol{x} = \alpha_{r+1}\boldsymbol{z}_{r+1} + \cdots + \alpha_m\boldsymbol{z}_m$ and so $\boldsymbol{x} \in T$. Thus, we also have $S^\perp \subseteq T$, and so this establishes that $T = S^\perp$. □

7. PROJECTION MATRICES

The orthogonal projection of an $m \times 1$ vector $\boldsymbol{x}$ onto a vector space S can be conveniently expressed in matrix form. Let $\{\boldsymbol{z}_1, \ldots, \boldsymbol{z}_r\}$ be any orthonormal basis for S while $\{\boldsymbol{z}_1, \ldots, \boldsymbol{z}_m\}$ is an orthonormal basis for R^m. Suppose $\alpha_1, \ldots, \alpha_m$ are the constants satisfying the relationship

$$\boldsymbol{x} = (\alpha_1\boldsymbol{z}_1 + \cdots + \alpha_r\boldsymbol{z}_r) + (\alpha_{r+1}\boldsymbol{z}_{r+1} + \cdots + \alpha_m\boldsymbol{z}_m) = \boldsymbol{u} + \boldsymbol{v},$$

where $\boldsymbol{u}$ and $\boldsymbol{v}$ are as previously defined. Write $\boldsymbol{\alpha} = (\boldsymbol{\alpha}_1', \boldsymbol{\alpha}_2')'$ and $Z = [Z_1 \quad Z_2]$, where $\boldsymbol{\alpha}_1 = (\alpha_1, \dots, \alpha_r)'$, $\boldsymbol{\alpha}_2 = (\alpha_{r+1}, \dots, \alpha_m)'$, $Z_1 = (\boldsymbol{z}_1, \dots, \boldsymbol{z}_r)$, and $Z_2 = (\boldsymbol{z}_{r+1}, \dots, \boldsymbol{z}_m)$. Then the expression for $\boldsymbol{x}$ given above can be written as

$$\boldsymbol{x} = Z\boldsymbol{\alpha} = Z_1\boldsymbol{\alpha}_1 + Z_2\boldsymbol{\alpha}_2;$$

that is, $\boldsymbol{u} = Z_1\boldsymbol{\alpha}_1$ and $\boldsymbol{v} = Z_2\boldsymbol{\alpha}_2$. Due to the orthonormality of the $\boldsymbol{z}_i$s, we have $Z_1'Z_1 = \mathrm{I}_r$ and $Z_1'Z_2 = (0)$, and so

$$Z_1Z_1'\boldsymbol{x} = Z_1Z_1'Z\boldsymbol{\alpha} = Z_1Z_1'[Z_1 \quad Z_2]\begin{bmatrix}\boldsymbol{\alpha}_1\\ \boldsymbol{\alpha}_2\end{bmatrix} = [Z_1 \quad (0)]\begin{bmatrix}\boldsymbol{\alpha}_1\\ \boldsymbol{\alpha}_2\end{bmatrix} = Z_1\boldsymbol{\alpha}_1 = \boldsymbol{u}$$

Thus, we have the following result.

Theorem 2.17. Suppose the columns of the $m \times r$ matrix Z_1 form an orthonormal basis for the vector space S which is a subspace of R^m. If $\boldsymbol{x} \in R^m$, the orthogonal projection of $\boldsymbol{x}$ onto S is given by $Z_1Z_1'\boldsymbol{x}$.

The matrix Z_1Z_1' appearing in Theorem 2.17 is called the projection matrix for the vector space S and sometimes will be denoted by P_S. Similarly, Z_2Z_2' is the projection matrix for $S^\perp$ and $ZZ' = \mathrm{I}_m$ is the projection matrix for R^m. Since $ZZ' = Z_1Z_1' + Z_2Z_2'$, we have the simple equation $Z_2Z_2' = \mathrm{I}_m - Z_1Z_1'$ relating the projection matrices of a vector subspace and its orthogonal complement. Although a vector space does not have a unique orthonormal basis, the projection matrix formed from these orthonormal bases is unique.

Theorem 2.18. Suppose the columns of the $m \times r$ matrices Z_1 and W_1 each form an orthonormal basis for the r-dimensional vector space S. Then $Z_1Z_1' = W_1W_1'$.

Proof. Each column of W_1 can be written as a linear combination of the columns of Z_1 since the columns of Z_1 span S and each column of W_1 is in S; that is, there exists an $r \times r$ matrix P such that $W_1 = Z_1P$. But $Z_1'Z_1 = W_1'W_1 = \mathrm{I}_r$, since each matrix has orthonormal columns. Thus,

$$\mathrm{I}_r = W_1'W_1 = P'Z_1'Z_1P = P'\mathrm{I}_rP = P'P,$$

so that P is an orthogonal matrix. Consequently, P also satisfies $PP' = \mathrm{I}_r$, and so

$$W_1W_1' = Z_1PP'Z_1' = Z_1\mathrm{I}_rZ_1' = Z_1Z_1' \qquad \square$$

We will take another look at the Gram–Schmidt orthonormalization procedure, this time utilizing projection matrices. The procedure takes an initial linearly independent set of vectors $\{\boldsymbol{x}_1,\ldots,\boldsymbol{x}_r\}$, which is transformed to an orthogonal set $\{\boldsymbol{y}_1,\ldots,\boldsymbol{y}_r\}$, which is then transformed to an orthonormal set $\{\boldsymbol{z}_1,\ldots,\boldsymbol{z}_r\}$. It is very easy to verify that for $i = 1,\ldots,r-1$, the vector $\boldsymbol{y}_{i+1}$ can be expressed as

$$\boldsymbol{y}_{i+1} = \left(\mathrm{I}_m - \sum_{j=1}^{i} \boldsymbol{z}_j \boldsymbol{z}_j'\right)\boldsymbol{x}_{i+1};$$

that is, $\boldsymbol{y}_{i+1} = (\mathrm{I}_m - Z_{(i)}Z_{(i)}')\boldsymbol{x}_{i+1}$, where $Z_{(i)} = (\boldsymbol{z}_1,\ldots,\boldsymbol{z}_i)$. Thus, the $(i+1)$th orthogonal vector $\boldsymbol{y}_{i+1}$ is obtained as the projection of the $(i+1)$th original vector onto the orthogonal complement of the vector space spanned by the first i orthogonal vectors, $\boldsymbol{y}_1,\ldots,\boldsymbol{y}_i$.

The Gram–Schmidt orthonormalization process represents one method of obtaining an orthonormal basis for a vector space S from a given basis $\{\boldsymbol{x}_1,\ldots,\boldsymbol{x}_r\}$. In general, if we define the $m \times r$ matrix $X_1 = (\boldsymbol{x}_1,\ldots,\boldsymbol{x}_r)$, the columns of

$$Z_1 = X_1 A \tag{2.6}$$

will form an orthonormal basis for S if A is any $r \times r$ matrix for which

$$Z_1'Z_1 = A'X_1'X_1A = \mathrm{I}_r$$

The matrix A must be nonsingular since we must have $\text{rank}(X_1) = \text{rank}(Z_1) = r$, so A^{-1} exists, and $X_1'X_1 = (A^{-1})'A^{-1}$ or $(X_1'X_1)^{-1} = AA'$; that is, A is a square root matrix of $(X_1'X_1)^{-1}$. Consequently, we can obtain an expression for the projection matrix P_S onto the vector space S in terms of X_1 as

$$P_S = Z_1Z_1' = X_1AA'X_1' = X_1(X_1'X_1)^{-1}X_1' \tag{2.7}$$

Note that the Gram–Schmidt equations given in (2.5) can be written in matrix form as $Y_1 = X_1T$, where $Y_1 = (\boldsymbol{y}_1,\ldots,\boldsymbol{y}_r)$, $X_1 = (\boldsymbol{x}_1,\ldots,\boldsymbol{x}_r)$, and T is an $r \times r$ upper triangular matrix with each diagonal element equal to 1. The normalization to produce Z_1 can then be written as $Z_1 = X_1TD^{-1}$, where D is the diagonal matrix with the positive square root of $\boldsymbol{y}_i'\boldsymbol{y}_i$ as its ith diagonal element. Consequently, the matrix $A = TD^{-1}$ is upper triangular with positive diagonal elements. Thus the Gram–Schmidt orthonormalization is the particular case of equation (2.6) in which the matrix A has been chosen to be the upper triangular square root matrix of $(X_1'X_1)^{-1}$ having positive diagonal elements.

Example 2.9. Using the basis $\{\boldsymbol{x}_1, \boldsymbol{x}_2, \boldsymbol{x}_3\}$ from Example 2.7, we form the X_1 matrix

$$X_1 = \begin{bmatrix} 1 & 1 & 3 \\ 1 & -2 & 1 \\ 1 & 1 & 1 \\ 1 & -2 & -1 \end{bmatrix},$$

and it is easy to verify that

$$X_1'X_1 = \begin{bmatrix} 4 & -2 & 4 \\ -2 & 10 & 4 \\ 4 & 4 & 12 \end{bmatrix}, \qquad (X_1'X_1)^{-1} = \frac{1}{36}\begin{bmatrix} 26 & 10 & -12 \\ 10 & 8 & -6 \\ -12 & -6 & 9 \end{bmatrix}$$

Thus, the projection matrix for the vector space S spanned by $\{\boldsymbol{x}_1, \boldsymbol{x}_2, \boldsymbol{x}_3\}$ is given by

$$P_S = X_1(X_1'X_1)^{-1}X_1' = \frac{1}{4}\begin{bmatrix} 3 & 1 & 1 & -1 \\ 1 & 3 & -1 & 1 \\ 1 & -1 & 3 & 1 \\ -1 & 1 & 1 & 3 \end{bmatrix}$$

This, of course, is the same as Z_1Z_1', where $Z_1 = (\boldsymbol{z}_1, \boldsymbol{z}_2, \boldsymbol{z}_3)$ and $\boldsymbol{z}_1, \boldsymbol{z}_2, \boldsymbol{z}_3$ are the vectors obtained by the Gram–Schmidt orthonormalization in the previous example. Now if $\boldsymbol{x} = (1, 2, -1, 0)'$, then the projection of $\boldsymbol{x}$ onto S is $X_1(X_1'X_1)^{-1}X_1'\boldsymbol{x} = \boldsymbol{x}$; the projection of $\boldsymbol{x}$ is equal to $\boldsymbol{x}$ since $\boldsymbol{x} = \boldsymbol{x}_3 - \boldsymbol{x}_1 - \boldsymbol{x}_2 \in S$. On the other hand, if $\boldsymbol{x} = (1, -1, 2, 1)'$, then the projection of $\boldsymbol{x}$ is given by $\boldsymbol{u} = X_1(X_1'X_1)^{-1}X_1'\boldsymbol{x} = (\frac{3}{4}, -\frac{3}{4}, \frac{9}{4}, \frac{3}{4})'$. The component of $\boldsymbol{x}$ orthogonal to S, or in other words, the orthogonal projection of $\boldsymbol{x}$ onto $S^{\perp}$, is $\{\mathrm{I} - X_1(X_1'X_1)^{-1}X_1'\}\boldsymbol{x} = \boldsymbol{x} - X_1(X_1'X_1)^{-1}X_1'\boldsymbol{x} = \boldsymbol{x} - \boldsymbol{u} = (\frac{1}{4}, -\frac{1}{4}, -\frac{1}{4}, \frac{1}{4})'$. This gives us the decomposition

$$\boldsymbol{x} = \begin{bmatrix} 1 \\ -1 \\ 2 \\ 1 \end{bmatrix} = \frac{1}{4}\begin{bmatrix} 3 \\ -3 \\ 9 \\ 3 \end{bmatrix} + \frac{1}{4}\begin{bmatrix} 1 \\ -1 \\ -1 \\ 1 \end{bmatrix} = \boldsymbol{u} + \boldsymbol{v}$$

of Theorem 2.14.

Example 2.10. We will generalize some of the ideas of Example 2.8 to the multiple regression model

$$y = \beta_0 + \beta_1 x_1 + \cdots + \beta_k x_k + \epsilon$$

relating a response variable y to k explanatory variables $x_1, \ldots, x_k$. If we have N observations, this model can be written as

$$y = X\boldsymbol{\beta} + \boldsymbol{\epsilon},$$

where $\boldsymbol{y}$ is $N \times 1$, X is $N \times (k+1)$, $\boldsymbol{\beta}$ is $(k+1) \times 1$, and $\boldsymbol{\epsilon}$ is $N \times 1$, while the vector of fitted values is given by

$$\hat{\boldsymbol{y}} = X\hat{\boldsymbol{\beta}},$$

where $\hat{\boldsymbol{\beta}}$ is an estimate of $\boldsymbol{\beta}$. Clearly, for any $\hat{\boldsymbol{\beta}}$, $\hat{\boldsymbol{y}}$ is a point in the subspace of R^N spanned by the columns of X. To be a least squares estimate of $\boldsymbol{\beta}$, $\hat{\boldsymbol{\beta}}$ must be such that $\hat{\boldsymbol{y}} = X\hat{\boldsymbol{\beta}}$ yields the point in this subspace closest to the vector $\boldsymbol{y}$, since this will have the sum of squared errors,

$$(\boldsymbol{y} - X\hat{\boldsymbol{\beta}})'(\boldsymbol{y} - X\hat{\boldsymbol{\beta}}),$$

minimized. Thus $X\hat{\boldsymbol{\beta}}$ must be the orthogonal projection of $\boldsymbol{y}$ onto the space spanned by the columns of X. If X has full column rank, then this space has projection matrix $X(X'X)^{-1}X'$, and so the required projection is

$$X\hat{\boldsymbol{\beta}} = X(X'X)^{-1}X'\boldsymbol{y}$$

Premultiplying this equation by $(X'X)^{-1}X'$, we obtain the least squares estimator

$$\hat{\boldsymbol{\beta}} = (X'X)^{-1}X'\boldsymbol{y}$$

In addition, we find that the sum of squared errors (SSE) for the fitted model $\hat{\boldsymbol{y}} = X\hat{\boldsymbol{\beta}}$ can be written as

$$\begin{aligned} \text{SSE}_1 &= (\boldsymbol{y} - X\hat{\boldsymbol{\beta}})'(\boldsymbol{y} - X\hat{\boldsymbol{\beta}}) = (\boldsymbol{y} - X(X'X)^{-1}X'\boldsymbol{y})'(\boldsymbol{y} - X(X'X)^{-1}X'\boldsymbol{y}) \\ &= \boldsymbol{y}'(\text{I}_N - X(X'X)^{-1}X')^2\boldsymbol{y} = \boldsymbol{y}'(\text{I}_N - X(X'X)^{-1}X')\boldsymbol{y}, \end{aligned}$$

and so this sum of squares represents the squared distance of the projection of $\boldsymbol{y}$ onto the orthogonal complement of the column space of X. Suppose now that $\boldsymbol{\beta}$ and X are partitioned as $\boldsymbol{\beta} = (\boldsymbol{\beta}_1', \boldsymbol{\beta}_2')'$ and $X = (X_1, X_2)$, where the number of columns of X_1 is the same as the number of elements in $\boldsymbol{\beta}_1$, and we wish to decide whether or not $\boldsymbol{\beta}_2 = \boldsymbol{0}$. If the columns of X_1 are orthogonal to the columns of X_2, then $X_1'X_2 = (0)$ and

$$(X'X)^{-1} = \begin{bmatrix} (X_1'X_1)^{-1} & (0) \\ (0) & (X_2'X_2)^{-1} \end{bmatrix},$$

and so $\hat{\boldsymbol{\beta}}$ can be partitioned as $\hat{\boldsymbol{\beta}} = (\hat{\boldsymbol{\beta}}_1', \hat{\boldsymbol{\beta}}_2')'$, where $\hat{\boldsymbol{\beta}}_1 = (X_1'X_1)^{-1}X_1'\boldsymbol{y}$ and $\hat{\boldsymbol{\beta}}_2 = (X_2'X_2)^{-1}X_2'\boldsymbol{y}$. Further, the sum of squared errors for the fitted model, $\hat{\boldsymbol{y}} = X\hat{\boldsymbol{\beta}}$, can be decomposed as

$$\begin{aligned}(\boldsymbol{y} - X\hat{\boldsymbol{\beta}})'(\boldsymbol{y} - X\hat{\boldsymbol{\beta}}) &= \boldsymbol{y}'(\mathrm{I}_N - X(X'X)^{-1}X')\boldsymbol{y} \\ &= \boldsymbol{y}'(\mathrm{I}_N - X_1(X_1'X_1)^{-1}X_1' - X_2(X_2'X_2)^{-1}X_2')\boldsymbol{y}\end{aligned}$$

On the other hand, the least squares estimator of $\boldsymbol{\beta}_1$ in the reduced model

$$\boldsymbol{y} = X_1\boldsymbol{\beta}_1 + \boldsymbol{\epsilon}$$

is $\hat{\boldsymbol{\beta}}_1 = (X_1'X_1)^{-1}X_1'\boldsymbol{y}$, while its sum of squared errors is given by

$$\mathrm{SSE}_2 = (\boldsymbol{y} - X_1\hat{\boldsymbol{\beta}}_1)'(\boldsymbol{y} - X_1\hat{\boldsymbol{\beta}}_1) = \boldsymbol{y}'(\mathrm{I}_N - X_1(X_1'X_1)^{-1}X_1')\boldsymbol{y}$$

Thus, the term $\mathrm{SSE}_2 - \mathrm{SSE}_1 = \boldsymbol{y}'X_2(X_2'X_2)^{-1}X_2'\boldsymbol{y}$ gives the reduction in the sum of squared errors attributable to the inclusion of the term $X_2\boldsymbol{\beta}_2$ in the model $\boldsymbol{y} = X\boldsymbol{\beta} + \boldsymbol{\epsilon} = X_1\boldsymbol{\beta}_1 + X_2\boldsymbol{\beta}_2 + \boldsymbol{\epsilon}$, and so its relative size will be helpful in deciding whether or not $\boldsymbol{\beta}_2 = \mathbf{0}$. If $\boldsymbol{\beta}_2 = \mathbf{0}$, then the N observations of y should be randomly clustered about the column space of X_1 in R^N with no tendency to deviate from this subspace in one direction more than in any other direction, while if $\boldsymbol{\beta}_2 \neq \mathbf{0}$, we would expect larger deviations in directions within the column space of X_2 than in directions orthogonal to the column space of X. Now, since the dimension of the column space of X is $k + 1$, SSE_1 is the sum of squared deviations in $N - k - 1$ orthogonal directions, while $\mathrm{SSE}_2 - \mathrm{SSE}_1$ gives the sum of squared deviations in k_2 orthogonal directions, where k_2 is the number of components in $\boldsymbol{\beta}_2$. Thus, $\mathrm{SSE}_1/(N - k - 1)$ and $(\mathrm{SSE}_2 - \mathrm{SSE}_1)/k_2$ should be of similar magnitudes if $\boldsymbol{\beta}_2 = \mathbf{0}$, while the latter should be larger than the former if $\boldsymbol{\beta}_2 \neq \mathbf{0}$. Consequently, a decision about $\boldsymbol{\beta}_2$ can be based on the value of the statistic

$$\mathrm{F} = \frac{(\mathrm{SSE}_2 - \mathrm{SSE}_1)/k_2}{\mathrm{SSE}_1/(N - k - 1)} \tag{2.8}$$

Using results that we will develop in Chapter 9, it can be shown that $\mathrm{F} \sim \mathrm{F}_{k_2, N-k-1}$ if $\boldsymbol{\epsilon} \sim \mathrm{N}_N(\mathbf{0}, \sigma^2\mathrm{I}_N)$ and $\boldsymbol{\beta}_2 = \mathbf{0}$.

When $X_1'X_2 \neq (0)$, the expression for $(\mathrm{SSE}_2 - \mathrm{SSE}_1)$ is not equal to $\boldsymbol{y}'X_2(X_2'X_2)^{-1}X_2'\boldsymbol{y}$ since, in this case, $\hat{\boldsymbol{y}}$ is not the sum of the projection of $\boldsymbol{y}$ onto the column space of X_1 and the projection of $\boldsymbol{y}$ onto the column space of X_2. To properly assess the effect of the inclusion of the term $X_2\boldsymbol{\beta}_2$ in the model, we must decompose $\hat{\boldsymbol{y}}$ into the sum of the projection of $\boldsymbol{y}$ onto the column space of X_1 and the projection of $\boldsymbol{y}$ onto the subspace of the column space of X_2 orthogonal to the column space of X_1. This latter subspace is spanned by the columns of

$$X_{2*} = (\mathrm{I}_N - X_1(X_1'X_1)^{-1}X_1')X_2,$$

since $(\mathrm{I}_N - X_1(X_1'X_1)^{-1}X_1')$ is the projection matrix of the orthogonal complement of the column space of X_1. Thus the vector of fitted values $\hat{\boldsymbol{y}} = X\hat{\boldsymbol{\beta}}$ can be written as

$$\hat{\boldsymbol{y}} = X_1(X_1'X_1)^{-1}X_1'\boldsymbol{y} + X_{2*}(X_{2*}'X_{2*})^{-1}X_{2*}'\boldsymbol{y}$$

Further, the sum of squared errors is given by

$$\boldsymbol{y}'(\mathrm{I}_N - X_1(X_1'X_1)^{-1}X_1' - X_{2*}(X_{2*}'X_{2*})^{-1}X_{2*}')\boldsymbol{y},$$

and the reduction in the sum of squared errors attributable to the inclusion of the term $X_2\boldsymbol{\beta}_2$ in the model $\boldsymbol{y} = X\boldsymbol{\beta} + \boldsymbol{\epsilon}$ is

$$\boldsymbol{y}'X_{2*}(X_{2*}'X_{2*})^{-1}X_{2*}'\boldsymbol{y}$$

Least squares estimators are not always unique as they have been throughout this example. For instance, let us return to the least squares estimation of $\boldsymbol{\beta}$ in the model $\boldsymbol{y} = X\boldsymbol{\beta} + \boldsymbol{\epsilon}$, where now X does not have full column rank. As before $\hat{\boldsymbol{y}} = X\hat{\boldsymbol{\beta}}$ will be given by the orthogonal projection of $\boldsymbol{y}$ onto the space spanned by the columns of X, but the necessary projection matrix can not be expressed as $X(X'X)^{-1}X'$, since $X'X$ is singular. If the projection matrix of the column space of X is denoted by $P_{R(X)}$, then a least squares estimator of $\boldsymbol{\beta}$ is any vector $\hat{\boldsymbol{\beta}}$ satisfying

$$X\hat{\boldsymbol{\beta}} = P_{R(X)}\boldsymbol{y}$$

Since X does not have full column rank, the dimension of its null space is at least one, and so we will be able to find a nonnull vector $\boldsymbol{a}$ satisfying $X\boldsymbol{a} = \mathbf{0}$. In this case, $\hat{\boldsymbol{\beta}} + \boldsymbol{a}$ is also a least squares estimator since

$$X(\hat{\boldsymbol{\beta}} + \boldsymbol{a}) = P_{R(X)}\boldsymbol{y},$$

and so the least squares estimator is not unique.

We have seen that if the columns of an $m \times r$ matrix Z_1 form an orthonormal basis for a vector space S, then the projection matrix of S is given by Z_1Z_1'. Clearly this projection matrix is symmetric and, since $Z_1'Z_1 = \mathrm{I}_r$, it is also idempotent; that is, every projection matrix is symmetric and idempotent. Our next result proves the converse. Every symmetric idempotent matrix is a projection matrix for some vector space.

Theorem 2.19. Let P be an $m \times m$ symmetric idempotent matrix of rank r. Then there is an r-dimensional vector space which has P as its projection matrix.

Proof. From Corollary 1.9.1, there exist an $m \times r$ matrix F and an $r \times m$ matrix G such that $\text{rank}(F) = \text{rank}(G) = r$ and $P = FG$. Since P is idempotent, we have

$$FGFG = FG,$$

which implies that

$$F'FGFGG' = F'FGG' \tag{2.9}$$

Since F and G' are full column rank, the matrices $F'F$ and GG' are nonsingular. Premultiplying (2.9) by $(F'F)^{-1}$ and postmultiplying by $(GG')^{-1}$, we obtain $GF = \mathrm{I}_r$. Using this and the symmetry of $P = FG$, we find that

$$F = FGF = (FG)'F = G'F'F,$$

which leads to $G' = F(F'F)^{-1}$. Thus, $P = FG = F(F'F)^{-1}F'$. Comparing this to equation (2.7), we see that P must be the projection matrix for the vector space spanned by the columns of F. This completes the proof. □

Example 2.11. Consider the 3×3 matrix

$$P = \frac{1}{6}\begin{bmatrix} 5 & -1 & 2 \\ -1 & 5 & 2 \\ 2 & 2 & 2 \end{bmatrix}$$

Clearly, P is symmetric and it is easily verified that P is idempotent, so P is a projection matrix. We will find the vector space S associated with this projection matrix. First note that the first two columns of P are linearly independent while the third column is the average of the first two columns. Thus, $\text{rank}(P) = 2$ and so the dimension of the vector space associated with P is 2. For any $\boldsymbol{x} \in R^3$, $P\boldsymbol{x}$ yields a vector in S. In particular, $P\boldsymbol{e}_1$ and $P\boldsymbol{e}_2$ are in S. These two vectors form a basis for S since they are linearly independent and the dimension of S is 2. Consequently, S contains all vectors of the form $(5a - b, 5b - a, 2a + 2b)'$.

8. LINEAR TRANSFORMATIONS AND SYSTEMS OF LINEAR EQUATIONS

If S is a vector subspace of R^m, with projection matrix P_S, then we have seen that for any $\boldsymbol{x} \in R^m$, $\boldsymbol{u} = \boldsymbol{u}(\boldsymbol{x}) = P_S\boldsymbol{x}$ is the orthogonal projection of $\boldsymbol{x}$ onto S; that is, each $\boldsymbol{x} \in R^m$ is transformed into a $\boldsymbol{u} \in S$. The function $\boldsymbol{u}(\boldsymbol{x}) = P_S\boldsymbol{x}$ is an example of a linear transformation of R^m into S.

Definition 2.11. Let $\boldsymbol{u}$ be a function defined for all $\boldsymbol{x}$ in the vector space T such that for any $\boldsymbol{x} \in T$, $\boldsymbol{u} = \boldsymbol{u}(\boldsymbol{x}) \in S$, where S is also a vector space. Then the transformation defined by $\boldsymbol{u}$ is a linear transformation of T into S if for any two scalars α_1 and α_2 and any two vectors $\boldsymbol{x}_1 \in T$ and $\boldsymbol{x}_2 \in T$,

$$\boldsymbol{u}(\alpha_1\boldsymbol{x}_1 + \alpha_2\boldsymbol{x}_2) = \alpha_1\boldsymbol{u}(\boldsymbol{x}_1) + \alpha_2\boldsymbol{u}(\boldsymbol{x}_2)$$

We will be interested in matrix transformations of the form $\boldsymbol{u} = A\boldsymbol{x}$, where $\boldsymbol{x}$ is in the subspace of R^n denoted by T, $\boldsymbol{u}$ is in the subspace of R^m denoted by S, and A is an $m \times n$ matrix. This defines a transformation of T into S, and the transformation is linear since for scalars α_1, α_2 and $n \times 1$ vectors $\boldsymbol{x}_1$ and $\boldsymbol{x}_2$, it follows immediately that

$$A(\alpha_1\boldsymbol{x}_1 + \alpha_2\boldsymbol{x}_2) = \alpha_1 A\boldsymbol{x}_1 + \alpha_2 A\boldsymbol{x}_2 \tag{2.10}$$

In fact, every linear transformation can be expressed as a matrix transformation. For the orthogonal projection described at the beginning of this section, $A = P_S$, so that $n = m$ and thus we have a linear transformation of R^m into R^m, or to be more specific, a linear transformation of R^m into S. In particular, for the multiple regression problem discussed in Example 2.10, we saw that for any $N \times 1$ vector of observations $\boldsymbol{y}$, the vector of estimated or fitted values was given by $\hat{\boldsymbol{y}} = X(X'X)^{-1}X'\boldsymbol{y}$. Thus, since $\boldsymbol{y} \in R^N$ and $\hat{\boldsymbol{y}} \in R(X)$, we have here a linear transformation of R^N into $R(X)$.

It should be obvious from (2.10) that if S is actually defined to be the set $\{\boldsymbol{u}\colon \boldsymbol{u} = A\boldsymbol{x}; \boldsymbol{x} \in T\}$, then T being a vector space guarantees that S will also be a vector space. In addition, if the vectors $\boldsymbol{x}_1, \dots, \boldsymbol{x}_r$ span T, then the vectors $A\boldsymbol{x}_1, \dots, A\boldsymbol{x}_r$ span S. In particular, if T is R^n, then since $\boldsymbol{e}_1, \dots, \boldsymbol{e}_n$ span R^n, we find that $(A)_{\cdot 1}, \dots, (A)_{\cdot n}$ span S; that is, S is the column space or range of A since it is spanned by the columns of A.

When the matrix A does not have full column rank then there will be vectors $\boldsymbol{x}$, other than the null vector, satisfying $A\boldsymbol{x} = \boldsymbol{0}$. The set of all such vectors is called the null space of the transformation $A\boldsymbol{x}$ or simply the null space of the matrix A.

Theorem 2.20. Let the linear transformation of R^n into S be given by $\boldsymbol{u} = A\boldsymbol{x}$, where $\boldsymbol{x} \in R^n$ and A is an $m \times n$ matrix. Then the null space of A, given by the set

$$N(A) = \{\boldsymbol{x}\colon A\boldsymbol{x} = \boldsymbol{0}, \boldsymbol{x} \in R^n\},$$

is a vector space.

Proof. Let $\boldsymbol{x}_1$ and $\boldsymbol{x}_2$ be in $N(A)$ so that $A\boldsymbol{x}_1 = A\boldsymbol{x}_2 = \boldsymbol{0}$. Then for any scalars α_1 and α_2, we have

$$A(\alpha_1\boldsymbol{x}_1 + \alpha_2\boldsymbol{x}_2) = \alpha_1 A\boldsymbol{x}_1 + \alpha_2 A\boldsymbol{x}_2 = \alpha_1(\boldsymbol{0}) + \alpha_2(\boldsymbol{0}) = \boldsymbol{0},$$

so that $(\alpha_1\boldsymbol{x}_1 + \alpha_2\boldsymbol{x}_2) \in N(A)$ and, hence, $N(A)$ is a vector space. □

The null space of a matrix A is related to the concept of orthogonal complements discussed in Section 2.6. In fact, the null space of the matrix A is the same as the orthogonal complement of the row space of A. Similarly, the null space of the matrix A' is the same as the orthogonal complement of the column space of A. The following result is an immediate consequence of Theorem 2.16.

Theorem 2.21. Let A be an $m \times n$ matrix. If the dimension of the row space of A is r_1 and the dimension of the null space of A is r_2, then $r_1 + r_2 = n$.

Since the rank of the matrix A is equal to the dimension of the row space of A, the result above can be equivalently expressed as

$$\text{rank}(A) = n - \dim\{N(A)\} \tag{2.11}$$

This connection between the rank of a matrix and the dimension of the null space of that matrix can be very useful in determining the rank of a matrix in certain situations.

Example 2.12. To illustrate the utility of (2.11), we will give an alternative proof of the identity $\text{rank}(A) = \text{rank}(A'A)$, which was given as Theorem 2.10(c). Suppose $\boldsymbol{x}$ is in the null space of A so that $A\boldsymbol{x} = \boldsymbol{0}$. Then, clearly, we must have $A'A\boldsymbol{x} = \boldsymbol{0}$, which implies that $\boldsymbol{x}$ is also in the null space of $A'A$, so it follows that $\dim\{N(A)\} \leq \dim\{N(A'A)\}$, or equivalently

$$\text{rank}(A) \geq \text{rank}(A'A) \tag{2.12}$$

On the other hand, if $\boldsymbol{x}$ is in the null space of $A'A$ then $A'A\boldsymbol{x} = \boldsymbol{0}$. Premultiplying by $\boldsymbol{x}'$ yields $\boldsymbol{x}'A'A\boldsymbol{x} = 0$, which is satisfied only if $A\boldsymbol{x} = \boldsymbol{0}$. Thus, $\boldsymbol{x}$ is also in the null space of A so that $\dim\{N(A)\} \geq \dim\{N(A'A)\}$, or

$$\text{rank}(A) \leq \text{rank}(A'A) \tag{2.13}$$

Combining (2.12) and (2.13), we get $\text{rank}(A) = \text{rank}(A'A)$.

When A is an $m \times m$ nonsingular matrix and $\boldsymbol{x} \in R^m$, then $\boldsymbol{u} = A\boldsymbol{x}$ defines a one-to-one transformation of R^m onto R^m. One way of viewing this transformation is as the movement of each point in R^m to another point in R^m. Alternatively, we can view the transformation as a change of coordinate axes. For instance, if we start with the standard coordinate axes which are given by the columns, $\boldsymbol{e}_1, \ldots, \boldsymbol{e}_m$ of the identity matrix I_m, then, since for any $\boldsymbol{x} \in R^m$, $\boldsymbol{x} = x_1\boldsymbol{e}_1 + \cdots + x_m\boldsymbol{e}_m$, the components of $\boldsymbol{x}$ give the coordinates of the point $\boldsymbol{x}$ relative to these standard coordinate axes. On the other hand, if $\boldsymbol{x}_1, \ldots, \boldsymbol{x}_m$ is another basis for R^m, then from Theorem 2.7 there exist scalars $u_1, \ldots, u_m$ so that with $\boldsymbol{u} = (u_1, \ldots, u_m)'$ and $X = (\boldsymbol{x}_1, \ldots, \boldsymbol{x}_m)$, we have

$$\boldsymbol{x} = \sum_{i=1}^{m} u_i\boldsymbol{x}_i = X\boldsymbol{u};$$

that is, $\boldsymbol{u} = (u_1, \ldots, u_m)'$ gives the coordinates of the point $\boldsymbol{x}$ relative to the coordinate axes $\boldsymbol{x}_1, \ldots, \boldsymbol{x}_m$. The transformation from the standard coordinate system to the one with axes $\boldsymbol{x}_1, \ldots, \boldsymbol{x}_m$ is then given by the matrix transformation $\boldsymbol{u} = A\boldsymbol{x}$, where $A = X^{-1}$. Note that the squared Euclidean distance of $\boldsymbol{u}$ from the origin,

$$\boldsymbol{u}'\boldsymbol{u} = (A\boldsymbol{x})'(A\boldsymbol{x}) = \boldsymbol{x}'A'A\boldsymbol{x},$$

will be the same as the squared Euclidean distance of $\boldsymbol{x}$ from the origin for every choice of $\boldsymbol{x}$ if and only if A, and hence also X, is an orthogonal matrix. In this case, $\boldsymbol{x}_1, \ldots, \boldsymbol{x}_m$ forms an orthonormal basis for R^m, and so the transformation has replaced the standard coordinate axes by a new set of orthogonal axes given by $\boldsymbol{x}_1, \ldots, \boldsymbol{x}_m$.

Example 2.13. Orthogonal transformations are of two types according to whether the determinant of A is +1 or -1. If $|A| = 1$, then the new axes can be obtained by a rotation of the standard axes. For example, for a fixed angle θ, let

$$A = \begin{bmatrix} \cos\theta & -\sin\theta & 0 \\ \sin\theta & \cos\theta & 0 \\ 0 & 0 & 1 \end{bmatrix},$$

so that $|A| = \cos^2\theta + \sin^2\theta = 1$. The transformation given by $\boldsymbol{u} = A\boldsymbol{x}$ transforms the standard axes $\boldsymbol{e}_1, \boldsymbol{e}_2, \boldsymbol{e}_3$ to the new axes $\boldsymbol{x}_1 = (\cos\theta, -\sin\theta, 0)'$,

$\boldsymbol{x}_2 = (\sin\theta, \cos\theta, 0)'$, $\boldsymbol{x}_3 = \boldsymbol{e}_3$, and this simply represents a rotation of $\boldsymbol{e}_1$ and $\boldsymbol{e}_2$ through an angle of θ. If instead we have

$$A = \begin{bmatrix} \cos\theta & -\sin\theta & 0 \\ \sin\theta & \cos\theta & 0 \\ 0 & 0 & -1 \end{bmatrix},$$

then $|A| = (\cos^2\theta + \sin^2\theta)\cdot(-1) = -1$. Now the transformation given by $\boldsymbol{u} = A\boldsymbol{x}$, transforms the standard axes to the new axes $\boldsymbol{x}_1 = (\cos\theta, -\sin\theta, 0)'$, $\boldsymbol{x}_2 = (\sin\theta, \cos\theta, 0)'$, and $\boldsymbol{x}_3 = -\boldsymbol{e}_3$; these axes are obtained by a rotation of $\boldsymbol{e}_1$ and $\boldsymbol{e}_2$ through an angle of θ followed by a reflection of $\boldsymbol{e}_3$ about the $\boldsymbol{x}_1$, $\boldsymbol{x}_2$ plane.

Although orthogonal transformations are very common, there are situations in which nonsingular nonorthogonal transformations are useful.

Example 2.14. Suppose we have several three-dimensional vectors $\boldsymbol{x}_1, \dots, \boldsymbol{x}_r$ that are observations from distributions, each having the same positive definite covariance matrix Ω. If we are interested in how these vectors differ from one another, then a plot of the points in R^3 may be useful. However, as discussed in Example 2.2, if Ω is not the identity matrix, then the Euclidean distance is not appropriate, and so it becomes difficult to compare and interpret the observed differences among the r points. This difficulty can be resolved by an appropriate transformation. We will see in a later chapter that since Ω is positive definite, there exists a nonsingular matrix T satisfying $\Omega = TT'$. If we let $\boldsymbol{u}_i = T^{-1}\boldsymbol{x}_i$, then the Mahalanobis distance, which was defined in Example 2.2, between $\boldsymbol{x}_i$ and $\boldsymbol{x}_j$ is

$$\begin{aligned} \mathrm{d}_\Omega(\boldsymbol{x}_i, \boldsymbol{x}_j) &= \{(\boldsymbol{x}_i - \boldsymbol{x}_j)'\Omega^{-1}(\boldsymbol{x}_i - \boldsymbol{x}_j)\}^{1/2} \\ &= \{(\boldsymbol{x}_i - \boldsymbol{x}_j)'T'^{-1}T^{-1}(\boldsymbol{x}_i - \boldsymbol{x}_j)\}^{1/2} \\ &= \{(T^{-1}\boldsymbol{x}_i - T^{-1}\boldsymbol{x}_j)'(T^{-1}\boldsymbol{x}_i - T^{-1}\boldsymbol{x}_j)\}^{1/2} \\ &= \{(\boldsymbol{u}_i - \boldsymbol{u}_j)'(\boldsymbol{u}_i - \boldsymbol{u}_j)\}^{1/2} = \mathrm{d}_I(\boldsymbol{u}_i, \boldsymbol{u}_j), \end{aligned}$$

while the variance of $\boldsymbol{u}_i$ is given by

$$\mathrm{var}(\boldsymbol{u}_i) = \mathrm{var}(T^{-1}\boldsymbol{x}_i) = T^{-1}\{\mathrm{var}(\boldsymbol{x}_i)\}T'^{-1} = T^{-1}\Omega T'^{-1} = I_3$$

That is, the transformation $\boldsymbol{u}_i = T^{-1}\boldsymbol{x}_i$ produces vectors for which the Euclidean distance function is an appropriate measure of distance between points.

In our next two examples, we discuss some transformations that are sometimes useful in regression analysis.

Example 2.15. A simple transformation that is useful in some situations is one that centers a collection of numbers at the origin. For instance, if $\bar{x}$ is the mean of the components of $\boldsymbol{x} = (x_1, \ldots, x_N)'$, then the average of the components of

$$\boldsymbol{v} = (\mathrm{I}_N - N^{-1}\mathbf{1}_N\mathbf{1}_N')\boldsymbol{x} = \begin{bmatrix} x_1 - \bar{x} \\ x_2 - \bar{x} \\ \vdots \\ x_N - \bar{x} \end{bmatrix}$$

is $\mathbf{0}$. This transformation is sometimes used in a regression analysis to center each of the explanatory variables. Thus the multiple regression model

$$\boldsymbol{y} = X\boldsymbol{\beta} + \boldsymbol{\epsilon} = [\mathbf{1}_N \quad X_1]\begin{bmatrix} \beta_0 \\ \boldsymbol{\beta}_1 \end{bmatrix} + \boldsymbol{\epsilon} = \beta_0\mathbf{1}_N + X_1\boldsymbol{\beta}_1 + \boldsymbol{\epsilon}$$

can be reexpressed as

$$\begin{aligned} \boldsymbol{y} &= \beta_0\mathbf{1}_N + \left\{N^{-1}\mathbf{1}_N\mathbf{1}_N' + (\mathrm{I}_N - N^{-1}\mathbf{1}_N\mathbf{1}_N')\right\}X_1\boldsymbol{\beta}_1 + \boldsymbol{\epsilon} \\ &= \gamma_0\mathbf{1}_N + V_1\boldsymbol{\beta}_1 + \boldsymbol{\epsilon} = V\boldsymbol{\gamma} + \boldsymbol{\epsilon}, \end{aligned}$$

where $V = [\mathbf{1}_N, V_1] = [\mathbf{1}_N, (\mathrm{I}_N - N^{-1}\mathbf{1}_N\mathbf{1}_N')X_1]$ and $\boldsymbol{\gamma} = (\gamma_0, \boldsymbol{\beta}_1')' = (\beta_0 + N^{-1}\mathbf{1}_N'X_1\boldsymbol{\beta}_1, \boldsymbol{\beta}_1')'$. Since the columns of V_1 are orthogonal to $\mathbf{1}_N$, the least squares estimator of $\boldsymbol{\gamma}$ simplifies to

$$\hat{\boldsymbol{\gamma}} = \begin{bmatrix} \hat{\gamma}_0 \\ \hat{\boldsymbol{\beta}}_1 \end{bmatrix} = (V'V)^{-1}V'\boldsymbol{y} = \begin{bmatrix} N^{-1} & \mathbf{0}' \\ \mathbf{0} & (V_1'V_1)^{-1} \end{bmatrix}\begin{bmatrix} \sum y_i \\ V_1'\boldsymbol{y} \end{bmatrix} = \begin{bmatrix} \bar{y} \\ (V_1'V_1)^{-1}V_1'\boldsymbol{y} \end{bmatrix}$$

Thus, $\hat{\gamma}_0 = \bar{y}$. The estimator, $\hat{\boldsymbol{\beta}}_1$, can be conveniently expressed in terms of the sample covariance matrix of the $N\,(k + 1) \times 1$ vectors that form the rows of the matrix $[\boldsymbol{y} \quad X_1]$. If we denote this covariance matrix by S and partition it as

$$S = \begin{bmatrix} s_{11} & \boldsymbol{s}_{21}' \\ \boldsymbol{s}_{21} & S_{22} \end{bmatrix},$$

then $(N - 1)^{-1}V_1'V_1 = S_{22}$ and, since $V_1'\mathbf{1}_N = \mathbf{0}$,

$$(N - 1)^{-1}V_1'\boldsymbol{y} = (N - 1)^{-1}V_1'(\boldsymbol{y} - \bar{y}\mathbf{1}_N) = \boldsymbol{s}_{21}$$

Consequently, $\hat{\boldsymbol{\beta}}_1 = S_{22}^{-1}\boldsymbol{s}_{21}$. Yet another adjustment to the original regression model involves the standardization of the explanatory variables. In this case, the model becomes

$$\boldsymbol{y} = \delta_0 \mathbf{1}_N + Z_1\boldsymbol{\delta}_1 + \boldsymbol{\epsilon} = Z\boldsymbol{\delta} + \boldsymbol{\epsilon},$$

where $\boldsymbol{\delta} = (\delta_0, \boldsymbol{\delta}_1')'$, $Z = (\mathbf{1}_N, Z_1)$, $\delta_0 = \gamma_0$, $Z_1 = V_1 D_{S_{22}}^{-1/2}$ and $\boldsymbol{\delta}_1 = D_{S_{22}}^{1/2}\boldsymbol{\beta}_1$. The least squares estimators are $\hat{\delta}_0 = \bar{y}$ and $\hat{\boldsymbol{\delta}}_1 = R_{22}^{-1}\boldsymbol{r}_{21}$, where we have partitioned the correlation matrix R in a fashion similar to that of S.

The centering of explanatory variables, discussed previously, involves a linear transformation on the columns of X_1. In some situations, it is advantageous to employ a linear transformation on the rows of X_1, V_1, or Z_1. For instance, suppose that T is a $k \times k$ nonsingular matrix, and we define $W_1 = Z_1 T$, $\alpha_0 = \delta_0$, and $\boldsymbol{\alpha}_1 = T^{-1}\boldsymbol{\delta}_1$, so that the model

$$\boldsymbol{y} = \delta_0 \mathbf{1}_N + Z_1\boldsymbol{\delta}_1 + \boldsymbol{\epsilon} = Z\boldsymbol{\delta} + \boldsymbol{\epsilon}$$

can be written as

$$\boldsymbol{y} = \alpha_0 \mathbf{1}_N + W_1\boldsymbol{\alpha}_1 + \boldsymbol{\epsilon} = W\boldsymbol{\alpha} + \boldsymbol{\epsilon}$$

This second model uses a different set of explanatory variables than the first; its ith explanatory variable is a linear combination of the explanatory variables of the first model with the coefficients given by the ith column of T. However, the two models yield equivalent results in terms of the fitted values. To see this, let

$$T_* = \begin{bmatrix} 1 & \mathbf{0}' \\ \mathbf{0} & T \end{bmatrix}$$

so that $W = ZT_*$, and note that the vector of fitted values from the second model,

$$\begin{aligned}\hat{\boldsymbol{y}} &= W\hat{\boldsymbol{\alpha}} = W(W'W)^{-1}W'\boldsymbol{y} = ZT_*(T_*'Z'ZT_*)^{-1}T_*'Z'\boldsymbol{y} \\ &= ZT_*T_*^{-1}(Z'Z)^{-1}T_*'^{-1}T_*'Z'\boldsymbol{y} = Z(Z'Z)^{-1}Z'\boldsymbol{y},\end{aligned}$$

is the same as that obtained from the first model.

Example 2.16. Consider the multiple regression model

$$\boldsymbol{y} = X\boldsymbol{\beta} + \boldsymbol{\epsilon},$$

where now $\text{var}(\boldsymbol{\epsilon}) \neq \sigma^2 I_N$. In this case, our previous estimator, $\hat{\boldsymbol{\beta}} = (X'X)^{-1}X'\boldsymbol{y}$, is still the least squares estimator of $\boldsymbol{\beta}$, but it doesn't possess certain optimality properties, one of which is illustrated later in Example 3.13, that hold when $\text{var}(\boldsymbol{\epsilon}) = \sigma^2 I_N$. In this example, we will consider the situation in which the ϵ_is are still uncorrelated, but their variances are not all the same. Thus, $\text{var}(\boldsymbol{\epsilon}) = \Omega = \sigma^2 C$, where $C = \text{diag}(c_1^2, \dots, c_N^2)$ and the c_is are known constants. This special regression problem is sometimes referred to as weighted least squares regression. The weighted least squares estimator of $\boldsymbol{\beta}$ is obtained by making a simple transformation so that ordinary least squares regression applies to the transformed model. Define the matrix $C^{-1/2} = \text{diag}(c_1^{-1}, \dots, c_N^{-1})$ and transform the original regression problem by premultiplying the model equation by $C^{-1/2}$; the new model equation is

$$C^{-1/2}\boldsymbol{y} = C^{-1/2}X\boldsymbol{\beta} + C^{-1/2}\boldsymbol{\epsilon}$$

or, equivalently,

$$\boldsymbol{y}_* = X_*\boldsymbol{\beta} + \boldsymbol{\epsilon}_*,$$

where $\boldsymbol{y}_* = C^{-1/2}\boldsymbol{y}$, $X_* = C^{-1/2}X$, and $\boldsymbol{\epsilon}_* = C^{-1/2}\boldsymbol{\epsilon}$. The covariance matrix of $\boldsymbol{\epsilon}_*$ is

$$\text{var}(\boldsymbol{\epsilon}_*) = \text{var}(C^{-1/2}\boldsymbol{\epsilon}) = C^{-1/2}\text{var}(\boldsymbol{\epsilon})C^{-1/2} = C^{-1/2}\{\sigma^2 C\}C^{-1/2} = \sigma^2 I_N$$

Thus, for the transformed model, ordinary least squares regression applies and so the least squares estimator of $\boldsymbol{\beta}$ can be expressed as

$$\hat{\boldsymbol{\beta}} = (X_*'X_*)^{-1}X_*'\boldsymbol{y}_*$$

Rewriting this in the original model terms X and $\boldsymbol{y}$, we get

$$\begin{aligned}\hat{\boldsymbol{\beta}} &= (X'C^{-1/2}C^{-1/2}X)^{-1}X'C^{-1/2}C^{-1/2}\boldsymbol{y} \\ &= (X'C^{-1}X)^{-1}X'C^{-1}\boldsymbol{y}.\end{aligned}$$

A common application related to linear transformations is one in which the matrix A and vector $\boldsymbol{u}$ consist of known constants, while $\boldsymbol{x}$ is a vector of variables, and we wish to determine all $\boldsymbol{x}$ for which $A\boldsymbol{x} = \boldsymbol{u}$; that is, we want to find the simultaneous solutions $x_1, \dots, x_n$ to the system of m equations

$$a_{11}x_1 + \cdots + a_{1n}x_n = u_1$$
$$\vdots$$
$$a_{m1}x_1 + \cdots + a_{mn}x_n = u_m$$

For instance, in Example 2.10, we saw that the least squares estimator of the parameter vector $\boldsymbol{\beta}$ in the multiple regression model satisfies the equation, $X\hat{\boldsymbol{\beta}} = X(X'X)^{-1}X'\boldsymbol{y}$; that is, here $A = X$, $\boldsymbol{u} = X(X'X)^{-1}X'\boldsymbol{y}$, and $\boldsymbol{x} = \hat{\boldsymbol{\beta}}$. In general, if $\boldsymbol{u} = \boldsymbol{0}$, then this system of equations is referred to as a homogeneous system, and the set of all solutions to $A\boldsymbol{x} = \boldsymbol{u}$, in this case, is simply given by the null space of A. Consequently, if A has full column rank, then $\boldsymbol{x} = \boldsymbol{0}$ is the only solution, whereas there are infinitely many solutions if A has less than full column rank. A nonhomogeneous system of linear equations is one which has $\boldsymbol{u} \neq \boldsymbol{0}$. While a homogeneous system always has at least one solution, $\boldsymbol{x} = \boldsymbol{0}$, a nonhomogeneous system may or may not have any solutions. A system of linear equations that has no solutions is called an inconsistent system of equations, while a system with solutions is referred to as a consistent system. If $\boldsymbol{u} \neq \boldsymbol{0}$ and $A\boldsymbol{x} = \boldsymbol{u}$ holds for some $\boldsymbol{x}$, then $\boldsymbol{u}$ must be a linear combination of the columns of A; that is, the nonhomogeneous system of equations $A\boldsymbol{x} = \boldsymbol{u}$ is consistent if and only if $\boldsymbol{u}$ is in the column space of A.

The mathematics involved in solving systems of linear equations is most conveniently handled using matrix methods. For example, consider one of the simplest nonhomogeneous systems of linear equations in which the matrix A is square and nonsingular. In this case, since A^{-1} exists, we find that the system $A\boldsymbol{x} = \boldsymbol{u}$ has a solution that is unique and is given by $\boldsymbol{x} = A^{-1}\boldsymbol{u}$. Similarly, when the matrix A is singular or not even square, matrix methods can be used to determine whether the system is consistent, and if so, the solutions can be given as matrix expressions. These results regarding the solution of a general system of linear equations will be developed in Chapter 6.

9. THE INTERSECTION AND SUM OF VECTOR SPACES

In this section, we discuss some common ways of forming a vector subspace from two or more given subspaces. The first of these utilizes a familiar operation from set theory.

Definition 2.12. Let S_1 and S_2 be vector subspaces of R^m. The intersection of S_1 and S_2, denoted by $S_1 \cap S_2$, is the vector subspace given as

$$S_1 \cap S_2 = \{\boldsymbol{x} \in R^m\colon \boldsymbol{x} \in S_1 \quad \text{and} \quad \boldsymbol{x} \in S_2\}$$

Note that this definition says that the set $S_1 \cap S_2$ is a vector subspace if S_1 and S_2 are vector subspaces. This follows from the fact that if $\boldsymbol{x}_1$ and $\boldsymbol{x}_2$ are in

$S_1 \cap S_2$, then $\boldsymbol{x}_1 \in S_1$, $\boldsymbol{x}_2 \in S_1$ and $\boldsymbol{x}_1 \in S_2$, $\boldsymbol{x}_2 \in S_2$. Thus, since S_1 and S_2 are vector spaces, for any scalars α_1 and α_2, $\alpha_1\boldsymbol{x}_1 + \alpha_2\boldsymbol{x}_2$ will be in S_1 and S_2 and hence also in $S_1 \cap S_2$. Definition 2.12 can be generalized in an obvious fashion to the intersection, $S_1 \cap \cdots \cap S_r$, of the r vector spaces $S_1, \ldots, S_r$.

A second set operation, which combines the elements of S_1 and S_2, is the union; that is, the union of S_1 and S_2 is given by

$$S_1 \cup S_2 = \{\boldsymbol{x} \in R^m: \boldsymbol{x} \in S_1 \quad \text{or} \quad \boldsymbol{x} \in S_2\}$$

If S_1 and S_2 are vector subspaces, then $S_1 \cup S_2$ will also be a vector subspace only if $S_1 \subseteq S_2$ or $S_2 \subseteq S_1$. It can be easily shown that the following combination of S_1 and S_2 yields the vector space with the smallest possible dimension containing $S_1 \cup S_2$.

Definition 2.13. If S_1 and S_2 are vector subspaces of R^m, then the sum of S_1 and S_2, denoted by $S_1 + S_2$, is the vector space given by

$$S_1 + S_2 = \{\boldsymbol{x}_1 + \boldsymbol{x}_2: \boldsymbol{x}_1 \in S_1,\ \boldsymbol{x}_2 \in S_2\}$$

Again our definition can be generalized to $S_1 + \cdots + S_r$, the sum of the r vector spaces $S_1, \ldots, S_r$. The proof of the following theorem has been left as an exercise.

Theorem 2.22. If S_1 and S_2 are vector subspaces of R^m, then

$$\dim(S_1 + S_2) = \dim(S_1) + \dim(S_2) - \dim(S_1 \cap S_2)$$

Example 2.17. Let S_1 and S_2 be subspaces of R^5 having bases $\{\boldsymbol{x}_1, \boldsymbol{x}_2, \boldsymbol{x}_3\}$ and $\{\boldsymbol{y}_1, \boldsymbol{y}_2\}$, respectively, where

$$\begin{aligned}
\boldsymbol{x}_1 &= (1, 0, 0, 1, 0)', \\
\boldsymbol{x}_2 &= (0, 0, 1, 0, 1)', \\
\boldsymbol{x}_3 &= (0, 1, 0, 0, 0)', \\
\boldsymbol{y}_1 &= (1, 0, 0, 1, 1)', \\
\boldsymbol{y}_2 &= (0, 1, 1, 0, 0)'
\end{aligned}$$

We wish to find bases for $S_1 + S_2$ and $S_1 \cap S_2$. Now, clearly, $S_1 + S_2$ is spanned by the set $\{\boldsymbol{x}_1, \boldsymbol{x}_2, \boldsymbol{x}_3, \boldsymbol{y}_1, \boldsymbol{y}_2\}$. Note that $\boldsymbol{y}_2 = \boldsymbol{x}_1 + \boldsymbol{x}_2 + \boldsymbol{x}_3 - \boldsymbol{y}_1$, and it can be easily verified that there are no constants $\alpha_1, \alpha_2, \alpha_3, \alpha_4$, except $\alpha_1 = \alpha_2 = \alpha_3 = \alpha_4 = 0$, satisfying $\alpha_1\boldsymbol{x}_1 + \alpha_2\boldsymbol{x}_2 + \alpha_3\boldsymbol{x}_3 + \alpha_4\boldsymbol{y}_1 = \boldsymbol{0}$. Thus, $\{\boldsymbol{x}_1, \boldsymbol{x}_2, \boldsymbol{x}_3, \boldsymbol{y}_1\}$ is a basis for $S_1 + S_2$, and so $\dim(S_1 + S_2) = 4$. From Theorem 2.22, we know that $\dim(S_1 \cap S_2) = 3 + 2 - 4 = 1$, and so any basis for $S_1 \cap S_2$ consists of one vector.

The dependency between the $\boldsymbol{x}$s and the $\boldsymbol{y}$s will indicate an appropriate vector; that is, a basis for $S_1 \cap S_2$ is given by the vector $\boldsymbol{y}_1 + \boldsymbol{y}_2 = (1, 1, 1, 1, 1)'$, since $\boldsymbol{y}_1 + \boldsymbol{y}_2 = \boldsymbol{x}_1 + \boldsymbol{x}_2 + \boldsymbol{x}_3$.

When S_1 and S_2 are such that $S_1 \cap S_2 = \{\mathbf{0}\}$, then the vector space obtained as the sum of S_1 and S_2 is sometimes referred to as the direct sum of S_1 and S_2 and written $S_1 \oplus S_2$. In this special case, each $\boldsymbol{x} \in S_1 \oplus S_2$ has a unique representation as $\boldsymbol{x} = \boldsymbol{x}_1 + \boldsymbol{x}_2$, where $\boldsymbol{x}_1 \in S_1$ and $\boldsymbol{x}_2 \in S_2$. A further special case is one in which S_1 and S_2 are orthogonal vector spaces; that is, for any $\boldsymbol{x}_1 \in S_1$ and $\boldsymbol{x}_2 \in S_2$, we have $\boldsymbol{x}_1'\boldsymbol{x}_2 = \mathbf{0}$. In this case, the unique representation $\boldsymbol{x} = \boldsymbol{x}_1 + \boldsymbol{x}_2$ for $\boldsymbol{x} \in S_1 \oplus S_2$ will have the vector x_1 given by the orthogonal projection of $\boldsymbol{x}$ onto S_1, while $\boldsymbol{x}_2$ will be given by the orthogonal projection of $\boldsymbol{x}$ onto S_2. For instance, for any vector subspace S of R^m, $R^m = S \oplus S^{\perp}$, and for any $\boldsymbol{x} \in R^m$,

$$\boldsymbol{x} = P_S\boldsymbol{x} + P_{S^{\perp}}\boldsymbol{x}$$

In general, if a vector space S is the sum of the r vector spaces $S_1, \dots, S_r$, and $S_i \cap S_j = \{\mathbf{0}\}$ for all $i \neq j$, then S is said to be the direct sum of $S_1, \dots, S_r$ and is written as $S = S_1 \oplus \cdots \oplus S_r$.

Example 2.18. Consider the vector spaces $S_1, \dots, S_m$, where S_i is spanned by $\{\boldsymbol{e}_i\}$ and, as usual, $\boldsymbol{e}_i$ is the ith column of the $m \times m$ identity matrix. Consider a second sequence of vector spaces, $T_1, \dots, T_m$, where T_i is spanned by $\{\boldsymbol{e}_i, \boldsymbol{e}_{i+1}\}$ if $i \leq m - 1$, while T_m is spanned by $\{\boldsymbol{e}_1, \boldsymbol{e}_m\}$. Then it follows that $R^m = S_1 + \cdots + S_m$, as well as $R^m = T_1 + \cdots + T_m$. However, although $R^m = S_1 \oplus \cdots \oplus S_m$, it does not follow that $R^m = T_1 \oplus \cdots \oplus T_m$, since it is not true that $T_i \cap T_j = \{\mathbf{0}\}$ for all $i \neq j$. Thus any $\boldsymbol{x} = (x_1, \dots, x_m)'$ in R^m can be expressed uniquely as a sum comprised of a vector from each of the spaces $S_1, \dots, S_m$; namely

$$\boldsymbol{x} = x_1\boldsymbol{e}_1 + \cdots + x_m\boldsymbol{e}_m,$$

where $\boldsymbol{e}_i \in S_i$. On the other hand, the decomposition corresponding to $T_1, \dots, T_m$ is not unique. For instance, we can get the same sum above by choosing $\boldsymbol{e}_1 \in T_1$, $\boldsymbol{e}_2 \in T_2, \dots, \boldsymbol{e}_m \in T_m$ or by choosing $\boldsymbol{e}_2 \in T_1$, $\boldsymbol{e}_3 \in T_2, \dots, \boldsymbol{e}_m \in T_{m-1}$, $\boldsymbol{e}_1 \in T_m$. In addition, the sum of the orthogonal projections of $\boldsymbol{x}$ onto the spaces $S_1, \dots, S_m$ yields $\boldsymbol{x}$, while the sum of the orthogonal projections of $\boldsymbol{x}$ onto the spaces $T_1, \dots, T_m$ yields $2\boldsymbol{x}$. Consider as a third sequence of vector spaces, $U_1, \dots, \dots U_m$, where U_i has the basis $\{\boldsymbol{\gamma}_i\}$ and $\boldsymbol{\gamma}_i = \boldsymbol{e}_1 + \cdots + \boldsymbol{e}_i$. Clearly, $U_i \cap U_j = \{\mathbf{0}\}$ if $i \neq j$, so $R^m = U_1 \oplus \cdots \oplus U_m$ and each $\boldsymbol{x} \in R^m$ has a unique decomposition $\boldsymbol{x} = \boldsymbol{x}_1 + \cdots + \boldsymbol{x}_m$ with $\boldsymbol{x}_i \in U_i$. However, in this case, since the U_is are not orthogonal vector spaces, this decomposition of $\boldsymbol{x}$ is not given by the sum of the orthogonal projections of $\boldsymbol{x}$ onto the spaces $U_1, \dots, U_m$.

10. CONVEX SETS

A special type of subset of a vector space is known as a convex set. Such a set has the property that it contains any point on the line segment connecting any other two points in the set. A formal definition follows.

Definition 2.14. A set $S \subseteq R^m$ is said to be a convex set if for any $\boldsymbol{x}_1 \in S$ and $\boldsymbol{x}_2 \in S$,

$$c\boldsymbol{x}_1 + (1-c)\boldsymbol{x}_2 \in S,$$

where c is any scalar satisfying $0 < c < 1$.

The condition for a convex set is very similar to the condition for a vector space; for S to be a vector space, we must have for any $\boldsymbol{x}_1 \in S$ and $\boldsymbol{x}_2 \in S$, $\alpha_1\boldsymbol{x}_1 + \alpha_2\boldsymbol{x}_2 \in S$ for all α_1 and α_2, while for S to be a convex set, this need only hold when α_1 and α_2 are nonnegative and $\alpha_1 + \alpha_2 = 1$. Thus, any vector space is a convex set. However, many familiar sets that are not vector spaces are, in fact, convex sets. For instance, intervals in R, rectangles in R^2, and ellipsoidal regions in R^m are all examples of convex sets. The linear combination of $\boldsymbol{x}_1$ and $\boldsymbol{x}_2$, $\alpha_1\boldsymbol{x}_1 + \alpha_2\boldsymbol{x}_2$, is called a convex combination when $\alpha_1 + \alpha_2 = 1$ and $\alpha_i \geq 0$ for each i. More generally, $\alpha_1\boldsymbol{x}_1 + \cdots + \alpha_r\boldsymbol{x}_r$ is called a convex combination of the vectors $\boldsymbol{x}_1, \ldots, \boldsymbol{x}_r$ when $\alpha_1 + \cdots + \alpha_r = 1$ and $\alpha_i \geq 0$ for each i. Thus, by a simple induction argument, we see that a set S is convex if and only if it is closed under all convex combinations of vectors in S.

The following result indicates that the intersection of convex sets and the sum of convex sets are themselves convex. The proof will be left as an exercise.

Theorem 2.23. Suppose that S_1 and S_2 are convex sets, where $S_i \subseteq R^m$ for each i. Then the set

(a) $S_1 \cap S_2$ is convex, and

(b) $S_1 + S_2 = \{\boldsymbol{x}_1 + \boldsymbol{x}_2\colon \boldsymbol{x}_1 \in S_1, \boldsymbol{x}_2 \in S_2\}$ is convex.

For any set S, the set C(S) defined as the intersection of all convex sets containing S is called the convex hull of S. Consequently, due to a generalization of Theorem 2.23(a), C(S) is the smallest convex set containing S.

A point $\boldsymbol{a}$ is a limit or accumulation point of a set $S \subseteq R^m$ if for any $\delta > 0$, the set $S_\delta = \{\boldsymbol{x}\colon \boldsymbol{x} \in R^m, (\boldsymbol{x} - \boldsymbol{a})'(\boldsymbol{x} - \boldsymbol{a}) < \delta\}$ contains at least one point of S distinct from $\boldsymbol{a}$. A closed set is one that contains all of its limit points. If S is a set, then $\overline{S}$ will denote its closure; that is, if S_0 is the set of all limit points of S, then $\overline{S} = S \cup S_0$. In our next theorem, we see that the convexity of S guarantees the convexity of $\overline{S}$.

Theorem 2.24. If $S \subseteq R^m$ is a convex set, then its closure $\overline{S}$ is also a convex set.

Proof. It is easily verified that the set $B_n = \{\boldsymbol{x}: \boldsymbol{x} \in R^m, \boldsymbol{x}'\boldsymbol{x} \leq n^{-1}\}$ is a convex set, where n is a positive integer. Consequently, it follows from Theorem 2.23(b) that $C_n = S + B_n$ is also convex. It also follows from a generalization of the result given in Theorem 2.23(a) that the set

$$A = \bigcap_{n=1}^{\infty} C_n$$

is convex. The result now follows by observing that $A = \overline{S}$. □

One of the most important results regarding convex sets is a theorem known as the separating hyperplane theorem. A hyperplane in R^m is a set of the form, $T = \{\boldsymbol{x}: \boldsymbol{x} \in R^m, \boldsymbol{a}'\boldsymbol{x} = c\}$, where $\boldsymbol{a}$ is an $m \times 1$ vector and c is a scalar. Thus, if $m = 2$, T represents a line in R^2 and if $m = 3$, T is a plane in R^3. We will see that the separating hyperplane theorem states that two convex sets S_1 and S_2 are separated by a hyperplane if their intersection is empty; that is, there is a hyperplane which partitions R^m into two parts so that S_1 is contained in one part, while S_2 is contained in the other. Before proving this result, we will need to obtain some preliminary results. Our first result is a special case of the separating hyperplane theorem in which one of the sets contains the single point $\boldsymbol{0}$.

Theorem 2.25. Let S be a nonempty closed convex subset of R^m and suppose that $\boldsymbol{0} \notin S$. Then there exists an $m \times 1$ vector $\boldsymbol{a}$ such that $\boldsymbol{a}'\boldsymbol{x} > 0$ for all $\boldsymbol{x} \in S$.

Proof. Let $\boldsymbol{a}$ be a point in S satisfying

$$\boldsymbol{a}'\boldsymbol{a} = \inf_{x \in S} \boldsymbol{x}'\boldsymbol{x},$$

where inf denotes the infimum or greatest lower bound. It is a consequence of the fact that S is closed and nonempty that such an $\boldsymbol{a} \in S$ exists. In addition, $\boldsymbol{a} \neq \boldsymbol{0}$ since $\boldsymbol{0} \notin S$. Now let c be an arbitrary scalar, $\boldsymbol{x}$ any vector in S except for $\boldsymbol{a}$, and consider the vector $c\boldsymbol{x} + (1 - c)\boldsymbol{a}$. The squared length of this vector as a function of c is given by

$$\begin{aligned} f(c) &= \{c\boldsymbol{x} + (1 - c)\boldsymbol{a}\}'\{c\boldsymbol{x} + (1 - c)\boldsymbol{a}\} = \{c(\boldsymbol{x} - \boldsymbol{a}) + \boldsymbol{a}\}'\{c(\boldsymbol{x} - \boldsymbol{a}) + \boldsymbol{a}\} \\ &= c^2(\boldsymbol{x} - \boldsymbol{a})'(\boldsymbol{x} - \boldsymbol{a}) + 2c\boldsymbol{a}'(\boldsymbol{x} - \boldsymbol{a}) + \boldsymbol{a}'\boldsymbol{a} \end{aligned}$$

Since the second derivative of this quadratic function $f(c)$ is positive, we find that it has a unique minimum at the point

$$c_* = \frac{\boldsymbol{a}'(\boldsymbol{x} - \boldsymbol{a})}{(\boldsymbol{x} - \boldsymbol{a})'(\boldsymbol{x} - \boldsymbol{a})}$$

Note that since S is convex, $\boldsymbol{x}_c = c\boldsymbol{x} + (1 - c)\boldsymbol{a} \in S$ when $0 \leq c \leq 1$, and so we must have $\boldsymbol{x}_c'\boldsymbol{x}_c = f(c) \geq f(0) = \boldsymbol{a}'\boldsymbol{a}$ for $0 \leq c \leq 1$ due to the way $\boldsymbol{a}$ was defined. But because of the quadratic structure of $f(c)$, this implies that $f(c) > f(0)$ for all $c > 0$. In other words, $c_* \leq 0$, and this leads to

$$\boldsymbol{a}'(\boldsymbol{x} - \boldsymbol{a}) \geq 0,$$

or

$$\boldsymbol{a}'\boldsymbol{x} \geq \boldsymbol{a}'\boldsymbol{a} > 0 \qquad \square$$

A point $\boldsymbol{x}_*$ is an interior point of S if there exists a $\delta > 0$ such that the set $S_\delta = \{\boldsymbol{x}: \boldsymbol{x} \in R^m, (\boldsymbol{x} - \boldsymbol{x}_*)'(\boldsymbol{x} - \boldsymbol{x}_*) < \delta\}$ is a subset of S. On the other hand, $\boldsymbol{x}_*$ is a boundary point of S if for each $\delta > 0$, the set S_δ contains at least one point in S and at least one point not in S. Our next result shows that the sets S and $\overline{S}$ have the same interior points if S is convex.

Theorem 2.26. Suppose that S is a convex subset of R^m, while T is an open subset of R^m. If $T \subset \overline{S}$, then $T \subset S$.

Proof. Let $\boldsymbol{x}_*$ be an arbitrary point in T and define the sets

$$S_* = \{\boldsymbol{x}: \boldsymbol{x} = \boldsymbol{y} - \boldsymbol{x}_*, \boldsymbol{y} \in S\}, \quad T_* = \{\boldsymbol{x}: \boldsymbol{x} = \boldsymbol{y} - \boldsymbol{x}_*, \boldsymbol{y} \in T\}$$

It follows from the conditions of the theorem that S_* is convex, T_* is open, and $T_* \subset \overline{S}_*$. The proof will be complete if we can show that $\boldsymbol{0} \in S_*$ since this will imply that $\boldsymbol{x}_* \in S$. Since $\boldsymbol{0} \in T_*$ and T_* is an open set, we can find an $\epsilon > 0$ such that each of the vectors, $\epsilon\boldsymbol{e}_1, \ldots, \epsilon\boldsymbol{e}_m, -\epsilon\boldsymbol{1}_m$ are in T_*. Since these vectors also must be in $\overline{S}_*$, we can find sequences, $\boldsymbol{x}_{i1}, \boldsymbol{x}_{i2}, \ldots$, for $i = 1, 2, \ldots, m + 1$, such that each $\boldsymbol{x}_{ij} \in S_*$ and $\boldsymbol{x}_{ij} \to \epsilon\boldsymbol{e}_i$ for $i = 1, \ldots, m$, and $\boldsymbol{x}_{ij} \to -\epsilon\boldsymbol{1}_m$ for $i = m+1$, as $j \to \infty$. Define the $m \times m$ matrix $X_j = (\boldsymbol{x}_{1j}, \ldots, \boldsymbol{x}_{mj})$ so that $X_j \to \epsilon \mathrm{I}_m$, as $j \to \infty$. It follows that there exists an integer N_1 such that X_j is nonsingular for all $j > N_1$. For $j > N_1$, define

$$\boldsymbol{y}_j = X_j^{-1}\boldsymbol{x}_{m+1,j}, \tag{2.14}$$

so that

$$y_j \to (\epsilon \mathrm{I}_m)^{-1}(-\epsilon \mathbf{1}_m) = -\mathbf{1}_m$$

Thus there exists some integer $N_2 \geq N_1$, such that for all $j > N_2$, all of the components of $\boldsymbol{y}_j$ are negative. But from (2.14) we have

$$\boldsymbol{x}_{m+1,j} - X_j \boldsymbol{y}_j = [X_j \quad \boldsymbol{x}_{m+1,j}] \begin{bmatrix} -\boldsymbol{y}_j \\ 1 \end{bmatrix} = \mathbf{0}$$

This same equation holds if we replace the vector $(-\boldsymbol{y}_j', \quad 1)'$ by the unit vector $(\boldsymbol{y}_j'\boldsymbol{y}_j + 1)^{-1/2}(-\boldsymbol{y}_j', \quad 1)'$. Thus $\mathbf{0}$ is a convex combination of the columns of $[X_j \quad \boldsymbol{x}_{m+1,j}]$, each of which is in S_*, so since S_* is convex, $\mathbf{0} \in S_*$. □

The next result is sometimes called the supporting hyperplane theorem. It states that for any boundary point of a convex set S, there exists a hyperplane passing through that point such that none of the points of S are on one side of the hyperplane.

Theorem 2.27. Suppose that S is a convex subset of R^m and that $\boldsymbol{x}_*$ either is not in S or is a boundary point of S if it is in S. Then there exists an $m \times 1$ vector $\boldsymbol{b} \neq \mathbf{0}$ such that $\boldsymbol{b}'\boldsymbol{x} \geq \boldsymbol{b}'\boldsymbol{x}_*$ for all $\boldsymbol{x} \in S$.

Proof. It follows from the previous theorem that $\boldsymbol{x}_*$ also is not in $\overline{S}$ or must be a boundary point of $\overline{S}$ if it is in $\overline{S}$. Consequently, there exists a sequence of vectors, $\boldsymbol{x}_1, \boldsymbol{x}_2, \ldots$ with each $\boldsymbol{x}_i \notin \overline{S}$ such that $\boldsymbol{x}_i \to \boldsymbol{x}_*$ as $i \to \infty$. Corresponding to each $\boldsymbol{x}_i$, define the set $S_i = \{\boldsymbol{y}\colon \boldsymbol{y} = \boldsymbol{x} - \boldsymbol{x}_i, \boldsymbol{x} \in S\}$, and note that $\mathbf{0} \notin \overline{S}_i$ since $\boldsymbol{x}_i \notin \overline{S}$. Thus, since $\overline{S}_i$ is closed and convex by Theorem 2.24, it follows from Theorem 2.25 that there exists an $m \times m$ vector $\boldsymbol{a}_i$ such that $\boldsymbol{a}_i'\boldsymbol{y} > 0$ for all $\boldsymbol{y} \in \overline{S}_i$ or, equivalently, $\boldsymbol{a}_i'(\boldsymbol{x} - \boldsymbol{x}_i) > 0$ for all $\boldsymbol{x} \in \overline{S}$. Alternatively, we can write this as $\boldsymbol{b}_i'(\boldsymbol{x} - \boldsymbol{x}_i) > 0$, where $\boldsymbol{b}_i = (\boldsymbol{a}_i'\boldsymbol{a}_i)^{-1/2}\boldsymbol{a}_i$. Now since $\boldsymbol{b}_i'\boldsymbol{b}_i = 1$, the sequence $\boldsymbol{b}_1, \boldsymbol{b}_2, \ldots$, is a bounded sequence and so it has a convergent subsequence; that is, there are positive integers $i_1 < i_2 < \cdots$, and some $m \times 1$ unit vector $\boldsymbol{b}$ such that $\boldsymbol{b}_{i_j} \to \boldsymbol{b}$ as $j \to \infty$. Consequently, $\boldsymbol{b}_{i_j}'(\boldsymbol{x} - \boldsymbol{x}_{i_j}) \to \boldsymbol{b}'(\boldsymbol{x} - \boldsymbol{x}_*)$ as $j \to \infty$, and we must have $\boldsymbol{b}'(\boldsymbol{x} - \boldsymbol{x}_*) \geq 0$ for all $\boldsymbol{x} \in S$ since $\boldsymbol{b}_{i_j}'(\boldsymbol{x} - \boldsymbol{x}_{i_j}) > 0$ for all $\boldsymbol{x} \in S$. This completes the proof. □

We are now ready to prove the separating hyperplane theorem.

Theorem 2.28. Let S_1 and S_2 be convex subsets of R^m with $S_1 \cap S_2 = \emptyset$. Then there exists an $m \times 1$ vector $\boldsymbol{b} \neq \mathbf{0}$ such that $\boldsymbol{b}'\boldsymbol{x}_1 \geq \boldsymbol{b}'\boldsymbol{x}_2$ for all $\boldsymbol{x}_1 \in S_1$ and all $\boldsymbol{x}_2 \in S_2$.

Proof. Clearly the set $S_{2*} = \{\boldsymbol{x}\colon -\boldsymbol{x} \in S_2\}$ is convex since S_2 is convex. Thus from Theorem 2.23 we know that the set

$$S = S_1 + S_{2*} = \{\boldsymbol{x}\colon \boldsymbol{x} = \boldsymbol{x}_1 - \boldsymbol{x}_2, \boldsymbol{x}_1 \in S_1, \boldsymbol{x}_2 \in S_2\}$$

is also convex. In addition, $\mathbf{0} \notin S$ since $S_1 \cap S_2 = \emptyset$. Consequently, using Theorem 2.27, we find that there is an $m \times 1$ vector $\boldsymbol{b} \neq \mathbf{0}$ for which $\boldsymbol{b}'\boldsymbol{x} \geq \mathbf{0}$ for all $\boldsymbol{x} \in S$. But this implies that $\boldsymbol{b}'(\boldsymbol{x}_1 - \boldsymbol{x}_2) \geq 0$ for all $\boldsymbol{x}_1 \in S_1$ and all $\boldsymbol{x}_2 \in S_2$, as is required. □

Suppose that $f(x)$ is a nonnegative function which is symmetric about $x = 0$ and has only one maximum, occurring at $x = 0$; in other words, $f(x) = f(-x)$ for all x and $f(x) \leq f(cx)$ if $0 \leq c \leq 1$. Clearly, the integral of $f(x)$ over an interval of fixed length will be maximized when the interval is centered at 0. This can be expressed as

$$\int_{-a}^{a} f(x + cy)\,\mathrm{d}x \geq \int_{-a}^{a} f(x + y)\,\mathrm{d}x,$$

for any y, $a > 0$, and $0 \leq c \leq 1$. This result has some important applications regarding probabilities of random variables. The following result, which is a generalization to a function $f(\boldsymbol{x})$ of the $m \times 1$ vector $\boldsymbol{x}$ replaces the interval in R^1 by a symmetric convex set in R^m. This generalization is due to Anderson (1955). For simple applications of the result to probabilities of random vectors, see Problem 2.44.

Theorem 2.29. Let S be a convex subset of R^m, symmetric about $\mathbf{0}$, so that if $\boldsymbol{x} \in S$, $-\boldsymbol{x} \in S$ also. Let $f(\boldsymbol{x}) \geq 0$ be a function for which $f(\boldsymbol{x}) = f(-\boldsymbol{x})$, $S_\alpha = \{\boldsymbol{x} : f(\boldsymbol{x}) \geq \alpha\}$ is convex for any positive α, and $\int_S f(\boldsymbol{x})\,\mathrm{d}\boldsymbol{x} < \infty$. Then

$$\int_S f(\boldsymbol{x} + c\boldsymbol{y})\,\mathrm{d}\boldsymbol{x} \geq \int_S f(\boldsymbol{x} + \boldsymbol{y})\,\mathrm{d}\boldsymbol{x},$$

for $0 \leq c \leq 1$ and $\boldsymbol{y} \in R^m$.

A more comprehensive discussion of convex sets can be found in Kelly and Weiss (1979), Lay (1982), and Rockafellar (1970), while some applications of the separating hyperplane theorem to statistical decision theory can be found in Ferguson (1967).

PROBLEMS

1. Determine whether each of the following sets of vectors is a vector space.
   (a) $\{(a, b, a + b, 1)'\colon -\infty < a < \infty, -\infty < b < \infty\}$.
   (b) $\{(a, b, c, a + b - 2c)'\colon -\infty < a < \infty, -\infty < b < \infty, -\infty < c < \infty\}$.
   (c) $\{(a, b, c, 1 - a - b - c)'\colon -\infty < a < \infty, -\infty < b < \infty, -\infty < c < \infty\}$.

2. Consider the vector space

$$S = \{(a, a+b, a+b, -b)' : -\infty < a < \infty, -\infty < b < \infty\}$$

Determine which of the following sets of vectors are spanning sets of S.
(a) $\{(1,0,0,1)', (1,2,2,-1)'\}$.
(b) $\{(1,1,0,0)', (0,0,1,-1)'\}$.
(c) $\{(2,1,1,1)', (3,1,1,2)', (3,2,2,1)'\}$.
(d) $\{(1,0,0,0)', (0,1,1,0)', (0,0,0,1)'\}$.

3. Is the vector $\boldsymbol{x} = (1,1,1,1)'$ in the vector space S given in Problem 2? Is the vector $\boldsymbol{y} = (4,1,1,3)'$ in S?

4. Let $\{\boldsymbol{x}_1, \ldots, \boldsymbol{x}_r\}$ be a set of vectors in a vector space S and let W be the vector subspace consisting of all possible linear combinations of these vectors. Prove that W is the smallest subspace of S that contains the vectors $\boldsymbol{x}_1, \ldots, \boldsymbol{x}_r$; that is, show that if V is another vector subspace containing $\boldsymbol{x}_1, \ldots, \boldsymbol{x}_r$, then W is a subspace of V.

5. Suppose that $\boldsymbol{x}$ is a random vector having a distribution with mean vector $\boldsymbol{\mu}$ and covariance matrix Ω given by

$$\boldsymbol{\mu} = \begin{bmatrix} 1 \\ 1 \end{bmatrix}, \qquad \Omega = \begin{bmatrix} 1 & -0.5 \\ -0.5 & 1 \end{bmatrix}$$

Let $\boldsymbol{x}_1 = (2,2)'$ and $\boldsymbol{x}_2 = (2,0)'$ be two observations from this distribution. Use the Mahalanobis distance function to determine which of these two observations is closer to the mean.

6. Show that the functions $\|\boldsymbol{x}\|_p$ and $\|\boldsymbol{x}\|_\infty$ defined in Section 2.2 are, in fact, vector norms.

7. Prove Theorem 2.3.

8. Show that the set of vectors $\{(1,2,2,2)', (1,2,1,2)', (1,1,1,1)'\}$ is a linearly independent set.

9. Consider the set of vectors

$$\{(2,1,4,3)', (3,0,5,2)', (0,3,2,5)', (4,2,8,6)'\}$$

(a) Show that this set of vectors is linearly dependent.

(b) From this set of four vectors find a subset of two vectors that is a linearly independent set.

10. Which of the following sets of vectors are bases for R^4?
   (a) $\{(0,1,0,1)', (1,1,0,0)', (0,0,1,1)'\}$.
   (b) $\{(2,2,2,1)', (2,1,1,1)', (3,2,1,1)', (1,1,1,1)'\}$.
   (c) $\{(2,0,1,1)', (3,1,2,2)', (2,1,1,2)', (2,1,2,1)'\}$.

11. Prove the results of Theorem 2.8.

12. Prove that if a set of orthogonal vectors does not contain the null vector, it is a linearly independent set.

13. Find a basis for the vector space given in Problem 2. What is the dimension of this vector space? Find a second different basis for this same vector space.

14. Show that the set of vectors $\{\boldsymbol{\gamma}_1, \ldots, \boldsymbol{\gamma}_m\}$, given in Example 2.4, is a basis for R^m.

15. Let A be an $m \times n$ matrix and B be an $n \times p$ matrix. Show that
   (a) $R(AB) \subseteq R(A)$.
   (b) $R(AB) = R(A)$ if $\text{rank}(AB) = \text{rank}(A)$.

16. Suppose A and B are $m \times n$ matrices. Show that there exists an $n \times n$ matrix C satisfying $AC = B$ if and only if $R(B) \subseteq R(A)$.

17. Prove the results of Theorem 2.11.

18. Let A, B, and C be $p \times n$, $m \times q$, and $m \times n$ matrices, respectively. Prove that

$$\text{rank}\left(\begin{bmatrix} C & B \\ A & (0) \end{bmatrix}\right) = \text{rank}(A) + \text{rank}(B)$$

if there exist an $m \times p$ matrix F and a $q \times n$ matrix G such that $C = FA + BG$.

19. Let A be an $m \times n$ matrix and B an $n \times p$ matrix with $\text{rank}(B) = n$. Show that $\text{rank}(A) = \text{rank}(AB)$.

20. Refer to Examples 2.7 and 2.9. Find the matrix A satisfying $Z_1 = X_1 A$, where $Z_1 = (z_1, z_2, z_3)$ and $X_1 = (\boldsymbol{x}_1, \boldsymbol{x}_2, \boldsymbol{x}_3)$. Show that $AA' = (X_1' X_1)^{-1}$.

21. Let S be the vector space spanned by the vectors $\boldsymbol{x}_1 = (1, 2, 1, 2)', \boldsymbol{x}_2 = (2, 3, 1, 2)', \boldsymbol{x}_3 = (3, 4, -1, 0)'$, and $\boldsymbol{x}_4 = (3, 4, 0, 1)'$.
    (a) Find a basis for S.
    (b) Use the Gram–Schmidt procedure on the basis found in (a) to determine an orthonormal basis for S.
    (c) Find the orthogonal projection of $\boldsymbol{x} = (1, 0, 0, 1)'$ onto S.
    (d) Find the component of $\boldsymbol{x}$ orthogonal to S.

22. Using equation (2.7), determine the projection matrix for the vector space S given in Problem 21. Use this to compute the orthogonal projection of $\boldsymbol{x} = (1, 0, 0, 1)'$ onto S.

23. Let S be the vector space spanned by the vectors $\boldsymbol{x}_1 = (1, 2, 3)'$ and $\boldsymbol{x}_2 = (1, 1, -1)'$. Find the point in S that is closest to the point $\boldsymbol{x} = (1, 1, 1)'$.

24. Suppose S is a vector subspace of R^4 having the projection matrix

$$P_S = \frac{1}{10}\begin{bmatrix} 6 & -2 & -2 & -4 \\ -2 & 9 & -1 & -2 \\ -2 & -1 & 9 & -2 \\ -4 & -2 & -2 & 6 \end{bmatrix}$$

    (a) What is the dimension of S?
    (b) Find a basis for S.

25. Consider the vector space $S = \{\boldsymbol{u}\colon \boldsymbol{u} = A\boldsymbol{x}, \boldsymbol{x} \in R^4\}$, where A is the 4×4 matrix given by

$$A = \begin{bmatrix} 1 & 2 & 0 & 1 \\ 1 & 1 & 2 & 2 \\ 1 & 0 & 4 & 3 \\ 1 & 3 & -2 & 0 \end{bmatrix}$$

    (a) Determine the dimension of S and find a basis.
    (b) Determine the dimension of the null space $N(A)$ and find a basis for it.
    (c) Is the vector $(3, 5, 2, 4)'$ in S?
    (d) Is the vector $(1, 1, 1, 1)'$ in $N(A)$?

26. Let $\boldsymbol{x} \in R^n$ and suppose that $\boldsymbol{u}(\boldsymbol{x})$ defines a linear transformation of R^n into R^m. Using the basis $\{\boldsymbol{e}_1, \ldots, \boldsymbol{e}_n\}$ for R^n and the $m \times 1$ vectors $\boldsymbol{u}(\boldsymbol{e}_1), \ldots, \boldsymbol{u}(\boldsymbol{e}_n)$, prove that there exists an $m \times n$ matrix A for which

$$\boldsymbol{u}(\boldsymbol{x}) = A\boldsymbol{x},$$

for every $\boldsymbol{x} \in R^n$.

27. Let T be a vector subspace of R^n and suppose that S is the subspace of R^m given by

$$S = \{\boldsymbol{u}(\boldsymbol{x}) \colon \boldsymbol{x} \in T\},$$

where the transformation defined by $\boldsymbol{u}$ is linear. Show that there exists an $m \times n$ matrix A satisfying

$$\boldsymbol{u}(\boldsymbol{x}) = A\boldsymbol{x},$$

for every $\boldsymbol{x} \in T$.

28. Let T be the vector space spanned by the two vectors $\boldsymbol{x}_1 = (1, 1, 0)'$ and $\boldsymbol{x}_2 = (0, 1, 1)'$. Let S be the vector space defined as $S = \{\boldsymbol{u}(\boldsymbol{x}) \colon \boldsymbol{x} \in T\}$, where the function $\boldsymbol{u}$ defines a linear transformation satisfying $\boldsymbol{u}(\boldsymbol{x}_1) = (2, 3, 1)'$ and $\boldsymbol{u}(\boldsymbol{x}_2) = (2, 5, 3)'$. Find a matrix A such that $\boldsymbol{u}(\boldsymbol{x}) = A\boldsymbol{x}$, for all $\boldsymbol{x} \in T$.

29. Consider the linear transformation defined by

$$\boldsymbol{u}(\boldsymbol{x}) = \begin{bmatrix} x_1 - \overline{x} \\ x_2 - \overline{x} \\ \vdots \\ x_m - \overline{x} \end{bmatrix},$$

for all $\boldsymbol{x} \in R^m$, where $\overline{x} = (1/m)\Sigma x_i$. Find the matrix A for which $\boldsymbol{u}(\boldsymbol{x}) = A\boldsymbol{x}$ and then determine the dimension of the range and null spaces.

30. In an introductory statistics course, students must take three 100-point exams followed by a 150-point final exam. We will identify the scores on these exams with the variables x_1, x_2, x_3, and y. We want to be able to estimate the value of y once x_1, x_2, and x_3 are known. A class of 32 students produced the following set of exam scores.

Student	x_1	x_2	x_3	y	Student	x_1	x_2	x_3	y
1	87	89	92	111	17	72	76	96	116
2	72	85	77	99	18	73	70	52	78
3	67	79	54	82	19	73	61	86	101
4	79	71	68	136	20	73	83	76	82
5	60	67	53	73	21	97	99	97	141
6	83	84	92	107	22	84	92	86	112
7	82	88	76	106	23	82	68	73	62
8	87	68	91	128	24	61	59	77	56
9	88	66	65	95	25	78	73	81	137
10	62	68	63	108	26	84	73	68	118
11	100	100	100	142	27	57	47	71	108
12	87	82	80	89	28	87	95	84	121
13	72	94	76	109	29	62	29	66	71
14	86	92	98	140	30	77	82	81	123
15	85	82	62	117	31	52	66	71	102
16	62	50	71	102	32	95	99	96	130

(a) Find the least squares estimator for $\boldsymbol{\beta} = (\beta_0, \beta_1, \beta_2, \beta_3)'$ in the multiple regression model

$$y = \beta_0 + \beta_1 x_1 + \beta_2 x_2 + \beta_3 x_3 + \epsilon$$

(b) Find the least squares estimator for $\boldsymbol{\beta}_1 = (\beta_0, \beta_1, \beta_2)'$ in the model

$$y = \beta_0 + \beta_1 x_1 + \beta_2 x_2 + \epsilon$$

(c) Compute the reduction in the sum of squared errors attributable to the inclusion of the variable x_3 in the model given in (a).

31. Suppose that we have independent samples of a response y corresponding to k different treatments with a sample size of n_i responses from the ith treatment. If the jth observation from the ith treatment is denoted, y_{ij}, then the model

$$y_{ij} = \mu_i + \epsilon_{ij}$$

is known as the one-way classification model. Here μ_i represents the expected value of a response from treatment i, while the ϵ_{ij}s are independent and identically distributed as $N(0, \sigma^2)$.

(a) If we let $\boldsymbol{\beta} = (\mu_1, \ldots, \mu_k)'$, write the model above in matrix form by defining $\boldsymbol{y}$, X, and $\boldsymbol{\epsilon}$ so that $\boldsymbol{y} = X\boldsymbol{\beta} + \boldsymbol{\epsilon}$.

(b) Find the least squares estimator of $\boldsymbol{\beta}$ and show that the sum of squared errors for the corresponding fitted model is given by

$$\text{SSE}_1 = \sum_{i=1}^{k} \sum_{j=1}^{n_i} (y_{ij} - \overline{y}_i)^2,$$

where

$$\overline{y}_i = \sum_{j=1}^{n_i} y_{ij}/n_i$$

(c) If $\mu_1 = \cdots = \mu_k = \mu$, then the reduced model

$$y_{ij} = \mu + \epsilon_{ij}$$

holds for all i and j. Find the least squares estimator of μ and the sum of squared errors SSE_2 for the fitted reduced model. Show that $\text{SSE}_2 - \text{SSE}_1$, referred to as the sum of squares for treatment and denoted SST, can be expressed as

$$\text{SST} = \sum_{i=1}^{k} n_i(\overline{y}_i - \overline{y})^2,$$

where

$$\overline{y} = \sum_{i=1}^{k} n_i \overline{y}_i / n, \qquad n = \sum_{i=1}^{k} n_i$$

(d) Show that the F statistic given in (2.8) takes the form

$$\text{F} = \frac{\text{SST}/(k-1)}{\text{SSE}_1/(n-k)}$$

32. Suppose that we have the model $\boldsymbol{y} = X\boldsymbol{\beta} + \boldsymbol{\epsilon}$ and wish to find the estimator $\hat{\boldsymbol{\beta}}$ which minimizes

$$(\boldsymbol{y} - X\hat{\boldsymbol{\beta}})'(\boldsymbol{y} - X\hat{\boldsymbol{\beta}}),$$

subject to the restriction that $\hat{\boldsymbol{\beta}}$ satisfies $A\hat{\boldsymbol{\beta}} = \boldsymbol{0}$, where X has full column rank and A has full row rank.

(a) Show that $S = \{\boldsymbol{y}: \boldsymbol{y} = X\hat{\boldsymbol{\beta}}, A\hat{\boldsymbol{\beta}} = \boldsymbol{0}\}$ is a vector space.

(b) Let C by any matrix whose columns form a basis for the null space of A; that is, C satisfies the identity $C(C'C)^{-1}C' = \mathrm{I} - A'(AA')^{-1}A$. Using the geometrical properties of least squares estimators, show that the restricted least squares estimator $\hat{\boldsymbol{\beta}}$ is given by

$$\hat{\boldsymbol{\beta}} = C(C'X'XC)^{-1}C'X'\boldsymbol{y}$$

33. Let S_1 and S_2 be vector subspaces of R^m. Show that $S_1 + S_2$ also must be a vector subspace of R^m.

34. Let S_1 and S_2 be vector subspaces of R^m. Show that $S_1 + S_2$ is the vector space of smallest dimension containing $S_1 \cup S_2$. In other words, show that if T is a vector space for which $S_1 \cup S_2 \subseteq T$, then $S_1 + S_2 \subseteq T$.

35. Prove Theorem 2.22.

36. Let S_1 and S_2 be vector subspaces of R^m. Suppose that $\{\boldsymbol{x}_1, \ldots, \boldsymbol{x}_r\}$ spans S_1 and $\{\boldsymbol{y}_1, \ldots, \boldsymbol{y}_h\}$ spans S_2. Show that $\{\boldsymbol{x}_1, \ldots, \boldsymbol{x}_r, \boldsymbol{y}_1, \ldots, \boldsymbol{y}_h\}$ spans the vector space $S_1 + S_2$.

37. Let S_1 be the vector space spanned by the vectors

$$\boldsymbol{x}_1 = \begin{bmatrix} 3 \\ 1 \\ 3 \\ 1 \end{bmatrix}, \qquad \boldsymbol{x}_2 = \begin{bmatrix} 1 \\ 1 \\ 1 \\ 1 \end{bmatrix}, \qquad \boldsymbol{x}_3 = \begin{bmatrix} 2 \\ 1 \\ 2 \\ 1 \end{bmatrix},$$

while the vector space S_2 is spanned by the vectors

$$\boldsymbol{y}_1 = \begin{bmatrix} 3 \\ 0 \\ 5 \\ -1 \end{bmatrix}, \qquad \boldsymbol{y}_2 = \begin{bmatrix} 1 \\ 2 \\ 3 \\ 1 \end{bmatrix}, \qquad \boldsymbol{y}_3 = \begin{bmatrix} 1 \\ -4 \\ -1 \\ -3 \end{bmatrix}$$

Find the following.

(a) Bases for S_1 and S_2.

(b) The dimension of $S_1 + S_2$.

(c) A basis for $S_1 + S_2$.

(d) The dimension of $S_1 \cap S_2$.

(e) A basis for $S_1 \cap S_2$.

38. Let S_1 and S_2 be vector subspaces of R^m with $\dim(S_1) = r_1$ and $\dim(S_2) = r_2$.
    (a) Find expressions in terms of m, r_1, and r_2 for the smallest and largest possible values of $\dim(S_1 + S_2)$.
    (b) Find the smallest and largest possible values of $\dim(S_1 \cap S_2)$.

39. Let T be the vector space spanned by the vectors $\{(1, 1, 1)', (2, 1, 2)'\}$. Find a vector space S_1 such that $R^3 = T \oplus S_1$. Find another vector space S_2 such that $R^3 = T \oplus S_2$ and $S_1 \cap S_2 = \{\mathbf{0}\}$.

40. Let S_1 be the vector space spanned by $\{(1, 1, -2, 0)', (2, 0, 1, -3)'\}$, while S_2 is spanned by $\{(1, 1, 1, -3)', (1, 1, 1, 1)'\}$. Show that $R^4 = S_1 + S_2$. Is this a direct sum? That is, can we write $S_1 \oplus S_2$? Are S_1 and S_2 orthogonal vector spaces?

41. Let S_1 and S_2 be vector subspaces of R^m and let $T = S_1 + S_2$. Show that this sum is a direct sum, that is, $T = S_1 \oplus S_2$ if and only if

$$\dim(T) = \dim(S_1) + \dim(S_2)$$

42. The concept of orthogonal projections and their associated projection matrices can be extended to projections that are not orthogonal. In the case of orthogonal projections onto the vector space $S \subseteq R^m$, we decompose R^m as $R^m = S \oplus S^{\perp}$. The projection matrix that projects orthogonally onto S is the matrix P satisfying $P\boldsymbol{y} \in S$ and $(\boldsymbol{y} - P\boldsymbol{y}) \in S^{\perp}$ for all $\boldsymbol{y} \in R^m$ and $P\boldsymbol{x} = \boldsymbol{x}$ for all $\boldsymbol{x} \in S$. If S is the column space of the full rank matrix X, then $S^{\perp}$ will be the null space of X', and the projection matrix described above is given by $P = X(X'X)^{-1}X'$. Suppose now that we decompose R^m as $R^m = S \oplus T$, where S is as before and T is the null space of the full rank matrix Y'. Note that S and T are not necessarily orthogonal vector spaces. We wish to find a projection matrix Q satisfying $Q\boldsymbol{y} \in S$ and $(\boldsymbol{y} - Q\boldsymbol{y}) \in T$ for all $\boldsymbol{y} \in R^m$ and $Q\boldsymbol{x} = \boldsymbol{x}$ for all $\boldsymbol{x} \in S$.
    (a) Show that Q is a projection matrix if and only if it is an idempotent matrix.
    (b) Show that Q can be expressed as $Q = X(Y'X)^{-1}Y'$.

43. Prove Theorem 2.23.

44. Show that if S_1 and S_2 are convex subsets of R^m, then $S_1 \cup S_2$ need not be convex.

45. Show that for any positive scalar n, the set $B_n = \{\boldsymbol{x}\colon \boldsymbol{x} \in R^m, \boldsymbol{x}'\boldsymbol{x} \leq n^{-1}\}$ is convex.

46. For any set $S \subseteq R^m$, show that its convex hull $\text{C}(S)$ consists of all convex combinations of the vectors in S.

47. Suppose that S is a nonempty subset of R^m. Show that every vector in the convex hull of S can be expressed as a convex combination of $m + 1$ or fewer vectors in S.

48. Let $\boldsymbol{x}$ be an $m \times 1$ random vector with density function $f(\boldsymbol{x})$ such that $f(\boldsymbol{x}) = f(-\boldsymbol{x})$ and the set $\{\boldsymbol{x}\colon f(\boldsymbol{x}) \geq \alpha\}$ is convex for all positive α. Suppose that S is a convex subset of R^m, symmetric about $\mathbf{0}$.
    (a) Show that $\text{P}(\boldsymbol{x} + c\boldsymbol{y} \in S) \geq \text{P}(\boldsymbol{x} + \boldsymbol{y} \in S)$ for any constant vector $\boldsymbol{y} \in S$ and $0 \leq c \leq 1$.
    (b) Show that the inequality in (a) also holds if $\boldsymbol{y}$ is an $m \times 1$ random vector distributed independently of $\boldsymbol{x}$.
    (c) Show that if $\boldsymbol{x} \sim N_m(\mathbf{0}, \Omega)$, its density function satisfies the conditions of this exercise.
    (d) Show that if $\boldsymbol{x}$ and $\boldsymbol{y}$ are independently distributed with $\boldsymbol{x} \sim N_m(\mathbf{0}, \Omega_1)$ and $\boldsymbol{y} \sim N_m(\mathbf{0}, \Omega_2)$ such that $\Omega_1 - \Omega_2$ is nonnegative definite, then $\text{P}(\boldsymbol{x} \in S) \leq \text{P}(\boldsymbol{y} \in S)$.

CHAPTER THREE

Eigenvalues and Eigenvectors

1. INTRODUCTION

Eigenvalues and eigenvectors are special implicitly defined functions of the elements of a square matrix. In many applications involving the analysis of a square matrix, the key information from the analysis is provided by some or all of these eigenvalues and eigenvectors. This is a consequence of some of the properties of eigenvalues and eigenvectors that we will develop in this chapter. But before we get to these properties, we must first understand how eigenvalues and eigenvectors are defined and how they are calculated.

2. EIGENVALUES, EIGENVECTORS, AND EIGENSPACES

If A is an $m \times m$ matrix, then any scalar λ satisfying the equation

$$A\boldsymbol{x} = \lambda \boldsymbol{x}, \tag{3.1}$$

for some $m \times 1$ vector $\boldsymbol{x} \neq \mathbf{0}$, is called an eigenvalue of A. The vector $\boldsymbol{x}$ is called an eigenvector of A corresponding to the eigenvalue λ, and equation (3.1) is called the eigenvalue–eigenvector equation of A. Eigenvalues and eigenvectors are also sometimes referred to as latent roots and vectors or characteristic roots and vectors. Equation (3.1) can be equivalently expressed as

$$(A - \lambda \mathrm{I})\boldsymbol{x} = \mathbf{0} \tag{3.2}$$

Note that if $|A - \lambda \mathrm{I}| \neq 0$, then $(A - \lambda \mathrm{I})^{-1}$ would exist and so premultiplication of equation (3.2) by this inverse would lead to a contradiction of the already stated assumption that $\boldsymbol{x} \neq \mathbf{0}$. Thus, any eigenvalue λ must satisfy the determinantal equation

$$|A - \lambda \mathrm{I}| = 0,$$

which is known as the characteristic equation of A. Using the definition of a determinant, we readily observe that the characteristic equation is an mth degree polynomial in λ; that is, there are scalars $\alpha_0, \dots, \alpha_{m-1}$ such that the characteristic equation above can be expressed equivalently as

$$(-\lambda)^m + \alpha_{m-1}(-\lambda)^{m-1} + \cdots + \alpha_1(-\lambda) + \alpha_0 = 0$$

Since an mth degree polynomial has m roots, it follows that an $m \times m$ matrix has m eigenvalues; that is, there are m scalars $\lambda_1, \dots, \lambda_m$, which satisfy the characteristic equation. When all of the eigenvalues of A are real, we will sometimes find it notationally convenient to identify the ith largest eigenvalue of the matrix A as $\lambda_i(A)$. In other words, in this case the ordered eigenvalues of A may be written as $\lambda_1(A) \geq \cdots \geq \lambda_m(A)$.

The characteristic equation can be used to obtain the eigenvalues of the matrix A. These can be then used in the eigenvalue–eigenvector equation to obtain corresponding eigenvectors.

Example 3.1. We will find the eigenvalues and eigenvectors of the 3×3 matrix A given by

$$A = \begin{bmatrix} 5 & -3 & 3 \\ 4 & -2 & 3 \\ 4 & -4 & 5 \end{bmatrix}$$

The characteristic equation of A is

$$\begin{aligned} |A - \lambda \mathrm{I}| &= \begin{vmatrix} 5-\lambda & -3 & 3 \\ 4 & -2-\lambda & 3 \\ 4 & -4 & 5-\lambda \end{vmatrix} \\ &= -(5-\lambda)^2(2+\lambda) - 3(4)^2 - 4(3)^2 \\ &\quad + 3(4)(2+\lambda) + 3(4)(5-\lambda) + 3(4)(5-\lambda) \\ &= -\lambda^3 + 8\lambda^2 - 17\lambda + 10 \\ &= -(\lambda - 5)(\lambda - 2)(\lambda - 1) = 0, \end{aligned}$$

so the three eigenvalues of A are 1, 2, and 5. To find an eigenvector of A corresponding to the eigenvalue $\lambda = 5$, we must solve the equation $A\boldsymbol{x} = 5\boldsymbol{x}$ for $\boldsymbol{x}$, which yields the system of equations

$$\begin{aligned} 5x_1 - 3x_2 + 3x_3 &= 5x_1 \\ 4x_1 - 2x_2 + 3x_3 &= 5x_2 \\ 4x_1 - 4x_2 + 5x_3 &= 5x_3 \end{aligned}$$

The first and third equations imply that $x_2 = x_3$ and $x_1 = x_2$, which when used in the second equation yields the identity $x_2 = x_2$. Thus, x_2 is arbitrary and so any $\boldsymbol{x}$ having $x_1 = x_2 = x_3$, such as the vector $(1, 1, 1)'$, is an eigenvector of A associated with the root 5. In a similar fashion, by solving the equation $A\boldsymbol{x} = \lambda\boldsymbol{x}$, when $\lambda = 2$ and $\lambda = 1$, we find that $(1, 1, 0)'$ is an eigenvector corresponding to the eigenvalue 2, and $(0, 1, 1)'$ is an eigenvector corresponding to the eigenvalue 1.

Note that if a nonnull vector $\boldsymbol{x}$ satisfies (3.1) for a given value of λ, then so will $(\alpha\boldsymbol{x})$ for any nonzero scalar α. Thus, eigenvectors are not uniquely defined unless we impose some scale constraint; for instance, we might only consider eigenvectors, $\boldsymbol{x}$, satisfying $\boldsymbol{x}'\boldsymbol{x} = 1$. In this case, for the previous example we would obtain the three normalized eigenvectors $(1/\sqrt{3}, 1/\sqrt{3}, 1/\sqrt{3})'$, $(1/\sqrt{2}, 1/\sqrt{2}, 0)'$ and $(0, 1/\sqrt{2}, 1/\sqrt{2})'$ corresponding to the eigenvalues 5, 2, and 1, respectively. These normalized eigenvectors are unique except for sign, since each of these eigenvectors, when multiplied by -1, yields another normalized eigenvector.

The following example illustrates the fact that a real matrix may have complex eigenvalues and eigenvectors.

Example 3.2. The matrix

$$A = \begin{bmatrix} 1 & 1 \\ -2 & -1 \end{bmatrix}$$

has the characteristic equation

$$|A - \lambda I| = \begin{vmatrix} 1-\lambda & 1 \\ -2 & -1-\lambda \end{vmatrix} = -(1-\lambda)(1+\lambda) + 2 = \lambda^2 + 1 = 0,$$

so that the eigenvalues of A are $i = \sqrt{-1}$ and $-i$. To find an eigenvector corresponding to the root i, write $\boldsymbol{x} = (x_1, x_2)' = (y_1 + iz_1, y_2 + iz_2)'$ and solve for y_1, y_2, z_1, z_2 using the equation $A\boldsymbol{x} = i\boldsymbol{x}$. From this we find that for any real scalar $\alpha \neq 0$, $\boldsymbol{x} = (\alpha + i\alpha, -2\alpha)'$ is an eigenvector corresponding to the eigenvalue i. In a similar manner, it can be shown that an eigenvector associated with the eigenvalue $-i$ has the form $\boldsymbol{x} = (\alpha - i\alpha, -2\alpha)'$.

The m eigenvalues of a matrix A need not all be different since the characteristic equation may have repeated roots. An eigenvalue that occurs as a single solution to the characteristic equation will be called a simple or distinct eigenvalue. Otherwise, an eigenvalue will be called a multiple eigenvalue, and its multiplicity will be given by the number of times this solution is repeated.

In some situations, we will find it useful to work with the set of all eigenvectors associated with a specific eigenvalue. This collection, $S_A(\lambda)$, of all eigenvectors corresponding to the particular eigenvalue λ, along with the trivial vector $\mathbf{0}$, is called the eigenspace of A associated with λ; that is, $S_A(\lambda)$ is given by $S_A(\lambda) = \{\boldsymbol{x}: \boldsymbol{x} \in R^m \text{ and } A\boldsymbol{x} = \lambda\boldsymbol{x}\}$.

Theorem 3.1. If $S_A(\lambda)$ is the eigenspace of the $m \times m$ matrix A corresponding to the root λ, then $S_A(\lambda)$ is a vector subspace of R^m.

Proof. By definition, if $\boldsymbol{x} \in S_A(\lambda)$, then $A\boldsymbol{x} = \lambda\boldsymbol{x}$. Thus, if $\boldsymbol{x} \in S_A(\lambda)$ and $\boldsymbol{y} \in S_A(\lambda)$, we have for any scalars α and β

$$A(\alpha\boldsymbol{x} + \beta\boldsymbol{y}) = \alpha A\boldsymbol{x} + \beta A\boldsymbol{y} = \alpha(\lambda\boldsymbol{x}) + \beta(\lambda\boldsymbol{y}) = \lambda(\alpha\boldsymbol{x} + \beta\boldsymbol{y})$$

Consequently, $(\alpha\boldsymbol{x} + \beta\boldsymbol{y}) \in S_A(\lambda)$, and so $S_A(\lambda)$ is a vector space. □

Example 3.3. The matrix

$$A = \begin{bmatrix} 2 & -1 & 0 \\ 0 & 1 & 0 \\ 0 & 0 & 1 \end{bmatrix}$$

has the characteristic equation

$$\begin{vmatrix} 2-\lambda & -1 & 0 \\ 0 & 1-\lambda & 0 \\ 0 & 0 & 1-\lambda \end{vmatrix} = (1-\lambda)^2(2-\lambda) = 0,$$

and so the eigenvalues of A are 1, with multiplicity 2, and 2. To find $S_A(1)$, the eigenspace corresponding to the eigenvalue 1, we solve the equation $A\boldsymbol{x} = \boldsymbol{x}$ for $\boldsymbol{x}$. We leave it to the reader to verify that this leads to two linearly independent solutions; any solution to $A\boldsymbol{x} = \boldsymbol{x}$ will be a linear combination of the two vectors $\boldsymbol{x}_1 = (0, 0, 1)'$ and $\boldsymbol{x}_2 = (1, 1, 0)'$. Thus, $S_A(1)$ is the subspace spanned by the basis $\{\boldsymbol{x}_1, \boldsymbol{x}_2\}$; that is, $S_A(1)$ is a plane in R^3. In a similar fashion, we may find the eigenspace $S_A(2)$. Solving $A\boldsymbol{x} = 2\boldsymbol{x}$, we find that $\boldsymbol{x}$ must be a scalar multiple of $(1, 0, 0)'$. Thus, $S_A(2)$ is the line in R^3 given by $\{(a, 0, 0)': -\infty < a < \infty\}$.

In the preceding example, for each value of λ, we have $\dim\{S(\lambda)\}$ being equal to the multiplicity of λ. This is not always the case; the following example illustrates that it is possible for $\dim\{S(\lambda)\}$ to be less than the multiplicity of the eigenvalue λ.

Example 3.4. Consider the 3×3 matrix given by

$$A = \begin{bmatrix} 1 & 2 & 3 \\ 0 & 1 & 0 \\ 0 & 2 & 1 \end{bmatrix}$$

Since $|A - \lambda I| = (1 - \lambda)^3$, A has the eigenvalue 1 repeated three times. The eigenvalue–eigenvector equation $A\boldsymbol{x} = \lambda\boldsymbol{x}$ yields the three scalar equations

$$\begin{aligned} x_1 + 2x_2 + 3x_3 &= x_1 \\ x_2 &= x_2 \\ 2x_2 + x_3 &= x_3, \end{aligned}$$

which have as a solution only vectors of the form $\boldsymbol{x} = (a, 0, 0)'$. Thus, although the multiplicity of the eigenvalue 1 is 3, the associated eigenspace $S_A(1) = \{(a, 0, 0)' : -\infty < a < \infty\}$ is only one-dimensional.

3. SOME BASIC PROPERTIES OF EIGENVALUES AND EIGENVECTORS

In this section, we establish some very useful results regarding eigenvalues. The proofs of the results in our first theorem, which are left to the reader as an exercise, are easily obtained by using the characteristic equation or the eigenvalue–eigenvector equation.

Theorem 3.2. Let A be an $m \times m$ matrix. Then

(a) The eigenvalues of A' are the same as the eigenvalues of A.
(b) A is singular if and only if at least one eigenvalue of A is equal to 0.
(c) The diagonal elements of A are the eigenvalues of A, if A is a triangular matrix.
(d) The eigenvalues of BAB^{-1} are the same as the eigenvalues of A, if B is a nonsingular $m \times m$ matrix.
(e) Each of the eigenvalues of A is either $+1$ or -1, if A is an orthogonal matrix.

We saw in Example 3.4 that it is possible for the dimension of an eigenspace associated with an eigenvalue λ to be less than the multiplicity of λ. The following theorem shows that if $\dim\{S_A(\lambda)\} \neq r$, where r denotes the multiplicity of λ, then $\dim\{S_A(\lambda)\} < r$.

Theorem 3.3. Suppose λ is an eigenvalue, with multiplicity $r \geq 1$, of the $m \times m$ matrix A. Then

$$1 \leq \dim\{S_A(\lambda)\} \leq r$$

Proof. If λ is an eigenvalue of A, by definition there exists an $\boldsymbol{x} \neq 0$ satisfying the eigenvalue–eigenvector equation $A\boldsymbol{x} = \lambda\boldsymbol{x}$ and so, clearly, $\dim\{S_A(\lambda)\} \geq 1$. Now let $k = \dim\{S_A(\lambda)\}$, and let $\boldsymbol{x}_1, \ldots, \boldsymbol{x}_k$ be linearly independent eigenvectors corresponding to λ. Form a nonsingular $m \times m$ matrix X which has these k vectors as its first k columns; that is, X has the form $X = [X_1 \quad X_2]$, where $X_1 = (\boldsymbol{x}_1, \ldots, \boldsymbol{x}_k)$ and X_2 is $m \times (m - k)$. Since each column of X_1 is an eigenvector of A corresponding to the eigenvalue λ, it follows that $AX_1 = \lambda X_1$, and

$$X^{-1}X_1 = \begin{bmatrix} I_k \\ (0) \end{bmatrix}$$

follows from the fact that $X^{-1}X = \mathrm{I}_m$. As a result we find that

$$\begin{aligned} X^{-1}AX &= X^{-1}[AX_1 \quad AX_2] = X^{-1}[\lambda X_1 \quad AX_2] \\ &= \begin{bmatrix} \lambda \mathrm{I}_k & B_1 \\ (0) & B_2 \end{bmatrix}, \end{aligned}$$

where B_1 and B_2 represent a partitioning of the matrix $X^{-1}AX_2$. If μ is an eigenvalue of $X^{-1}AX$, then

$$\begin{aligned} 0 = |X^{-1}AX - \mu\mathrm{I}_m| &= \begin{vmatrix} (\lambda - \mu)\mathrm{I}_k & B_1 \\ (0) & B_2 - \mu\mathrm{I}_{m-k} \end{vmatrix} \\ &= (\lambda - \mu)^k |B_2 - \mu\mathrm{I}_{m-k}|, \end{aligned}$$

where the last equality can be obtained by repeated use of the cofactor expansion formula for a determinant. Thus, λ must be an eigenvalue of $X^{-1}AX$ with multiplicity of at least k. The result now follows since, from Theorem 3.2(d), the eigenvalues of $X^{-1}AX$ are the same as those of A. □

We now prove the following theorem involving both the eigenvalues and the eigenvectors of a matrix.

Theorem 3.4. Let λ be an eigenvalue of the $m \times m$ matrix A and let $\boldsymbol{x}$ be a corresponding eigenvector. Then,

(a) If n is an integer ≥ 1, λ^n is an eigenvalue of A^n corresponding to the eigenvector $\boldsymbol{x}$.

(b) If A is nonsingular, λ^{-1} is an eigenvalue of A^{-1} corresponding to the eigenvector $\boldsymbol{x}$.

Proof. Part (a) is proven by repeatedly using the relationship $A\boldsymbol{x} = \lambda\boldsymbol{x}$; that is, we have

$$A^n\boldsymbol{x} = A^{n-1}(A\boldsymbol{x}) = A^{n-1}(\lambda\boldsymbol{x}) = \lambda A^{n-1}\boldsymbol{x} = \cdots = \lambda^n\boldsymbol{x}$$

To prove part (b), premultiply the eigenvalue–eigenvector equation

$$A\boldsymbol{x} = \lambda\boldsymbol{x}$$

by A^{-1}, yielding the equation

$$\boldsymbol{x} = \lambda A^{-1}\boldsymbol{x} \tag{3.3}$$

Since A is nonsingular, we know from Theorem 3.2(b) that $\lambda \neq 0$, and so dividing both sides of (3.3) by λ yields

$$A^{-1}\boldsymbol{x} = \lambda^{-1}\boldsymbol{x},$$

which is the eigenvalue–eigenvector equation for A^{-1}, with eigenvalue λ^{-1} and eigenvector $\boldsymbol{x}$. □

The determinant and trace of a matrix have very simple and useful relationships with the eigenvalues of that matrix. These relationships are established in the next theorem.

Theorem 3.5. Let A be an $m \times m$ matrix with eigenvalues $\lambda_1, \ldots, \lambda_m$. Then

(a) $\text{tr}(A) = \sum_{i=1}^m \lambda_i$,
(b) $|A| = \prod_{i=1}^m \lambda_i$.

Proof. Recall that the characteristic equation, $|A - \lambda \text{I}| = 0$, can be expressed in the polynomial form

$$(-\lambda)^m + \alpha_{m-1}(-\lambda)^{m-1} + \cdots + \alpha_1(-\lambda) + \alpha_0 = 0 \tag{3.4}$$

We will first identify the coefficients α_0 and α_{m-1}. We can determine α_0 by evaluating the left-hand side of equation (3.4) at $\lambda = 0$; thus, $\alpha_0 = |A - (0)\text{I}| = |A|$. In order to find α_{m-1}, recall that, from its definition, the determinant is expressed as a sum of terms over all permutations of the integers $(1, 2, \ldots, m)$. Since α_{m-1} is the coefficient of $(-\lambda)^{m-1}$, to evaluate this term we only need to consider the terms in the sum which involve at least $m - 1$ of the diagonal elements of $(A - \lambda\text{I})$. But each term in the sum is the product of m elements from the matrix $(A - \lambda\text{I})$, multiplied by the appropriate sign, with one element chosen from each row and each column of $(A - \lambda I)$. Consequently, the only term in the sum involving at least $m - 1$ of the diagonal elements of $(A - \lambda I)$ is the term that involves the product of all of the diagonal elements. Since this term involves an even permutation, the sign term will equal +1, and so α_{m-1} will be the coefficient of $(-\lambda)^{m-1}$ in

$$(a_{11} - \lambda)(a_{22} - \lambda) \cdots (a_{mm} - \lambda),$$

which clearly is $a_{11} + a_{22} + \cdots + a_{mm}$ or simply $\text{tr}(A)$. Now to relate $\alpha_0 = |A|$ and $\alpha_{m-1} = \text{tr}(A)$ to the eigenvalues of A, note that since $\lambda_1, \ldots, \lambda_m$ are the roots to the characteristic equation, which is an mth degree polynomial in λ, it follows that

$$(\lambda_1 - \lambda)(\lambda_2 - \lambda) \cdots (\lambda_m - \lambda) = 0$$

Multiplying out the left-hand side of this equation and then matching corresponding terms with those in (3.4), we find that

$$|A| = \prod_{i=1}^{m} \lambda_i, \qquad \text{tr}(A) = \sum_{i=1}^{m} \lambda_i \qquad \square$$

The following theorem gives a sufficient condition for a set of eigenvectors to be linearly independent.

Theorem 3.6. Suppose $\boldsymbol{x}_1, \ldots, \boldsymbol{x}_r$ are eigenvectors of the $m \times m$ matrix A, where $r \leq m$. If the corresponding eigenvalues $\lambda_1, \ldots, \lambda_r$ are such that $\lambda_i \neq \lambda_j$ for all $i \neq j$, then the vectors $\boldsymbol{x}_1, \ldots, \boldsymbol{x}_r$ are linearly independent.

Proof. Our proof is by contradiction; that is, we begin by assuming that the vectors $\boldsymbol{x}_1, \ldots, \boldsymbol{x}_r$ are linearly dependent. Let h be the largest integer for which $\boldsymbol{x}_1, \ldots, \boldsymbol{x}_h$ are linearly independent. Such a set can be found since $\boldsymbol{x}_1$, being an eigenvector, cannot equal $\mathbf{0}$, and so it is linearly independent. The vectors $\boldsymbol{x}_1, \ldots, \boldsymbol{x}_{h+1}$ must be linearly dependent, so there exist scalars $\alpha_1, \ldots, \alpha_{h+1}$ with at least two not equal to zero since no eigenvector can be the null vector, such that

$$\alpha_1 \boldsymbol{x}_1 + \cdots + \alpha_{h+1} \boldsymbol{x}_{h+1} = \mathbf{0}$$

Premultiplying the left-hand side of this equation by $(A - \lambda_{h+1} I)$, we find that

$$\begin{aligned}\alpha_1(A - \lambda_{h+1}\mathrm{I})\boldsymbol{x}_1 + \cdots + \alpha_{h+1}(A - \lambda_{h+1}\mathrm{I})\boldsymbol{x}_{h+1} \\ = \alpha_1(A\boldsymbol{x}_1 - \lambda_{h+1}\boldsymbol{x}_1) + \cdots + \alpha_{h+1}(A\boldsymbol{x}_{h+1} - \lambda_{h+1}\boldsymbol{x}_{h+1}) \\ = \alpha_1(\lambda_1 - \lambda_{h+1})\boldsymbol{x}_1 + \cdots + \alpha_h(\lambda_h - \lambda_{h+1})\boldsymbol{x}_h\end{aligned}$$

also must be equal to $\mathbf{0}$. But $\boldsymbol{x}_1, \ldots, \boldsymbol{x}_h$ are linearly independent so it follows that

$$\alpha_1(\lambda_1 - \lambda_{h+1}) = \cdots = \alpha_h(\lambda_h - \lambda_{h+1}) = 0$$

We know that at least one of the scalars $\alpha_1, \ldots, \alpha_h$ is not equal to zero, and if, for instance, α_i is one of these nonzero scalars, we then must have $\lambda_i = \lambda_{h+1}$. This contradicts the conditions of the theorem, so the vectors $\boldsymbol{x}_1, \ldots, \boldsymbol{x}_r$ must be linearly independent. □

If the eigenvalues $\lambda_1, \ldots, \lambda_m$ of an $m \times m$ matrix A are all distinct, then it follows from Theorem 3.6 that the matrix $X = (\boldsymbol{x}_1, \ldots, \boldsymbol{x}_m)$, where $\boldsymbol{x}_i$ is an eigenvector corresponding to λ_i, is nonsingular. It also follows from the eigenvalue–eigenvector equation $A\boldsymbol{x}_i = \lambda_i\boldsymbol{x}_i$ that if we define the diagonal matrix $\Lambda = \mathrm{diag}(\lambda_1, \ldots, \lambda_m)$, then $AX = X\Lambda$. Premultiplying this equation by X^{-1} yields the identity $X^{-1}AX = \Lambda$. Any square matrix that can be transformed to a diagonal matrix through the postmultiplication by a nonsingular matrix and premultiplication by its inverse is said to be diagonalizable. Thus, a square matrix with distinct eigenvalues is diagonalizable.

Clearly, when a matrix is diagonalizable, its rank equals the number of its nonzero eigenvalues, since

$$\mathrm{rank}(A) = \mathrm{rank}(X^{-1}AX) = \mathrm{rank}(\Lambda)$$

follows from Theorem 1.8. This relationship between the number of nonzero eigenvalues and the rank of a square matrix does not necessarily hold if the matrix is not diagonalizable.

Example 3.5. Consider the 2×2 matrices

$$A = \begin{bmatrix} 1 & 1 \\ 0 & 0 \end{bmatrix} \quad \text{and} \quad B = \begin{bmatrix} 0 & 1 \\ 0 & 0 \end{bmatrix}$$

Clearly, both A and B have rank of 1. Now the characteristic equation of A simplifies to $\lambda(1 - \lambda) = 0$ so that the eigenvalues of A are 0 and 1, and thus, in this case, rank(A) equals the number of nonzero eigenvalues. The characteristic equation for B simplifies to $\lambda^2 = 0$, so B has the eigenvalue 0 repeated twice. Hence, the rank of B exceeds its number of nonzero eigenvalues.

Our final theorem, known as the Cayley–Hamilton Theorem, states that a matrix satisfies its own characteristic equation. A proof of this result can be found in Hammarling (1970).

Theorem 3.7. Let A be an $m \times m$ matrix with eigenvalues $\lambda_1, \dots, \lambda_m$. Then

$$\prod_{i=1}^{m} (A - \lambda_i \mathrm{I}) = (0);$$

that is, if $(-\lambda)^m + \alpha_{m-1}(-\lambda)^{m-1} + \cdots + \alpha_1(-\lambda) + \alpha_0 = 0$ is the characteristic equation of A, then

$$(-A)^m + \alpha_{m-1}(-A)^{m-1} + \cdots + \alpha_1(-A) + \alpha_0 \mathrm{I} = (0)$$

4. SYMMETRIC MATRICES

Many of the applications involving eigenvalues and eigenvectors in statistics are ones that deal with a symmetric matrix, and symmetric matrices have some especially nice properties regarding eigenvalues and eigenvectors. In this section, we will develop some of these properties.

We have seen that a matrix may have complex eigenvalues even when the matrix itself is real. This is not the case for symmetric matrices.

Theorem 3.8. Let A be an $m \times m$ real symmetric matrix. Then the eigenvalues of A are real, and corresponding to any eigenvalue there exist eigenvectors that are real.

Proof. Let $\lambda = \alpha + i\beta$ be an eigenvalue of A and $\boldsymbol{x} = \boldsymbol{y} + i\boldsymbol{z}$ a corresponding eigenvector, where $i = \sqrt{-1}$. We will first show that $\beta = 0$. Substitution of these expressions for λ and $\boldsymbol{x}$ in the eigenvalue–eigenvector equation $A\boldsymbol{x} = \lambda\boldsymbol{x}$ yields

$$A(\boldsymbol{y} + i\boldsymbol{z}) = (\alpha + i\beta)(\boldsymbol{y} + i\boldsymbol{z}) \tag{3.5}$$

Premultiplying (3.5) by $(\boldsymbol{y} - i\boldsymbol{z})'$, we get

$$(\boldsymbol{y} - i\boldsymbol{z})'A(\boldsymbol{y} + i\boldsymbol{z}) = (\alpha + i\beta)(\boldsymbol{y} - i\boldsymbol{z})'(\boldsymbol{y} + i\boldsymbol{z}),$$

which simplifies to

$$\boldsymbol{y}'A\boldsymbol{y} + \boldsymbol{z}'A\boldsymbol{z} = (\alpha + i\beta)(\boldsymbol{y}'\boldsymbol{y} + \boldsymbol{z}'\boldsymbol{z}),$$

since $y'Az = z'Ay$ follows from the symmetry of A. Now $x \neq 0$ implies that $(y'y + z'z) > 0$ and, consequently, we must have $\beta = 0$ since the left-hand side of the equation above is real. Substituting $\beta = 0$ in (3.5), we find that

$$Ay + iAz = \alpha y + i\alpha z$$

Thus, $x = y + iz$ will be an eigenvector of A corresponding to $\lambda = \alpha$ as long as y and z satisfy $Ay = \alpha y$, $Az = \alpha z$, and at least one is not $\mathbf{0}$ so that $x \neq \mathbf{0}$. A real eigenvector is then constructed by selecting $y \neq \mathbf{0}$, such that $Ay = \alpha y$ and $z = \mathbf{0}$. □

We have seen that a set of eigenvectors of an $m \times m$ matrix A is linearly independent if the associated eigenvalues are all different from one another. We will now show that, if A is symmetric, we can say a bit more. Suppose that x and y are eigenvectors of A corresponding to the eigenvalues λ and γ, where $\lambda \neq \gamma$. Then, since A is symmetric, it follows that

$$\lambda x'y = (\lambda x)'y = (Ax)'y = x'A'y = x'(Ay) = x'(\gamma y) = \gamma x'y$$

Since $\lambda \neq \gamma$, we must have $x'y = 0$; that is, eigenvectors corresponding to different eigenvalues must be orthogonal. Thus, if the m eigenvalues of A are distinct, then the set of corresponding eigenvectors will form a group of mutually orthogonal vectors. We will show that this is still possible when A has multiple eigenvalues. Before we prove this, we will need the following result.

Theorem 3.9. Let A be an $m\times m$ symmetric matrix and let x be any nonzero $m \times 1$ vector. Then for some $r \geq 1$, the vector space spanned by the vectors $x, Ax, \dots, A^{r-1}x$, contains an eigenvector of A.

Proof. Let r be the smallest integer for which $x, Ax, \dots, A^r x$ form a linearly dependent set. Then there exist scalars, $\alpha_0, \dots, \alpha_r$, not all of which are zero, such that

$$\alpha_0 x + \alpha_1 Ax + \cdots + \alpha_r A^r x = (\alpha_0 \mathrm{I}_m + \alpha_1 A + \cdots + A^r)x = \mathbf{0},$$

where without loss of generality we have taken $\alpha_r = 1$, since the way r was chosen guarantees that α_r is not zero. The expression in the parentheses is an rth-degree matrix polynomial in A. This can be factored in a fashion similar to the way scalar polynomials are factored; that is, it can be written as

$$(A - \gamma_1 \mathrm{I}_m)(A - \gamma_2 \mathrm{I}_m) \cdots (A - \gamma_r \mathrm{I}_m),$$

where $\gamma_1, \dots, \gamma_r$ are the roots of the polynomial satisfying $\alpha_0 = (-1)^r \gamma_1 \cdot \gamma_2 \cdots \gamma_r, \dots, \alpha_{r-1} = -(\gamma_1 + \gamma_2 + \cdots + \gamma_r)$. Let

$$\begin{aligned} \boldsymbol{y} &= (A - \gamma_2 \mathrm{I}_m) \cdots (A - \gamma_r \mathrm{I}_m)\boldsymbol{x}, \\ &= (-1)^{r-1}\gamma_2 \cdots \gamma_r \boldsymbol{x} + \cdots + A^{r-1}\boldsymbol{x}, \end{aligned}$$

and note that $\boldsymbol{y} \neq \mathbf{0}$ since, otherwise, $\boldsymbol{x}, A\boldsymbol{x}, \ldots, A^{r-1}\boldsymbol{x}$ would be a linearly dependent set, contradicting the definition of r. Thus, $\boldsymbol{y}$ is in the space spanned by $\boldsymbol{x}, A\boldsymbol{x}, \ldots, A^{r-1}\boldsymbol{x}$ and

$$(A - \gamma_1 \mathrm{I}_m)\boldsymbol{y} = (A - \gamma_1 \mathrm{I}_m)(A - \gamma_2 \mathrm{I}_m) \cdots (A - \gamma_r \mathrm{I}_m)\boldsymbol{x} = \mathbf{0}$$

Consequently, $\boldsymbol{y}$ is an eigenvector of A corresponding to the eigenvalue γ_1, and so the proof is complete. □

Theorem 3.10. If the $m \times m$ matrix A is symmetric, then it is possible to construct a set of m eigenvectors of A such that the set is orthonormal.

Proof. We first show that if we have an orthonormal set of eigenvectors, $\boldsymbol{x}_1, \ldots, \boldsymbol{x}_h$, where $1 \leq h < m$, then we can find another normalized eigenvector $\boldsymbol{x}_{h+1}$ orthogonal to each of these vectors. Select any vector $\boldsymbol{x}$ which is orthogonal to each of the vectors $\boldsymbol{x}_1, \ldots, \boldsymbol{x}_h$. Note that for any positive integer $k, A^k\boldsymbol{x}$ is also orthogonal to $\boldsymbol{x}_1, \ldots, \boldsymbol{x}_h$ since, if λ_i is the eigenvalue corresponding to $\boldsymbol{x}_i$, it follows from the symmetry of A and Theorem 3.4(a) that

$$\boldsymbol{x}_i' A^k \boldsymbol{x} = \{(A^k)'\boldsymbol{x}_i\}'\boldsymbol{x} = (A^k\boldsymbol{x}_i)'\boldsymbol{x} = \lambda_i^k \boldsymbol{x}_i'\boldsymbol{x} = 0$$

From the previous theorem we know that, for some r, the space spanned by $\boldsymbol{x}, A\boldsymbol{x}, \ldots, A^{r-1}\boldsymbol{x}$ contains an eigenvector, say $\boldsymbol{y}$, of A. This vector $\boldsymbol{y}$ also must be orthogonal to $\boldsymbol{x}_1, \ldots, \boldsymbol{x}_h$ since it is from a vector space spanned by a set of vectors orthogonal to $\boldsymbol{x}_1, \ldots, \boldsymbol{x}_h$. Thus, we can take $\boldsymbol{x}_{h+1} = (\boldsymbol{y}'\boldsymbol{y})^{-1/2}\boldsymbol{y}$. The theorem now follows by starting with any eigenvector of A, and then using the previous argument $m - 1$ times. □

If we let the $m \times m$ matrix $X = (\boldsymbol{x}_1, \ldots, \boldsymbol{x}_m)$, where $\boldsymbol{x}_1, \ldots, \boldsymbol{x}_m$ are the orthonormal vectors described in the proof, and $\Lambda = \mathrm{diag}(\lambda_1, \ldots, \lambda_m)$, then the eigenvalue–eigenvector equation $A\boldsymbol{x}_i = \lambda_i \boldsymbol{x}_i$ can be expressed collectively as the matrix equation $AX = X\Lambda$. Since the columns of X are orthonormal vectors, X is an orthogonal matrix. Premultiplication of our matrix equation by X yields the relationship $X'AX = \Lambda$, or equivalently

$$A = X\Lambda X',$$

which is known as the spectral decomposition of A. We will see in Section 4.2 that there is a very useful generalization of this decomposition, known as the singular value decomposition, which holds for any $m \times n$ matrix A; in particular, there exist $m \times m$ and $n \times n$ orthogonal matrices P and Q and an $m \times n$ matrix D with $d_{ij} = 0$ if $i \neq j$, such that $A = PDQ'$.

Note that it follows from Theorem 3.2(d) that the eigenvalues of A are the same as the eigenvalues of Λ, which are the diagonal elements of Λ. Thus, if λ is a multiple root of A with multiplicity $r > 1$, then r of the diagonal elements of Λ are equal to λ and r of the eigenvectors, say $\boldsymbol{x}_1, \ldots, \boldsymbol{x}_r$, correspond to this root λ. Consequently, the dimension of the eigenspace of $A, S_A(\lambda)$, corresponding to λ, is equal to the multiplicity r. The set of orthonormal eigenvectors corresponding to this root is not unique. Any orthonormal basis for $S_A(\lambda)$ will be a set of r orthonormal vectors associated with the eigenvalue λ. For example, if we let $X_1 = (\boldsymbol{x}_1, \ldots, \boldsymbol{x}_r)$ and let Q be any $r \times r$ orthogonal matrix, the columns of $Y_1 = X_1 Q$ also form a set of orthonormal eigenvectors corresponding to λ.

Example 3.6. One application of an eigenanalysis in statistics involves overcoming difficulties associated with a regression analysis in which the explanatory variables are nearly linearly dependent. This situation is often referred to as multicollinearity. In this case, some of the explanatory variables are providing redundant information about the response variable. As a result, the least squares estimator of $\boldsymbol{\beta}$ in the model $\boldsymbol{y} = X\boldsymbol{\beta} + \boldsymbol{\epsilon}$

$$\hat{\boldsymbol{\beta}} = (X'X)^{-1}X'\boldsymbol{y}$$

will be imprecise since its covariance matrix

$$\begin{aligned} \text{var}(\hat{\boldsymbol{\beta}}) &= (X'X)^{-1}X'\{\text{var}(\boldsymbol{y})\}X(X'X)^{-1} \\ &= (X'X)^{-1}X'\{\sigma^2 \text{I}\}X(X'X)^{-1} = \sigma^2(X'X)^{-1} \end{aligned}$$

will tend to have some large elements due to the near singularity of $X'X$. If the near linear dependence is simply because one of the explanatory variables, say x_j, is nearly a scalar multiple of another, say x_l, one could simply eliminate one of these variables from the model. However, in most cases, the near linear dependence is not this straightforward. We will see that an eigenanalysis will help reveal any of these dependencies. Suppose that we standardize the explanatory variables so that we have the model

$$\boldsymbol{y} = \delta_0 \mathbf{1}_N + Z_1 \boldsymbol{\delta}_1 + \boldsymbol{\epsilon}$$

discussed in Example 2.15. Let $\Lambda = \text{diag}(\lambda_1, \ldots, \lambda_k)$ contain the eigenvalues of $Z_1'Z_1$ in descending order of magnitude, and let U be an orthogonal matrix that has corresponding normalized eigenvectors of $Z_1'Z_1$ as its columns, so that $Z_1'Z_1 = U\Lambda U'$. It was shown in Example 2.15 that the estimation of $\boldsymbol{y}$ is unaffected by a nonsingular transformation of the explanatory variables; that is, we could just as well work with the model

$$y = \alpha_0 \mathbf{1}_N + W_1 \boldsymbol{\alpha}_1 + \boldsymbol{\epsilon},$$

where $\alpha_0 = \delta_0, \boldsymbol{\alpha}_1 = T^{-1}\boldsymbol{\delta}_1, W_1 = Z_1 T$, and T is a nonsingular matrix. A method, referred to as principal components regression, deals with the problems associated with multicollinearity by utilizing the orthogonal transformations $W_1 = Z_1 U$ and $\boldsymbol{\alpha}_1 = U'\boldsymbol{\delta}_1$ of the standardized explanatory variables and parameter vector. The k new explanatory variables are called the principal components; the variable corresponding to the ith column of W_1 is called the ith principal component. Since $W_1'W_1 = U'Z_1'Z_1U = \Lambda$ and $\mathbf{1}_N'W_1 = \mathbf{1}_N'Z_1U = \mathbf{0}'U = \mathbf{0}'$, the least squares estimate of $\boldsymbol{\alpha}_1$ is

$$\hat{\boldsymbol{\alpha}}_1 = (W_1'W_1)^{-1}W_1'y = \Lambda^{-1}W_1'y,$$

while its covariance matrix simplifies to

$$\text{var}(\hat{\boldsymbol{\alpha}}_1) = \sigma^2(W_1'W_1)^{-1} = \sigma^2\Lambda^{-1}$$

If $Z_1'Z_1$ and hence also $W_1'W_1$ is nearly singular, then at least one of the λ_is will be very small, while the variances of the corresponding α_is will be very large. Since the explanatory variables have been standardized, $W_1'W_1$ is $N-1$ times the sample correlation matrix of the principal components computed from the N observations. Thus, if $\lambda_i \simeq 0$, then the ith principal component is nearly constant from observation to observation, and so it contributes very little to the estimation of y. If $\lambda_i \simeq 0$ for $i = k-r+1, \ldots, k$, then the problems associated with multicollinearity can be avoided by eliminating the last r principal components from the model; in other words, the principal components regression model is

$$y = \alpha_0 \mathbf{1}_N + W_{11}\boldsymbol{\alpha}_{11} + \boldsymbol{\epsilon},$$

where W_{11} and $\boldsymbol{\alpha}_{11}'$ are obtained from W_1 and $\boldsymbol{\alpha}_1'$ by deleting their last r columns. If we let $\Lambda_1 = \text{diag}(\lambda_1, \ldots, \lambda_{k-r})$, then the least squares estimate of $\boldsymbol{\alpha}_{11}$ can be written as

$$\hat{\boldsymbol{\alpha}}_{11} = (W_{11}'W_{11})^{-1}W_{11}'y = \Lambda_1^{-1}W_{11}'y$$

Note that due to the orthogonality of the principal components, $\hat{\boldsymbol{\alpha}}_{11}$ is identical to the first $k-r$ components of $\hat{\boldsymbol{\alpha}}_1$. The estimate $\hat{\boldsymbol{\alpha}}_{11}$ can be used to find the principal components estimate of $\boldsymbol{\delta}_1$ in the original standardized model. Recall that $\boldsymbol{\delta}_1$ and $\boldsymbol{\alpha}_1$ are related through the identity $\boldsymbol{\delta}_1 = U\boldsymbol{\alpha}_1$. By eliminating the last r principal components, we are replacing this identity with the identity

$\boldsymbol{\delta}_1 = U_1\boldsymbol{\alpha}_{11}$, and so the principal components regression estimate of $\boldsymbol{\delta}_1$ is given by $\hat{\boldsymbol{\delta}}_{11} = U_1\hat{\boldsymbol{\alpha}}_{11} = U_1\Lambda_1^{-1}W_{11}'\boldsymbol{y}$.

A set of orthonormal eigenvectors of a matrix A can be used to find what are known as the eigenprojections of A.

Definition 3.1. Let λ be an eigenvalue of the $m \times m$ symmetric matrix A with multiplicity $r \geq 1$. If $\boldsymbol{x}_1, \ldots, \boldsymbol{x}_r$ is a set of orthonormal eigenvectors corresponding to λ, then the eigenprojection of A associated with the eigenvalue λ is given by

$$P_A(\lambda) = \sum_{i=1}^{r} \boldsymbol{x}_i\boldsymbol{x}_i'$$

The eigenprojection $P_A(\lambda)$ is simply the projection matrix for the vector space $S_A(\lambda)$. Thus, for any $\boldsymbol{x} \in R^m, \boldsymbol{y} = P_A(\lambda)\boldsymbol{x}$ gives the orthogonal projection of $\boldsymbol{x}$ onto the eigenspace $S_A(\lambda)$. If we define X_1 as before, that is $X_1 = (\boldsymbol{x}_1, \ldots, \boldsymbol{x}_r)$, then $P_A(\lambda) = X_1X_1'$. Note that $P_A(\lambda)$ is unique even though the set of eigenvectors $\boldsymbol{x}_1, \ldots, \boldsymbol{x}_r$ is not unique; for instance, if $Y_1 = X_1Q$, where Q is an arbitrary $r \times r$ orthogonal matrix, then the columns of Y_1 form another set of orthonormal eigenvectors corresponding to λ, but

$$Y_1Y_1' = (X_1Q)(X_1Q)' = X_1QQ'X_1' = X_1\mathrm{I}_mX_1' = X_1X_1' = P_A(\lambda)$$

The term spectral decomposition comes from the use of the term spectral set of A for the set of all eigenvalues of A excluding repetitions of the same value. Suppose the $m \times m$ matrix A has the spectral set $\{\mu_1, \ldots, \mu_k\}$, where $k \leq m$, since some of the μ_i may correspond to multiple eigenvalues. The set of μ_i may be different from our set of λ_i in that we do not repeat values for the μ_i. Thus, if A is 4×4 with eigenvalues $\lambda_1 = 3, \lambda_2 = 2, \lambda_3 = 2$, and $\lambda_4 = 1$, then the spectral set of A is $\{3, 2, 1\}$. Using X and Λ as previously defined, the spectral decomposition states that

$$A = X\Lambda X' = \sum_{i=1}^{m} \lambda_i\boldsymbol{x}_i\boldsymbol{x}_i' = \sum_{i=1}^{k} \mu_iP_A(\mu_i),$$

so that A has been decomposed into a sum of terms, one corresponding to each value in the spectral set.

Example 3.7. It can be easily verified by solving the characteristic equation for the 3×3 symmetric matrix

$$A = \begin{bmatrix} 5 & -1 & -1 \\ -1 & 5 & -1 \\ -1 & -1 & 5 \end{bmatrix}$$

that A has the simple eigenvalue 3 and the multiple eigenvalue 6, with multiplicity 2. The unique (except for sign) unit eigenvector associated with the eigenvalue 3 can be shown to equal $(1/\sqrt{3}, 1/\sqrt{3}, 1/\sqrt{3})'$, while a set of orthonormal eigenvectors associated with 6 is given by $(-2/\sqrt{6}, 1/\sqrt{6}, 1/\sqrt{6})'$ and $(0, 1/\sqrt{2}, -1/\sqrt{2})'$. Thus, the spectral decomposition of A is given by

$$\begin{bmatrix} 5 & -1 & -1 \\ -1 & 5 & -1 \\ -1 & -1 & 5 \end{bmatrix} = \begin{bmatrix} 1/\sqrt{3} & -2/\sqrt{6} & 0 \\ 1/\sqrt{3} & 1/\sqrt{6} & 1/\sqrt{2} \\ 1/\sqrt{3} & 1/\sqrt{6} & -1/\sqrt{2} \end{bmatrix} \begin{bmatrix} 3 & 0 & 0 \\ 0 & 6 & 0 \\ 0 & 0 & 6 \end{bmatrix}$$

$$\times \begin{bmatrix} 1/\sqrt{3} & 1/\sqrt{3} & 1/\sqrt{3} \\ -2/\sqrt{6} & 1/\sqrt{6} & 1/\sqrt{6} \\ 0 & 1/\sqrt{2} & -1/\sqrt{2} \end{bmatrix},$$

and the two eigenprojections of A are

$$P_A(3) = \begin{bmatrix} 1/\sqrt{3} \\ 1/\sqrt{3} \\ 1/\sqrt{3} \end{bmatrix} [1/\sqrt{3} \quad 1/\sqrt{3} \quad 1/\sqrt{3}] = \frac{1}{3} \begin{bmatrix} 1 & 1 & 1 \\ 1 & 1 & 1 \\ 1 & 1 & 1 \end{bmatrix},$$

$$P_A(6) = \begin{bmatrix} -2/\sqrt{6} & 0 \\ 1/\sqrt{6} & 1/\sqrt{2} \\ 1/\sqrt{6} & -1/\sqrt{2} \end{bmatrix} \begin{bmatrix} -2/\sqrt{6} & 1/\sqrt{6} & 1/\sqrt{6} \\ 0 & 1/\sqrt{2} & -1/\sqrt{2} \end{bmatrix}$$

$$= \frac{1}{3} \begin{bmatrix} 2 & -1 & -1 \\ -1 & 2 & -1 \\ -1 & -1 & 2 \end{bmatrix}$$

The relationship between the rank of a matrix and the number of its nonzero eigenvalues becomes an exact one for symmetric matrices.

Theorem 3.11. Suppose that the $m \times m$ matrix A has r nonzero eigenvalues. Then, if A is symmetric, $\text{rank}(A) = r$.

Proof. If $A = X\Lambda X'$ is the spectral decomposition of A, then the diagonal matrix Λ has r nonzero diagonal elements and

$$\text{rank}(A) = \text{rank}(X\Lambda X') = \text{rank}(\Lambda),$$

since the multiplication of a matrix by nonsingular matrices does not affect the rank. Clearly, the rank of a diagonal matrix equals the number of its nonzero diagonal elements, so the result follows. □

Some of the most important applications of eigenvalues and eigenvectors in statistics involve the analysis of covariance and correlation matrices.

Example 3.8. In some situations, a matrix has some special structure that when recognized can be used to expedite the calculation of eigenvalues and eigenvectors. In this example we consider a structure sometimes possessed by an $m \times m$ covariance matrix. This structure is one in which we have equal variances and equal correlations; that is, the covariance matrix has the form

$$\Omega = \sigma^2 \begin{bmatrix} 1 & \rho & \cdots & \rho \\ \rho & 1 & \cdots & \rho \\ \vdots & \vdots & & \vdots \\ \rho & \rho & \cdots & 1 \end{bmatrix}$$

Alternatively, Ω can be expressed as $\Omega = \sigma^2\{(1-\rho)\text{I}_m + \rho\mathbf{1}_m\mathbf{1}_m'\}$ so that it is a function of the vector $\mathbf{1}_m$. This vector also plays a crucial role in the eigenanalysis of Ω since

$$\Omega\mathbf{1}_m = \sigma^2\{(1-\rho)\mathbf{1}_m + \rho\mathbf{1}_m\mathbf{1}_m'\mathbf{1}_m\} = \sigma^2\{(1-\rho) + m\rho\}\mathbf{1}_m$$

Thus, $\mathbf{1}_m$ is an eigenvector of Ω corresponding to the eigenvalue $\sigma^2\{(1-\rho) + m\rho\}$. The remaining eigenvalues of Ω can be identified by noting that if $\boldsymbol{x}$ is any $m \times 1$ vector orthogonal to $\mathbf{1}_m$, then

$$\Omega\boldsymbol{x} = \sigma^2\{(1-\rho)\boldsymbol{x} + \rho\mathbf{1}_m\mathbf{1}_m'\boldsymbol{x}\} = \sigma^2(1-\rho)\boldsymbol{x},$$

and so $\boldsymbol{x}$ is an eigenvector of Ω corresponding to the eigenvalue $\sigma^2(1-\rho)$. Since there are $m-1$ linearly independent vectors orthogonal to $\mathbf{1}_m$, the eigenvalue $\sigma^2(1-\rho)$ is repeated $m-1$ times. The order of these two distinct eigenvalues depends on the value of ρ; $\sigma^2\{(1-\rho)+m\rho\}$ will be larger than $\sigma^2(1-\rho)$ only if ρ is positive.

Example 3.9. A covariance matrix can be any symmetric nonnegative definite matrix. Consequently, for a given set of m nonnegative numbers and a given set of m orthonormal $m\times 1$ vectors, it is possible to construct an $m\times m$ covariance matrix with these numbers and vectors as its eigenvalues and eigenvectors. On the other hand, a correlation matrix has the additional constraint that its diagonal elements must each equal 1, and this extra restriction has an impact on the eigenanalysis of correlation matrices; that is, there is a much more limited set of possible eigenvalues and eigenvectors for correlation matrices. For the most extreme case, consider a 2×2 correlation matrix that must have the form

$$P = \begin{bmatrix} 1 & \rho \\ \rho & 1 \end{bmatrix}$$

with $-1 \leq \rho \leq 1$, since P must be nonnegative definite. The characteristic equation $|P - \lambda I_2| = 0$ readily admits the two eigenvalues $1 + \rho$ and $1 - \rho$. Using these in the eigenvalue–eigenvector equation $P\boldsymbol{x} = \lambda\boldsymbol{x}$ we find that regardless of the value of ρ, $(1/\sqrt{2}, 1/\sqrt{2})'$ must be an eigenvector corresponding to $1 + \rho$, while $(1/\sqrt{2}, -1/\sqrt{2})'$ must be an eigenvector corresponding to $1 - \rho$. Thus, ignoring sign changes, there is only one set of orthonormal eigenvectors possible for a 2×2 correlation matrix if $\rho \neq 0$. This number of possible sets of orthonormal eigenvectors increases as the order m increases. In some situations, such as simulation studies of analyses of correlation matrices, one may wish to construct a correlation matrix with some particular structure with regard to its eigenvalues or eigenvectors. For example, suppose that we want to construct an $m \times m$ correlation matrix that has three distinct eigenvalues with one of them being repeated $m - 2$ times. Thus, this correlation matrix has the form

$$P = \lambda_1 \boldsymbol{x}_1 \boldsymbol{x}_1' + \lambda_2 \boldsymbol{x}_2 \boldsymbol{x}_2' + \sum_{i=3}^{m} \lambda_i \boldsymbol{x}_i \boldsymbol{x}_i',$$

where λ_1, λ_2, and λ are the distinct eigenvalues of P, and $\boldsymbol{x}_1, \ldots, \boldsymbol{x}_m$ are corresponding normalized eigenvectors. Since P is nonnegative definite, we must have $\lambda_1 \geq 0$, $\lambda_2 \geq 0$, and $\lambda \geq 0$, while $\text{tr}(P) = m$ implies that $\lambda = (m - \lambda_1 - \lambda_2)/(m - 2)$. Note that P can be written as

$$P = (\lambda_1 - \lambda)\boldsymbol{x}_1 \boldsymbol{x}_1' + (\lambda_2 - \lambda)\boldsymbol{x}_2 \boldsymbol{x}_2' + \lambda I_m,$$

so that the constraint $(P)_{ii} = 1$ implies that

$$(\lambda_1 - \lambda)x_{i1}^2 + (\lambda_2 - \lambda)x_{i2}^2 + \lambda = 1$$

or, equivalently,

$$x_{i2}^2 = \frac{1 - \lambda - (\lambda_1 - \lambda)x_{i1}^2}{(\lambda_2 - \lambda)}$$

The constraints described can then be used to construct a particular matrix. For instance, suppose that we want to construct a 4×4 correlation matrix with eigenvalues $\lambda_1 = 2$, $\lambda_2 = 1$, and $\lambda = 0.5$ repeated twice. If we choose $\boldsymbol{x}_1 = (\frac{1}{2}, \frac{1}{2}, \frac{1}{2}, \frac{1}{2})'$, then we must have $x_{i2}^2 = \frac{1}{4}$, and so because of the orthogonality of $\boldsymbol{x}_1$ and $\boldsymbol{x}_2$, $\boldsymbol{x}_2$ can be any vector obtained from $\boldsymbol{x}_1$ by negating two of its components. For example, if we take $\boldsymbol{x}_2 = (\frac{1}{2}, -\frac{1}{2}, \frac{1}{2}, -\frac{1}{2})'$, then

$$P = \begin{bmatrix} 1 & 0.25 & 0.50 & 0.25 \\ 0.25 & 1 & 0.25 & 0.50 \\ 0.50 & 0.25 & 1 & 0.25 \\ 0.25 & 0.50 & 0.25 & 1 \end{bmatrix}$$

5. CONTINUITY OF EIGENVALUES AND EIGENPROJECTIONS

Our first result of this section is one which bounds the absolute difference between eigenvalues of two matrices by a function of the absolute differences of the elements of the two matrices. A proof of this theorem can be found in Ostrowski (1973). For some other similar bounds see Elsner (1982).

Theorem 3.12. Let A and B be $m \times m$ matrices possessing eigenvalues $\lambda_1, \ldots, \lambda_m$ and $\gamma_1, \ldots, \gamma_m$, respectively. Define

$$M = \max_{1 \le i \le m,\, 1 \le j \le m} (|a_{ij}|, |b_{ij}|),$$

and

$$\delta(A, B) = \frac{1}{m} \sum_{i=1}^{m} \sum_{j=1}^{m} |a_{ij} - b_{ij}|$$

Then

$$\max_{1 \le i \le m} \min_{1 \le j \le m} |\lambda_i - \gamma_j| \le (m+2) M^{1-1/m} \delta(A, B)^{1/m}$$

Theorem 3.12 will allow us to establish a very useful result regarding the eigenvalues of any matrix A. Let $B_1, B_2, \ldots,$ be a sequence of $m \times m$ matrices such that $B_n \to A$, as $n \to \infty$, and let $\delta(A, B_n)$ be as defined in Theorem 3.12. It follows from the fact that $B_n \to A$, as $n \to \infty$, that $\delta(A, B_n) \to 0$, as $n \to \infty$. Hence, if $\gamma_{1,n}, \ldots, \gamma_{m,n}$ are the eigenvalues of B_n, then Theorem 3.12 tells us that

$$\max_{1 \le i \le m} \min_{1 \le j \le m} |\lambda_i - \gamma_{j,n}| \to 0,$$

as $n \to \infty$. In other words, if B_n is very close to A, then for each i, there exists some j such that $\gamma_{j,n}$ is very close to λ_i, or more precisely, as $B_n \to A$, the eigenvalues of B_n are converging to those of A. This leads to the following important result.

Theorem 3.13. Let $\lambda_1, \dots, \lambda_m$ be the eigenvalues of the $m \times m$ matrix A. Then, for each i, λ_i is a continuous function of the elements of A.

Our next result addresses the continuity of the eigenprojection $P_A(\lambda)$ of a symmetric matrix A. A detailed treatment of this problem, as well as the more general problem of the continuity of the eigenprojections of nonsymmetric matrices, can be found in Kato (1982).

Theorem 3.14. Suppose that A is an $m \times m$ symmetric matrix and λ is one of its eigenvalues. Then $P_A(\lambda)$, the eigenprojection associated with the eigenvalue λ, is a continuous function of the elements of A.

Example 3.10. Consider the matrix A

$$A = \begin{bmatrix} 2 & 0 & 0 \\ 0 & 1 & 0 \\ 0 & 0 & 1 \end{bmatrix},$$

which clearly has the simple eigenvalue 2 and the repeated eigenvalue 1. Suppose that $B_1, B_2, \dots$ is a sequence of 3×3 matrices such that $B_n \to A$, as $n \to \infty$. Let $\gamma_{1,n} \geq \gamma_{2,n} \geq \gamma_{3,n}$ be the eigenvalues of B_n, while $\boldsymbol{x}_{1,n}$, $\boldsymbol{x}_{2,n}$, and $\boldsymbol{x}_{3,n}$ is a corresponding set of orthonormal eigenvectors. Theorem 3.13 implies that, as $n \to \infty$,

$$\gamma_{1,n} \to 2, \qquad \text{and} \qquad \gamma_{i,n} \to 1, \quad \text{for} \quad i = 2, 3$$

On the other hand, Theorem 3.14 implies that, as $n \to \infty$,

$$P_{1,n} \to P_A(2), \qquad P_{2,n} \to P_A(1),$$

where

$$P_{1,n} = \boldsymbol{x}_{1,n}\boldsymbol{x}'_{1,n}, \qquad P_{2,n} = \boldsymbol{x}_{2,n}\boldsymbol{x}'_{2,n} + \boldsymbol{x}_{3,n}\boldsymbol{x}'_{3,n}$$

For instance, suppose that

$$B_n = \begin{bmatrix} 2 & 0 & n^{-1} \\ 0 & 1 & 0 \\ n^{-1} & 0 & 1 \end{bmatrix},$$

so that, clearly, $B_n \to A$. The characteristic equation of B_n simplifies to

$$\lambda^3 - 4\lambda^2 + (5 - n^{-2})\lambda - 2 + n^{-2} = (\lambda - 1)(\lambda^2 - 3\lambda + 2 - n^{-2}) = 0,$$

so that the eigenvalues of B_n are

$$1, \frac{3}{2} - \frac{\sqrt{1 + 4n^{-2}}}{2}, \qquad \frac{3}{2} + \frac{\sqrt{1 + 4n^{-2}}}{2},$$

which do converge to 1, 1, and 2, respectively. It is left as an exercise for the reader to verify that

$$P_{1,n} \to \begin{bmatrix} 1 & 0 & 0 \\ 0 & 0 & 0 \\ 0 & 0 & 0 \end{bmatrix} = P_A(2), \qquad P_{2,n} \to \begin{bmatrix} 0 & 0 & 0 \\ 0 & 1 & 0 \\ 0 & 0 & 1 \end{bmatrix} = P_A(1)$$

6. EXTREMAL PROPERTIES OF EIGENVALUES

One of the reasons that eigenvalues play a prominent role in many applications is because they can be expressed as maximum or minimum values of certain functions involving a quadratic form. In this section, we derive some of these extremal properties of eigenvalues.

Let A be a fixed $m \times m$ symmetric matrix and consider the quadratic form $\boldsymbol{x}'A\boldsymbol{x}$ as a function of $\boldsymbol{x}$. If α is a nonzero scalar, then $(\alpha\boldsymbol{x})'A(\alpha\boldsymbol{x}) = \alpha^2\boldsymbol{x}'A\boldsymbol{x}$, so that the quadratic form can be made arbitrarily small or large, depending on whether $\boldsymbol{x}'A\boldsymbol{x}$ is negative or positive, through the proper choice of α. Thus, any meaningful study of the variational properties of $\boldsymbol{x}'A\boldsymbol{x}$ as we change $\boldsymbol{x}$ will require the removal of the effect of scale changes in $\boldsymbol{x}$. One way of doing this is through the construction of what is commonly called the Rayleigh quotient given by

$$R(\boldsymbol{x}, A) = \frac{\boldsymbol{x}'A\boldsymbol{x}}{\boldsymbol{x}'\boldsymbol{x}}$$

Note that $R(\alpha\boldsymbol{x}, A) = R(\boldsymbol{x}, A)$. Our first result involves the global maximization and minimization of $R(\boldsymbol{x}, A)$.

Theorem 3.15. Let A be a symmetric $m\times m$ matrix with ordered eigenvalues $\lambda_1 \geq \cdots \geq \lambda_m$. For any $m \times 1$ vector $\boldsymbol{x} \neq \boldsymbol{0}$,

$$\lambda_m \leq \frac{\boldsymbol{x}'A\boldsymbol{x}}{\boldsymbol{x}'\boldsymbol{x}} \leq \lambda_1, \tag{3.6}$$

and, in particular,

$$\lambda_m = \min_{\boldsymbol{x}\neq\boldsymbol{0}} \frac{\boldsymbol{x}'A\boldsymbol{x}}{\boldsymbol{x}'\boldsymbol{x}}, \qquad \lambda_1 = \max_{\boldsymbol{x}\neq\boldsymbol{0}} \frac{\boldsymbol{x}'A\boldsymbol{x}}{\boldsymbol{x}'\boldsymbol{x}} \tag{3.7}$$

Proof. Let $A = X\Lambda X'$ be the spectral decomposition of A, where the columns of $X = (\boldsymbol{x}_1,\ldots,\boldsymbol{x}_m)$ are normalized eigenvectors of A and $\Lambda = \text{diag}(\lambda_1,\ldots,\lambda_m)$. Then, if $\boldsymbol{y} = X'\boldsymbol{x}$, we have

$$\frac{\boldsymbol{x}'A\boldsymbol{x}}{\boldsymbol{x}'\boldsymbol{x}} = \frac{\boldsymbol{x}'X\Lambda X'\boldsymbol{x}}{\boldsymbol{x}'XX'\boldsymbol{x}} = \frac{\boldsymbol{y}'\Lambda\boldsymbol{y}}{\boldsymbol{y}'\boldsymbol{y}} = \frac{\sum_{i=1}^m \lambda_i y_i^2}{\sum_{i=1}^m y_i^2},$$

so that (3.6) follows from the fact that

$$\lambda_m \sum_{i=1}^m y_i^2 \leq \sum_{i=1}^m \lambda_i y_i^2 \leq \lambda_1 \sum_{i=1}^m y_i^2$$

Now (3.7) is verified by choices of $\boldsymbol{x}$ for which the bounds in (3.6) are attained; for instance, the lower bound is attained with $\boldsymbol{x} = \boldsymbol{x}_m$, while the upper bound holds with $\boldsymbol{x} = \boldsymbol{x}_1$. □

Note that, since for any nonnull $\boldsymbol{x}$, $\boldsymbol{z} = (\boldsymbol{x}'\boldsymbol{x})^{-1/2}\boldsymbol{x}$ is a unit vector, the minimization and maximization of $\boldsymbol{z}'A\boldsymbol{z}$ over all unit vectors $\boldsymbol{z}$ will also yield λ_m and λ_1, respectively; that is,

$$\lambda_m = \min_{\boldsymbol{z}'\boldsymbol{z}=1} \boldsymbol{z}'A\boldsymbol{z}, \qquad \lambda_1 = \max_{\boldsymbol{z}'\boldsymbol{z}=1} \boldsymbol{z}'A\boldsymbol{z}$$

The following theorem shows that each eigenvalue of a symmetric matrix A can be expressed as a constrained maximum or minimum of the Rayleigh quotient, $R(\boldsymbol{x}, A)$.

Theorem 3.16. Let A be an $m \times m$ symmetric matrix having eigenvalues $\lambda_1 \geq \lambda_2 \geq \cdots \geq \lambda_m$ with $\boldsymbol{x}_1,\ldots,\boldsymbol{x}_m$ being a corresponding set of orthonormal eigenvectors. For $h = 1,\ldots,m$, define S_h and T_h to be the vector spaces spanned by the columns of $X_h = (\boldsymbol{x}_1, \ldots, \boldsymbol{x}_h)$ and $Y_h = (\boldsymbol{x}_h, \ldots, \boldsymbol{x}_m)$, respectively. Then

$$\lambda_h = \min_{\boldsymbol{x} \in S_h} \frac{\boldsymbol{x}'A\boldsymbol{x}}{\boldsymbol{x}'\boldsymbol{x}} = \min_{Y'_{h+1}\boldsymbol{x}=\boldsymbol{0}} \frac{\boldsymbol{x}'A\boldsymbol{x}}{\boldsymbol{x}'\boldsymbol{x}},$$

and

$$\lambda_h = \max_{\boldsymbol{x} \in T_h} \frac{\boldsymbol{x}'A\boldsymbol{x}}{\boldsymbol{x}'\boldsymbol{x}} = \max_{X'_{h-1}\boldsymbol{x}=\boldsymbol{0}} \frac{\boldsymbol{x}'A\boldsymbol{x}}{\boldsymbol{x}'\boldsymbol{x}},$$

where the vector $\boldsymbol{x} = \boldsymbol{0}$ has been excluded from the maximization and minimization processes.

Proof. We will prove the result concerning the minimum; the proof for the maximum is similar. Let $X = (\boldsymbol{x}_1, \ldots, \boldsymbol{x}_m)$ and $\Lambda = \text{diag}(\lambda_1, \ldots, \lambda_m)$. Note that, since $X'AX = \Lambda$ and $X'X = \text{I}_m$, it follows that $X'_h X_h = \text{I}_h$ and $X'_h AX_h = \Lambda_h$, where $\Lambda_h = \text{diag}(\lambda_1, \ldots, \lambda_h)$. Now $\boldsymbol{x} \in S_h$ if and only if there exists an $h \times 1$ vector $\boldsymbol{y}$ such that $\boldsymbol{x} = X_h\boldsymbol{y}$. Consequently,

$$\min_{\boldsymbol{x} \in S_h} \frac{\boldsymbol{x}'A\boldsymbol{x}}{\boldsymbol{x}'\boldsymbol{x}} = \min_{\boldsymbol{y} \neq \boldsymbol{0}} \frac{\boldsymbol{y}'X'_hAX_h\boldsymbol{y}}{\boldsymbol{y}'X'_hX_h\boldsymbol{y}} = \min_{\boldsymbol{y} \neq \boldsymbol{0}} \frac{\boldsymbol{y}'\Lambda_h\boldsymbol{y}}{\boldsymbol{y}'\boldsymbol{y}} = \lambda_h,$$

where the last equality follows from Theorem 3.15. The second version of the minimization follows immediately from the first and the fact that the null space of Y'_{h+1} is S_h. □

The next two examples give some indication of how the extremal properties of eigenvalues make them important features in many applications.

Example 3.11. Suppose that the same m variables are measured on individuals from k different groups with the goal being to identify differences in the means for the k groups. Let the $m \times 1$ vectors $\boldsymbol{\mu}_1, \ldots, \boldsymbol{\mu}_k$ represent the k group mean vectors, and let $\boldsymbol{\mu} = (\boldsymbol{\mu}_1 + \cdots + \boldsymbol{\mu}_k)/k$ be the average of these mean vectors. To investigate the differences in group means, we will utilize the deviations $(\boldsymbol{\mu}_i - \boldsymbol{\mu})$ from the average mean; in particular, we form the sum of squares and cross products matrix given by

$$A = \sum_{i=1}^{k} (\boldsymbol{\mu}_i - \boldsymbol{\mu})(\boldsymbol{\mu}_i - \boldsymbol{\mu})'$$

Note that for a particular unit vector $\boldsymbol{x}$, $\boldsymbol{x}'A\boldsymbol{x}$ will give a measure of the differences among the k groups in the direction $\boldsymbol{x}$; a value of zero indicates the groups have identical means in this direction, while increasingly large values of $\boldsymbol{x}'A\boldsymbol{x}$ indicate increasingly widespread differences in this same direction. If $\boldsymbol{x}_1,\ldots,\boldsymbol{x}_m$ are normalized eigenvectors of A corresponding to its ordered eigenvalues $\lambda_1 \geq \cdots \geq \lambda_m$, then it follows from Theorems 3.15 and 3.16 that the greatest difference among the k groups, in terms of deviations from the overall mean, occurs in the direction given by $\boldsymbol{x}_1$. Of all directions orthogonal to $\boldsymbol{x}_1,\boldsymbol{x}_2$ gives the direction with the greatest difference among the k groups, and so on. If some of the eigenvalues are very small relative to the rest, then we will be able to effectively reduce the dimension of the problem. For example, suppose that $\lambda_3,\ldots,\lambda_m$ are all very small relative to λ_1 and λ_2. Then all substantial differences among the group means will be observed in the plane spanned by $\boldsymbol{x}_1$ and $\boldsymbol{x}_2$. In Example 4.11 we will discuss the statistical procedure, called canonical variate analysis, that utilizes this sort of dimension reducing process.

Example 3.12. In Example 3.11, the focus was on means. In this example, we will look at a procedure that concentrates on variances. This technique, called principal component analysis, was developed by Hotelling (1933). Some good references on this subject are Jackson (1991) and Jolliffe (1986). Let $\boldsymbol{x}$ be an $m \times 1$ random vector having the covariance matrix Ω. Suppose that we wish to find the $m \times 1$ vector $\boldsymbol{a}_1$ so as to make the variance of $\boldsymbol{a}_1'\boldsymbol{x}$ as large as possible. But from Section 1.13, we know that

$$\text{var}(\boldsymbol{a}_1'\boldsymbol{x}) = \boldsymbol{a}_1'\{\text{var}(\boldsymbol{x})\}\boldsymbol{a}_1 = \boldsymbol{a}_1'\Omega\boldsymbol{a}_1 \tag{3.8}$$

Clearly, we can make this arbitrarily large by taking $\boldsymbol{a}_1 = \alpha\boldsymbol{c}$ for some scalar α and some vector $\boldsymbol{c} \neq \boldsymbol{0}$, and then let $\alpha \to \infty$. We will remove this effect of the scale of $\boldsymbol{a}_1$ by imposing a constraint. For example, we may consider maximizing (3.8) over all choices of $\boldsymbol{a}_1$ satisfying $\boldsymbol{a}_1'\boldsymbol{a}_1 = 1$. In this case, we are searching for the one direction in R^m, that is, the line, for which the variability of observations of $\boldsymbol{x}$ projected onto that line is maximized. It follows from Theorem 3.15 that this direction is given by the normalized eigenvector of Ω corresponding to its largest eigenvalue. Suppose we also wish to find a second direction, given by $\boldsymbol{a}_2$ and orthogonal to $\boldsymbol{a}_1$, where $\boldsymbol{a}_2'\boldsymbol{a}_2 = 1$ and $\text{var}(\boldsymbol{a}_2'\boldsymbol{x})$ is maximized. From Theorem 3.16, this second direction is given by the normalized eigenvector of Ω corresponding to its second largest eigenvalue. Continuing in this fashion, we would obtain m directions identified by the set $\boldsymbol{a}_1,\ldots,\boldsymbol{a}_m$ of orthonormal eigenvectors of Ω. Effectively, what we will have done is to find a rotation of the original axes to a new set of orthogonal axes, where each successive axis is selected so as to maximize the dispersion among the $\boldsymbol{x}$ observations along that axis. Note that the components of the transformed vector $(\boldsymbol{a}_1'\boldsymbol{x},\ldots,\boldsymbol{a}_m'\boldsymbol{x})'$, which are called the principal components of Ω, are uncorrelated since for $i \neq j$,

$$\text{cov}(\boldsymbol{a}_i'\boldsymbol{x}, \boldsymbol{a}_j'\boldsymbol{x}) = \boldsymbol{a}_i'\Omega\boldsymbol{a}_j = 0$$

For some specific examples, first consider the 4×4 covariance matrix given by

$$\Omega = \begin{bmatrix} 4.65 & 4.35 & 0.55 & 0.45 \\ 4.35 & 4.65 & 0.45 & 0.55 \\ 0.55 & 0.45 & 4.65 & 4.35 \\ 0.45 & 0.55 & 4.35 & 4.65 \end{bmatrix}$$

The eigenvalues of Ω are 10, 8, 0.4, and 0.2, so the first two eigenvalues account for a large proportion, actually $18/18.6 = 0.97$, of the total variability of $\boldsymbol{x}$. This means that although observations of $\boldsymbol{x}$ would appear as points in R^4, almost all of the dispersion among these points will be confined to a plane. This plane is spanned by the first two normalized eigenvectors of Ω, $(0.25, 0.25, 0.25, 0.25)'$ and $(0.25, 0.25, -0.25, -0.25)'$. As a second illustration, consider a covariance matrix such as

$$\Omega = \begin{bmatrix} 59 & 5 & 2 \\ 5 & 35 & -10 \\ 2 & -10 & 56 \end{bmatrix},$$

which has a repeated eigenvalue; specifically the eigenvalues are 60 and 30 with multiplicities 2 and 1, respectively. Since the largest eigenvalue of Ω is repeated, there is no one direction $\boldsymbol{a}_1$ that maximizes $\text{var}(\boldsymbol{a}_1'\boldsymbol{x})$. Instead, the dispersion of $\boldsymbol{x}$ observations is the same in all directions in the plane given by the eigenspace $S_\Omega(60)$, which is spanned by the vectors $(1, 1, -2)'$ and $(2, 0, 1)'$. Consequently, a scatter plot of $\boldsymbol{x}$ observations would produce a circular pattern of points in this plane.

Our final result, known as the Courant–Fischer min-max theorem, gives alternative expressions for the intermediate eigenvalues of A as constrained minima and maxima of the Rayleigh quotient $R(\boldsymbol{x}, A)$.

Theorem 3.17. Let A be an $m \times m$ symmetric matrix having eigenvalues $\lambda_1 \geq \lambda_2 \geq \cdots \geq \lambda_m$. For $h = 1, \ldots, m$, let B_h be any $m \times (h-1)$ matrix and C_h any $m \times (m-h)$ matrix satisfying $B_h'B_h = \text{I}_{h-1}$ and $C_h'C_h = \text{I}_{m-h}$. Then

$$\lambda_h = \min_{B_h} \max_{B_h'\boldsymbol{x}=\boldsymbol{0}} \frac{\boldsymbol{x}'A\boldsymbol{x}}{\boldsymbol{x}'\boldsymbol{x}}, \tag{3.9}$$

as well as

$$\lambda_h = \max_{C_h} \min_{C_h'x=0} \frac{x'Ax}{x'x} \tag{3.10}$$

where the vector $x = 0$ has been excluded.

Proof. We first prove the min-max result given by (3.9). Let $X_h = (x_1, \ldots, x_h)$, where $x_1, \ldots, x_h$ is a set of orthonormal eigenvectors of A, corresponding to the eigenvalues $\lambda_1, \ldots, \lambda_h$. Since X_{h-1} is an $m \times (h-1)$ matrix satisfying $X_{h-1}' X_{h-1} = I_{h-1}$, it follows that

$$\min_{B_h} \max_{B_h'x=0} \frac{x'Ax}{x'x} \leq \max_{X_{h-1}'x=0} \frac{x'Ax}{x'x} = \lambda_h, \tag{3.11}$$

where the equality follows from Theorem 3.16. Now for arbitrary B_h satisfying $B_h'B_h = I_{h-1}$, the matrix $B_h'X_h$ is $(h-1) \times h$, so that the columns must be linearly dependent. Consequently, we can find an $h \times 1$ nonnull vector y such that $B_h'X_h y = 0$. Since $X_h y$ is one choice for x, we find that

$$\max_{B_h'x=0} \frac{x'Ax}{x'x} \geq \frac{y'X_h'AX_h y}{y'X_h'X_h y} = \frac{y'\Lambda_h y}{y'y} \geq \lambda_h, \tag{3.12}$$

where $\Lambda_h = \text{diag}(\lambda_1, \ldots, \lambda_h)$, and the last inequality follows from (3.6). Minimizing (3.12) over all choices of B_h gives

$$\min_{B_h} \max_{B_h'x=0} \frac{x'Ax}{x'x} \geq \lambda_h$$

This, along with (3.11), proves (3.9). The proof of (3.10) is along the same lines. Let $Y_h = (x_h, \ldots, x_m)$, where $x_h, \ldots, x_m$ is a set of orthonormal eigenvectors of A, corresponding to the eigenvalues $\lambda_h, \ldots, \lambda_m$. Since Y_{h+1} is an $m \times (m-h)$ matrix satisfying $Y_{h+1}' Y_{h+1} = I_{m-h}$, it follows that

$$\max_{C_h} \min_{C_h'x=0} \frac{x'Ax}{x'x} \geq \min_{Y_{h+1}'x=0} \frac{x'Ax}{x'x} = \lambda_h, \tag{3.13}$$

where the equality follows from Theorem 3.16. For an arbitrary C_h satisfying $C_h'C_h = I_{m-h}$, the matrix $C_h'Y_h$ is $(m-h) \times (m-h+1)$, so the columns of $C_h'Y_h$ must be linearly dependent. Thus, there exists an $(m-h+1) \times 1$ nonnull vector

$\boldsymbol{y}$ satisfying $C_h'Y_h\boldsymbol{y} = \mathbf{0}$. Since $Y_h\boldsymbol{y}$ is one choice for $\boldsymbol{x}$, we have

$$\min_{C_h'\boldsymbol{x}=\mathbf{0}} \frac{\boldsymbol{x}'A\boldsymbol{x}}{\boldsymbol{x}'\boldsymbol{x}} \leq \frac{\boldsymbol{y}'Y_h'AY_h\boldsymbol{y}}{\boldsymbol{y}'Y_h'Y_h\boldsymbol{y}} = \frac{\boldsymbol{y}'\Delta_h\boldsymbol{y}}{\boldsymbol{y}'\boldsymbol{y}} \leq \lambda_h, \tag{3.14}$$

where $\Delta_h = \text{diag}(\lambda_h, \dots, \lambda_m)$ and the last inequality follows from (3.6). Maximizing (3.14) over all choices of C_h yields

$$\max_{C_h} \min_{C_h'\boldsymbol{x}=\mathbf{0}} \frac{\boldsymbol{x}'A\boldsymbol{x}}{\boldsymbol{x}'\boldsymbol{x}} \leq \lambda_h$$

This together with (3.13) establishes (3.10). □

Corollary 3.17.1. Let A be an $m \times m$ symmetric matrix having eigenvalues $\lambda_1 \geq \lambda_2 \geq \cdots \geq \lambda_m$. For $h = 1, \dots, m$, let B_h be any $m \times (h-1)$ matrix and C_h be any $m \times (m-h)$ matrix. Then

$$\lambda_h \leq \max_{B_h'\boldsymbol{x}=\mathbf{0}} \frac{\boldsymbol{x}'A\boldsymbol{x}}{\boldsymbol{x}'\boldsymbol{x}},$$

and

$$\lambda_h \geq \min_{C_h'\boldsymbol{x}=\mathbf{0}} \frac{\boldsymbol{x}'A\boldsymbol{x}}{\boldsymbol{x}'\boldsymbol{x}}$$

Proof. If $B_h'B_h = \mathrm{I}_{h-1}$ and $C_h'C_h = \mathrm{I}_{m-h}$, then the two inequalities follow directly from Theorem 3.17. We need to establish them for arbitrary B_h and C_h. When $B_h'B_h = \mathrm{I}_{h-1}$, the set $S_{B_h} = \{\boldsymbol{x}: \boldsymbol{x} \in R^m, B_h'\boldsymbol{x} = \mathbf{0}\}$ is the orthogonal complement of the vector space which has the columns of B_h as an orthonormal basis. Thus, the first inequality holds when maximizing over all $\boldsymbol{x} \neq \mathbf{0}$ in any $(m-h+1)$-dimensional vector subspace of R^m. Consequently, this inequality also will hold for any $m \times (h-1)$ matrix B_h since, in this case, $\text{rank}(B_h) \leq h-1$ guarantees that the maximization is over a vector subspace of dimension at least $m-h+1$. A similar argument applies to the second inequality. □

The proof of the following result is left to the reader as an exercise.

Theorem 3.18. Suppose that A and B are $m \times m$ symmetric matrices and $A - B$ is nonnegative definite. Then $\lambda_i(A) \geq \lambda_i(B)$ for $i = 1, \dots, m$.

Some additional extremal properties of eigenvalues can be found in Bellman (1970) and Horn and Johnson (1985).

7. SOME ADDITIONAL RESULTS CONCERNING EIGENVALUES

Let A be an $m \times m$ symmetric matrix and H, an $m \times h$ matrix satisfying $H'H = I_h$. In some situations it is of interest to compare the eigenvalues of A to those of $H'AH$. Some comparisons follow immediately from Theorem 3.17. For instance, it is easily verified that from (3.9), we have

$$\lambda_1(H'AH) \geq \lambda_{m-h+1}(A),$$

and from (3.10) we have

$$\lambda_h(H'AH) \leq \lambda_h(A)$$

The following result, known as the Poincaré separation theorem [Poincaré, (1890); see also Fan (1949)], provides some inequalities involving the eigenvalues of A and $H'AH$ in addition to the two given above.

Theorem 3.19. Let A be an $m \times m$ symmetric matrix and H be an $m \times h$ matrix satisfying $H'H = I_h$. Then, for $i = 1, \ldots, h$, it follows that

$$\lambda_{m-h+i}(A) \leq \lambda_i(H'AH) \leq \lambda_i(A)$$

Proof. To establish the lower bound on $\lambda_i(H'AH)$, let $Y_n = (\boldsymbol{x}_n, \ldots, \boldsymbol{x}_m)$, where $n = m - h + i + 1$, and $\boldsymbol{x}_1, \ldots, \boldsymbol{x}_m$ is a set of orthonormal eigenvectors of A corresponding to the eigenvalues $\lambda_1(A) \geq \cdots \geq \lambda_m(A)$. Then it follows that

$$\lambda_{m-h+i}(A) = \lambda_{n-1}(A) = \min_{Y_n'\boldsymbol{x}=\boldsymbol{0}} \frac{\boldsymbol{x}'A\boldsymbol{x}}{\boldsymbol{x}'\boldsymbol{x}} \leq \min_{\substack{Y_n'\boldsymbol{x}=\boldsymbol{0} \\ \boldsymbol{x}=H\boldsymbol{y}}} \frac{\boldsymbol{x}'A\boldsymbol{x}}{\boldsymbol{x}'\boldsymbol{x}}$$

$$= \min_{Y_n'H\boldsymbol{y}=\boldsymbol{0}} \frac{\boldsymbol{y}'H'AH\boldsymbol{y}}{\boldsymbol{y}'\boldsymbol{y}} \leq \lambda_{h-(m-n+1)}(H'AH) = \lambda_i(H'AH),$$

where the second equality follows from Theorem 3.16. The last inequality follows from Corollary 3.17.1, after noting that the order of $H'AH$ is h and $Y_n'H$ is $(m-n+1) \times h$. To prove the upper bound for $\lambda_i(H'AH)$, let $X_{i-1} = (\boldsymbol{x}_1, \ldots, \boldsymbol{x}_{i-1})$, and note that

$$\lambda_i(A) = \max_{X'_{i-1}x=0} \frac{x'Ax}{x'x} \geq \max_{\substack{X'_{i-1}x=0\\ x=Hy}} \frac{x'Ax}{x'x}$$

$$= \max_{X'_{i-1}Hy=0} \frac{y'H'AHy}{y'y} \geq \lambda_i(H'AH),$$

where the first equality follows from Theorem 3.16 and the final inequality follows from Corollary 3.17.1. □

Theorem 3.19 can be used to prove the following useful result.

Theorem 3.20. Let A be an $m \times m$ symmetric matrix and let A_k be its leading $k \times k$ principal submatrix; that is, A_k is the matrix obtained by deleting the last $m - k$ rows and columns of A. Then, for $i = 1, \ldots, k$,

$$\lambda_{m-i+1}(A) \leq \lambda_{k-i+1}(A_k) \leq \lambda_{k-i+1}(A)$$

In Chapter 1, the conditions for a symmetric matrix A to be a positive definite or positive semidefinite matrix were given in terms of the possible values of the quadratic form $x'Ax$. We now show that these conditions also can be expressed in terms of the eigenvalues of A.

Theorem 3.21. Let A be an $m \times m$ symmetric matrix with eigenvalues $\lambda_1, \ldots, \lambda_m$. Then

(a) A is positive definite if and only if $\lambda_i > 0$ for all i.
(b) A is positive semidefinite if and only if $\lambda_i \geq 0$ for all i and $\lambda_i = 0$ for at least one i.

Proof. Let the columns of $X = (x_1, \ldots, x_m)$ be a set of orthonormal eigenvectors of A corresponding to the eigenvalues $\lambda_1, \ldots, \lambda_m$, so that $A = X\Lambda X'$, where $\Lambda = \text{diag}(\lambda_1, \ldots, \lambda_m)$. If A is positive definite, then $x'Ax > 0$ for all $x \neq 0$, so in particular, choosing $x = x_i$, we have

$$x_i'Ax_i = \lambda_i > 0$$

Conversely, if $\lambda_i > 0$ for all i, then for any $x \neq 0$ define $y = X'x$, and note that

$$x'Ax = x'X\Lambda X'x = y'\Lambda y = \sum_{i=1}^{m} y_i^2\lambda_i \tag{3.15}$$

has to be positive because the λ_i are positive and at least one of the y_i^2 is positive since $\boldsymbol{y} \neq \mathbf{0}$. This proves (a). By a similar argument, we find that A is nonnegative definite if and only if $\lambda_i \geq 0$ for all i. Thus, to prove (b) we only need to prove that $\boldsymbol{x}'A\boldsymbol{x} = 0$ for some $\boldsymbol{x} \neq \mathbf{0}$ if and only if at least one $\lambda_i = 0$. It follows from (3.15) that if $\boldsymbol{x}'A\boldsymbol{x} = 0$, then $\lambda_i = 0$ for every i for which $y_i^2 > 0$. On the other hand, if for some i, $\lambda_i = 0$, then $\boldsymbol{x}_i'A\boldsymbol{x}_i = \lambda_i = 0$. □

Since a square matrix is singular if and only if it has a zero eigenvalue, it follows immediately from Theorem 3.21 that positive definite matrices are nonsingular, while positive semidefinite matrices are singular.

Example 3.13. Consider the ordinary least squares estimator $\hat{\boldsymbol{\beta}} = (X'X)^{-1}X'\boldsymbol{y}$ of $\boldsymbol{\beta}$ in the model

$$\boldsymbol{y} = X\boldsymbol{\beta} + \boldsymbol{\epsilon},$$

where $E(\boldsymbol{\epsilon}) = \mathbf{0}$ and $\text{var}(\boldsymbol{\epsilon}) = \sigma^2 I_N$. For an arbitrary $(k+1) \times 1$ vector $\boldsymbol{c}$, we will prove that $\boldsymbol{c}'\hat{\boldsymbol{\beta}}$ is the best linear unbiased estimator of $\boldsymbol{c}'\boldsymbol{\beta}$; an estimator t is an unbiased estimator of $\boldsymbol{c}'\boldsymbol{\beta}$ if $E(t) = \boldsymbol{c}'\boldsymbol{\beta}$. Clearly, $\boldsymbol{c}'\hat{\boldsymbol{\beta}}$ is unbiased since $E(\boldsymbol{\epsilon}) = \mathbf{0}$ implies that

$$E(\boldsymbol{c}'\hat{\boldsymbol{\beta}}) = \boldsymbol{c}'(X'X)^{-1}X'E(\boldsymbol{y}) = \boldsymbol{c}'(X'X)^{-1}X'X\boldsymbol{\beta} = \boldsymbol{c}'\boldsymbol{\beta}$$

To show that it is the best linear unbiased estimator, we must show that it has variance at least as small as the variance of any other linear unbiased estimator of $\boldsymbol{c}'\boldsymbol{\beta}$. Let $\boldsymbol{a}'\boldsymbol{y}$ be an arbitrary linear unbiased estimator of $\boldsymbol{c}'\boldsymbol{\beta}$, so that

$$\boldsymbol{c}'\boldsymbol{\beta} = E(\boldsymbol{a}'\boldsymbol{y}) = \boldsymbol{a}'E(\boldsymbol{y}) = \boldsymbol{a}'X\boldsymbol{\beta},$$

regardless of the value of the vector $\boldsymbol{\beta}$. But this implies that

$$\boldsymbol{c}' = \boldsymbol{a}'X$$

Now

$$\text{var}(\boldsymbol{c}'\hat{\boldsymbol{\beta}}) = \boldsymbol{c}'\{\text{var}(\hat{\boldsymbol{\beta}})\}\boldsymbol{c} = \boldsymbol{c}'\{\sigma^2(X'X)^{-1}\}\boldsymbol{c} = \sigma^2\boldsymbol{a}'X(X'X)^{-1}X'\boldsymbol{a},$$

while

$$\text{var}(\boldsymbol{a}'\boldsymbol{y}) = \boldsymbol{a}'\{\text{var}(\boldsymbol{y})\}\boldsymbol{a} = \boldsymbol{a}'\{\sigma^2 I_N\}\boldsymbol{a} = \sigma^2\boldsymbol{a}'\boldsymbol{a}$$

Thus, the difference in their variances is

$$\begin{aligned}\text{var}(\boldsymbol{a}'\boldsymbol{y}) - \text{var}(\boldsymbol{c}'\hat{\boldsymbol{\beta}}) &= \sigma^2\boldsymbol{a}'\boldsymbol{a} - \sigma^2\boldsymbol{a}'X(X'X)^{-1}X'\boldsymbol{a}\\ &= \sigma^2\boldsymbol{a}'(\mathrm{I}_N - X(X'X)^{-1}X')\boldsymbol{a}\end{aligned}$$

But

$$\{\mathrm{I}_N - X(X'X)^{-1}X'\}^2 = \{\mathrm{I}_N - X(X'X)^{-1}X'\},$$

and so using Theorem 3.4, we find that each of the eigenvalues of $\mathrm{I}_N - X(X'X)^{-1}X'$ must be 0 or 1. Thus, from Theorem 3.21, we see that $\mathrm{I}_N - X(X'X)^{-1}X'$ is nonnegative definite and so

$$\text{var}(\boldsymbol{a}'\boldsymbol{y}) - \text{var}(\boldsymbol{c}'\hat{\boldsymbol{\beta}}) \geq 0$$

as is required.

Symmetric matrices are often obtained as the result of a transpose product; that is, if T is an $m \times n$ matrix, then both $T'T$ and TT' are symmetric matrices. The following two theorems show that their eigenvalues are nonnegative and their positive eigenvalues are equal.

Theorem 3.22. Let T be an $m \times n$ matrix, with rank$(T) = r$. Then $T'T$ has r positive eigenvalues. It is positive definite if $r = n$ and positive semidefinite if $r < n$.

Proof. For any nonnull $n \times 1$ vector $\boldsymbol{x}$, let $\boldsymbol{y} = T\boldsymbol{x}$. Then clearly

$$\boldsymbol{x}'T'T\boldsymbol{x} = \boldsymbol{y}'\boldsymbol{y} = \sum_{i=1}^{m} y_i^2$$

is nonnegative, so $T'T$ is nonnegative definite and, thus, by Theorem 3.21 all of its eigenvalues are nonnegative. If $\boldsymbol{x}$ is an eigenvector of $T'T$ corresponding to a zero eigenvalue, then the equation above must equal zero, and this can only happen if $\boldsymbol{y} = T\boldsymbol{x} = \boldsymbol{0}$. Since rank$(T) = r$, we can find a set of $n - r$ linearly independent $\boldsymbol{x}$s satisfying $T\boldsymbol{x} = \boldsymbol{0}$, that is, any basis of the null space of T, and so the number of zero eigenvalues of $T'T$ is equal to $n - r$. The result now follows. □

Theorem 3.23. Let T be an $m \times n$ matrix, with rank$(T) = r$. Then the positive eigenvalues of $T'T$ are equal to the positive eigenvalues of TT'.

Proof. Let $\lambda > 0$ be an eigenvalue of $T'T$ with multiplicity h. Since the $n \times n$ matrix $T'T$ is symmetric, we can find an $n \times h$ matrix X, whose columns are orthonormal, satisfying

$$T'TX = \lambda X$$

Let $Y = TX$ and observe that

$$TT'Y = TT'TX = T(\lambda X) = \lambda TX = \lambda Y,$$

so that λ is also an eigenvalue of TT'. Its multiplicity is also h since

$$\begin{aligned} \text{rank}(Y) &= \text{rank}(TX) = \text{rank}((TX)'TX) = \text{rank}(X'T'TX) \\ &= \text{rank}(\lambda X'X) = \text{rank}(\lambda \text{I}_h) = h \end{aligned}$$

□

Next we will use the Courant–Fischer min-max theorem to prove the following important monotonicity property of the eigenvalues of symmetric matrices.

Theorem 3.24. Let A be an $m \times m$ symmetric matrix and B be an $m \times m$ nonnegative definite matrix. Then, for $h = 1, \dots, m$, we have

$$\lambda_h(A + B) \geq \lambda_h(A),$$

where the inequality is strict if B is positive definite.

Proof. For an arbitrary $m \times (h - 1)$ matrix B_h satisfying $B_h'B_h = I_{h-1}$, we have

$$\begin{aligned} \max_{B_h'x=0} \frac{x'(A+B)x}{x'x} &= \max_{B_h'x=0} \left(\frac{x'Ax}{x'x} + \frac{x'Bx}{x'x} \right) \geq \max_{B_h'x=0} \frac{x'Ax}{x'x} \\ &\quad + \min_{B_h'x=0} \frac{x'Bx}{x'x} \\ &\geq \max_{B_h'x=0} \frac{x'Ax}{x'x} + \min_{x \neq 0} \frac{x'Bx}{x'x} \\ &= \max_{B_h'x=0} \frac{x'Ax}{x'x} + \lambda_m(B) \geq \max_{B_h'x=0} \frac{x'Ax}{x'x}, \end{aligned}$$

where the last equality follows from Theorem 3.15. The final inequality above is strict if B is positive definite since, in this case, $\lambda_m(B) > 0$. Now minimizing both sides of the equation above over all choices of B_h satisfying $B_h'B_h = \text{I}_h$ and using (3.9) of Theorem 3.17, we get

$$\lambda_h(A+B) = \min_{B_h} \max_{B_h'x=0} \frac{x'(A+B)x}{x'x} \geq \min_{B_h} \max_{B_h'x=0} \frac{x'Ax}{x'x} = \lambda_h(A)$$

This completes the proof. □

Note that there is not a general bounding relationship between $\lambda_h(A+B)$ and $\lambda_h(A)+\lambda_h(B)$. For instance, if $A = \text{diag}(1, 2, 3, 4)$ and $B = \text{diag}(8, 6, 4, 2)$, then

$$\lambda_2(A+B) = 8 < \lambda_2(A) + \lambda_2(B) = 3 + 6 = 9,$$

while

$$\lambda_3(A+B) = 7 > \lambda_3(A) + \lambda_3(B) = 2 + 4 = 6$$

In Example 3.11 we discussed a situation in which the eigenvalues and eigenvectors of

$$A = \sum_{i=1}^{k} (\boldsymbol{\mu}_i - \boldsymbol{\mu})(\boldsymbol{\mu}_i - \boldsymbol{\mu})'$$

were utilized in analyzing differences among the group means $\boldsymbol{\mu}_1, \ldots, \boldsymbol{\mu}_k$. For instance, an eigenvector $\boldsymbol{x}_1$, corresponding to the largest eigenvalue of A, gives the direction of maximum dispersion among the group means in that

$$\frac{\boldsymbol{x}_1'A\boldsymbol{x}_1}{\boldsymbol{x}_1'\boldsymbol{x}_1}$$

is maximized. The division here by $\boldsymbol{x}_1'\boldsymbol{x}_1$, which removes the effect of scale, may not be appropriate if the groups have covariance matrices other than the identity matrix. Suppose, for example, that each group has the same covariance matrix B. If $\boldsymbol{y}$ is a random vector with covariance matrix B, then the variability of $\boldsymbol{y}$ in the direction given by $\boldsymbol{x}$ will be $\text{var}(\boldsymbol{x}'\boldsymbol{y}) = \boldsymbol{x}'B\boldsymbol{x}$. Since differences among the groups in a direction with high variability will not be as important as similar differences in another direction with low variability, we will adjust for these differences in variability by constructing the ratio

$$\frac{\boldsymbol{x}'A\boldsymbol{x}}{\boldsymbol{x}'B\boldsymbol{x}}$$

The vector $\boldsymbol{x}_1$ that maximizes this ratio will then identify the one-dimensional subspace of R^m in which the group means differ the most, when adjusting for differences in variability. The next step after finding $\boldsymbol{x}_1$ would be to find the vector $\boldsymbol{x}_2$ that maximizes the ratio but has $\boldsymbol{x}_2'\boldsymbol{y}$ uncorrelated with $\boldsymbol{x}_1'\boldsymbol{y}$; this would be the vector $\boldsymbol{x}_2$ that maximizes the ratio in the equation subject to the constraint that $\boldsymbol{x}_1'B\boldsymbol{x}_2 = \boldsymbol{0}$. Continuing in this fashion, we would determine the m vectors $\boldsymbol{x}_1, \ldots, \boldsymbol{x}_m$ that yield the m extremal values $\lambda_1, \ldots, \lambda_m$ of the ratio. These extremal values are identified in the following theorem.

Theorem 3.25. Let A and B be $m \times m$ matrices, with A being nonnegative definite and B positive definite. For $h = 1, \ldots, m$, define $X_h = (\boldsymbol{x}_1, \ldots, \boldsymbol{x}_h)$ and $Y_h = (\boldsymbol{x}_h, \ldots, \boldsymbol{x}_m)$, where $\boldsymbol{x}_1, \ldots, \boldsymbol{x}_m$ are linearly independent eigenvectors of $B^{-1}A$ corresponding to the eigenvalues $\lambda_1(B^{-1}A) \geq \cdots \geq \lambda_m(B^{-1}A)$. Then

$$\lambda_h(B^{-1}A) = \min_{Y_{h+1}'B\boldsymbol{x}=\boldsymbol{0}} \frac{\boldsymbol{x}'A\boldsymbol{x}}{\boldsymbol{x}'B\boldsymbol{x}},$$

and

$$\lambda_h(B^{-1}A) = \max_{X_{h-1}'B\boldsymbol{x}=\boldsymbol{0}} \frac{\boldsymbol{x}'A\boldsymbol{x}}{\boldsymbol{x}'B\boldsymbol{x}},$$

where $\boldsymbol{x} = \boldsymbol{0}$ is excluded, and the min and max are over all $\boldsymbol{x} \neq \boldsymbol{0}$ when $h = m$ and $h = 1$, respectively.

Proof. We will prove the result involving the minimum; the proof for the maximum is similar. Let $B = PDP'$ be the spectral decomposition of B, so that $D = \text{diag}(d_1, \ldots, d_m)$, where the eigenvalues of B, $d_1, \ldots, d_m$, are all positive due to Theorem 3.19. If we let $T = PD^{1/2}P'$, where $D^{1/2} = \text{diag}(d_1^{1/2}, \ldots, d_m^{1/2})$, then $B = TT = T^2$ and T, like B, is symmetric and nonsingular. Putting $\boldsymbol{y} = T\boldsymbol{x}$, we find that

$$\begin{aligned}\min_{Y_{h+1}'B\boldsymbol{x}=\boldsymbol{0}} \frac{\boldsymbol{x}'A\boldsymbol{x}}{\boldsymbol{x}'B\boldsymbol{x}} &= \min_{Y_{h+1}'TT\boldsymbol{x}=\boldsymbol{0}} \frac{\boldsymbol{x}'TT^{-1}AT^{-1}T\boldsymbol{x}}{\boldsymbol{x}'TT\boldsymbol{x}} \\ &= \min_{Y_{h+1}'T\boldsymbol{y}=\boldsymbol{0}} \frac{\boldsymbol{y}'T^{-1}AT^{-1}\boldsymbol{y}}{\boldsymbol{y}'\boldsymbol{y}} \end{aligned} \tag{3.16}$$

Note that if we write $\lambda_i = \lambda_i(B^{-1}A)$, then $B^{-1}A\boldsymbol{x}_i = \lambda_i\boldsymbol{x}_i$, so that

$$T^{-1}T^{-1}A\boldsymbol{x}_i = \lambda_i\boldsymbol{x}_i,$$

which implies

$$T^{-1}AT^{-1}T\boldsymbol{x}_i = \lambda_i T\boldsymbol{x}_i$$

Thus, $T\boldsymbol{x}_i$ is an eigenvector of $T^{-1}AT^{-1}$ corresponding to the eigenvalue

$\lambda_i = \lambda_i(T^{-1}AT^{-1})$; that is, the eigenvalues of $B^{-1}A$ are the same as those of $T^{-1}AT^{-1}$. Since the rows of $Y'_{h+1}T$ are the transposes of the eigenvectors $T\boldsymbol{x}_{h+1}, \dots, T\boldsymbol{x}_m$, it follows from Theorem 3.16 that (3.16) equals $\lambda_h(T^{-1}AT^{-1})$, which we have already established as being the same as $\lambda_h(B^{-1}A)$. □

Since $\boldsymbol{x}_i$ is an eigenvector of $B^{-1}A$ corresponding to the eigenvalue $\lambda_i = \lambda_i(B^{-1}A)$, we know that

$$B^{-1}A\boldsymbol{x}_i = \lambda_i \boldsymbol{x}_i$$

or, equivalently,

$$A\boldsymbol{x}_i = \lambda_i B\boldsymbol{x}_i \tag{3.17}$$

Equation (3.17) is similar to the eigenvalue–eigenvector equation of A, except for the multiplication of $\boldsymbol{x}_i$ by B on the right-hand side of the equation. The eigenvalues satisfying (3.17) are sometimes referred to as the eigenvalues of A in the metric of B. Note that if we premultiply (3.17) by $\boldsymbol{x}_i'$ and then solve for λ_i, we get

$$\lambda_i(B^{-1}A) = \frac{\boldsymbol{x}_i' A\boldsymbol{x}_i}{\boldsymbol{x}_i' B\boldsymbol{x}_i};$$

that is, the extremal values given in Theorem 3.25 are attained at the eigenvectors of $B^{-1}A$.

The proof of the previous theorem suggests a way of simultaneously diagonalizing the matrices A and B. Since $T^{-1}AT^{-1}$ is symmetric, it can be expressed in the form $Q\Lambda Q'$, where Q is orthogonal and $\Lambda = \text{diag}(\lambda_1(T^{-1}AT^{-1}), \dots, \lambda_m(T^{-1}AT^{-1}))$. The matrix $C = Q'T^{-1}$ is nonsingular since Q and T^{-1} are nonsingular and

$$CAC' = Q'T^{-1}AT^{-1}Q = Q'Q\Lambda Q'Q = \Lambda,$$
$$CBC' = Q'T^{-1}TTT^{-1}Q = Q'Q = \text{I}_m$$

Equivalently, if $G = C^{-1}$ we have $A = G\Lambda G'$ and $B = GG'$. This simultaneous diagonalization is useful in proving our next result. For some other related results see Olkin and Tomsky (1981).

Theorem 3.26. Let A be an $m \times m$ nonnegative definite matrix and B be an $m \times m$ positive definite matrix. If F is any $m \times h$ matrix with full column rank, then for $i = 1, \dots, h$

$$\lambda_i((F'AF)(F'BF)^{-1}) \leq \lambda_i(AB^{-1}),$$

and further

$$\max_F \lambda_i((F'AF)(F'BF)^{-1}) = \lambda_i(AB^{-1})$$

Proof. Note that the second equation implies the first, so our proof simply involves the verification of the second equation. Let the nonsingular $m \times m$ matrix G be such that $B = GG'$ and $A = G\Lambda G'$, where $\Lambda = \text{diag}(\lambda_1(B^{-1}A), \dots, \lambda_m(B^{-1}A))$. Then

$$\begin{aligned}\max_F \lambda_i((F'AF)(F'BF)^{-1}) &= \max_F \lambda_i((F'G\Lambda G'F)(F'GG'F)^{-1}) \\ &= \max_E \lambda_i((E'\Lambda E)(E'E)^{-1}),\end{aligned}$$

where this last maximization is also over all $m \times h$ matrices of rank h, since $E = G'F$ must have the same rank as F. Note that since E has rank h, the $h \times h$ matrix $E'E$ is a nonsingular symmetric matrix. As was seen in the previous proof, such a matrix can be expressed as $E'E = TT$ for some nonsingular symmetric $h \times h$ matrix T. It then follows that

$$\begin{aligned}\max_E \lambda_i((E'\Lambda E)(E'E)^{-1}) &= \max_E \lambda_i((E'\Lambda E)(TT)^{-1}) \\ &= \max_E \lambda_i(T^{-1}E'\Lambda ET^{-1}),\end{aligned}$$

where this last equality follows from Theorem 3.2(d). Now if we define the $m \times h$ rank h matrix $H = ET^{-1}$, then $H'H = T^{-1}E'ET^{-1} = T^{-1}TTT^{-1} - \text{I}_h$. Thus,

$$\max_E \lambda_i(T^{-1}E'\Lambda ET^{-1}) = \max_H \lambda_i(H'\Lambda H) = \lambda_i(B^{-1}A),$$

where the final equality follows from Theorem 3.19 and the fact that equality is actually achieved with the choice of $H' = [\text{I}_h \quad (0)]$. □

Example 3.14. Many multivariate analyses are simply generalizations or extensions of corresponding univariate analyses. In this example, we begin with what is known as the univariate one-way classification model in which we have independent samples of a response y from k different populations or treatments, with a sample size of n_i from the ith population. The jth observation from the ith sample can be expressed as

$$y_{ij} = \mu_i + \epsilon_{ij},$$

where the μ_is are constants and the ϵ_{ij}s are independent and identically distributed as $N(0, \sigma^2)$. Our goal is to determine whether or not the μ_is are all the same; that is, we wish to test the null hypothesis $H_0: \mu_1 = \cdots = \mu_k$ against the alternative hypothesis H_1: at least two of the μ_is differ. An analysis of variance compares (see Problem 2.31) the variability between treatments,

$$\text{SST} = \sum_{i=1}^{k} n_i(\bar{y}_i - \bar{y})^2,$$

to the variability within treatments,

$$\text{SSE} = \sum_{i=1}^{k} \sum_{j=1}^{n_i} (y_{ij} - \bar{y}_i)^2,$$

where

$$\bar{y}_i = \sum_{j=1}^{n_i} y_{ij}/n_i, \qquad \bar{y} = \sum_{i=1}^{k} n_i \bar{y}_i/n, \qquad n = \sum_{i=1}^{k} n_i$$

SST is referred to as the sum of squares for treatment while SSE is called the sum of squares for error. The hypothesis H_0 is rejected if the statistic

$$\text{F} = \frac{\text{SST}/(k-1)}{\text{SSE}/(n-k)}$$

exceeds the appropriate quantile of the F distribution with $k-1$ and $n-k$ degrees of freedom. Now suppose that instead of obtaining the value of one response variable for each observation, we obtain the values of m different response variables for each observation. If $\boldsymbol{y}_{ij}$ is the $m \times 1$ vector of responses obtained as the jth observation from the ith treatment, then we have the multivariate one-way classification model given by

$$\boldsymbol{y}_{ij} = \boldsymbol{\mu}_i + \boldsymbol{\epsilon}_{ij},$$

where $\boldsymbol{\mu}_i$ is an $m \times 1$ vector of constants and $\boldsymbol{\epsilon}_{ij} \sim N_m(\mathbf{0}, \Omega)$, independently. Measures of the between treatment variability and within treatment variability are now given by the matrices,

$$B = \sum_{i=1}^{k} n_i(\bar{\boldsymbol{y}}_i - \bar{\boldsymbol{y}})(\bar{\boldsymbol{y}}_i - \bar{\boldsymbol{y}})', \qquad W = \sum_{i=1}^{k} \sum_{j=1}^{n_i} (\boldsymbol{y}_{ij} - \bar{\boldsymbol{y}}_i)(\boldsymbol{y}_{ij} - \bar{\boldsymbol{y}}_i)'$$

One approach to testing the null hypothesis H_0: $\boldsymbol{\mu}_1 = \cdots = \boldsymbol{\mu}_k$ against the alternative hypothesis, H_1: at least two of the $\boldsymbol{\mu}_i$s differ, is by a method called the union–intersection procedure. This technique is based on the following decomposition of the hypotheses H_0 and H_1 into univariate hypotheses. If $\boldsymbol{c}$ is any $m \times 1$ vector, and we define the hypothesis $H_0(\boldsymbol{c})$: $\boldsymbol{c}'\boldsymbol{\mu}_1 = \cdots = \boldsymbol{c}'\boldsymbol{\mu}_k$, then the intersection of $H_0(\boldsymbol{c})$ over all $\boldsymbol{c} \in R^m$ is the hypothesis H_0. In addition, if we define the hypothesis $H_1(\boldsymbol{c})$: at least two of the $\boldsymbol{c}'\boldsymbol{\mu}_i$s differ, then the union of the hypotheses $H_1(\boldsymbol{c})$ over all $\boldsymbol{c} \in R^m$ is the hypothesis H_1. Thus, we should reject the hypothesis H_0 if and only if we reject $H_0(\boldsymbol{c})$ for at least one $\boldsymbol{c}$. Now the null hypothesis $H_0(\boldsymbol{c})$ involves the univariate one-way classification model in which $\boldsymbol{c}'\boldsymbol{y}_{ij}$ is the response, and so we would reject $H_0(\boldsymbol{c})$ for large values of the F statistic

$$\mathrm{F}(\boldsymbol{c}) = \frac{\mathrm{SST}(\boldsymbol{c})/(k-1)}{\mathrm{SSE}(\boldsymbol{c})/(n-k)},$$

where $\mathrm{SST}(\boldsymbol{c})$ and $\mathrm{SSE}(\boldsymbol{c})$ are the sums of squares for treatments and errors, respectively, computed for the responses $\boldsymbol{c}'\boldsymbol{y}_{ij}$. Since H_0 is rejected if $H_0(\boldsymbol{c})$ is rejected for at least one $\boldsymbol{c}$, we will reject H_0 is $\mathrm{F}(\boldsymbol{c})$ is sufficiently large for at least one $\boldsymbol{c}$ or, equivalently, if

$$\max_{\boldsymbol{c} \neq \boldsymbol{0}} \mathrm{F}(\boldsymbol{c})$$

is sufficiently large. Omitting the constants $(k-1)$ and $(n-k)$ and noting that the sums of squares $\mathrm{SST}(\boldsymbol{c})$ and $\mathrm{SSE}(\boldsymbol{c})$ can be expressed using B and W as

$$\mathrm{SST}(\boldsymbol{c}) = \boldsymbol{c}'B\boldsymbol{c}, \qquad \mathrm{SSE}(\boldsymbol{c}) = \boldsymbol{c}'W\boldsymbol{c},$$

we find that we reject H_0 for large values of

$$\max_{\boldsymbol{c} \neq \boldsymbol{0}} \frac{\boldsymbol{c}'B\boldsymbol{c}}{\boldsymbol{c}'W\boldsymbol{c}} = \lambda_1(W^{-1}B), \tag{3.18}$$

where the right-hand side follows from Theorem 3.25. Thus, if $u_{1-\alpha}$ is the $(1-\alpha)$th quantile of the distribution of the largest eigenvalue $\lambda_1(W^{-1}B)$ [see, for example, Morrison (1990)] so that

$$P[\lambda_1(W^{-1}B) \leq u_{1-\alpha} | H_0] = 1 - \alpha, \tag{3.19}$$

then we would reject H_0 if $\lambda_1(W^{-1}B) > u_{1-\alpha}$. One advantage of the union–intersection procedure is that it naturally leads to simultaneous confidence intervals. It follows immediately from (3.18) and (3.19) that for any mean vectors $\boldsymbol{\mu}_1, \ldots, \boldsymbol{\mu}_k$, with probability $1 - \alpha$, the inequality

$$\frac{\sum_{i=1}^{k} n_i \boldsymbol{c}'\{(\bar{\boldsymbol{y}}_i - \bar{\boldsymbol{y}}) - (\boldsymbol{\mu}_i - \boldsymbol{\mu})\}\{(\bar{\boldsymbol{y}}_i - \bar{\boldsymbol{y}}) - (\boldsymbol{\mu}_i - \boldsymbol{\mu})\}'\boldsymbol{c}}{\boldsymbol{c}'W\boldsymbol{c}} \leq u_{1-\alpha}, \qquad (3.20)$$

holds for all $m \times 1$ vectors $\boldsymbol{c}$, where

$$\boldsymbol{\mu} = \sum_{i=1}^{k} n_i \boldsymbol{\mu}_i / n$$

Scheffé's method [see Scheffé (1953) or Miller (1981)] can then be used on (3.20) to yield the inequalities

$$\sum_{i=1}^{k}\sum_{j=1}^{m} a_i c_j \bar{x}_{ij} - \sqrt{u_{1-\alpha}\boldsymbol{c}'W\boldsymbol{c}\left(\sum_{i=1}^{k} a_i^2/n_i\right)}$$
$$\leq \sum_{i=1}^{k}\sum_{j=1}^{m} a_i c_j \mu_{ij}$$
$$\leq \sum_{i=1}^{k}\sum_{j=1}^{m} a_i c_j \bar{x}_{ij} + \sqrt{u_{1-\alpha}\boldsymbol{c}'W\boldsymbol{c}\left(\sum_{i=1}^{k} a_i^2/n_i\right)}$$

which hold with probability $1 - \alpha$, for all $m \times 1$ vectors $\boldsymbol{c}$ and all $k \times 1$ vectors $\boldsymbol{a}$ satisfying $\boldsymbol{a}'\boldsymbol{1}_k = 0$.

PROBLEMS

1. Consider the 3×3 matrix

$$A = \begin{bmatrix} 9 & -3 & -4 \\ 12 & -4 & -6 \\ 8 & -3 & -3 \end{bmatrix}$$

   (a) Find the eigenvalues of A.
   (b) Find a normalized eigenvector corresponding to each eigenvalue.

2. Find the eigenvalues of A', where A is the matrix given in Problem 1. Determine the eigenspaces for A' and compare these to those of A.

3. Let the 3×3 matrix A be given by

$$A = \begin{bmatrix} 1 & -2 & 0 \\ 1 & 4 & 0 \\ 0 & 0 & 2 \end{bmatrix}$$

(a) Find the eigenvalues of A.
(b) For each different value of λ, determine the associated eigenspace $S_A(\lambda)$.
(c) Describe the eigenspaces obtained in part (b).

4. If the $m \times m$ matrix A has eigenvalues $\lambda_1, \ldots, \lambda_m$ and corresponding eigenvectors $\boldsymbol{x}_1, \ldots, \boldsymbol{x}_m$, show that the matrix $(A + \gamma\text{I})$ has eigenvalues $\lambda_1 + \gamma, \ldots, \lambda_m + \gamma$ and corresponding eigenvectors $\boldsymbol{x}_1, \ldots, \boldsymbol{x}_m$.

5. In Example 3.6, we discussed the use of principal components regression as a way of overcoming the difficulties associated with multicollinearity. Another approach, called ridge regression, replaces the ordinary least squares estimator in the standardized model $\hat{\boldsymbol{\delta}}_1 = (Z_1'Z_1)^{-1}Z_1'\boldsymbol{y}$ by $\hat{\boldsymbol{\delta}}_{1\gamma} = (Z_1'Z_i + \gamma\text{I})^{-1}Z_1'\boldsymbol{y}$, where γ is a small positive number. This adjustment will reduce the impact of the near singularity of $Z_1'Z_1$ since the addition of γI increases each of the eigenvalues of $Z_1'Z_1$ by γ.
(a) Show that if $N > 2k + 1$, there is an $N \times k$ matrix W such that $\hat{\boldsymbol{\delta}}_{1\gamma}$ is the ordinary least squares estimate of $\boldsymbol{\delta}_1$ in the model

$$\boldsymbol{y} = \delta_0 \mathbf{1}_N + (Z_1 + W)\boldsymbol{\delta}_1 + \boldsymbol{\epsilon};$$

that is, $\hat{\boldsymbol{\delta}}_{1\gamma}$ can be viewed as the ordinary least squares estimator of $\boldsymbol{\delta}_1$ after we have perturbed the matrix of values for the explanatory variables Z_1 by W.
(b) Show that there exists a $k \times k$ matrix U such that $\hat{\boldsymbol{\delta}}_{1\gamma}$ is the ordinary least squares estimate of $\boldsymbol{\delta}_1$ in the model

$$\begin{bmatrix} \boldsymbol{y} \\ \mathbf{0} \end{bmatrix} = \begin{bmatrix} \delta_0 \mathbf{1}_N \\ \mathbf{0} \end{bmatrix} + \begin{bmatrix} Z_1 \\ U \end{bmatrix} \boldsymbol{\delta}_1 + \begin{bmatrix} \boldsymbol{\epsilon} \\ \boldsymbol{\epsilon}_* \end{bmatrix},$$

where $\mathbf{0}$ is a $k \times 1$ vector of zeros and $\boldsymbol{\epsilon}_* \sim N_k(\mathbf{0}, \sigma^2\text{I})$, independently of $\boldsymbol{\epsilon}$. Thus, the ridge regression estimator also can be viewed as the least squares estimator obtained after adding k observations, each having zero for the response variable and the small values in U as the values for the explanatory variables.

6. Refer to Example 3.6 and the previous exercise.
   (a) Find the expected values of the principal components regression estimator, $\hat{\boldsymbol{\delta}}_{1*}$ and the ridge regression estimator $\hat{\boldsymbol{\delta}}_{1\gamma}$, thereby showing that each is a biased estimator of $\boldsymbol{\delta}_1$.
   (b) Find the covariance matrix of $\hat{\boldsymbol{\delta}}_{1*}$ and show that $\text{var}(\hat{\boldsymbol{\delta}}_1) - \text{var}(\hat{\boldsymbol{\delta}}_{1*})$ is a nonnegative definite matrix, where $\hat{\boldsymbol{\delta}}_1$ is the ordinary least squares estimator of $\boldsymbol{\delta}_1$.
   (c) Find the covariance matrix of $\hat{\boldsymbol{\delta}}_{1\gamma}$ and show that $\text{tr}\{\text{var}(\hat{\boldsymbol{\delta}}_1) - \text{var}(\hat{\boldsymbol{\delta}}_{1\gamma})\}$ is nonnegative.

7. If A and B are $m \times m$ matrices and at least one of them is nonsingular, show that the eigenvalues of AB and BA are the same.

8. If λ is a real eigenvalue of the $m \times m$ real matrix A, show that there exist real eigenvectors of A corresponding to the eigenvalue λ.

9. Prove the results given in Theorem 3.2.

10. Suppose that λ is a simple eigenvalue of the $m \times m$ matrix A. Show that rank $(A - \lambda \mathrm{I}) = m - 1$.

11. If A is an $m \times m$ matrix and $\text{rank}(A - \lambda \mathrm{I}) = m - 1$, show that λ is an eigenvalue of A with multiplicity of at least one.

12. Consider the $m \times m$ matrix

$$A = \begin{bmatrix} 1 & 1 & 0 & \cdots & 0 \\ 0 & 1 & 1 & \cdots & 0 \\ \vdots & \vdots & \vdots & & \vdots \\ 0 & 0 & 0 & \cdots & 1 \\ 0 & 0 & 0 & 0 & 1 \end{bmatrix},$$

which has each element on and directly above the diagonal equal to 1. Find the eigenvalues and eigenvectors of A.

13. Let A be an $m \times m$ nonsingular matrix with eigenvalues $\lambda_1, \dots, \lambda_m$ and corresponding eigenvectors $\boldsymbol{x}_1, \dots, \boldsymbol{x}_m$. If $\mathrm{I} + A$ is nonsingular, find the eigenvalues and eigenvectors of
   (a) $(\mathrm{I} + A)^{-1}$,
   (b) $A + A^{-1}$,
   (c) $(\mathrm{I} + A^{-1})$.

14. Let the $m \times m$ nonsingular matrix A be such that $\mathrm{I} + A$ is nonsingular, and define

$$B = (\mathrm{I} + A)^{-1} + (\mathrm{I} + A^{-1})^{-1}$$

(a) Show that if $\boldsymbol{x}$ is an eigenvector of A corresponding to the eigenvalue λ, then $\boldsymbol{x}$ is an eigenvector of B corresponding to the eigenvalue 1.
(b) Use Theorem 1.7 to show that $B = \mathrm{I}$.

15. Consider the 2×2 matrix

$$A = \begin{bmatrix} 4 & 2 \\ 3 & 5 \end{bmatrix}$$

(a) Find the characteristic equation of A.
(b) Illustrate Theorem 3.7 by substituting A for λ in the characteristic equation obtained in (a) and then showing that the resulting matrix is the null matrix.
(c) Rearrange the matrix polynomial equation in (b) to obtain an expression for A^2 as a linear combination of A and I.
(d) In a similar fashion, write A^3 and A^{-1} as linear combinations of A and I.

16. Consider the general 2×2 matrix

$$A = \begin{bmatrix} a_{11} & a_{12} \\ a_{21} & a_{22} \end{bmatrix}$$

(a) Find the characteristic equation of A.
(b) Obtain expressions for the two eigenvalues of A in terms of the elements of A.
(c) When will these eigenvalues be real?

17. Find the eigenvalues and eigenvectors of the matrix $\mathbf{1}_m\mathbf{1}_m'$.

18. Consider the $m \times m$ matrix $A = \alpha \mathrm{I}_m + \beta \mathbf{1}_m\mathbf{1}_m'$, where α and β are scalars.
(a) Find the eigenvalues and eigenvectors of A.
(b) Determine the eigenspaces and associated eigenprojections of A.
(c) For which values of α and β will A be nonsingular?
(d) Using (a), show that when A is nonsingular, then

$$A^{-1} = \alpha^{-1}\mathrm{I}_m - \frac{\beta}{\alpha(\alpha + m\beta)} \mathbf{1}_m \mathbf{1}_m'$$

(e) Show that the determinant of A is $\alpha^{m-1}(\alpha + m\beta)$.

19. Consider the $m \times m$ matrix $A = \alpha\mathrm{I}_m + \beta\boldsymbol{cc}'$, where α and β are scalars and $\boldsymbol{c} \neq \mathbf{0}$ is an $m \times 1$ vector.
    (a) Find the eigenvalues and eigenvectors of A.
    (b) Find the determinant of A.
    (c) Give conditions for A to be nonsingular and find an expression for the inverse of A.

20. Let A be the 3×3 matrix given by

$$A = \begin{bmatrix} 2 & -1 & 0 \\ -1 & 1 & 1 \\ 0 & 1 & 2 \end{bmatrix}$$

    (a) Find the eigenvalues and associated normalized eigenvectors of A.
    (b) What is the rank of A?
    (c) Find the eigenspaces and associated eigenprojections of A.

21. Construct a 3×3 symmetric matrix having eigenvalues 18, 21, and 28, and corresponding eigenvectors $(1, 1, 2)'$, $(4, -2, -1)'$, and $(1, 3, -2)'$.

22. Show that if A is an $m \times m$ symmetric matrix with eigenvalues $\lambda_1, \dots, \lambda_m$, then

$$\sum_{i=1}^{m} \sum_{j=1}^{m} a_{ij}^2 = \sum_{i=1}^{m} \lambda_i^2$$

23. Show that if A is an $m \times m$ symmetric matrix with its eigenvalues equal to its diagonal elements, then A must be a diagonal matrix.

24. Use Theorem 3.17 to prove Theorem 3.18. Show that the converse is not true; that is, find symmetric matrices A and B for which $\lambda_i(A) \geq \lambda_i(B)$ for $i = 1, \dots, m$, yet $A - B$ is not nonnegative definite.

25. Let A be an $m \times n$ matrix with $\operatorname{rank}(A) = r$. Use the spectral decomposition of $A'A$ to show that there exists an $n \times (n - r)$ matrix X such that

$$AX = (0) \quad \text{and} \quad X'X = \mathrm{I}_{n-r}$$

In a similar fashion, show that there exists an $(m - r) \times m$ matrix Y such that

$$YA = (0) \quad \text{and} \quad YY' = \mathrm{I}_{m-r}$$

26. Let A be the 2×3 matrix given by

$$A = \begin{bmatrix} 6 & 4 & 4 \\ 3 & 2 & 2 \end{bmatrix}$$

Find matrices X and Y satisfying the conditions given in the previous exercise.

27. An $m \times m$ matrix A is said to be nilpotent if $A^k = (0)$ for some positive integer k.
    (a) Show that all of the eigenvalues of a nilpotent matrix are equal to 0.
    (b) Find a matrix, other than the null matrix, that is nilpotent.

28. Complete the details of Example 3.10 by showing that

$$P_{1,n} \to \begin{bmatrix} 1 & 0 & 0 \\ 0 & 0 & 0 \\ 0 & 0 & 0 \end{bmatrix}, \qquad P_{2,n} \to \begin{bmatrix} 0 & 0 & 0 \\ 0 & 1 & 0 \\ 0 & 0 & 1 \end{bmatrix}$$

29. Let A and B be $m \times m$ symmetric matrices. Show that

$$\lambda_1(A + B) \le \lambda_1(A) + \lambda_1(B),$$
$$\lambda_m(A + B) \ge \lambda_m(A) + \lambda_m(B)$$

30. Prove Theorem 3.20.

31. Our proof of Theorem 3.24 utilized (3.9) of Theorem 3.17. Obtain an alternative proof of Theorem 3.24 by using (3.10) of Theorem 3.17.

32. Let A be an $m \times m$ nonnegative definite matrix and B be an $m \times m$ positive definite matrix. If F is any $m \times h$ matrix with full column rank, then show the following:
    (a) $\lambda_{h-i+1}((F'AF)(F'BF)^{-1}) \ge \lambda_{m-i+1}(AB^{-1})$, for $i = 1, \ldots, h$.
    (b) $\min_F \lambda_1((F'AF)(F'BF)^{-1}) = \lambda_{m-h+1}(AB^{-1})$.
    (c) $\min_F \lambda_h((F'AF)(F'BF)^{-1}) = \lambda_m(AB^{-1})$.

33. Suppose A is an $m \times m$ matrix with eigenvalues $\lambda_1, \dots, \lambda_m$ and associated eigenvectors $\boldsymbol{x}_1, \dots, \boldsymbol{x}_m$, while B is $n \times n$ with eigenvalues $\gamma_1, \dots, \gamma_n$ and eigenvectors $\boldsymbol{y}_1, \dots, \boldsymbol{y}_n$. What are the eigenvalues and eigenvectors of the $(m+n) \times (m+n)$ matrix

$$C = \begin{bmatrix} A & (0) \\ (0) & B \end{bmatrix}?$$

Generalize this result by giving the eigenvalues and eigenvectors of the matrix

$$C = \begin{bmatrix} C_1 & (0) & \cdots & (0) \\ (0) & C_2 & \cdots & (0) \\ \vdots & \vdots & & \vdots \\ (0) & (0) & \cdots & C_r \end{bmatrix}$$

in terms of the eigenvalues and eigenvectors of the matrices $C_1, \dots, C_r$.

34. Let

$$T = \begin{bmatrix} 1 & -1 & 2 \\ 2 & 1 & 1 \end{bmatrix}$$

(a) Find the eigenvalues and corresponding eigenvectors of TT'.

(b) Find the eigenvalues and corresponding eigenvectors of $T'T$.

35. Show that if A is a nonnegative definite matrix and $a_{ii} = 0$ for some i, then $a_{ij} = a_{ji} = 0$ for all j.

36. Suppose that A is an $m \times m$ symmetric matrix with eigenvalues $\lambda_1, \dots, \lambda_m$ and associated eigenvectors $\boldsymbol{x}_1, \dots, \boldsymbol{x}_m$, while B is an $m \times m$ symmetric matrix with eigenvalues $\gamma_1, \dots, \gamma_m$ and associated eigenvectors $\boldsymbol{x}_1, \dots, \boldsymbol{x}_m$; that is, A and B have common eigenvectors.

(a) Find the eigenvalues and eigenvectors of $A + B$.

(b) Find the eigenvalues and eigenvectors of AB.

(c) Show that $AB = BA$.

37. Suppose that $\boldsymbol{x}_1, \dots, \boldsymbol{x}_r$ is a set of orthonormal eigenvectors corresponding to the r largest eigenvalues $\gamma_1, \dots, \gamma_r$ of the $m \times m$ symmetric matrix A and assume that $\gamma_r > \gamma_{r+1}$. Let P be the total eigenprojection of A associated

with the eigenvalues $\gamma_1, \ldots, \gamma_r$; that is,

$$P = \sum_{i=1}^{r} \boldsymbol{x}_i \boldsymbol{x}_i'$$

Let B be another $m \times m$ symmetric matrix with its r largest eigenvalues given by $\mu_1, \ldots, \mu_r$, where $\mu_r > \mu_{r+1}$, and a corresponding set of orthonormal eigenvectors given by $\boldsymbol{y}_1, \ldots, \boldsymbol{y}_r$. Let Q be the total eigenprojection of B associated with the eigenvalues $\mu_1, \ldots, \mu_r$ so that

$$Q = \sum_{i=1}^{r} \boldsymbol{y}_i \boldsymbol{y}_i'$$

(a) Show that $P = Q$ if and only if

$$\sum_{i=1}^{r} \{\gamma_i + \mu_i - \lambda_i(A + B)\} = 0$$

(b) Let $X = (\boldsymbol{x}_1, \ldots, \boldsymbol{x}_m)$, where $\boldsymbol{x}_{r+1}, \ldots, \boldsymbol{x}_m$ is a set of orthonormal eigenvectors corresponding to the smallest $m - r$ eigenvalues of A. Show that if $P = Q$, then $X'BX$ has the block diagonal form

$$\begin{bmatrix} U & (0) \\ (0) & V \end{bmatrix},$$

where U is $r \times r$ and V is $(m - r) \times (m - r)$. Show that the converse is not true.

38. Let $\lambda_1 \geq \cdots \geq \lambda_m$ be the eigenvalues of the $m \times m$ symmetric matrix A and let $\boldsymbol{x}_1, \ldots, \boldsymbol{x}_m$ be a set of corresponding orthonormal eigenvectors. For some k, define the total eigenprojection associated with the eigenvalues $\lambda_k, \ldots, \lambda_m$ as

$$P = \sum_{i=k}^{m} \boldsymbol{x}_i \boldsymbol{x}_i'$$

Show that $\lambda_k = \cdots = \lambda_m = \lambda$ if and only if

$$P(A - \lambda \mathrm{I})P = (0)$$

39. Let $A_1, \ldots, A_k$ be $m \times m$ symmetric matrices and let τ_i be one of the eigenvalues of A_i. Let $\boldsymbol{x}_1, \ldots, \boldsymbol{x}_r$ be a set of orthonormal $m \times 1$ vectors, and define

$$P = \sum_{i=1}^{r} \boldsymbol{x}_i \boldsymbol{x}_i'$$

Show that if each of the eigenvalues τ_i has multiplicity r and has $\boldsymbol{x}_1, \ldots, \boldsymbol{x}_r$ as associated eigenvectors, then

$$P \left\{ \sum_{i=1}^{k} (A_i - \tau_i \mathrm{I})^2 \right\} P = (0)$$

CHAPTER FOUR

Matrix Factorizations and Matrix Norms

1. INTRODUCTION

In this chapter, we take a look at some useful ways of expressing a given matrix A in the form of a product of other matrices having some special structure or canonical form. In many applications such a decomposition of A may reveal to us the key features of A that are of interest to us. These factorizations are particularly useful in multivariate distribution theory in that they can expedite the mathematical development and often simplify the generalization of results from a special case to a more general situation. Our focus here will be on conditions for the existence of these factorizations as well as mathematical properties and consequences of the factorizations. Details on the numerical computation of the component matrices in these factorizations can be found in texts on numerical methods. Some useful references are Golub and Van Loan (1989) and Press, Flannery, Teukolsky, and Vetterling (1992).

2. THE SINGULAR VALUE DECOMPOSITION

The first factorization that we consider, the singular value decomposition, could be described as the most useful because this is a factorization for a matrix of any size; the subsequent decompositions will only apply to square matrices. We will find this decomposition particularly useful in the next chapter when we generalize the concept of an inverse of a nonsingular square matrix to any matrix.

Theorem 4.1. If A is an $m \times n$ matrix of rank $r > 0$, there exist orthogonal $m \times m$ and $n \times n$ matrices P and Q, such that $A = PDQ'$ and $D = P'AQ$, where the $m \times n$ matrix D is given by

$$\text{(a) } \Delta \quad \text{if } r = m = n, \qquad \text{(b) } [\Delta \quad (0)] \quad \text{if } r = m < n,$$

$$\text{(c) } \begin{bmatrix} \Delta \\ (0) \end{bmatrix} \quad \text{if } r = n < m, \qquad \text{(d) } \begin{bmatrix} \Delta & (0) \\ (0) & (0) \end{bmatrix} \quad \text{if } r < m, r < n,$$

and Δ is an $r \times r$ diagonal matrix with positive diagonal elements. The diagonal elements of Δ^2 are the positive eigenvalues of $A'A$ and AA'.

Proof. We will prove the result for the case $r < m$ and $r < n$. The proofs of (a)–(c) only require notational changes. Let Δ^2 be the $r \times r$ diagonal matrix whose diagonal elements are the r positive eigenvalues of $A'A$ which are identical to the positive eigenvalues of AA' by Theorem 3.23. Define Δ to be the diagonal matrix whose diagonal elements are the positive square roots of the corresponding diagonal elements of Δ^2. Since $A'A$ is an $n \times n$ symmetric matrix, we can find an $n \times n$ orthogonal matrix Q such that

$$Q'A'AQ = \begin{bmatrix} \Delta^2 & (0) \\ (0) & (0) \end{bmatrix}$$

Partitioning Q as $Q = [Q_1 \quad Q_2]$, where Q_1 is $n \times r$, the identity above implies that

$$Q_1'A'AQ_1 = \Delta^2, \tag{4.1}$$

and

$$Q_2'A'AQ_2 = (0) \tag{4.2}$$

Note that from (4.2) it follows that

$$AQ_2 = (0) \tag{4.3}$$

Now let $P = [P_1 \quad P_2]$ be an $m \times m$ orthogonal matrix, where the $m \times r$ matrix $P_1 = AQ_1\Delta^{-1}$ and the $m \times (m - r)$ matrix P_2 is any matrix which makes P orthogonal. Consequently, we must have $P_2'P_1 = P_2'AQ_1\Delta^{-1} = (0)$ or, equivalently,

$$P_2'AQ_1 = (0) \tag{4.4}$$

By using (4.1), (4.3), and (4.4), we find that

$$P'AQ = \begin{bmatrix} P_1'AQ_1 & P_1'AQ_2 \\ P_2'AQ_1 & P_2'AQ_2 \end{bmatrix} = \begin{bmatrix} \Delta^{-1}Q_1'A'AQ_1 & \Delta^{-1}Q_1'A'AQ_2 \\ P_2'AQ_1 & P_2'AQ_2 \end{bmatrix}$$

$$= \begin{bmatrix} \Delta^{-1}\Delta^2 & \Delta^{-1}Q_1'A'(0) \\ (0) & P_2'(0) \end{bmatrix} = \begin{bmatrix} \Delta & (0) \\ (0) & (0) \end{bmatrix} \qquad \square$$

The diagonal elements of Δ, that is, the positive square roots of the positive eigenvalues of $A'A$ and AA', are called the singular values of A. It is obvious from the proof of Theorem 4.1 that the columns of Q form an orthonormal set of eigenvectors of $A'A$ and so $A'A = QD'DQ'$. It is important to note also that the columns of P form an orthonormal set of eigenvectors of AA' since $AA' = PDQ'QD'P' = PDD'P'$.

If we again partition P and Q as $P = [P_1 \quad P_2]$ and $Q = [Q_1 \quad Q_2]$, where P_1 is $m \times r$ and Q_1 is $n \times r$, then the singular value decomposition can be restated as follows.

Corollary 4.1.1. If A is an $m \times n$ matrix of rank $r > 0$, then there exist $m \times r$ and $n \times r$ matrices P_1 and Q_1, such that $P_1'P_1 = Q_1'Q_1 = I_r$ and $A = P_1 \Delta Q_1'$, where Δ is an $r \times r$ diagonal matrix with positive diagonal elements.

Quite a bit of information about the structure of a matrix A can be obtained from its singular value decomposition. The number of singular values gives the rank of A, while the columns of P_1 and Q_1 are orthonormal bases for the column space and row space of A, respectively. Similarly, the columns of P_2 span the null space of A' and the columns of Q_2 span the null space of A.

Theorem 4.1 and Corollary 4.1.1 are related to Theorem 1.9 and its corollary, Corollary 1.9.1, which were stated as consequences of the properties of elementary transformations. It is easily verified that Theorem 1.9 and Corollary 1.9.1 also follow directly from Theorem 4.1 and Corollary 4.1.1.

Example 4.1. We will find a singular value decomposition for the 4×3 matrix

$$A = \begin{bmatrix} 2 & 0 & 1 \\ 3 & -1 & 1 \\ -2 & 4 & 1 \\ 1 & 1 & 1 \end{bmatrix}$$

First an eigenanalysis of the matrix

$$A'A = \begin{bmatrix} 18 & -10 & 4 \\ -10 & 18 & 4 \\ 4 & 4 & 4 \end{bmatrix}$$

reveals that it has eigenvalues 28, 12, and 0 with associated normalized eigenvectors $(1/\sqrt{2}, -1/\sqrt{2}, 0)'$, $(1/\sqrt{3}, 1/\sqrt{3}, 1/\sqrt{3})'$, and $(1/\sqrt{6}, 1/\sqrt{6}, -2/\sqrt{6})'$, respectively. Let these be the columns of the 3×3 orthogonal matrix Q. Clearly, $\text{rank}(A) = 2$ and the two singular values of A are $\sqrt{28}$ and $\sqrt{12}$. Thus, the 4×2 matrix P_1 is given by

$$P_1 = AQ_1\Delta^{-1} = \begin{bmatrix} 2 & 0 & 1 \\ 3 & -1 & 1 \\ -2 & 4 & 1 \\ 1 & 1 & 1 \end{bmatrix} \begin{bmatrix} 1/\sqrt{2} & 1/\sqrt{3} \\ -1/\sqrt{2} & 1/\sqrt{3} \\ 0 & 1/\sqrt{3} \end{bmatrix} \begin{bmatrix} 1/\sqrt{28} & 0 \\ 0 & 1/\sqrt{12} \end{bmatrix}$$

$$= \begin{bmatrix} 1/\sqrt{14} & 1/2 \\ 2/\sqrt{14} & 1/2 \\ -3/\sqrt{14} & 1/2 \\ 0 & 1/2 \end{bmatrix}$$

The 4×2 matrix P_2 can be any matrix satisfying $P_1'P_2 = (0)$ and $P_2'P_2 = I_2$; for instance, we can take $(1/\sqrt{12}, 1/\sqrt{12}, 1/\sqrt{12}, -3/\sqrt{12})'$ and $(-5/\sqrt{42}, 4/\sqrt{42}, 1/\sqrt{42}, 0)'$ as the columns of P_2. Then our singular value decomposition of A is given by

$$\begin{bmatrix} 1/\sqrt{14} & 1/2 & 1/\sqrt{12} & -5/\sqrt{42} \\ 2/\sqrt{14} & 1/2 & 1/\sqrt{12} & 4/\sqrt{42} \\ -3/\sqrt{14} & 1/2 & 1/\sqrt{12} & 1/\sqrt{42} \\ 0 & 1/2 & -3/\sqrt{12} & 0 \end{bmatrix} \begin{bmatrix} \sqrt{28} & 0 & 0 \\ 0 & \sqrt{12} & 0 \\ 0 & 0 & 0 \\ 0 & 0 & 0 \end{bmatrix}$$

$$\cdot \begin{bmatrix} 1/\sqrt{2} & -1/\sqrt{2} & 0 \\ 1/\sqrt{3} & 1/\sqrt{3} & 1/\sqrt{3} \\ 1/\sqrt{6} & 1/\sqrt{6} & -2/\sqrt{6} \end{bmatrix},$$

or in the form of Corollary 4.1.1,

$$\begin{bmatrix} 1/\sqrt{14} & 1/2 \\ 2/\sqrt{14} & 1/2 \\ -3/\sqrt{14} & 1/2 \\ 0 & 1/2 \end{bmatrix} \begin{bmatrix} \sqrt{28} & 0 \\ 0 & \sqrt{12} \end{bmatrix} \begin{bmatrix} 1/\sqrt{2} & -1/\sqrt{2} & 0 \\ 1/\sqrt{3} & 1/\sqrt{3} & 1/\sqrt{3} \end{bmatrix}$$

Alternatively, we could have determined the matrix P by using the fact that its columns are eigenvectors of the matrix

$$AA' = \begin{bmatrix} 5 & 7 & -3 & 3 \\ 7 & 11 & -9 & 3 \\ -3 & -9 & 21 & 3 \\ 3 & 3 & 3 & 3 \end{bmatrix}$$

However, when constructing P this way, one will have to check the decomposition $A = P_1\Delta Q_1'$ to determine the correct sign for each of the columns of P_1.

The singular value decomposition of a vector is very easy to construct. We illustrate this in the next example.

Example 4.2. Let $\boldsymbol{x}$ be an $m \times 1$ nonnull vector. Its singular value decomposition will be of the form

$$\boldsymbol{x} = P\boldsymbol{d}q,$$

where P is an $m \times m$ orthogonal matrix, $\boldsymbol{d}$ is an $m \times 1$ vector having only its first component nonzero, and q is a scalar satisfying $q^2 = 1$. The single singular value of $\boldsymbol{x}$ is given by $\lambda^{1/2}$, where $\lambda = \boldsymbol{x}'\boldsymbol{x}$. If we define $\boldsymbol{x}_* = \lambda^{-1/2}\boldsymbol{x}$, note that $\boldsymbol{x}_*'\boldsymbol{x}_* = 1$, and

$$\boldsymbol{x}\boldsymbol{x}'\boldsymbol{x}_* = \boldsymbol{x}\boldsymbol{x}'(\lambda^{-1/2}\boldsymbol{x}) = (\lambda^{-1/2}\boldsymbol{x})\boldsymbol{x}'\boldsymbol{x} = \lambda\boldsymbol{x}_*$$

so that $\boldsymbol{x}_*$ is a normalized eigenvector of $\boldsymbol{x}\boldsymbol{x}'$ corresponding to its single positive eigenvalue λ. Any vector orthogonal to $\boldsymbol{x}_*$ is an eigenvector of $\boldsymbol{x}\boldsymbol{x}'$ corresponding to the repeated eigenvalue 0. Thus, if we let $\boldsymbol{d} = (\lambda^{1/2}, 0, \dots, 0)'$, $q = 1$, and $P = [\boldsymbol{x}_*, \boldsymbol{p}_2, \dots, \boldsymbol{p}_m]$ be any orthogonal matrix with $\boldsymbol{x}_*$ as its first column, then

$$P\boldsymbol{d}q = [\boldsymbol{x}_*, \boldsymbol{p}_2, \dots, \boldsymbol{p}_m] \begin{bmatrix} \lambda^{1/2} \\ 0 \\ \vdots \\ 0 \end{bmatrix} 1 = \lambda^{1/2}\boldsymbol{x}_* = \boldsymbol{x}$$

as is required.

When A is $m \times m$ and symmetric, the singular values of A are directly related to the eigenvalues of A. This follows from the fact that $AA' = A^2$, and the eigenvalues of A^2 are the squares of the eigenvalues of A. Thus, the singular values of A will be given by the absolute values of the eigenvalues of A. If we let the columns of P be a set of orthonormal eigenvectors of A, then the Q matrix in Theorem 4.1 will be identical to P except that any column of Q that is associated with a negative eigenvalue will be -1 times the corresponding column of P. If A is nonnegative definite, then the singular values of A will be the same as the positive eigenvalues of A and, in fact, the singular value decomposition of A is simply the spectral decomposition of A discussed in the next section. This nice relationship between the eigenvalues and singular values of a symmetric matrix does not carry over to general square matrices.

Example 4.3. Consider the 2×2 matrix

$$A = \begin{bmatrix} 6 & 6 \\ -1 & 1 \end{bmatrix},$$

which has

$$AA' = \begin{bmatrix} 72 & 0 \\ 0 & 2 \end{bmatrix}, \quad A'A = \begin{bmatrix} 37 & 35 \\ 35 & 37 \end{bmatrix}$$

Clearly, the singular values of A are $\sqrt{72} = 6\sqrt{2}$ and $\sqrt{2}$. Normalized eigenvectors corresponding to 72 and 2 are $(1,0)'$ and $(0,1)'$ for AA', while $A'A$ has $(1/\sqrt{2}, 1/\sqrt{2})'$ and $(-1/\sqrt{2}, 1/\sqrt{2})'$. Thus, the singular value decomposition of A can be written as

$$\begin{bmatrix} 1 & 0 \\ 0 & 1 \end{bmatrix} \begin{bmatrix} 6\sqrt{2} & 0 \\ 0 & \sqrt{2} \end{bmatrix} \begin{bmatrix} 1/\sqrt{2} & 1/\sqrt{2} \\ -1/\sqrt{2} & 1/\sqrt{2} \end{bmatrix}$$

On the other hand, an eigenanalysis of A yields the eigenvalues 4 and 3. Associated normalized eigenvectors are $(3/\sqrt{10}, -1/\sqrt{10})'$ and $(2/\sqrt{5}, -1/\sqrt{5})'$.

We end this section with an example which illustrates an application of the singular value decomposition to least squares regression. For more discussion of this and other uses of the singular value decomposition in statistics, the reader is referred to Mandel (1982), Eubank and Webster (1985), and Nelder (1985).

Example 4.4. In this example, we will take a closer look at the multicollinearity problem, which we first discussed in Example 3.6. Suppose we have the standardized regression model

$$\boldsymbol{y} = \delta_0 \boldsymbol{1}_N + Z_1 \boldsymbol{\delta}_1 + \boldsymbol{\epsilon}$$

We have seen in Example 2.15 that the least squares estimator of δ_0 is $\bar{y}$. The fitted model $\hat{\boldsymbol{y}} = \bar{y}\boldsymbol{1}_N + Z_1\hat{\boldsymbol{\delta}}_1$ gives points on a hyperplane in R^{k+1}, where the $(k + 1)$ axes correspond to the k standardized explanatory variables and the fitted response variable. Now let $Z_1 = VDU'$ be the singular value decomposition of the $N \times k$ matrix Z_1. Thus, V is an $N \times N$ orthogonal matrix, U is a $k \times k$ orthogonal matrix, and D is an $N \times k$ matrix that has the square roots of the eigenvalues of $Z_1'Z_1$ as its diagonal elements and zeros elsewhere. We can rewrite the model $\boldsymbol{y} = \delta_0 \boldsymbol{1}_N + Z_1 \boldsymbol{\delta}_1 + \boldsymbol{\epsilon}$ as we did in Example 2.15 by defining $\alpha_0 = \delta_0$, $\boldsymbol{\alpha}_1 = U'\boldsymbol{\delta}_1$, and $W_1 = VD$, so that $\boldsymbol{y} = \alpha_0 \boldsymbol{1}_N + W_1\boldsymbol{\alpha}_1 + \boldsymbol{\epsilon}$. Suppose that exactly r of the diagonal elements of D, specifically the last r diagonal elements, are zeros, and so by partitioning U, V, and D appropriately, we get $Z_1 = V_1 D_1 U_1'$, where D_1 is a $(k - r) \times (k - r)$ diagonal matrix. This means that the row space of Z_1 is a $(k - r)$-dimensional subspace of R^k, and this subspace is spanned by the columns of U_1; that is, the points on the fitted regression hyperplane described above, when projected onto the k-dimensional standardized explanatory variable space, are actually confined to a $(k-r)$-dimensional subspace. Also, the model $\boldsymbol{y} = \alpha_0 \mathbf{1}_N + W_1\boldsymbol{\alpha}_1 + \boldsymbol{\epsilon}$ simplifies to

$$\boldsymbol{y} = \alpha_0 \mathbf{1}_N + W_{11}\boldsymbol{\alpha}_{11} + \boldsymbol{\epsilon}, \tag{4.5}$$

where $W_{11} = V_1 D_1$ and $\boldsymbol{\alpha}_{11} = U_1'\boldsymbol{\delta}_1$, and the least squares estimator of the $(k-r) \times 1$ vector $\boldsymbol{\alpha}_{11}$ is given by $\hat{\boldsymbol{\alpha}}_{11} = (W_{11}'W_{11})^{-1}W_{11}'\boldsymbol{y} = D_1^{-1}V_1'\boldsymbol{y}$. This can be used to find a least squares estimator of $\boldsymbol{\delta}_1$ since we must have $\hat{\boldsymbol{\alpha}}_{11} = U_1'\hat{\boldsymbol{\delta}}_1$. Partitioning $\hat{\boldsymbol{\delta}}_1 = (\hat{\boldsymbol{\delta}}_{11}', \hat{\boldsymbol{\delta}}_{12}')'$ and $U_1' = (U_{11}', U_{12}')$, where $\hat{\boldsymbol{\delta}}_{11}$ is $(k-r) \times 1$, we obtain the relationship

$$\hat{\boldsymbol{\alpha}}_{11} = U_{11}'\hat{\boldsymbol{\delta}}_{11} + U_{12}'\hat{\boldsymbol{\delta}}_{12}$$

Premultiplying this by $U_{11}'^{-1}$ (if U_{11} is not nonsingular, then $\boldsymbol{\delta}_1$ and U_1 can be rearranged so that it is), we find that

$$\hat{\boldsymbol{\delta}}_{11} = U_{11}'^{-1}\hat{\boldsymbol{\alpha}}_{11} - U_{11}'^{-1}U_{12}'\hat{\boldsymbol{\delta}}_{12};$$

that is, the least squares estimator of $\boldsymbol{\delta}_1$ is not unique since $\hat{\boldsymbol{\delta}}_1 = (\hat{\boldsymbol{\delta}}_{11}', \hat{\boldsymbol{\delta}}_{12}')'$ is a least squares estimator for any choice of $\hat{\boldsymbol{\delta}}_{12}$, as long as $\hat{\boldsymbol{\delta}}_{11}$ satisfies the identity given. Now suppose that we wish to estimate the response variable y corresponding to an observation that has the standardized explanatory variables at the values given in the $k \times 1$ vector $\boldsymbol{z}$. Using a least squares estimate $\hat{\boldsymbol{\delta}}_1$ we obtain the estimate $\hat{y} = \bar{y} + \boldsymbol{z}'\hat{\boldsymbol{\delta}}_1$. This estimated response, like $\hat{\boldsymbol{\delta}}_1$, may not be unique since, if we partition $\boldsymbol{z}$ as $\boldsymbol{z}' = (\boldsymbol{z}_1', \boldsymbol{z}_2')$ with $\boldsymbol{z}_1$ $(k-r) \times 1$,

$$\begin{aligned}\hat{y} &= \bar{y} + \boldsymbol{z}'\hat{\boldsymbol{\delta}}_1 = \bar{y} + \boldsymbol{z}_1'\hat{\boldsymbol{\delta}}_{11} + \boldsymbol{z}_2'\hat{\boldsymbol{\delta}}_{12} \\ &= \bar{y} + \boldsymbol{z}_1'U_{11}'^{-1}\hat{\boldsymbol{\alpha}}_{11} + (\boldsymbol{z}_2' - \boldsymbol{z}_1'U_{11}'^{-1}U_{12}')\hat{\boldsymbol{\delta}}_{12}\end{aligned}$$

Thus, $\hat{y}$ does not depend on the arbitrary $\hat{\boldsymbol{\delta}}_{12}$ and is therefore unique, only if

$$(\boldsymbol{z}_2' - \boldsymbol{z}_1'U_{11}'^{-1}U_{12}') = \mathbf{0}', \tag{4.6}$$

in which case the unique estimated value is given by $\hat{y} = \bar{y} + \boldsymbol{z}_1'U_{11}'^{-1}\hat{\boldsymbol{\alpha}}_{11}$. It is easily shown that the set of all vectors $\boldsymbol{z} = (\boldsymbol{z}_1', \boldsymbol{z}_2')'$ satisfying (4.6) is simply the column space of U_1. Thus, $y = \delta_0 + \boldsymbol{z}'\boldsymbol{\delta}_1$ is uniquely estimated only if the vector of standardized explanatory variables $\boldsymbol{z}$ falls within the space spanned by the collection of all vectors of standardized explanatory variables available to compute $\hat{\boldsymbol{\delta}}_1$.

In the typical multicollinearity problem, Z_1 is full rank so that the matrix D has no zero diagonal elements but instead has r of its diagonal elements very small relative to the others. In this case, the row space of Z_1 is all of R^k, but the points corresponding to the rows of Z_1 all lie very close to a $(k-r)$-dimensional subspace S of R^k, specifically, the space spanned by the columns of U_1. Small changes in the values of the response variables corresponding to these points can substantially alter the position of the fitted regression hyperplane $\hat{y} = \bar{y} + z'\hat{\boldsymbol{\delta}}_1$ for vectors z lying outside of and, in particular, far from S. For instance, if $k = 2$ and $r = 1$, the points corresponding to the rows of Z_1 all lie very close to S, which in this case is a line in the z_1, z_2 plane, and $\hat{y} = \bar{y} + z'\hat{\boldsymbol{\delta}}_1$ will be given by a plane in R^3 extended over the z_1, z_2 plane. The fitted regression plane $\hat{y} = \bar{y} + z'\hat{\boldsymbol{\delta}}_1$ can be identified by the line formed as the intersection of this plane and the plane perpendicular to the z_1, z_2 plane and passing through the line S, along with the tilt of the fitted regression plane. Small changes in the values of the response variables will produce small changes in both the location of this line of intersection and the tilt of the plane. However, even a slight change in the tilt of the regression plane will yield large changes on the surface of this plane for vectors z far from S. The adverse effect of this tilting can be eliminated by the use of principal components regression. As we saw in Example 3.6, principal components regression utilizes the regression model (4.5), and so an estimated response will be given by $\hat{y} = \bar{y} + z'U_1D_1^{-1}V_1'\boldsymbol{y}$. Since this regression model technically holds only for $z \in S$, by using this model for $z \in S$ we will introduce bias into our estimate of y. The advantage of principal components regression is that this may be compensated for by a large enough reduction in the variance of our estimate so as to reduce the mean squared error (see Problem 4.9). However, it should be apparent that the predicted values of y obtained from both ordinary least squares regression and principal components regression will be poor if the vector z is far from S.

3. THE SPECTRAL DECOMPOSITION AND SQUARE ROOT MATRICES OF A SYMMETRIC MATRIX

The spectral decomposition of a symmetric matrix, briefly discussed in the previous chapter, is nothing more than a special case of the singular value decomposition. We summarize this result in the following theorem.

Theorem 4.2. Let A be an $m \times m$ symmetric matrix with eigenvalues $\lambda_1, \dots, \lambda_m$ and suppose that $\boldsymbol{x}_1, \dots, \boldsymbol{x}_m$ is a set of orthonormal eigenvectors corresponding to these eigenvalues. Then, if $\Lambda = \text{diag}(\lambda_1, \dots, \lambda_m)$ and $X = (\boldsymbol{x}_1, \dots, \boldsymbol{x}_m)$, it follows that

$$A = X\Lambda X'$$

We can use the spectral decomposition of a nonnegative definite matrix A to find a square root matrix of A; that is, we wish to find an $m \times m$ matrix $A^{1/2}$ for which $A = A^{1/2}A^{1/2}$. If Λ and X are defined as in the theorem above, and we let $\Lambda^{1/2} = \text{diag}(\lambda_1^{1/2}, \ldots, \lambda_m^{1/2})$ and $A^{1/2} = X\Lambda^{1/2}X'$, then since $X'X = \text{I}$,

$$A^{1/2}A^{1/2} = X\Lambda^{1/2}X'X\Lambda^{1/2}X' = X\Lambda^{1/2}\Lambda^{1/2}X' = X\Lambda X' = A,$$

as is required. Note that $(A^{1/2})' = (X\Lambda^{1/2}X')' = X\Lambda^{1/2}X' = A^{1/2}$; consequently, $X\Lambda^{1/2}X'$ is referred to as the symmetric square root of A. Note also that if we did not require A to be nonnegative definite, then $A^{1/2}$ would be a complex matrix if some of the eigenvalues of A are negative.

We can expand the set of square root matrices if we do not insist that $A^{1/2}$ be symmetric; that is, now let us consider any matrix $A^{1/2}$ satisfying $A = A^{1/2}(A^{1/2})'$. If Q is any $m \times m$ orthogonal matrix, then $A^{1/2} = X\Lambda^{1/2}Q'$ is such a square root matrix since

$$A^{1/2}A^{1/2\prime} = X\Lambda^{1/2}Q'Q\Lambda^{1/2}X' = X\Lambda^{1/2}\Lambda^{1/2}X' = X\Lambda X' = A$$

If $A^{1/2}$ is a lower triangular matrix with nonnegative diagonal elements, then the factorization $A = A^{1/2}A^{1/2\prime}$ is known as the Cholesky decomposition of A. The following theorem establishes the existence of such a decomposition.

Theorem 4.3. Let A be an $m \times m$ nonnegative definite matrix. Then there exists an $m\times m$ lower triangular matrix T having nonnegative diagonal elements such that $A = TT'$. Further, if A is positive definite, the matrix T is unique and has positive diagonal elements.

Proof. We will prove the result for positive definite matrices. Our proof is by induction. The result clearly holds if $m = 1$, since in this case A is a positive scalar, and so the unique T would be given by the positive square root of A. Now assume that the result holds for all positive definite $(m - 1) \times (m - 1)$ matrices. Partition A as

$$A = \begin{bmatrix} A_{11} & \boldsymbol{a}_{12} \\ \boldsymbol{a}'_{12} & a_{22} \end{bmatrix},$$

where A_{11} is $(m - 1) \times (m - 1)$. Since A_{11} must be positive definite if A is, we know there exists a unique $(m - 1) \times (m - 1)$ lower triangular matrix T_{11} having positive diagonal elements and satisfying $A_{11} = T_{11}T'_{11}$. Our proof will be complete if we can show that there is a unique $(m - 1)\times 1$ vector $\boldsymbol{t}_{12}$ and a unique positive scalar t_{22} such that

$$\begin{bmatrix} A_{11} & \boldsymbol{a}_{12} \\ \boldsymbol{a}'_{12} & a_{22} \end{bmatrix} = \begin{bmatrix} T_{11} & \boldsymbol{0} \\ \boldsymbol{t}'_{12} & t_{22} \end{bmatrix} \begin{bmatrix} T'_{11} & \boldsymbol{t}_{12} \\ \boldsymbol{0}' & t_{22} \end{bmatrix} = \begin{bmatrix} T_{11}T'_{11} & T_{11}\boldsymbol{t}_{12} \\ \boldsymbol{t}'_{12}T'_{11} & \boldsymbol{t}'_{12}\boldsymbol{t}_{12} + t_{22}^2 \end{bmatrix};$$

that is, we must have $\boldsymbol{a}_{12} = T_{11}\boldsymbol{t}_{12}$ and $a_{22} = \boldsymbol{t}_{12}'\boldsymbol{t}_{12} + t_{22}^2$. Since T_{11} must be nonsingular, the unique choice of $\boldsymbol{t}_{12}$ is given by $\boldsymbol{t}_{12} = T_{11}^{-1}\boldsymbol{a}_{12}$, and so t_{22}^2 must satisfy

$$\begin{aligned} t_{22}^2 &= a_{22} - \boldsymbol{t}_{12}'\boldsymbol{t}_{12} = a_{22} - \boldsymbol{a}_{12}'(T_{11}^{-1})'T_{11}^{-1}\boldsymbol{a}_{12} \\ &= a_{22} - \boldsymbol{a}_{12}'(T_{11}T_{11}')^{-1}\boldsymbol{a}_{12} = a_{22} - \boldsymbol{a}_{12}'A_{11}^{-1}\boldsymbol{a}_{12} \end{aligned}$$

Note that since A is positive definite, $a_{22} - \boldsymbol{a}_{12}'A_{11}^{-1}\boldsymbol{a}_{12}$ will be positive since, if we let $\boldsymbol{x} = (\boldsymbol{x}_1', -1)' = (\boldsymbol{a}_{12}'A_{11}^{-1}, -1)'$, then

$$\begin{aligned} \boldsymbol{x}'A\boldsymbol{x} &= \boldsymbol{x}_1'A_{11}\boldsymbol{x}_1 - 2\boldsymbol{x}_1'\boldsymbol{a}_{12} + a_{22} \\ &= \boldsymbol{a}_{12}'A_{11}^{-1}A_{11}A_{11}^{-1}\boldsymbol{a}_{12} - 2\boldsymbol{a}_{12}'A_{11}^{-1}\boldsymbol{a}_{12} + a_{22} \\ &= a_{22} - \boldsymbol{a}_{12}'A_{11}^{-1}\boldsymbol{a}_{12} \end{aligned}$$

Consequently, the unique $t_{22} > 0$ is given by $t_{22} = (a_{22} - \boldsymbol{a}_{12}'A_{11}^{-1}\boldsymbol{a}_{12})^{1/2}$. □

The following decomposition, commonly known as the *QR* factorization, can be used to establish the triangular factorization of Theorem 4.3 for positive semidefinite matrices.

Theorem 4.4. Let A be an $m \times n$ matrix, where $m \geq n$. There exist an $n \times n$ upper triangular matrix R and an $m \times n$ matrix Q satisfying $Q'Q = \mathrm{I}_n$, such that $A = QR$.

For a proof of Theorem 4.4 see Horn and Johnson (1985). If A is a positive semidefinite matrix and $A = A^{1/2}(A^{1/2})'$, then the triangular factorization of Theorem 4.3 for positive semidefinite matrices can be proven by using the *QR* factorization of $(A^{1/2})'$.

Example 4.5. Suppose that the $m \times 1$ random vector $\boldsymbol{x}$ has mean vector $\boldsymbol{\mu}$ and the positive definite covariance matrix Ω. By using a square root matrix of Ω, we can determine a linear transformation of $\boldsymbol{x}$ so that the transformed random vector is standardized; that is, it has mean vector $\boldsymbol{0}$ and covariance matrix I_m. If we let $\Omega^{1/2}$ be any matrix satisfying $\Omega = \Omega^{1/2}(\Omega^{1/2})'$ and put $\boldsymbol{z} = \Omega^{-1/2}(\boldsymbol{x} - \boldsymbol{\mu})$, where $\Omega^{-1/2} = (\Omega^{1/2})^{-1}$, then by using (1.8) and (1.9) of Section 1.13, we find that

$$\mathrm{E}(\boldsymbol{z}) = \mathrm{E}\{\Omega^{-1/2}(\boldsymbol{x} - \boldsymbol{\mu})\} = \Omega^{-1/2}\{\mathrm{E}(\boldsymbol{x} - \boldsymbol{\mu})\} = \Omega^{-1/2}(\boldsymbol{\mu} - \boldsymbol{\mu}) = \boldsymbol{0},$$

and

$$\begin{aligned}\text{var}(z) &= \text{var}\{\Omega^{-1/2}(\boldsymbol{x} - \boldsymbol{\mu})\} = \Omega^{-1/2}\{\text{var}(\boldsymbol{x} - \boldsymbol{\mu})\}(\Omega^{-1/2})' \\ &= \Omega^{-1/2}\{\text{var}(\boldsymbol{x})\}(\Omega^{-1/2})' = \Omega^{-1/2}\Omega(\Omega^{-1/2})' = \text{I}_m\end{aligned}$$

Since the covariance matrix of z is the identity matrix, the Euclidean distance function will give a meaningful measure of the distance between observations from this distribution. By making use of the linear transformation defined above, we can relate distances between z observations to distances between $\boldsymbol{x}$ observations. For example, the Euclidean distance between an observation z and its expected value $\mathbf{0}$ is

$$\begin{aligned}d_{\text{I}}(z, \mathbf{0}) &= \{(z - \mathbf{0})'(z - \mathbf{0})\}^{1/2} = (z'z)^{1/2} \\ &= \{(\boldsymbol{x} - \boldsymbol{\mu})'(\Omega^{-1/2})'\Omega^{-1/2}(\boldsymbol{x} - \boldsymbol{\mu})\}^{1/2} \\ &= \{(\boldsymbol{x} - \boldsymbol{\mu})'\Omega^{-1}(\boldsymbol{x} - \boldsymbol{\mu})\}^{1/2} \\ &= d_\Omega(\boldsymbol{x}, \boldsymbol{\mu}),\end{aligned}$$

where d_Ω is the Mahalanobis distance function defined in Section 2.2. Similarly, if $\boldsymbol{x}_1$ and $\boldsymbol{x}_2$ are two observations from the distribution of $\boldsymbol{x}$ and z_1 and z_2 are the corresponding transformed vectors, then $d_{\text{I}}(z_1, z_2) = d_\Omega(\boldsymbol{x}_1, \boldsymbol{x}_2)$. This relationship between the Mahalanobis distance and the Euclidean distance makes the construction of the Mahalanobis distance function more apparent. It is nothing more than a two-stage computation of distance; the first stage transforms points so as to remove the effect of correlations and differing variances, while the second stage simply computes the Euclidean distance for these transformed points.

Example 4.6. In Example 2.16, we obtained the weighted least squares estimator of $\boldsymbol{\beta}$ in the multiple regression model

$$\boldsymbol{y} = X\boldsymbol{\beta} + \boldsymbol{\epsilon},$$

where $\text{var}(\boldsymbol{\epsilon}) = \sigma^2\text{diag}(c_1^2, \ldots, c_N^2)$ and $c_1^2, \ldots, c_N^2$ are known constants. We now consider a more general regression problem, sometimes referred to as generalized least squares regression, in which $\text{var}(\boldsymbol{\epsilon}) = \sigma^2 C$, where C is a known $N \times N$ positive definite matrix. Thus, the random errors not only may have different variances but also may be correlated, and weighted least squares regression is simply a special case of generalized least squares regression. As with weighted least squares regression, the approach here is to transform the problem to ordinary least squares regression; that is, we wish to transform the model so that the vector of random errors in the transformed model has $\sigma^2\text{I}_N$ as its covariance matrix. This can be done by utilizing any square root matrix of C. Let T be any $N \times N$ matrix satisfying $TT' = C$ or, equivalently, $T'^{-1}T^{-1} = C^{-1}$. Now transform our original regression model to the model

$$y_* = X_*\boldsymbol{\beta} + \boldsymbol{\epsilon}_*,$$

where $\boldsymbol{y}_* = T^{-1}\boldsymbol{y}$, $X_* = T^{-1}X$, and $\boldsymbol{\epsilon}_* = T^{-1}\boldsymbol{\epsilon}$, and note that $\mathrm{E}(\boldsymbol{\epsilon}_*) = T^{-1}\mathrm{E}(\boldsymbol{\epsilon}) = \mathbf{0}$ and

$$\begin{aligned}\mathrm{var}(\boldsymbol{\epsilon}_*) &= \mathrm{var}(T^{-1}\boldsymbol{\epsilon}) = T^{-1}\{\mathrm{var}(\boldsymbol{\epsilon})\}T'^{-1} \\ &= T^{-1}(\sigma^2 C)T'^{-1} = \sigma^2 T^{-1}TT'T'^{-1} = \sigma^2 \mathrm{I}_N\end{aligned}$$

Thus, the generalized least squares estimator $\hat{\boldsymbol{\beta}}_*$ of $\boldsymbol{\beta}$ in the model $\boldsymbol{y} = X\boldsymbol{\beta}+\boldsymbol{\epsilon}$ is given by the ordinary least squares estimator of $\boldsymbol{\beta}$ in the model $\boldsymbol{y}_* = X_*\boldsymbol{\beta} + \boldsymbol{\epsilon}_*$ and so can be expressed as

$$\begin{aligned}\hat{\boldsymbol{\beta}}_* &= (X_*'X_*)^{-1}X_*'\boldsymbol{y}_* = (X'T'^{-1}T^{-1}X)^{-1}X'T'^{-1}T^{-1}\boldsymbol{y} \\ &= (X'C^{-1}X)^{-1}X'C^{-1}\boldsymbol{y}\end{aligned}$$

In some situations a matrix A can be expressed in the form of the transpose product, BB', where the $m \times r$ matrix B has $r < m$, so that unlike a square root matrix, B is not square. This is the subject of our next theorem, the proof of which will be left to the reader as an exercise.

Theorem 4.5. Let A be an $m \times m$ nonnegative definite matrix with $\mathrm{rank}(A) = r$. Then there exists an $m \times r$ matrix B, having rank of r, such that $A = BB'$.

The transpose product form $A = BB'$ of the nonnegative definite matrix A is not unique. However, if C is another matrix of order $m \times n$ where $n \geq r$ and $A = CC'$, there is an explicit relationship between the matrices B and C. This is established in the next theorem.

Theorem 4.6. Suppose that B is an $m \times h$ matrix and C is an $m \times n$ matrix, where $h \leq n$. Then $BB' = CC'$ if and only if there exists an $h \times n$ matrix Q such that $QQ' = \mathrm{I}_h$ and $C = BQ$.

Proof. If $C = BQ$ with $QQ' = \mathrm{I}_h$, then clearly

$$CC' = BQ(BQ)' = BQQ'B' = BB'$$

Conversely, now suppose that $BB' = CC'$. We will assume that $h = n$ since if $h < n$, we can form the matrix $B_* = [B \quad (0)]$ so that B_* is $m \times n$ and $B_*B_*' = BB'$; then proving that there exists an $n \times n$ orthogonal matrix Q_* such that $C = B_*Q_*$ will yield $C = BQ$, if we take Q to be the first h rows of Q_*.

Now since BB' is symmetric, there exists an orthogonal matrix X such that

$$BB' = CC' = X\begin{bmatrix} \Lambda & (0) \\ (0) & (0) \end{bmatrix} X' = X_1 \Lambda X_1',$$

where $\text{rank}(BB') = r$ and the $r \times r$ diagonal matrix Λ contains the positive eigenvalues of the nonnegative definite matrix BB'. Here X has been partitioned as $X = [X_1 \quad X_2]$, where X_1 is $m \times r$. Form the matrices

$$E = \begin{bmatrix} \Lambda^{-1/2} & (0) \\ (0) & I_{m-r} \end{bmatrix} X'B = \begin{bmatrix} \Lambda^{-1/2} X_1'B \\ X_2'B \end{bmatrix} = \begin{bmatrix} E_1 \\ E_2 \end{bmatrix}, \tag{4.7}$$

$$F = \begin{bmatrix} \Lambda^{-1/2} & (0) \\ (0) & I_{m-r} \end{bmatrix} X'C = \begin{bmatrix} \Lambda^{-1/2} X_1'C \\ X_2'C \end{bmatrix} = \begin{bmatrix} F_1 \\ F_2 \end{bmatrix}, \tag{4.8}$$

so that

$$EE' = FF' = \begin{bmatrix} I_r & (0) \\ (0) & (0) \end{bmatrix};$$

that is, $E_1E_1' = F_1F_1' = I_r$, $E_2E_2' = F_2F_2' = (0)$, and so $E_2 = F_2 = (0)$. Now let E_3 and F_3 be any $(h-r) \times h$ matrices such that $E_* = [E_1' \quad E_3']'$ and $F_* = [F_1' \quad F_3']'$ are both orthogonal matrices. Consequently, if $Q = E_*'F_*$, then $QQ' = E_*'F_*F_*'E_* = E_*'E_* = I_h$, so Q is orthogonal. Since E_* is orthogonal, we have $E_1E_3' = (0)$, and so

$$\begin{aligned} EQ = EE_*'F_* &= \begin{bmatrix} E_1 \\ (0) \end{bmatrix} [E_1' \quad E_3'] \begin{bmatrix} F_1 \\ F_3 \end{bmatrix} \\ &= \begin{bmatrix} I_r & (0) \\ (0) & (0) \end{bmatrix} \begin{bmatrix} F_1 \\ F_3 \end{bmatrix} = \begin{bmatrix} F_1 \\ (0) \end{bmatrix} = F \end{aligned}$$

But using (4.7) and (4.8), $EQ = F$ can be written as

$$\begin{bmatrix} \Lambda^{-1/2} & (0) \\ (0) & I_{m-r} \end{bmatrix} X'BQ = \begin{bmatrix} \Lambda^{-1/2} & (0) \\ (0) & I_{m-r} \end{bmatrix} X'C$$

The result now follows by premultiplying this equation by

$$X\begin{bmatrix} \Lambda^{1/2} & (0) \\ (0) & \mathrm{I}_{m-r} \end{bmatrix},$$

since $XX' = \mathrm{I}_m$. □

4. THE DIAGONALIZATION OF A SQUARE MATRIX

From the spectral decomposition theorem, we know that every symmetric matrix can be transformed to a diagonal matrix by postmultiplying by an appropriately chosen orthogonal matrix and premultiplying by its transpose. This result gives us a very useful and simple relationship between a symmetric matrix and its eigenvalues and eigenvectors. In this section, we investigate a generalization of this relationship to square matrices in general. We begin with the following definition.

Definition 4.1. The $m \times m$ matrices A and B are said to be similar matrices if there exists a nonsingular matrix C such that $A = CBC^{-1}$.

It follows from Theorem 3.2(d) that similar matrices have identical eigenvalues. However, the converse is not true. For instance, if we have

$$A = \begin{bmatrix} 0 & 1 \\ 0 & 0 \end{bmatrix}, \qquad B = \begin{bmatrix} 0 & 0 \\ 0 & 0 \end{bmatrix},$$

then A and B have identical eigenvalues since each has 0 with multiplicity 2. Clearly, however, there is no nonsingular matrix C satisfying $A = CBC^{-1}$.

The spectral decomposition theorem given as Theorem 4.2 tells us that every symmetric matrix is similar to a diagonal matrix. Unfortunately, the same statement does not hold for all square matrices. If the diagonal elements of the diagonal matrix Λ are the eigenvalues of A, and the columns of X are corresponding eigenvectors, then the eigenvalue–eigenvector equation $AX = X\Lambda$ immediately leads to the identity $X^{-1}AX = \Lambda$, if X is nonsingular; that is, the diagonability of an $m \times m$ matrix simply depends on the existence of a set of m linearly independent eigenvectors. Consequently, we have the following result, previously mentioned in Section 3.3, which follows immediately from Theorem 3.6.

Theorem 4.7. Suppose that the $m \times m$ matrix A has the eigenvalues $\lambda_1, \dots, \lambda_m$ which are distinct. If $\Lambda = \mathrm{diag}(\lambda_1, \dots, \lambda_m)$ and $X = (\boldsymbol{x}_1, \dots, \boldsymbol{x}_m)$, where $\boldsymbol{x}_1, \dots, \boldsymbol{x}_m$ are eigenvectors of A corresponding to $\lambda_1, \dots, \lambda_m$, then

$$X^{-1}AX = \Lambda \tag{4.9}$$

The theorem above gives a sufficient but not necessary condition for the diagonalization of a general square matrix; that is, some nonsymmetric matrices that have multiple eigenvalues are similar to a diagonal matrix. The next theorem gives a necessary and sufficient condition for a matrix to be diagonalizable.

Theorem 4.8. Suppose the eigenvalues $\lambda_1, \ldots, \lambda_m$ of the $m \times m$ matrix A consist of h distinct values $\mu_1, \ldots, \mu_h$ having multiplicities $r_1, \ldots, r_h$, so that $r_1 + \cdots + r_h = m$. Then A has a set of m linearly independent eigenvectors and, thus, is diagonalizable if and only if $\text{rank}(A - \mu_i \text{I}_m) = m - r_i$ for $i = 1, \ldots, h$.

Proof. First, suppose that A is diagonalizable, so that using the usual notation, we have $X^{-1}AX = \Lambda$, or equivalently $A = X\Lambda X^{-1}$. Thus,

$$\begin{aligned}\text{rank}(A - \mu_i \text{I}_m) &= \text{rank}(X\Lambda X^{-1} - \mu_i \text{I}_m) = \text{rank}\{X(\Lambda - \mu_i \text{I}_m)X^{-1}\} \\ &= \text{rank}(\Lambda - \mu_i \text{I}_m),\end{aligned}$$

where the last equality follows from the fact that the rank of a matrix is unaltered by its multiplication by a nonsingular matrix. Now, since μ_i has multiplicity r_i, the diagonal matrix $(\Lambda - \mu_i \text{I}_m)$ has exactly $m - r_i$ nonzero diagonal elements which then guarantees that $\text{rank}(A - \mu_i \text{I}_m) = m - r_i$. Conversely, now suppose that $\text{rank}(A - \mu_i \text{I}_m) = m - r_i$, for $i = 1, \ldots, h$. This implies that the dimension of the null space of $(A - \mu_i \text{I}_m)$ is $m - (m - r_i) = r_i$, and so we can find r_i linearly independent vectors satisfying the equation

$$(A - \mu_i \text{I}_m)\boldsymbol{x} = \mathbf{0}$$

But any such $\boldsymbol{x}$ is an eigenvector of A corresponding to the eigenvalue μ_i. Consequently, we can find a set of r_i linearly independent eigenvectors associated with the eigenvalue μ_i. From Theorem 3.6, we know that eigenvectors corresponding to different eigenvalues are linearly independent. As a result, any set of m eigenvectors of A, which has r_i linearly independent eigenvectors corresponding to μ_i for each i, will also be linearly independent. Therefore, A is diagonalizable and so the proof is complete. □

We saw in Chapter 3 that the rank of a symmetric matrix is equal to the number of its nonzero eigenvalues. The diagonal factorization given in (4.9) immediately yields the following generalization of this result.

Theorem 4.9. Let A be an $m \times m$ matrix. If A is diagonalizable, then the rank of A is equal to the number of nonzero eigenvalues of A.

The converse of Theorem 4.9 is not true; that is, a matrix need not be diagonalizable for its rank to equal the number of its nonzero eigenvalues.

Example 4.7. Let A, B, and C be the 2×2 matrices given by

$$A = \begin{bmatrix} 1 & 1 \\ 4 & 1 \end{bmatrix}, \qquad B = \begin{bmatrix} 0 & 1 \\ 0 & 0 \end{bmatrix}, \qquad C = \begin{bmatrix} 1 & 1 \\ 0 & 1 \end{bmatrix}$$

The characteristic equation of A simplifies to $(\lambda-3)(\lambda+1) = 0$, so its eigenvalues are $\lambda = 3, -1$. Since the eigenvalues are simple, A is diagonalizable. Eigenvectors corresponding to these two eigenvalues are $\boldsymbol{x}_1 = (1, 2)'$ and $\boldsymbol{x}_2 = (1, -2)'$, so the diagonalization of A is given by

$$\begin{bmatrix} 1/2 & 1/4 \\ 1/2 & -1/4 \end{bmatrix} \begin{bmatrix} 1 & 1 \\ 4 & 1 \end{bmatrix} \begin{bmatrix} 1 & 1 \\ 2 & -2 \end{bmatrix} = \begin{bmatrix} 3 & 0 \\ 0 & -1 \end{bmatrix}$$

Clearly, the rank of A is 2, which is the same as the number of nonzero eigenvalues of A. The characteristic equation of B reduces to $\lambda^2 = 0$, so B has the eigenvalue $\lambda = 0$ with multiplicity $r = 2$. Since $\text{rank}(B - \lambda \text{I}_2) = \text{rank}(B) = 1 \neq 0 = m - r$, B will not have two linearly independent eigenvectors. The equation $B\boldsymbol{x} = \lambda \boldsymbol{x} = \boldsymbol{0}$ has only one linearly independent solution for $\boldsymbol{x}$, namely, vectors of the form $(a, 0)'$. Thus, B is not diagonalizable. Note also that the rank of B is 1, which is greater than the number of its nonzero eigenvalues. Finally, turning to C, we see that it has the eigenvalue $\lambda = 1$ with multiplicity $r = 2$, since its characteristic equation simplifies to $(1 - \lambda)^2 = 0$. This matrix is not diagonalizable since $\text{rank}(C - \lambda \text{I}_2) = \text{rank}(C - \text{I}_2) = \text{rank}(B) = 1 \neq 0 = m - r$. Any eigenvector of C is a scalar multiple of the vector $\boldsymbol{x} = (1, 0)'$. However, notice that even though C is not diagonalizable, it has rank of 2, which is the same as the number of its nonzero eigenvalues.

The next result shows that the connection between the rank and the number of nonzero eigenvalues of a matrix A hinges on the dimension of the eigenspace associated with the eigenvalue 0.

Theorem 4.10. Let A be an $m \times m$ matrix and let k be the dimension of the eigenspace associated with the eigenvalue 0 if 0 is an eigenvalue of A, and let $k = 0$ otherwise. Then

$$\text{rank}(A) = m - k$$

Proof. From Theorem 2.21, we know that

$$\text{rank}(A) = m - \dim\{N(A)\},$$

where $N(A)$ is the null space of A. But since the null space of A consists of all vectors $\boldsymbol{x}$ satisfying $A\boldsymbol{x} = \boldsymbol{0}$, we see that $N(A)$ is the same as $S_A(0)$, and so the result follows. □

We have seen that the number of nonzero eigenvalues of a matrix A equals the rank of A if A is similar to a diagonal matrix; that is, A being diagonalizable is a sufficient condition for this exact relationship between rank and the number of nonzero eigenvalues. The following necessary and sufficient condition for this relationship to exist is an immediate consequence of Theorem 4.10.

Corollary 4.10.1. Let A be an $m \times m$ matrix and let m_0 denote the multiplicity of the eigenvalue 0. Then the rank of A is equal to the number of nonzero eigenvalues of A if and only if

$$\dim\{S_A(0)\} = m_0$$

Example 4.8. We saw in Example 4.7 that the two matrices

$$B = \begin{bmatrix} 0 & 1 \\ 0 & 0 \end{bmatrix}, \qquad C = \begin{bmatrix} 1 & 1 \\ 0 & 1 \end{bmatrix}$$

are not diagonalizable since each has only one linearly independent eigenvector associated with its single eigenvalue, which has multiplicity two. This eigenvalue is 0 for B, so

$$\text{rank}(B) = 2 - \dim\{S_B(0)\} = 2 - 1 = 1$$

On the other hand, since 0 is not an eigenvalue of C, $\dim\{S_C(0)\} = 0$, and so the rank of C equals the number of its nonzero eigenvalues, 2.

5. THE JORDAN DECOMPOSITION

Our next factorization of a square matrix A is one that could be described as an attempt to find a matrix similar to A, which, if not diagonal, is as diagonal as is possible. We begin with the following definition.

Definition 4.2. For $h > 1$, the $h \times h$ matrix $J_h(\lambda)$ is said to be a Jordan block matrix if it has the form

$$J_h(\lambda) = \lambda \mathrm{I}_h + \sum_{i=1}^{h-1} \boldsymbol{e}_i \boldsymbol{e}_{i+1}' = \begin{bmatrix} \lambda & 1 & 0 & \cdots & 0 \\ 0 & \lambda & 1 & \cdots & 0 \\ 0 & 0 & \lambda & \cdots & 0 \\ \vdots & \vdots & \vdots & & \vdots \\ 0 & 0 & 0 & \cdots & \lambda \end{bmatrix},$$

where $\boldsymbol{e}_i$ is the ith column of I_h. If $h = 1, J_1(\lambda) = \lambda$.

The matrices B and C from Examples 4.7 and 4.8 are both 2×2 Jordan block matrices; in particular, $B = J_2(0)$ and $C = J_2(1)$. We saw that neither of these matrices is similar to a diagonal matrix. This is true for Jordan block matrices in general; if $h > 1$, then $J_h(\lambda)$ is not diagonalizable. To see this, note that since $J_h(\lambda)$ is a triangular matrix, its diagonal elements are its eigenvalues, and so it has the one value, λ, repeated h times. However, the solution to $J_h(\lambda)\boldsymbol{x} = \lambda \boldsymbol{x}$ has x_1 arbitrary while $x_2 = \cdots = x_h = 0$; that is, $J_h(\lambda)$ has only one linearly independent eigenvector, which is of the form $\boldsymbol{x} = (x_1, 0, \dots, 0)'$.

We now state the Jordan decomposition theorem. For a proof of this result see Horn and Johnson (1985).

Theorem 4.11. Let A be an $m \times m$ matrix. Then there exists a nonsingular matrix B such that

$$B^{-1}AB = J = \mathrm{diag}(J_{h_1}(\lambda_1), \dots, J_{h_r}(\lambda_r))$$
$$= \begin{bmatrix} J_{h_1}(\lambda_1) & (0) & \cdots & (0) \\ (0) & J_{h_2}(\lambda_2) & \cdots & (0) \\ \vdots & \vdots & & \vdots \\ (0) & (0) & \cdots & J_{h_r}(\lambda_r) \end{bmatrix},$$

where $h_1 + \cdots + h_r = m$ and $\lambda_1, \dots, \lambda_r$ are the not necessarily distinct eigenvalues of A.

The matrix J will be diagonal if $h_i = 1$ for all i. Since the $h_i \times h_i$ matrix $J_{h_i}(\lambda_i)$ has only one linearly independent eigenvector, it follows that the Jordan canonical form $J = \mathrm{diag}(J_{h_1}(\lambda_1), \dots, J_{h_r}(\lambda_r))$ has r linearly independent eigenvectors. Thus, if $h_i > 1$ for at least one i, then J will not be diagonal; in fact, J will not be diagonalizable. The vector $\boldsymbol{x}_i$ is an eigenvector of J corresponding to the eigenvalue λ_i if and only if the vector $\boldsymbol{y}_i = B\boldsymbol{x}_i$ is an eigenvector of A corresponding to λ_i; for instance, if $\boldsymbol{x}_i$ satisfies $J\boldsymbol{x}_i = \lambda_i \boldsymbol{x}_i$, then

$$A\boldsymbol{y}_i = (BJB^{-1})B\boldsymbol{x}_i = BJ\boldsymbol{x}_i = \lambda_i B\boldsymbol{x}_i = \lambda_i \boldsymbol{y}_i$$

Thus, r also gives the number of linearly independent eigenvectors of A, and A is diagonalizable only if J is diagonal.

Example 4.9. Suppose that A is a 4×4 matrix with the eigenvalue λ having multiplicity 4. Then A will be similar to one of the following five Jordan canonical forms:

$$\text{diag}(J_1(\lambda), J_1(\lambda), J_1(\lambda), J_1(\lambda)) = \begin{bmatrix} \lambda & 0 & 0 & 0 \\ 0 & \lambda & 0 & 0 \\ 0 & 0 & \lambda & 0 \\ 0 & 0 & 0 & \lambda \end{bmatrix},$$

$$\text{diag}(J_2(\lambda), J_1(\lambda), J_1(\lambda)) = \begin{bmatrix} \lambda & 1 & 0 & 0 \\ 0 & \lambda & 0 & 0 \\ 0 & 0 & \lambda & 0 \\ 0 & 0 & 0 & \lambda \end{bmatrix},$$

$$\text{diag}(J_3(\lambda), J_1(\lambda)) = \begin{bmatrix} \lambda & 1 & 0 & 0 \\ 0 & \lambda & 1 & 0 \\ 0 & 0 & \lambda & 0 \\ 0 & 0 & 0 & \lambda \end{bmatrix},$$

$$\text{diag}(J_2(\lambda), J_2(\lambda)) = \begin{bmatrix} \lambda & 1 & 0 & 0 \\ 0 & \lambda & 0 & 0 \\ 0 & 0 & \lambda & 1 \\ 0 & 0 & 0 & \lambda \end{bmatrix},$$

$$J_4(\lambda) = \begin{bmatrix} \lambda & 1 & 0 & 0 \\ 0 & \lambda & 1 & 0 \\ 0 & 0 & \lambda & 1 \\ 0 & 0 & 0 & \lambda \end{bmatrix}$$

The first form given is diagonal so this corresponds to the case in which A has four linearly independent eigenvectors associated with the eigenvalue λ. The second and last forms correspond to A having three and one linearly independent eigenvectors, respectively. If A has two linearly independent eigenvectors, then it will be similar to either the third or the fourth matrix given.

6. THE SCHUR DECOMPOSITION

Our next result can be viewed as another generalization of the spectral decomposition theorem to any square matrix, A. The diagonalization theorem and the Jordan decomposition were generalizations of the spectral decomposition in which our goal was to obtain a diagonal or "nearly" diagonal matrix. Now, instead we focus on the orthogonal matrix employed in the spectral decomposition theorem. Specifically, if we restrict attention only to orthogonal matrices, X, what is the simplest structure that we can get for $X'AX$? It turns out that for the general case of any real square matrix A, we can find an X such that X^*AX is a triangular matrix, where we have broadened the choice of X to include all unitary matrices. Recall that a real unitary matrix is an orthogonal matrix and, in general, X is unitary if $X^*X = \mathrm{I}$, where X^* is the transpose of the complex conjugate of X. This decomposition, sometimes referred to as the Schur decomposition, is given in the following theorem.

Theorem 4.12. Let A be an $m \times m$ matrix. Then there exists an $m \times m$ unitary matrix X such that

$$X^*AX = T,$$

where T is an upper triangular matrix with the eigenvalues of A as its diagonal elements.

Proof. Let $\lambda_1, \dots, \lambda_m$ be the eigenvalues of A, and let $\boldsymbol{y}_1$ be an eigenvector of A corresponding to λ_1 and normalized so that $\boldsymbol{y}_1^*\boldsymbol{y}_1 = 1$. Let Y be any $m \times m$ unitary matrix having $\boldsymbol{y}_1$ as its first column. Writing Y in partitioned form as $Y = [\boldsymbol{y}_1 \quad Y_2]$, we see that, since $A\boldsymbol{y}_1 = \lambda_1\boldsymbol{y}_1$ and $Y_2^*\boldsymbol{y}_1 = \boldsymbol{0}$,

$$Y^*AY = \begin{bmatrix} \boldsymbol{y}_1^*A\boldsymbol{y}_1 & \boldsymbol{y}_1^*AY_2 \\ Y_2^*A\boldsymbol{y}_1 & Y_2^*AY_2 \end{bmatrix} = \begin{bmatrix} \lambda_1\boldsymbol{y}_1^*\boldsymbol{y}_1 & \boldsymbol{y}_1^*AY_2 \\ \lambda_1 Y_2^*\boldsymbol{y}_1 & Y_2^*AY_2 \end{bmatrix}$$

$$= \begin{bmatrix} \lambda_1 & \boldsymbol{y}_1^*AY_2 \\ \boldsymbol{0} & B \end{bmatrix},$$

where the $(m-1) \times (m-1)$ matrix $B = Y_2^*AY_2$. Using the identity above and the cofactor expansion formula for a determinant, it follows that the characteristic equation of Y^*AY is

$$(\lambda_1 - \lambda)|B - \lambda\mathrm{I}_{m-1}| = 0,$$

and, since by Theorem 3.2(d) the eigenvalues of Y^*AY are the same as those of A, the eigenvalues of B must be $\lambda_2, \dots, \lambda_m$. Now if $m = 2$, then the scalar B must equal λ_2 and Y^*AY is upper triangular, so the proof is complete. For $m > 2$, we proceed by induction; that is, we show that if our result holds for $(m-1) \times (m-1)$ matrices, then it must also hold for $m \times m$ matrices. Since B is $(m-1) \times (m-1)$ we may assume that there exists a unitary matrix W such that $W^*BW = T_2$, where T_2 is an upper triangular matrix with diagonal elements $\lambda_2, \dots, \lambda_m$. Define the $m \times m$ matrix U by

$$U = \begin{bmatrix} 1 & \mathbf{0}' \\ \mathbf{0} & W \end{bmatrix},$$

and note that U is unitary since W is. If we let $X = YU$, then X is also unitary and

$$\begin{aligned} X^*AX = U^*Y^*AYU &= \begin{bmatrix} 1 & \mathbf{0}' \\ \mathbf{0} & W^* \end{bmatrix} \begin{bmatrix} \lambda_1 & \boldsymbol{y}_1^*AY_2 \\ \mathbf{0} & B \end{bmatrix} \begin{bmatrix} 1 & \mathbf{0}' \\ \mathbf{0} & W \end{bmatrix} \\ &= \begin{bmatrix} \lambda_1 & \boldsymbol{y}_1^*AY_2W \\ \mathbf{0} & W^*BW \end{bmatrix} = \begin{bmatrix} \lambda_1 & \boldsymbol{y}_1^*AY_2W \\ \mathbf{0} & T_2 \end{bmatrix}, \end{aligned}$$

where this final matrix is upper triangular with $\lambda_1, \dots, \lambda_m$ as its diagonal elements. Thus, the proof is complete. □

If all of the eigenvalues of A are real, then there exist corresponding real eigenvectors. In this case, a real matrix X satisfying the conditions of Theorem 4.12 can be found. Consequently, we have the following result.

Corollary 4.12.1. If the $m \times m$ matrix A has real eigenvalues, then there exists an $m \times m$ orthogonal matrix X such that $X'AX = T$, where T is an upper triangular matrix.

Example 4.10. Consider the 3×3 matrix given by

$$A = \begin{bmatrix} 5 & -3 & 3 \\ 4 & -2 & 3 \\ 4 & -4 & 5 \end{bmatrix}$$

In Example 3.1, the eigenvalues of A were shown to be $\lambda_1 = 1$, $\lambda_2 = 2$, and $\lambda_3 = 5$, with eigenvectors, $\boldsymbol{x}_1 = (0, 1, 1)'$, $\boldsymbol{x}_2 = (1, 1, 0)'$, and $\boldsymbol{x}_3 = (1, 1, 1)'$, respectively. We will find an orthogonal matrix X and an upper triangular matrix T so that $A = XTX'$. First, we construct an orthogonal matrix Y having a normalized version of $\boldsymbol{x}_1$ as its first column; for instance, by inspection we set

$$Y = \begin{bmatrix} 0 & 0 & 1 \\ 1/\sqrt{2} & 1/\sqrt{2} & 0 \\ 1/\sqrt{2} & -1/\sqrt{2} & 0 \end{bmatrix}$$

Thus, our first stage yields

$$Y'AY = \begin{bmatrix} 1 & -7 & 4\sqrt{2} \\ 0 & 2 & 0 \\ 0 & -3\sqrt{2} & 5 \end{bmatrix}$$

The 2×2 matrix

$$B = \begin{bmatrix} 2 & 0 \\ -3\sqrt{2} & 5 \end{bmatrix}$$

has a normalized eigenvector $(1/\sqrt{3}, \sqrt{2}/\sqrt{3})'$, and so we can construct an orthogonal matrix

$$W = \begin{bmatrix} 1/\sqrt{3} & -\sqrt{2}/\sqrt{3} \\ \sqrt{2}/\sqrt{3} & 1/\sqrt{3} \end{bmatrix}$$

for which

$$W'BW = \begin{bmatrix} 2 & 3\sqrt{2} \\ 0 & 5 \end{bmatrix}$$

Putting it all together, we have

$$X = Y \begin{bmatrix} 1 & \mathbf{0}' \\ \mathbf{0} & W \end{bmatrix} = \frac{1}{\sqrt{6}} \begin{bmatrix} 0 & 2 & \sqrt{2} \\ \sqrt{3} & 1 & -\sqrt{2} \\ \sqrt{3} & -1 & \sqrt{2} \end{bmatrix},$$

and

$$T = X'AX = \begin{bmatrix} 1 & 1/\sqrt{3} & 22/\sqrt{6} \\ 0 & 2 & 3\sqrt{2} \\ 0 & 0 & 5 \end{bmatrix}$$

The matrices X and T in the Schur decomposition are not unique; that is, if $A = XTX^*$ is a Schur decomposition of A, then $A = X_0T_0X_0^*$ is also, where $X_0 = XP$ and P is any unitary matrix for which $P^*TP = T_0$ is upper triangular. The triangular matrices T and T_0 must have the same diagonal elements, possibly ordered differently. Otherwise, however, the two matrices T and T_0 may be quite different. For example, it can be easily verified that the matrices

$$X_0 = \begin{bmatrix} 1/\sqrt{3} & 2/\sqrt{6} & 0 \\ 1/\sqrt{3} & -1/\sqrt{6} & -1/\sqrt{2} \\ 1/\sqrt{3} & -1/\sqrt{6} & 1/\sqrt{2} \end{bmatrix}, \qquad T_0 = \begin{bmatrix} 5 & 8/\sqrt{2} & 20/\sqrt{6} \\ 0 & 1 & -1/\sqrt{3} \\ 0 & 0 & 2 \end{bmatrix}$$

give another Schur decomposition of the matrix A of Example 4.10.

In Chapter 3, by utilizing the characteristic equation of the $m \times m$ matrix A, we were able to prove that the determinant of A equals the product of its eigenvalues, while the trace of A equals the sum of its eigenvalues. These results are also very easily proven using the Schur decomposition of A. If the eigenvalues of A are $\lambda_1, \dots, \lambda_m$ and $A = XTX^*$ is a Schur decomposition of A, then it follows that

$$|A| = |XTX^*| = |X^*X|\,|T| = |T| = \prod_{i=1}^{m} \lambda_i,$$

since $|X^*X| = 1$ follows from the fact that X is a unitary matrix, and the determinant of a triangular matrix is the product of its diagonal elements. Also, using properties of the trace of a matrix, we have

$$\text{tr}(A) = \text{tr}(XTX^*) = \text{tr}(X^*XT) = \text{tr}(T) = \sum_{i=1}^{m} \lambda_i$$

The Schur decomposition also provides a method of easily establishing the fact that the number of nonzero eigenvalues of a matrix serves as a lower bound for the rank of that matrix. This is the subject of our next theorem.

Theorem 4.13. Suppose the $m \times m$ matrix A has r nonzero eigenvalues. Then $\text{rank}(A) \geq r$.

Proof. Let X be a unitary matrix and T be an upper triangular matrix such that $A = XTX^*$. Since the eigenvalues of A are the diagonal elements of T, T must have exactly r nonzero diagonal elements. The $r \times r$ submatrix of T, formed by deleting the columns and rows occupied by the zero diagonal elements of T, will be upper triangular with nonzero diagonal elements. This submatrix will be nonsingular since the determinant of a triangular matrix is the product of its diagonal elements, so we must have $\text{rank}(T) \geq r$. The result then follows from the fact that since X is unitary, it must be nonsingular, so

$$\text{rank}(A) = \text{rank}(XTX^*) = \text{rank}(T) \geq r$$

□

7. THE SIMULTANEOUS DIAGONALIZATION OF TWO SYMMETRIC MATRICES

We have already discussed in Section 3.7 one manner in which two symmetric matrices can be simultaneously diagonalized. We restate this result in the following theorem.

Theorem 4.14. Let A and B be $m \times m$ symmetric matrices, with A being nonnegative definite and B, positive definite. Let $\Lambda = \text{diag}(\lambda_1, \dots, \lambda_m)$, where $\lambda_1, \dots, \lambda_m$ are the eigenvalues of $B^{-1}A$. Then there exists a nonsingular matrix C such that

$$CAC' = \Lambda, \qquad CBC' = I_m$$

Example 4.11. One application of the simultaneous diagonalization described in the theorem above is in a multivariate analysis commonly referred to as canonical variate analysis (see Krzanowski, 1988 or Mardia, Kent, and Bibby, 1979). This analysis involves data from the multivariate one-way classification model discussed in Example 3.14, so that we have independent random samples from k different groups or treatments, with the ith sample of $m \times 1$ vectors given by $\boldsymbol{y}_{i1}, \dots, \boldsymbol{y}_{in_i}$. The model is

$$\boldsymbol{y}_{ij} = \boldsymbol{\mu}_i + \boldsymbol{\epsilon}_{ij},$$

where $\boldsymbol{\mu}_i$ is an $m \times 1$ vector of constants and $\boldsymbol{\epsilon}_{ij} \sim N_m(\mathbf{0}, \Omega)$. In Example 3.14, we saw how the matrices

$$B = \sum_{i=1}^{k} n_i(\bar{\boldsymbol{y}}_i - \bar{\boldsymbol{y}})(\bar{\boldsymbol{y}}_i - \bar{\boldsymbol{y}})', \qquad W = \sum_{i=1}^{k} \sum_{j=1}^{n_i} (\boldsymbol{y}_{ij} - \bar{\boldsymbol{y}}_i)(\boldsymbol{y}_{ij} - \bar{\boldsymbol{y}}_i)',$$

where

$$\bar{\boldsymbol{y}}_i = \sum_{j=1}^{n_i} \frac{\boldsymbol{y}_{ij}}{n_i}, \qquad \bar{\boldsymbol{y}} = \sum_{i=1}^{k} \frac{n_i \bar{\boldsymbol{y}}_i}{n}, \qquad n = \sum_{i=1}^{k} n_i,$$

could be used to test the hypothesis, H_0: $\boldsymbol{\mu}_1 = \cdots = \boldsymbol{\mu}_k$. Canonical variate analysis is an analysis of the differences in the mean vectors, performed when this hypothesis is rejected. This analysis is particularly useful when the differences between the vectors $\boldsymbol{\mu}_1, \dots, \boldsymbol{\mu}_k$ are confined, or nearly confined, to some lower dimensional subspace of R^m. Note that if these vectors span an r-dimensional subspace of R^m, then the population version of B,

$$\Phi = \sum_{i=1}^{k} n_i(\boldsymbol{\mu}_i - \boldsymbol{\mu})(\boldsymbol{\mu}_i - \boldsymbol{\mu})',$$

where $\boldsymbol{\mu} = \Sigma n_i \boldsymbol{\mu}_i / n$, will have rank r; in fact, the eigenvectors of Φ corresponding to its positive eigenvalues will span this r-dimensional space. Thus, a plot of the projections of $\boldsymbol{\mu}_1, \ldots, \boldsymbol{\mu}_k$ onto this subspace will yield a reduced-dimension diagram of the population means. Unfortunately, if $\Omega \neq I_m$, it will be difficult to interpret the differences in these mean vectors since Euclidean distance would not be appropriate. This difficulty can be resolved by analyzing the transformed data $\Omega^{-1/2}\mathbf{y}_{ij}$, where $\Omega'^{-1/2}\Omega^{-1/2} = \Omega^{-1}$, since $\Omega^{-1/2}\mathbf{y}_{ij} \sim N_m(\Omega^{-1/2}\boldsymbol{\mu}_i, I_m)$. Thus, we would plot the projections of $\Omega^{-1/2}\boldsymbol{\mu}_1, \ldots, \Omega^{-1/2}\boldsymbol{\mu}_k$ onto the subspace spanned by the eigenvectors of $\Omega^{-1/2}\Phi\Omega'^{-1/2}$ corresponding to its r positive eigenvalues; that is, if the spectral decomposition of $\Omega^{-1/2}\Phi\Omega'^{-1/2}$ is given by $P_1\Lambda_1 P_1'$, where P_1 is an $m \times r$ matrix satisfying $P_1'P_1 = I_r$ and Λ_1 is an $r \times r$ diagonal matrix, then we could simply plot the vectors $P_1'\Omega^{-1/2}\boldsymbol{\mu}_1, \ldots, P_1'\Omega^{-1/2}\boldsymbol{\mu}_k$ in R^r. The r components of the vector $\boldsymbol{v}_i = P_1'\Omega^{-1/2}\boldsymbol{\mu}_i$ in this r-dimensional space are called the canonical variates means for the ith population. Note that in obtaining these canonical variates we have essentially used the simultaneous diagonalization of Φ and Ω, since if $C' = (C_1', C_2')$ satisfies

$$\begin{bmatrix} C_1 \\ C_2 \end{bmatrix} \Phi [C_1' \quad C_2'] = \begin{bmatrix} \Lambda_1 & (0) \\ (0) & (0) \end{bmatrix}, \qquad \begin{bmatrix} C_1 \\ C_2 \end{bmatrix} \Omega [C_1' \quad C_2'] = \begin{bmatrix} I_r & (0) \\ (0) & I_{m-r} \end{bmatrix},$$

then we can take $C_1 = P_1'\Omega^{-1/2}$. When $\boldsymbol{\mu}_1, \ldots, \boldsymbol{\mu}_k$ are unknown, the canonical variate means can be estimated by the sample canonical variate means, which are computed using the samples means $\bar{\mathbf{y}}_1, \ldots, \bar{\mathbf{y}}_k$ and the corresponding simultaneous diagonalization of B and W.

The matrix C that diagonalizes A and B in Theorem 4.14 is nonsingular but not necessarily orthogonal. Further, the diagonal elements of the two diagonal matrices are not the eigenvalues of A nor B. This sort of diagonalization, one which will be useful in our study of quadratic forms in normal random vectors ins Chapter 9, is what we consider next; we would like to know whether or not there exists an orthogonal matrix that diagonalizes both A and B. The following result gives a necessary and sufficient condition for such an orthogonal matrix to exist.

Theorem 4.15. Suppose that A and B are $m \times m$ symmetric matrices. Then there exists an orthogonal matrix P such that $P'AP$ and $P'BP$ are both diagonal if and only if A and B commute; that is, if and only if $AB = BA$.

Proof. First suppose that such an orthogonal matrix does exist; that is, there is an orthogonal matrix P such $P'AP = \Lambda_1$ and $P'BP = \Lambda_2$, where Λ_1 and

Λ_2 are diagonal matrices. Then since Λ_1 and Λ_2 are diagonal matrices, clearly $\Lambda_1\Lambda_2 = \Lambda_2\Lambda_1$, so we have

$$AB = P\Lambda_1 P' P\Lambda_2 P' = P\Lambda_1\Lambda_2 P' = P\Lambda_2\Lambda_1 P' = P\Lambda_2 P' P\Lambda_1 P' = BA$$

and, hence, A and B do commute. Conversely, now assuming that $AB = BA$, we need to show that such an orthogonal matrix P does exist. Let $\mu_1, \ldots, \mu_h$ be the distinct values of the eigenvalues of A having multiplicities $r_1, \ldots, r_h$, respectively. Since A is symmetric there exists an orthogonal matrix Q satisfying

$$Q'AQ = \Lambda_1 = \text{diag}(\mu_1 \text{I}_{r_1}, \ldots, \mu_h \text{I}_{r_h})$$

Performing this same transformation on B and partitioning the resulting matrix in the same way that $Q'AQ$ has been partitioned, we get

$$C = Q'BQ = \begin{bmatrix} C_{11} & C_{12} & \cdots & C_{1h} \\ C_{21} & C_{22} & \cdots & C_{2h} \\ \vdots & \vdots & & \vdots \\ C_{h1} & C_{h2} & \cdots & C_{hh} \end{bmatrix},$$

where C_{ij} is $r_i \times r_j$. Note that since $AB = BA$, we must have

$$\Lambda_1 C = Q'AQQ'BQ = Q'ABQ = Q'BAQ = Q'BQQ'AQ = C\Lambda_1$$

Equating the (i,j)th submatrix of $\Lambda_1 C$ to the (i,j)th submatrix of $C\Lambda_1$ yields the identity $\mu_i C_{ij} = \mu_j C_{ij}$. Since $\mu_i \neq \mu_j$ if $i \neq j$, we must have $C_{ij} = (0)$ if $i \neq j$; that is, the matrix $C = \text{diag}(C_{11}, \ldots, C_{hh})$ is block diagonal. Now since C is symmetric so also is C_{ii} for each i, and, thus, we can find an $r_i \times r_i$ orthogonal matrix X_i satisfying

$$X_i' C_{ii} X_i = \Delta_i,$$

where Δ_i is diagonal. Let $P = QX$, where X is the block diagonal matrix $X = \text{diag}(X_1, \ldots, X_h)$, and note that

$$\begin{aligned} P'P = X'Q'QX = X'X &= \text{diag}(X_1'X_1, \ldots, X_h'X_h) \\ &= \text{diag}(\text{I}_{r_1}, \ldots, \text{I}_{r_h}) = \text{I}_m, \end{aligned}$$

so that P is orthogonal. Finally, the matrix $\Delta = \text{diag}(\Delta_1, \ldots, \Delta_h)$ is diagonal and

$$\begin{aligned} P'AP &= X'Q'AQX = X'\Lambda_1 X \\ &= \text{diag}(X_1', \ldots, X_h')\,\text{diag}(\mu_1 \mathrm{I}_{r_1}, \ldots, \mu_h \mathrm{I}_{r_h})\,\text{diag}(X_1, \ldots, X_h) \\ &= \text{diag}(\mu_1 X_1' X_1, \ldots, \mu_h X_h' X_h) = \text{diag}(\mu_1 \mathrm{I}_{r_1}, \ldots, \mu_h \mathrm{I}_{r_h}) = \Lambda_1, \end{aligned}$$

and

$$\begin{aligned} P'BP &= X'Q'BQX = X'CX \\ &= \text{diag}(X_1', \ldots, X_h')\,\text{diag}(C_{11}, \ldots, C_{hh})\,\text{diag}(X_1, \ldots, X_h) \\ &= \text{diag}(X_1' C_{11} X_1, \ldots, X_h' C_{hh} X_h) = \text{diag}(\Delta_1, \ldots, \Delta_h) = \Delta, \end{aligned}$$

and so the proof is complete. □

The columns of the matrix P are eigenvectors of A as well as B; that is, A and B commute if and only if the two matrices have common eigenvectors. Also, note that since A and B are symmetric, $(AB)' = B'A' = BA$, and so $AB = BA$ if and only if AB is symmetric. The previous theorem easily generalizes to a collection of symmetric matrices.

Theorem 4.16. Let $A_1, \ldots, A_k$ be $m \times m$ symmetric matrices. Then there exists an orthogonal matrix P such that $P'A_iP = \Lambda_i$ is diagonal for each i if and only if $A_iA_j = A_jA_i$ for all pairs (i, j).

The two previous theorems involving symmetric matrices are special cases of more general results regarding diagonalizable matrices. For instance, Theorem 4.16 is a special case of the following result. The proof, which is similar to that given for Theorem 4.15, is left as an exercise.

Theorem 4.17. Suppose that each of the $m \times m$ matrices $A_1, \ldots, A_k$ is diagonalizable. Then there exists a nonsingular matrix X such that $X^{-1}A_iX = \Lambda_i$ is diagonal for each i if and only if $A_iA_j = A_jA_i$ for all pairs (i, j).

8. MATRIX NORMS

In Chapter 2, we saw that vector norms can be used to measure the size of a vector. Similarly, we may be interested in measuring the size of an $m \times m$ matrix A or measuring the closeness of A to another $m \times m$ matrix B. Matrix norms will provide the means to do this. In a later chapter, we will need to apply some of our results on matrix norms to matrices that are possibly complex matrices. Consequently, throughout this section, we will not be restricting attention only to real matrices.

Definition 4.3. A function $||A||$ defined on all $m \times m$ matrices A, real or complex, is a matrix norm if the following conditions hold for all $m \times m$ matrices A and B.

(a) $||A|| \geq 0$.
(b) $||A|| = 0$ if and only if $A = (0)$.
(c) $||cA|| = |c|\,||A||$ for any complex scalar c.
(d) $||A + B|| \leq ||A|| + ||B||$.
(e) $||AB|| \leq ||A||\,||B||$.

Any vector norm defined on $m^2 \times 1$ vectors, when applied to the $m^2 \times 1$ vector formed by stacking the columns of A, one on top of the other, will satisfy conditions (a)–(d) since these are the conditions of a vector norm. However, condition (e), which relates the sizes of A and B to that of AB, will not necessarily hold for vector norms; that is, not all vector norms can be used as matrix norms.

We now give examples of some commonly encountered matrix norms. We will leave it to the reader to verify that these functions, in fact, satisfy the conditions of Definition 4.3. The Euclidean matrix norm is simply the Euclidean vector norm computed on the stacked columns of A, and so is given by

$$||A||_E = \left(\sum_{i=1}^{m} \sum_{j=1}^{m} |a_{ij}|^2 \right)^{1/2} = \{\text{tr}(A^*A)\}^{1/2}$$

The maximum column sum matrix norm is given by

$$||A||_1 = \max_{1 \leq j \leq m} \sum_{i=1}^{m} |a_{ij}|,$$

while the maximum row sum matrix norm is given by

$$||A||_\infty = \max_{1 \leq i \leq m} \sum_{j=1}^{m} |a_{ij}|$$

The spectral norm utilizes the eigenvalues of A^*A; in particular, if $\mu_1, \ldots, \mu_m$ are the eigenvalues of A^*A, then the spectral norm is given by

$$||A||_2 = \max_{1 \leq i \leq m} \sqrt{\mu_i}$$

We will find the following theorem useful. The proof, which simply involves the verification of the conditions of Definition 4.3, is left to the reader as an exercise.

Theorem 4.18. Let $||A||$ be any matrix norm defined on $m \times m$ matrices. If C is an $m \times m$ nonsingular matrix, then the function defined by

$$||A||_C = ||C^{-1}AC||$$

is also a matrix norm.

The eigenvalues of a matrix A play an important role in the study of matrix norms of A. Particularly important is the maximum modulus of this set of eigenvalues.

Definition 4.4. Let $\lambda_1, \ldots, \lambda_m$ be the eigenvalues of the $m \times m$ matrix A. The spectral radius of A, denoted $\rho(A)$, is defined to be

$$\rho(A) = \max_{1 \leq i \leq m} |\lambda_i|$$

Although $\rho(A)$ does give us some information about the size of A, it is not a matrix norm itself. To see this, consider the case in which $m = 2$ and

$$A = \begin{bmatrix} 0 & 1 \\ 0 & 0 \end{bmatrix}$$

Both of the eigenvalues of A are 0, so $\rho(A) = 0$ even though A is not the null matrix; that is, $\rho(A)$ violates condition (b) of Definition 4.3. The following result shows that $\rho(A)$ actually serves as a lower bound for any matrix norm of A.

Theorem 4.19. For any $m \times m$ matrix A and any matrix norm $||A||$, $\rho(A) \leq ||A||$.

Proof. Suppose that λ is an eigenvalue of A for which $|\lambda| = \rho(A)$, and let $\boldsymbol{x}$ be a corresponding eigenvector, so that $A\boldsymbol{x} = \lambda\boldsymbol{x}$. Then $\boldsymbol{x}\boldsymbol{1}_m'$ is an $m \times m$ matrix satisfying $A\boldsymbol{x}\boldsymbol{1}_m' = \lambda\boldsymbol{x}\boldsymbol{1}_m'$, and so using properties (c) and (e) of matrix norms, we find that

$$\rho(A)||\boldsymbol{x}\boldsymbol{1}_m'|| = |\lambda|\,||\boldsymbol{x}\boldsymbol{1}_m'|| = ||\lambda\boldsymbol{x}\boldsymbol{1}_m'|| = ||A\boldsymbol{x}\boldsymbol{1}_m'|| \leq ||A||\,||\boldsymbol{x}\boldsymbol{1}_m'||.$$

The result now follows by dividing the equation above by $||\boldsymbol{x}\boldsymbol{1}_m'||$. □

Although the spectral radius of A is at least as small as every norm of A, our next result shows that we can always find a matrix norm so that $\|A\|$ is arbitrarily close to $\rho(A)$.

Theorem 4.20. For any $m \times m$ matrix A and any scalar $\epsilon > 0$, there exists a matrix norm, $\|A\|_{A,\epsilon}$, such that

$$\|A\|_{A,\epsilon} - \rho(A) < \epsilon$$

Proof. Let $A = XTX^*$ be the Schur decomposition of A, so that X is a unitary matrix and T is an upper triangular matrix with the eigenvalues of A, $\lambda_1, \ldots, \lambda_m$, as its diagonal elements. For any scalar $c > 0$, define the matrix $D_c = \text{diag}(c, c^2, \ldots, c^m)$ and note that the diagonal elements of the upper triangular matrix $D_c T D_c^{-1}$ are also $\lambda_1, \ldots, \lambda_m$. Further, the ith column sum of $D_c T D_c^{-1}$ is given by

$$\lambda_i + \sum_{j=1}^{i-1} c^{-(i-j)} t_{ji}$$

Clearly, by choosing c large enough, we can guarantee that

$$\sum_{j=1}^{i-1} |c^{-(i-j)} t_{ji}| < \epsilon,$$

for each i. In this case, since $|\lambda_i| \le \rho(A)$, we must have

$$\|D_c T D_c^{-1}\|_1 < \rho(A) + \epsilon,$$

where $\|A\|_1$ denotes the maximum column sum matrix norm previously defined. For any $m \times m$ matrix B, define $\|B\|_{A,\epsilon}$ as

$$\|B\|_{A,\epsilon} = \|(XD_c^{-1})^{-1} B (XD_c^{-1})\|_1$$

The result now follows from Theorem 4.18 and the fact that

$$\|A\|_{A,\epsilon} = \|(XD_c^{-1})^{-1} A (XD_c^{-1})\|_1 = \|D_c T D_c^{-1}\|_1 \qquad \square$$

Often we will be interested in the limit of a sequence of vectors or the limit of a sequence of matrices. The sequence of $m \times 1$ vectors, $\boldsymbol{x}_1, \boldsymbol{x}_2, \ldots$ converges to the $m \times 1$ vector $\boldsymbol{x}$ if the jth component of $\boldsymbol{x}_k$ converges to the jth component of $\boldsymbol{x}$, as $k \to \infty$, for each j; that is, $|x_{jk} - x_j| \to 0$, as $k \to \infty$, for each j. Similarly, a sequence of $m \times m$ matrices, $A_1, A_2, \ldots$ converges to the $m \times m$ matrix A if each element of A_k converges to the corresponding element of A as $k \to \infty$. Alternatively, we can consider the notion of the convergence of a sequence with respect to a specific norm. Thus, the sequence of vectors $\boldsymbol{x}_1, \boldsymbol{x}_2, \ldots$ converges to $\boldsymbol{x}$, with respect to the vector norm $\|\boldsymbol{x}\|$, if $\|\boldsymbol{x}_k - \boldsymbol{x}\| \to 0$ as $k \to \infty$. The following result indicates that the actual choice of a norm is not important. For a proof of this result, see Horn and Johnson (1985).

Theorem 4.21. Let $\|\boldsymbol{x}\|_a$ and $\|\boldsymbol{x}\|_b$ be any two vector norms defined on any $m \times 1$ vector $\boldsymbol{x}$. If $\boldsymbol{x}_1, \boldsymbol{x}_2, \ldots$ is a sequence of $m \times 1$ vectors, then $\boldsymbol{x}_k$ converges to $\boldsymbol{x}$, as $k \to \infty$, with respect to $\|\boldsymbol{x}\|_a$ if and only if $\boldsymbol{x}_k$ converges to $\boldsymbol{x}$, as $k \to \infty$, with respect to $\|\boldsymbol{x}\|_b$.

Since the first four conditions of a matrix norm are the conditions of a vector norm, the previous theorem immediately leads to the following.

Corollary 4.21.1. Let $\|A\|_a$ and $\|A\|_b$ be any two matrix norms defined on any $m \times m$ matrix A. If $A_1, A_2, \ldots$ is a sequence of $m \times m$ matrices, then A_k converges to A, as $k \to \infty$, with respect to $\|A\|_a$ if and only if A_k converges to A, as $k \to \infty$, with respect to $\|A\|_b$.

A sequence of matrices that is sometimes of interest is the sequence, $A, A^2, A^3, \ldots$, formed from a fixed $m \times m$ matrix A. A sufficient condition for this sequence of matrices to converge to the null matrix is given next.

Theorem 4.22. Let A be an $m \times m$ matrix, and suppose that for some matrix norm, $\|A\| < 1$. Then $\lim A^k = (0)$, as $k \to \infty$.

Proof. By repeatedly using condition (e) of a matrix norm, we find that $\|A^k\| \leq \|A\|^k$, and so $\|A^k\| \to 0$, as $k \to \infty$, since $\|A\| < 1$. Thus, A^k converges to (0) with respect to the norm $\|A\|$. But by Corollary 4.21.1, A^k also converges to (0) with respect to the matrix norm (see Problem 4.37)

$$\|A\|_* = m\left(\max_{1 \leq i,j \leq m} |a_{ij}|\right)$$

But this implies that $|a_{ij}^k| \to 0$, as $k \to \infty$, for each (i, j) and so the proof is complete. □

Our next result relates the convergence of A^k to (0), to the size of the spectral radius of A.

Theorem 4.23. Suppose that A is an $m \times m$ matrix. Then A^k converges to (0), as $k \to \infty$, if and only if $\rho(A) < 1$.

Proof. Suppose that $A^k \to (0)$, in which case, $A^k\boldsymbol{x} \to \mathbf{0}$ for any $m \times 1$ vector $\boldsymbol{x}$. Now if $\boldsymbol{x}$ is an eigenvector of A corresponding to the eigenvalue λ, we must also have $\lambda^k\boldsymbol{x} \to \mathbf{0}$, since $A^k\boldsymbol{x} = \lambda^k\boldsymbol{x}$. This can only happen if $|\lambda| < 1$, and so $\rho(A) < 1$, since λ was an arbitrary eigenvalue of A. On the other hand, if $\rho(A) < 1$, then we know from Theorem 4.20 that there is a matrix norm satisfying $||A|| < 1$. Hence, it follows from Theorem 4.22 that $A^k \to (0)$. □

Our final result shows that the spectral radius of A is the limit of a particular sequence that can be computed from any matrix norm.

Theorem 4.24. Let A be an $m \times m$ matrix. Then for any matrix norm $||A||$

$$\lim_{k \to \infty} ||A^k||^{1/k} = \rho(A)$$

Proof. λ is an eigenvalue of A if and only if λ^k is an eigenvalue of A^k. Further, $|\lambda|^k = |\lambda^k|$, so $\rho(A)^k = \rho(A^k)$. This, along with Theorem 4.19, yields $\rho(A)^k \leq ||A^k||$, or equivalently, $\rho(A) \leq ||A^k||^{1/k}$. Thus, the proof will be complete if we can show that for arbitrary $\epsilon > 0$, there exists an integer N_ϵ such that $||A^k||^{1/k} < \rho(A) + \epsilon$ for all $k > N_\epsilon$. But this is the same as showing that there exists an integer N_ϵ such that for all $k > N_\epsilon$, $||A^k|| < \{\rho(A) + \epsilon\}^k$, or equivalently,

$$||B^k|| < 1, \tag{4.10}$$

where $B = \{\rho(A) + \epsilon\}^{-1}A$. Now (4.10) follows immediately from Theorem 4.23 since

$$\rho(B) = \frac{\rho(A)}{\rho(A) + \epsilon} < 1$$

□

PROBLEMS

1. Obtain a singular value decomposition for the matrix

$$A = \begin{bmatrix} 1 & 2 & 2 & 1 \\ 1 & 1 & 1 & -1 \end{bmatrix}$$

2. Let A be an $m \times n$ matrix.
   (a) Show that the singular values of A are the same as those of A'.
   (b) Show that the singular values of A are the same as those of FAG, if F and G are orthogonal matrices.

(c) If $\alpha \neq 0$ is a scalar, how do the singular values of αA compare to those of A?

3. Let A be an $m \times m$ matrix. Show that A has a zero eigenvalue if and only if it has fewer than m singular values.

4. Let A be $m \times n$ and B be $n \times m$. We will see in Chapter 7 that the nonzero eigenvalues of AB are the same as those of BA. This is not necessarily true for the singular values. Give an example of matrices A and B for which the nonzero singular values of AB are not the same as those of BA.

5. Let A be an $m \times n$ matrix having rank r and singular values $\mu_1, \ldots, \mu_r$. Show that the $(m+n) \times (m+n)$ matrix

$$B = \begin{bmatrix} (0) & A \\ A' & (0) \end{bmatrix}$$

has eigenvalues $\mu_1, \ldots, \mu_r, -\mu_1, \ldots, -\mu_r$, with the remaining eigenvalues being zero.

6. Find a singular value decomposition for the vector $\boldsymbol{x} = (1, 5, 7, 5)'$.

7. Let $\boldsymbol{x}$ be an $m \times 1$ nonnull vector and $\boldsymbol{y}$ be an $n \times 1$ nonnull vector. Obtain a singular value decomposition of $\boldsymbol{x}\boldsymbol{y}'$ in terms of $\boldsymbol{x}$ and $\boldsymbol{y}$.

8. Let A be an $m \times n$ matrix and let $A = P_1 \Delta Q_1'$ be the decomposition given in Corollary 4.1.1. Define the $n \times m$ matrix B as $B = Q_1 \Delta^{-1} P_1'$. Simplify, as much as possible, the expressions for ABA and BAB.

9. If t is an estimator of θ, then the mean squared error (MSE) of t is defined by

$$\text{MSE}(t) = \text{var}(t) + \{\text{E}(t) - \theta\}^2$$

Consider the multicollinearity problem discussed in Example 4.4 in which r of the singular values of Z_1 are very small relative to the others. Suppose that we want to estimate the response variable corresponding to an observation which has the standardized explanatory variables at the values given in the $k \times 1$ vector $\boldsymbol{z}$. Let $\hat{y} = \bar{y} + \boldsymbol{z}'(Z_1'Z_1)^{-1}Z_1'\boldsymbol{y}$ be the estimate obtained using ordinary least squares, while $\tilde{y} = \bar{y} + \boldsymbol{z}'U_1 D_1^{-1} V_1'\boldsymbol{y}$ is the estimate obtained using principal components regression. Assume throughout that $\boldsymbol{\epsilon} \sim \text{N}_N(\boldsymbol{0}, \sigma^2 \text{I}_N)$.

(a) Show that if the vector $\boldsymbol{v} = (v_1, \ldots, v_N)'$ satisfies $\boldsymbol{z}' = \boldsymbol{v}'DU'$, then

$$\text{MSE}(\hat{y}) = \sigma^2 \left(N^{-1} + \sum_{i=1}^{k} v_i^2 \right)$$

(b) Show that

$$\text{MSE}(\tilde{y}) = \sigma^2 \left(N^{-1} + \sum_{i=1}^{k-r} v_i^2 \right) + \left(\sum_{i=k-r+1}^{k} d_i v_i \alpha_i \right)^2,$$

where d_i is the ith diagonal element of D.

(c) If $r = 1$, when will $\text{MSE}(\tilde{y}) < \text{MSE}(\hat{y})$?

10. Suppose that ten observations are obtained in a process involving two explanatory variables and a response variable resulting in the following data:

x_1	x_2	y
−2.49	6.49	28.80
0.85	4.73	21.18
−0.78	4.24	24.73
−0.75	5.54	25.34
1.16	4.74	28.50
−1.52	5.86	27.19
−0.51	5.65	26.22
−0.05	4.50	20.71
−1.01	5.75	25.47
0.13	5.69	29.83

(a) Obtain the matrix of standardized explanatory variables Z_1, use ordinary least squares to estimate the parameters in the model $\boldsymbol{y} = \delta_0 \mathbf{1}_N + Z_1 \boldsymbol{\delta}_1 + \boldsymbol{\epsilon}$, and obtain the fitted values $\hat{\boldsymbol{y}} = \hat{\delta}_0 \mathbf{1}_N + Z_1 \hat{\boldsymbol{\delta}}_1$.

(b) Compute the singular value decomposition of Z_1. Then use principal components regression to obtain an alternative vector of fitted values.

(c) Use both models of (a) and (b) to estimate the response variable for an observation having $x_1 = -2$ and $x_2 = 4$.

11. Consider the 3×3 symmetric matrix given by

$$A = \begin{bmatrix} 3 & 1 & -1 \\ 1 & 3 & 1 \\ -1 & 1 & 3 \end{bmatrix}$$

(a) Find the spectral decomposition of A.
(b) Find a symmetric square root matrix for A.
(c) Find a nonsymmetric square root matrix for A.

12. Use the spectral decomposition theorem to prove Theorem 4.5.

13. Find a 3×2 matrix T such that $TT' = A$, where

$$A = \begin{bmatrix} 5 & 4 & 0 \\ 4 & 5 & 3 \\ 0 & 3 & 5 \end{bmatrix}$$

14. Suppose $\boldsymbol{x} \sim N_3(\mathbf{0}, \Omega)$, where

$$\Omega = \begin{bmatrix} 2 & 1 & 1 \\ 1 & 2 & 1 \\ 1 & 1 & 2 \end{bmatrix}$$

Find a 3×3 matrix A such that the components of $\boldsymbol{z} = A\boldsymbol{x}$ are independently distributed.

15. Let the matrices A, B, and C be given by

$$A = \begin{bmatrix} 1 & 2 & 5 \\ 2 & 1 & 4 \\ -1 & 1 & 1 \end{bmatrix}, \qquad B = \begin{bmatrix} 1 & 1 & -1 \\ -2 & 2 & 2 \\ -1 & 3 & 1 \end{bmatrix}, \qquad C = \begin{bmatrix} 2 & 1 & -1 \\ 2 & 5 & 3 \\ -2 & -1 & 1 \end{bmatrix}$$

(a) Which of these matrices are diagonalizable?
(b) Which of these matrices have their rank equal to the number of nonzero eigenvalues?

16. Let A be an $m \times m$ matrix and B be an $n \times n$ matrix. Prove that the matrix

$$C = \begin{bmatrix} A & (0) \\ (0) & B \end{bmatrix}$$

is diagonalizable if and only if the matrices A and B are diagonalizable.

Using induction, show that the square matrices $A_1, \ldots, A_k$ are diagonalizable if and only if $\text{diag}(A_1, \ldots, A_k)$ is diagonalizable.

17. Find a 4×4 matrix A having eigenvalues 0 and 1 with multiplicities 3 and 1, respectively, such that
    (a) the rank of A is 1,
    (b) the rank of A is 2,
    (c) the rank of A is 3.

18. Repeat Example 4.9 for 5×5 matrices; that is, obtain a collection of 5×5 matrices in Jordan canonical form such that every 5×5 matrix having the eigenvalue λ with multiplicity 5 is similar to one of the matrices in this set.

19. Consider the 6×6 matrix

$$J = \begin{bmatrix} 2 & 1 & 0 & 0 & 0 & 0 \\ 0 & 2 & 0 & 0 & 0 & 0 \\ 0 & 0 & 2 & 1 & 0 & 0 \\ 0 & 0 & 0 & 2 & 0 & 0 \\ 0 & 0 & 0 & 0 & 3 & 1 \\ 0 & 0 & 0 & 0 & 0 & 3 \end{bmatrix},$$

    which is in Jordan canonical form.
    (a) Find the eigenvalues of J and their multiplicities.
    (b) Find the eigenspaces of J.

20. An $m \times m$ matrix B is said to be nilpotent if $B^k = (0)$ for some positive integer k.
    (a) Show that $J_h(\lambda) = \lambda I_h + B_h$, where B_h is nilpotent. In particular, show that $B_h^h = (0)$.
    (b) Let $J = \text{diag}(J_{h_1}(\lambda_1), \ldots, J_{h_r}(\lambda_r))$ be a Jordan canonical form. Show that J can be written as $J = D + B$, where D is diagonal and B is nilpotent. What is the smallest h such that $B^h = (0)$?
    (c) Use part (b) to show that if A is similar to J, then A can be expressed as $A = F + G$, where F is diagonalizable and G is nilpotent.

21. Let A be an $m \times m$ nilpotent matrix. In Problem 3.27, it was shown that all of the eigenvalues of A are 0. Use this and the Jordan canonical form of A to show that there must be a positive integer $h \leq m$ satisfying $A^h = (0)$.

22. Let A be an $m \times m$ matrix. Show that the rank of A is equal to the number of nonzero eigenvalues of A if and only if $\text{rank}(A^2) = \text{rank}(A)$.

23. Suppose that λ is an eigenvalue of A with multiplicity r. Show that there are r linearly independent eigenvectors of A corresponding to λ if and only if $\text{rank}(A - \lambda \text{I}) = \text{rank}\{(A - \lambda \text{I})^2\}$.

24. Let A and B be $m \times m$ matrices. Suppose that there exists an $m \times m$ unitary matrix X such X^*AX and X^*BX are both upper triangular matrices. Show then that the eigenvalues of $AB - BA$ are all equal to 0.

25. Let T and U be $m \times m$ upper triangular matrices. In addition, suppose that for some positive integer $r < m$, $t_{ij} = 0$ for $1 \leq i \leq r$, $1 \leq j \leq r$, and $u_{r+1,r+1} = 0$. Show that the upper triangular matrix $V = TU$ is such that $v_{ij} = 0$ for $1 \leq i \leq r+1$, $1 \leq j \leq r+1$.

26. Use the Schur decomposition of a matrix A and the result of the previous exercise to prove the Cayley–Hamilton theorem given as Theorem 3.7; that is, if $\lambda_1, \ldots, \lambda_m$ are the eigenvalues of A, show that

$$(A - \lambda_1 \text{I})(A - \lambda_2 \text{I}) \cdots (A - \lambda_m \text{I}) = (0).$$

27. Obtain a Schur decomposition for the matrix C given in Problem 15.

28. Repeat Problem 27 by obtaining a different Schur decomposition of C.

29. Let A be $m \times n$, with $m \leq n$. Show that there exist an $m \times m$ nonnegative definite matrix B and an $m \times n$ matrix H such that $HH' = \text{I}_m$ and $A = BH$.

30. Suppose that A and B are $m \times m$ and diagonalizable. Show that A and B commute; that is, $AB = BA$ if and only if they are simultaneously diagonalizable; in other words, $AB = BA$, if and only if there exists a nonsingular matrix X such that both $X^{-1}AX$ and $X^{-1}BX$ are diagonal matrices. This proves Theorem 4.17 when $k = 2$.

31. Let

$$A = \begin{bmatrix} 1 & 0 \\ 0 & 1 \end{bmatrix}, \qquad B = \begin{bmatrix} 0 & 1 \\ 0 & 0 \end{bmatrix}$$

(a) Show that $AB = BA$.
(b) Show that AB is not diagonalizable.
(c) Why does this not contradict the result of Problem 30?

32. Suppose that the $m \times m$ matrices A and B are diagonalizable and $AB = BA$. Denote the eigenvalues of A by $\lambda_1, \ldots, \lambda_m$ and those of B by $\mu_1, \ldots, \mu_m$.

If the eigenvalues of $A + B$ are $\gamma_1, \ldots, \gamma_m$, show that for $k = 1, \ldots, m$,

$$\gamma_k = \lambda_{i_k} + \mu_{j_k},$$

where $(i_1, \ldots, i_m)$ and $(j_1, \ldots, j_m)$ are permutations of $(1, \ldots, m)$.

33. The following is a generalization of Theorem 4.14 to arbitrary nonnegative definite matrices. Let A and B be $m \times m$ nonnegative definite matrices with $\text{rank}(A) = r \leq s = \text{rank}(B)$. Show that there exists a nonsingular matrix C such that

$$CAC' = \begin{bmatrix} D_1 & (0) \\ (0) & (0) \end{bmatrix}, \quad CBC' = \begin{bmatrix} D_2 & (0) \\ (0) & (0) \end{bmatrix},$$

where D_1 and D_2 are $t \times t$ diagonal matrices, $t \geq s$, and

$$D_1 = \begin{bmatrix} \mathrm{I}_r & (0) \\ (0) & (0) \end{bmatrix}$$

34. Let A and B be $m \times m$ matrices and suppose that A and B commute.
   (a) If A and B are nonsingular, show that A^{-1} and B^{-1} commute.
   (b) If i and j are positive integers, show that A^i and B^j commute.

35. Suppose that A and B are $m \times m$ positive definite matrices. Show that $A - B$ is positive definite if and only if $B^{-1} - A^{-1}$ is positive definite.

36. Show that the functions, $\|A\|_E$, $\|A\|_1$, $\|A\|_\infty$, and $\|A\|_2$ given in Section 4.8 are, in fact, matrix norms.

37. Let A be an $m \times m$ matrix and consider the function

$$\|A\|_* = m(\max_{1 \leq i,j \leq m} |a_{ij}|)$$

Show that $\|A\|_*$ is a matrix norm.

38. Prove Theorem 4.18.

39. For any matrix norm defined on $m \times m$ matrices, show that
   (a) $\|\mathrm{I}_m\| \geq 1$,
   (b) $\|A^{-1}\| \geq \|A\|^{-1}$, if A is an $m \times m$ nonsingular matrix.

40. Show that if for some matrix norm $\|\mathrm{I}_m - A\| < 1$ then A is a nonsingular matrix.

41. Consider the 2×2 matrix of the form

$$A = \begin{bmatrix} a & 1 \\ 0 & a \end{bmatrix}$$

(a) Determine A^k for general positive integer k.
(b) Find $\rho(A)$ and $\rho(A^k)$.
(c) For which values of a does A^k converge to (0) as $k \to \infty$? In this case, show how to construct a norm so that $||A|| < 1$.

42. Let A be an $m \times m$ matrix. Show that if for some matrix norm $||A|| < 1$ then the matrix $I_m - A$ has an inverse and

$$(I_m - A)^{-1} = I_m + \sum_{k=1}^{\infty} A^k$$

43. In this problem, we consider a factorization of an $m \times m$ matrix A of the form $A = LU$, where L is an $m \times m$ lower triangular matrix and U is an $m \times m$ upper triangular matrix.
(a) Let A_j be the $j \times j$ submatrix of A consisting of the first j rows and j columns of A. Show that if $r = \text{rank}(A)$ and $|A_j| \neq 0$, $j = 1, \dots, r$, then A_r can be factored as $A_r = L_* U_*$, where L_* is an $r \times r$ nonsingular lower triangular matrix and U_* is an $r \times r$ nonsingular upper triangular matrix. Use this to then show that A may be factored as $A = LU$, where L is an $m \times m$ lower triangular matrix and U is an $m \times m$ upper triangular matrix.
(b) Show that not every $m \times m$ matrix has an LU factorization by finding a 2×2 matrix that cannot be factored in this way.
(c) Show how the LU factorization of A can be used to simplify the computation of a solution $\boldsymbol{x}$, to the system of equations $A\boldsymbol{x} = \boldsymbol{c}$.

44. Suppose that A is an $m \times m$ matrix. Show that there exist an $m \times m$ lower triangular matrix L, an $m \times m$ upper triangular matrix U, and $m \times m$ permutation matrices P and Q, such that $A = PLUQ$.

45. Suppose that A is an $m \times m$ matrix for which $|A_j| \neq 0, j = 1, \dots, m$, where A_j denotes the $j \times j$ submatrix of A consisting of the first j rows and j columns of A.
(a) Show that there exist $m \times m$ lower triangular matrices L and M having all diagonal elements equal to one and an $m \times m$ diagonal matrix D, such that $A = LDM'$.
(b) Show that if A is also symmetric, then $M = L$ so that $A = LDL'$.

CHAPTER FIVE

Generalized Inverses

1. INTRODUCTION

The inverse of a matrix is defined for all square matrices that are nonsingular. There are some situations in which we may have a rectangular matrix or a square singular matrix, A, and still be in need of another matrix that in some ways behaves like the inverse of A. One such situation, which is often encountered in the study of statistics as well as many other fields of application, involves finding solutions to a system of linear equations. A system of linear equations can be written in matrix form as

$$A\boldsymbol{x} = \boldsymbol{c},$$

where A is an $m \times n$ matrix of constants, $\boldsymbol{c}$ is an $m \times 1$ vector of constants, and $\boldsymbol{x}$ is an $n \times 1$ vector of variables for which we need to find solutions. If $m = n$ and A is nonsingular, then A^{-1} exists and so by premultiplying our system of equations by A^{-1}, we see that the system is satisfied only if $\boldsymbol{x} = A^{-1}\boldsymbol{c}$; that is, the system has a solution, the solution is unique, and it is given by $\boldsymbol{x} = A^{-1}\boldsymbol{c}$. When A^{-1} does not exist, how do we determine whether the system has any solutions, and if solutions exist, how many solutions are there, and how do we find them? We will see in the next chapter that the answers to all of these questions can be conveniently expressed in terms of the generalized inverses discussed in this chapter.

A second application of generalized inverses in statistics involves quadratic forms and chi-squared distributions. Suppose we have an m-dimensional random vector $\boldsymbol{x}$ which has a mean vector of zero and covariance matrix Ω. A useful transformation in some situations is one that transforms $\boldsymbol{x}$ to another random vector, $\boldsymbol{z}$, having the identity matrix as its covariance matrix. For instance, in Chapter 9 we will see that if $\boldsymbol{z}$ has a normal distribution, then the sum of squares of the components of $\boldsymbol{z}$, that is $\boldsymbol{z}'\boldsymbol{z}$, has a chi-squared distribution. We saw in Example 4.5 that if T is any $m \times m$ matrix satisfying $\Omega^{-1} = TT'$, then $\boldsymbol{z} = T'\boldsymbol{x}$ will have I_m as its covariance matrix. Then

$$z'z = x'(T')'T'x = x'(TT')x = x'\Omega^{-1}x$$

This, of course, will be possible only if Ω is positive definite. If Ω is positive semidefinite with rank r, then it will be possible to find $m \times m$ matrices A and B, with $A = BB'$, such that when z is defined by $z = B'x$,

$$\text{var}(z) = \begin{bmatrix} I_r & (0) \\ (0) & (0) \end{bmatrix},$$

and $z'z = x'Ax$. We will see later that A is a generalized inverse of Ω and $z'z$ still has a chi-squared distribution if z has a normal distribution.

2. THE MOORE–PENROSE GENERALIZED INVERSE

A very useful generalized inverse in statistical applications is one developed by Moore (1920, 1935) and Penrose (1955). This inverse is defined so as to possess four properties that the inverse of a square nonsingular matrix has.

Definition 5.1. The Moore–Penrose inverse of the $m \times n$ matrix A is the $n \times m$ matrix, denoted by A^+, which satisfies the conditions

$$AA^+A = A \tag{5.1}$$

$$A^+AA^+ = A^+ \tag{5.2}$$

$$(AA^+)' = AA^+ \tag{5.3}$$

$$(A^+A)' = A^+A \tag{5.4}$$

One of the most important features of the Moore–Penrose inverse, one which distinguishes it from other generalized inverses that we will discuss in this chapter, is that it is uniquely defined. This fact, along with the existence of the Moore–Penrose inverse, is established in the following theorem.

Theorem 5.1. Corresponding to each $m \times n$ matrix A, there exists one and only one $n \times m$ matrix A^+ satisfying conditions (5.1)–(5.4).

Proof. First we will prove the existence of A^+. If A is the $m \times n$ null matrix, then it is easily verified that the four conditions in Definition 5.1 are satisfied with $A^+ = (0)$, the $n \times m$ null matrix. If $A \neq (0)$, so that $\text{rank}(A) = r > 0$, then from Corollary 4.1.1, we know there exist $m \times r$ and $n \times r$ matrices P and Q such that $P'P = Q'Q = \mathrm{I}_r$ and

$$A = P\Delta Q',$$

where Δ is a diagonal matrix with positive diagonal elements. Define $A^+ = Q\Delta^{-1}P'$, and note that

$$AA^+A = P\Delta Q'Q\Delta^{-1}P'P\Delta Q' = P\Delta\Delta^{-1}\Delta Q' = P\Delta Q' = A$$
$$A^+AA^+ = Q\Delta^{-1}P'P\Delta Q'Q\Delta^{-1}P' = Q\Delta^{-1}\Delta\Delta^{-1}P' = Q\Delta^{-1}P' = A^+$$
$$AA^+ = P\Delta Q'Q\Delta^{-1}P' = PP' \quad \text{is symmetric}$$
$$A^+A = Q\Delta^{-1}P'P\Delta Q' = QQ' \quad \text{is symmetric}$$

Thus, $A^+ = Q\Delta^{-1}P'$ is a Moore–Penrose inverse of A, and so we have established the existence of the Moore–Penrose inverse. Next, suppose that B and C are any two matrices satisfying conditions (5.1)–(5.4) for A^+. Then using these four conditions we find that

$$AB = (AB)' = B'A' = B'(ACA)' = B'A'(AC)' = (AB)'AC = ABAC = AC,$$

and

$$BA = (BA)' = A'B' = (ACA)'B' = (CA)'A'B' = CA(BA)' = CABA = CA$$

Now using these two identities, we see that

$$B = BAB = BAC = CAC = C$$

Since B and C are identical, the Moore–Penrose inverse is unique. □

We saw in the proof of Theorem 5.1 that the Moore–Penrose inverse of a matrix A is explicitly related to the singular value decomposition of A; that is, this inverse is nothing more than a very simple function of the component matrices making up the singular value decomposition of A.

Definition 5.1 is the definition of a generalized inverse given by Penrose (1955). The following alternative definition, which we will find useful on some occasions, is the original definition given by Moore (1935). This definition utilizes the concept of projection matrices that were discussed in Chapter 2. Recall that if S is a vector subspace of R^m and P_S is its projection matrix, then for any $\boldsymbol{x} \in R^m$, $P_S\boldsymbol{x}$ gives the orthogonal projection of $\boldsymbol{x}$ onto S, while $\boldsymbol{x} - P_S\boldsymbol{x}$ is the component of $\boldsymbol{x}$ orthogonal to S; further, this unique matrix

P_S is given by $\boldsymbol{x}_1\boldsymbol{x}_1' + \cdots + \boldsymbol{x}_r\boldsymbol{x}_r'$, where $\{\boldsymbol{x}_1, \ldots, \boldsymbol{x}_r\}$ is any orthonormal basis for S.

Definition 5.2. Let A be an $m\times n$ matrix. Then the Moore–Penrose inverse of A is the unique $n \times m$ matrix A^+, satisfying

(a) $AA^+ = P_{R(A)}$, (b) $A^+A = P_{R(A^+)}$,

where $P_{R(A)}$ and $P_{R(A^+)}$ are the projection matrices of the range spaces of A and A^+, respectively.

The equivalence of Definitions 5.1 and 5.2 is not immediately obvious. Consequently, we will establish it in the next theorem.

Theorem 5.2. Definition 5.2 is equivalent to Definition 5.1.

Proof. We first show that a matrix A^+ satisfying Definition 5.2 must also satisfy Definition 5.1. Conditions (5.3) and (5.4) follow immediately since by definition, a projection matrix is symmetric, while (5.1) and (5.2) follow since the columns of A are in $R(A)$ imply that

$$AA^+A = P_{R(A)}A = A,$$

and the columns of A^+ are in $R(A^+)$ imply that

$$A^+AA^+ = P_{R(A^+)}A^+ = A^+$$

Conversely, now suppose that A^+ satisfies Definition 5.1. Premultiplying (5.2) by A yields the identity

$$AA^+AA^+ = (AA^+)^2 = AA^+,$$

which along with (5.3) shows that AA^+ is idempotent and symmetric and thus by Theorem 2.19 is a projection matrix. To show that it is the projection matrix of the range space of A, note that for any matrices B and C, for which BC is defined, $R(BC) \subseteq R(B)$. Using this twice along with (5.1), we find that

$$R(A) = R(AA^+A) \subseteq R(AA^+) \subseteq R(A),$$

so that $R(AA^+) = R(A)$. This proves that $P_{R(A)} = AA^+$. A proof of $P_{R(A^+)} = A^+A$ is obtained in a similar fashion using (5.1) and (5.4). □

3. SOME BASIC PROPERTIES OF THE MOORE–PENROSE INVERSE

In this section, we will establish some of the basic properties of the Moore–Penrose inverse, while in some of the subsequent sections, we will look at some more specialized results. First, we have the following theorem.

Theorem 5.3. Let A be an $m \times n$ matrix. Then

(a) $(\alpha A)^+ = \alpha^{-1}A^+$, if $\alpha \neq 0$ is a scalar,
(b) $(A')^+ = (A^+)'$,
(c) $(A^+)^+ = A$,
(d) $A^+ = A^{-1}$, if A is square and nonsingular,
(e) $(A'A)^+ = A^+A^{+\prime}$ and $(AA')^+ = A^{+\prime}A^+$,
(f) $(AA^+)^+ = AA^+$ and $(A^+A)^+ = A^+A$,
(g) $A^+ = (A'A)^+A' = A'(AA')^+$,
(h) $A^+ = (A'A)^{-1}A'$ and $A^+A = \mathrm{I}_n$, if $\mathrm{rank}(A) = n$,
(i) $A^+ = A'(AA')^{-1}$ and $AA^+ = \mathrm{I}_m$, if $\mathrm{rank}(A) = m$,
(j) $A^+ = A'$, if the columns of A are orthogonal, that is, $A'A = \mathrm{I}_n$.

Proof. Each part is proven by simply verifying that the stated inverse satisfies conditions (5.1)–(5.4). Here, we will only verify that $(A'A)^+ = A^+A^{+\prime}$, given in (e), and leave the remaining proofs to the reader. Since A^+ satisfies the four conditions of a Moore–Penrose inverse, we find that

$$\begin{aligned} A'A(A'A)^+A'A &= A'AA^+A^{+\prime}A'A = A'AA^+(AA^+)'A = A'AA^+AA^+A \\ &= A'AA^+A = A'A, \\ (A'A)^+A'A(A'A)^+ &= A^+A^{+\prime}A'AA^+A^{+\prime} = A^+(AA^+)'AA^+A^{+\prime} = A^+AA^+AA^+A^{+\prime} \\ &= A^+AA^+A^{+\prime} = A^+A^{+\prime} = (A'A)^+, \end{aligned}$$

so that $A^+A^{+\prime}$ satisfies conditions (5.1) and (5.2) of the Moore–Penrose inverse $(A'A)^+$. In addition, note that

$$A'A(A'A)^+ = A'AA^+A^{+\prime} = A'(A^+(AA^+)')' = A'(A^+AA^+)' = A'A^{+\prime} = (A^+A)',$$

and A^+A must be symmetric by definition, so condition (5.3) is satisfied for $(A'A)^+ = A^+A^{+\prime}$. Likewise condition (5.4) holds since

$$(A'A)^+A'A = A^+A^{+\prime}A'A = A^+(AA^+)'A = A^+AA^+A = A^+A$$

This then proves that $(A'A)^+ = A^+A^{+\prime}$. □

Example 5.1. Properties (h) and (i) of Theorem 5.3 give useful ways of computing the Moore–Penrose inverse of matrices that have full column rank or full row rank. We will demonstrate this by finding the Moore–Penrose inverses of

$$\boldsymbol{a} = \begin{bmatrix} 1 \\ 1 \end{bmatrix} \quad \text{and} \quad A = \begin{bmatrix} 1 & 2 & 1 \\ 2 & 1 & 0 \end{bmatrix}$$

From property (h), for any vector $\boldsymbol{a} \neq \mathbf{0}$, $\boldsymbol{a}^+$ will be given by $(\boldsymbol{a}'\boldsymbol{a})^{-1}\boldsymbol{a}'$, so here we find that

$$\boldsymbol{a}^+ = [0.5 \quad 0.5]$$

For A, we can use property (i) since $\text{rank}(A) = 2$. Computing AA' and $(AA')^{-1}$, we get

$$AA' = \begin{bmatrix} 6 & 4 \\ 4 & 5 \end{bmatrix}, \qquad (AA')^{-1} = \frac{1}{14}\begin{bmatrix} 5 & -4 \\ -4 & 6 \end{bmatrix},$$

and so

$$A^+ = A'(AA')^{-1} = \frac{1}{14}\begin{bmatrix} 1 & 2 \\ 2 & 1 \\ 1 & 0 \end{bmatrix}\begin{bmatrix} 5 & -4 \\ -4 & 6 \end{bmatrix} = \frac{1}{14}\begin{bmatrix} -3 & 8 \\ 6 & -2 \\ 5 & -4 \end{bmatrix}$$

Our next result establishes a relationship between the rank of a matrix and the rank of its Moore–Penrose inverse.

Theorem 5.4. For any $m \times n$ matrix A,

$$\text{rank}(A) = \text{rank}(A^+) = \text{rank}(AA^+) = \text{rank}(A^+A)$$

Proof. Using condition (5.1) and the fact that the rank of a matrix product cannot exceed the rank of any of the matrices in the product, we find that

$$\text{rank}(A) = \text{rank}(AA^+A) \leq \text{rank}(AA^+) \leq \text{rank}(A^+) \tag{5.5}$$

In a similar fashion, using condition (5.2), we get

$$\text{rank}(A^+) = \text{rank}(A^+AA^+) \leq \text{rank}(A^+A) \leq \text{rank}(A) \tag{5.6}$$

The result follows immediately from (5.5) and (5.6). □

We have seen through Definition 5.2 and Theorem 5.2 that A^+A is the projection matrix of the range of A^+. It also will be the projection matrix of the range of any matrix B satisfying $\text{rank}(B) = \text{rank}(A^+)$ and $A^+AB = B$. For instance, from Theorem 5.4 we have $\text{rank}(A') = \text{rank}(A^+)$ and

$$A^+AA' = (A^+A)'A' = A'A^{+\prime}A' = A',$$

so A^+A is also the projection matrix of the range of A'; that is $P_{R(A')} = A^+A$.

Our next result summarizes some of the special properties possessed by the Moore–Penrose inverse of a symmetric matrix.

Theorem 5.5 Let A be an $m \times m$ symmetric matrix. Then

(a) A^+ is also symmetric,
(b) $AA^+ = A^+A$,
(c) $A^+ = A$, if A is idempotent.

Proof. Using Theorem 5.3(b) and the fact that $A = A'$, we have

$$A^+ = (A')^+ = (A^+)',$$

which then proves (a). To prove (b), note that it follows from condition (5.3) of the Moore–Penrose inverse of a matrix, along with the symmetry of both A and A^+, that

$$AA^+ = (AA^+)' = A^{+\prime}A' = A^+A$$

Finally, (c) is established by verifying the four conditions of the Moore–Penrose inverse for $A^+ = A$, when $A^2 = A$. For instance, both conditions (5.1) and (5.2) hold since

$$AAA = A^2A = AA = A^2 = A,$$

while conditions (5.3) and (5.4) hold because

$$(AA)' = A'A' = AA \qquad \square$$

In the proof of Theorem 5.1, we saw that the Moore–Penrose inverse of any matrix can be conveniently expressed in terms of the components involved in the singular value decomposition of that matrix. Likewise, in the special case of a symmetric matrix, we will be able to write the Moore–Penrose inverse in terms of the components of the spectral decomposition of that matrix; that is, in terms of its eigenvalues and eigenvectors. Before identifying this relationship, we first consider the Moore–Penrose inverse of a diagonal matrix. The proof of this result, which simply involves the verification of conditions (5.1)–(5.4), is left to the reader.

Theorem 5.6. Let Λ be the $m \times m$ diagonal matrix $\text{diag}(\lambda_1, \dots, \lambda_m)$. Then the Moore–Penrose inverse Λ^+ of Λ, is the diagonal matrix $\text{diag}(\phi_1, \dots, \phi_m)$, where

$$\phi_i = \begin{cases} \lambda_i^{-1}, & \text{if } \lambda_i \neq 0, \\ 0, & \text{if } \lambda_i = 0 \end{cases}$$

Theorem 5.7. Let $\boldsymbol{x}_1, \dots, \boldsymbol{x}_m$ be a set of orthonormal eigenvectors corresponding to the eigenvalues, $\lambda_1, \dots, \lambda_m$, of the $m \times m$ symmetric matrix A. If we define $\Lambda = \text{diag}(\lambda_1, \dots, \lambda_m)$ and $X = (\boldsymbol{x}_1, \dots, \boldsymbol{x}_m)$, then

$$A^+ = X\Lambda^+ X'$$

Proof. Let $r = \text{rank}(A)$, and suppose that we have ordered the λ_is so that $\lambda_{r+1} = \cdots = \lambda_m = 0$. Partition X as $X = [X_1 \quad X_2]$, where X_1 is $m \times r$, and partition Λ in block diagonal form as $\Lambda = \text{diag}(\Lambda_1, (0))$, where $\Lambda_1 = \text{diag}(\lambda_1, \dots, \lambda_r)$. Then, the spectral decomposition of A is given by

$$A = [X_1 \quad X_2] \begin{bmatrix} \Lambda_1 & (0) \\ (0) & (0) \end{bmatrix} \begin{bmatrix} X_1' \\ X_2' \end{bmatrix} = X_1 \Lambda_1 X_1',$$

and similarly the expression above for A^+ reduces to $A^+ = X_1 \Lambda_1^{-1} X_1'$. Thus, since $X_1' X_1 = \text{I}_r$, we have

$$AA^+ = X_1 \Lambda_1 X_1' X_1 \Lambda_1^{-1} X_1' = X_1 \Lambda_1 \Lambda_1^{-1} X_1' = X_1 X_1',$$

which is clearly symmetric, so condition (5.3) is satisfied. Similarly, $A^+ A = X_1 X_1'$ and so (5.4) also holds. Conditions (5.1) and (5.2) hold since

$$AA^+ A = (AA^+)A = X_1 X_1' X_1 \Lambda_1 X_1' = X_1 \Lambda_1 X_1' = A$$

and

$$A^+ AA^+ = A^+(AA^+) = X_1 \Lambda_1^{-1} X_1' X_1 X_1' = X_1 \Lambda_1^{-1} X_1' = A^+,$$

and so the proof is complete. □

Example 5.2. Consider the symmetric matrix

$$A = \begin{bmatrix} 32 & 16 & 16 \\ 16 & 14 & 2 \\ 16 & 2 & 14 \end{bmatrix}$$

It is easily verified that an eigenanalysis of A reveals that it can be expressed as

$$A = \begin{bmatrix} 2/\sqrt{6} & 0 \\ 1/\sqrt{6} & -1/\sqrt{2} \\ 1/\sqrt{6} & 1/\sqrt{2} \end{bmatrix} \begin{bmatrix} 48 & 0 \\ 0 & 12 \end{bmatrix} \begin{bmatrix} 2/\sqrt{6} & 1/\sqrt{6} & 1/\sqrt{6} \\ 0 & -1/\sqrt{2} & 1/\sqrt{2} \end{bmatrix}$$

Thus, using Theorem 5.7, we find that

$$\begin{aligned} A^+ &= \begin{bmatrix} 2/\sqrt{6} & 0 \\ 1/\sqrt{6} & -1/\sqrt{2} \\ 1/\sqrt{6} & 1/\sqrt{2} \end{bmatrix} \begin{bmatrix} 1/48 & 0 \\ 0 & 1/12 \end{bmatrix} \begin{bmatrix} 2/\sqrt{6} & 1/\sqrt{6} & 1/\sqrt{6} \\ 0 & -1/\sqrt{2} & 1/\sqrt{2} \end{bmatrix} \\ &= \frac{1}{288} \begin{bmatrix} 4 & 2 & 2 \\ 2 & 13 & -11 \\ 2 & -11 & 13 \end{bmatrix} \end{aligned}$$

In Section 2.7, we saw that if the columns of an $m \times r$ matrix X form a basis for a vector space S, then the projection matrix of S is given by $X(X'X)^{-1}X'$; that is

$$P_{R(X)} = X(X'X)^{-1}X'$$

Definition 5.2 indicates how this can be generalized to the situation in which X is not full column rank. Thus, using Definition 5.2 and Theorem 5.3(g), we have

$$P_{R(X)} = XX^+ = X(X'X)^+X' \tag{5.7}$$

Example 5.3. We will utilize (5.7) to obtain the projection matrix of the range of

$$X = \begin{bmatrix} 4 & 1 & 3 \\ -4 & -3 & -1 \\ 0 & -2 & 2 \end{bmatrix}$$

The Moore–Penrose inverse of

$$X'X = \begin{bmatrix} 32 & 16 & 16 \\ 16 & 14 & 2 \\ 16 & 2 & 14 \end{bmatrix}$$

was obtained in the previous exercise. Using this we find that

$$\begin{aligned} P_{R(X)} &= X(X'X)^{+}X' \\ &= \frac{1}{288}\begin{bmatrix} 4 & 1 & 3 \\ -4 & -3 & -1 \\ 0 & -2 & 2 \end{bmatrix}\begin{bmatrix} 4 & 2 & 2 \\ 2 & 13 & -11 \\ 2 & -11 & 13 \end{bmatrix}\begin{bmatrix} 4 & -4 & 0 \\ 1 & -3 & -2 \\ 3 & -1 & 2 \end{bmatrix} \\ &= \frac{1}{3}\begin{bmatrix} 2 & -1 & 1 \\ -1 & 2 & 1 \\ 1 & 1 & 2 \end{bmatrix} \end{aligned}$$

This illustrates the use of (5.7). Actually, $P_{R(X)}$ can be computed without ever formally computing any Moore–Penrose inverse since $P_{R(X)}$ is the total eigenprojection corresponding to the positive eigenvalues of XX'. Here we have

$$XX' = \begin{bmatrix} 26 & -22 & 4 \\ -22 & 26 & 4 \\ 4 & 4 & 8 \end{bmatrix},$$

which has the normalized eigenvectors $z_1 = (1/\sqrt{2}, -1/\sqrt{2}, 0)'$ and $z_2 = (1/\sqrt{6}, 1/\sqrt{6}, 2/\sqrt{6})'$ corresponding to its two positive eigenvalues. Thus, if we let $Z = (z_1, z_2)$, then

$$P_{R(X)} = ZZ' = \frac{1}{3}\begin{bmatrix} 2 & -1 & 1 \\ -1 & 2 & 1 \\ 1 & 1 & 2 \end{bmatrix}$$

Example 5.4. The Moore–Penrose inverse is useful in constructing quadratic forms, in normal random vectors, so that they have chi-squared distributions. This is a topic that we will investigate in more detail in Chapter 9; here we will look at a simple illustration. A common situation encountered in inferential statistics is one in which one has a sample statistic, $\boldsymbol{t} \sim N_m(\boldsymbol{\theta}, \Omega)$, and it is desired to determine whether or not the $m \times 1$ parameter vector $\boldsymbol{\theta} = \mathbf{0}$; formally, we want to test the null hypothesis $H_0\colon \boldsymbol{\theta} = \mathbf{0}$ versus the alternative hypothesis $H_1\colon \boldsymbol{\theta} \neq \mathbf{0}$. One approach to this problem, if Ω is positive definite, is to base the decision between H_0 and H_1 on the statistic

$$v_1 = \boldsymbol{t}'\Omega^{-1}\boldsymbol{t}$$

Now if T is any $m \times m$ matrix satisfying $TT' = \Omega$, and we define $\boldsymbol{u} = T^{-1}\boldsymbol{t}$, then $\text{E}(\boldsymbol{u}) = T^{-1}\boldsymbol{\theta}$ and

$$\text{var}(\boldsymbol{u}) = T^{-1}\,\{\text{var}(\boldsymbol{t})\}\,T'^{-1} = T^{-1}(TT')T'^{-1} = \text{I}_m,$$

so $\boldsymbol{u} \sim \text{N}_m(T^{-1}\boldsymbol{\theta}, \text{I}_m)$. Consequently, $u_1, \ldots, u_m$ are independently distributed normal random variables, and so

$$v_1 = \boldsymbol{t}'\Omega^{-1}\boldsymbol{t} = \boldsymbol{u}'\boldsymbol{u} = \sum_{i=1}^{m} u_i^2$$

has a chi-squared distribution with m degrees of freedom. This chi-squared distribution is central if $\boldsymbol{\theta} = \boldsymbol{0}$ and noncentral if $\boldsymbol{\theta} \neq \boldsymbol{0}$, so we would choose H_1 over H_0 is v_1 is sufficiently large. When Ω is positive semidefinite, the construction of v_1 above can be generalized by using the Moore–Penrose inverse of Ω. In this case, if $\text{rank}(\Omega) = r$, and we write $\Omega = X_1\Lambda_1X_1'$ and $\Omega^+ = X_1\Lambda_1^{-1}X_1'$, where the $m \times r$ matrix X_1 and the $r \times r$ diagonal matrix Λ_1 are defined as in the proof of Theorem 5.7, then $\boldsymbol{w} = \Lambda_1^{-1/2}X_1'\boldsymbol{t} \sim \text{N}_r\,(\Lambda_1^{-1/2}X_1'\boldsymbol{\theta}, \text{I}_r)$, since

$$\text{var}(\boldsymbol{w}) = \Lambda_1^{-1/2}X_1'\,\{\text{var}(\boldsymbol{t})\}\,X_1\Lambda_1^{-1/2} = \Lambda_1^{-1/2}X_1'(X_1\Lambda_1X_1')X_1\Lambda_1^{-1/2} = \text{I}_r$$

Thus, since the w_is are independently distributed normal random variables,

$$v_2 = \boldsymbol{t}'\Omega^+\boldsymbol{t} = \boldsymbol{w}'\boldsymbol{w} = \sum_{i=1}^{r} w_i^2$$

has a chi-squared distribution, which is central if $\Lambda_1^{-1/2}X_1'\boldsymbol{\theta} = \boldsymbol{0}$, with r degrees of freedom.

4. THE MOORE–PENROSE INVERSE OF A MATRIX PRODUCT

If A and B each is an $m \times m$ nonsingular matrix, then it follows that $(AB)^{-1} = B^{-1}A^{-1}$. This property of the matrix inverse does not immediately generalize to the Moore–Penrose inverse of a matrix; that is, if A is $m \times p$ and B is $p \times n$, then we cannot, in general, be assured that $(AB)^+ = B^+A^+$. In this section, we look at some results regarding this sort of factorization of the Moore–Penrose inverse of a product.

Example 5.5. Here we look at a very simple example, given by Greville (1966), that illustrates a situation in which the factorization does not hold. Define the 2×1 vectors

$$\boldsymbol{a} = \begin{bmatrix} 1 \\ 0 \end{bmatrix}, \qquad \boldsymbol{b} = \begin{bmatrix} 1 \\ 1 \end{bmatrix}$$

so that

$$\boldsymbol{a}^+ = (\boldsymbol{a}'\boldsymbol{a})^{-1}\boldsymbol{a}' = [1 \quad 0], \qquad \boldsymbol{b}^+ = (\boldsymbol{b}'\boldsymbol{b})^{-1}\boldsymbol{b}' = [0.5 \quad 0.5]$$

Thus, we have

$$(\boldsymbol{a}'\boldsymbol{b})^+ = (1)^+ = 1 \neq \boldsymbol{b}^+\boldsymbol{a}^{+\prime} = 0.5$$

Actually, in the previous section, we have already given a few situations in which the identity $(AB)^+ = B^+A^+$ does hold. For example in Theorem 5.3 we saw that

$$(A'A)^+ = A^+A^{+\prime} = A^+A'^+,$$

and

$$(AA^+)^+ = AA^+ = (A^+)^+A^+$$

The next theorem gives yet another situation.

Theorem 5.8. Let A be an $m \times n$ matrix, while P and Q are $h \times m$ and $n \times p$ matrices satisfying $P'P = \mathrm{I}_m$ and $QQ' = \mathrm{I}_n$. Then

$$(PAQ)^+ = Q^+A^+P^+ = Q'A^+P'$$

The proof of Theorem 5.8, which we leave to the reader, simply involves the verification of conditions (5.1)–(5.4). Note that Theorem 5.7, regarding the Moore–Penrose inverse of a symmetric matrix, is a special case of the theorem above.

Our next result gives a sufficient condition on the matrices A and B to guarantee that $(AB)^+ = B^+A^+$.

Theorem 5.9. Let A be an $m \times p$ matrix and B be a $p \times n$ matrix. If rank$(A) =$ rank$(B) = p$, then $(AB)^+ = B^+A^+$.

Proof. Since A is full column rank and B is full row rank, we know from Theorem 5.3 that $A^+ = (A'A)^{-1}A'$ and $B^+ = B'(BB')^{-1}$. Consequently, we find that

$$\begin{aligned} ABB^+A^+AB &= ABB'(BB')^{-1}(A'A)^{-1}A'AB = AB, \\ B^+A^+ABB^+A^+ &= B'(BB')^{-1}(A'A)^{-1}A'ABB'(BB')^{-1}(A'A)^{-1}A' \\ &= B'(BB')^{-1}(A'A)^{-1}A' = B^+A^+, \end{aligned}$$

so conditions (5.1) and (5.2) are satisfied. In addition,

$$\begin{aligned} ABB^+A^+ &= ABB'(BB')^{-1}(A'A)^{-1}A' = A(A'A)^{-1}A', \\ B^+A^+AB &= B'(BB')^{-1}(A'A)^{-1}A'AB = B'(BB')^{-1}B \end{aligned}$$

are symmetric, so B^+A^+ is the Moore–Penrose inverse of AB. □

While Theorem 5.9 is useful, its major drawback is that it only gives a sufficient condition for the factorization of $(AB)^+$. The following result, due to Greville (1966), gives several necessary and sufficient conditions for this factorization to hold.

Theorem 5.10. Let A be an $m \times p$ matrix and B be a $p \times n$ matrix. Then each of the following conditions are necessary and sufficient for $(AB)^+ = B^+A^+$.

(a) $A^+ABB'A' = BB'A'$ and $BB^+A'AB = A'AB$.
(b) A^+ABB' and $A'ABB^+$ are symmetric matrices.
(c) $A^+ABB'A'ABB^+ = BB'A'A$.
(d) $A^+AB = B(AB)^+AB$ and $BB^+A' = A'AB(AB)^+$.

Proof. We will prove that the conditions given in (a) are necessary and sufficient; the proofs for (b)–(d) will be left to the reader as an exercise. First assume that the conditions of (a) hold. Premultiplying the first identity by B^+ while postmultiplying by $(AB)'^+$ yields

$$B^+A^+AB(AB)'(AB)'^+ = B^+BB'A'(AB)'^+ \tag{5.8}$$

Now for any matrix C,

$$C^+CC' = (C^+C)'C' = C'C^{+\prime}C' = C'C'^+C' = C' \tag{5.9}$$

Using this identity, when $C = B$, on the right-hand side of (5.8) and its transpose on the left-hand side, when $C = AB$, we obtain the equation

$$B^+A^+AB = (AB)'(AB)'^+,$$

which, due to condition (5.4), is equivalent to

$$B^+A^+AB = (AB)^+(AB) = P_{R((AB)^+)} \tag{5.10}$$

The final equality in (5.10) follows from the definition of the Moore–Penrose inverse in terms of projection matrices, as given in Definition 5.2. In a similar fashion, if we take the transpose of the second identity in (a), which yields

$$B'A'ABB^+ = B'A'A$$

and premultiply this by $(AB)'^+$ and postmultiply by A^+, then, after simplifying by using (5.9) on the left-hand side with $C = (AB)'$ and the transpose of (5.9) on the right-hand side with $C = A'$, we obtain the equation

$$ABB^+A^+ = (AB)(AB)^+ = P_{R(AB)} \tag{5.11}$$

But from Definition 5.2, $(AB)^+$ is the only matrix satisfying both (5.10) and (5.11). Consequently, we must have $(AB)^+ = B^+A^+$. Conversely, now suppose that $(AB)^+ = B^+A^+$. Using this in (5.9), when $C = AB$, gives

$$(AB)' = B^+A^+(AB)(AB)'$$

Premultiplying this by $ABB'B$, we obtain

$$ABB'BB'A' = ABB'BB^+A^+ABB'A',$$

which, after using the transpose of (5.9) with $C = B'$ and then rearranging, simplifies to

$$ABB'(\mathrm{I} - A^+A)BB'A' = (0)$$

Note that since $D = (\mathrm{I} - A^+A)$ is symmetric and idempotent, the equation above is in the form $E'D'DE = (0)$, where $E = BB'A'$. This then implies that $ED = (0)$; that is,

$$(\mathrm{I} - A^+A)BB'A' = (0),$$

which is equivalent to the first identity in (a). In a similar fashion, using $(AB)^+ = B^+A^+$ in (5.9) with $C = (AB)'$ yields

$$AB = A^{+\prime}B^{+\prime}B'A'AB$$

This, when premultiplied by $B'A'AA'$, can be simplified to an equation that is equivalent to the second identity of (a). □

Our next step is to find a general expression for $(AB)^+$ which holds for all A and B for which the product AB is defined. Our approach involves transforming A to a matrix A_1 and transforming B to B_1, such that $AB = A_1B_1$ and $(A_1B_1)^+ = B_1^+A_1^+$. The result, due to Cline (1964a), is given in the next theorem.

Theorem 5.11. Let A be an $m \times p$ matrix and B be a $p \times n$ matrix. If we define $B_1 = A^+AB$ and $A_1 = AB_1B_1^+$, then $AB = A_1B_1$ and $(AB)^+ = B_1^+A_1^+$.

Proof. Note that

$$AB = AA^+AB = AB_1 = AB_1B_1^+B_1 = A_1B_1,$$

so the first result holds. To verify the second statement, we will show that the two conditions given in Theorem 5.10(a) are satisfied for A_1 and B_1. First note that

$$A^+A_1 = A^+AB_1B_1^+ = A^+A(A^+AB)B_1^+ = A^+ABB_1^+ = B_1B_1^+, \tag{5.12}$$

and

$$A_1^+A_1 = A_1^+AB_1B_1^+ = A_1^+A(B_1B_1^+B_1)B_1^+ = A_1^+A_1B_1B_1^+ \tag{5.13}$$

Taking the transpose of (5.13) and using (5.12), along with conditions (5.3) and (5.4), we get

$$A_1^+A_1 = B_1B_1^+A_1^+A_1 = A^+A_1A_1^+A_1 = A^+A_1 = B_1B_1^+,$$

and so

$$A_1^+A_1B_1B_1'A_1' = B_1B_1^+B_1B_1'A_1' = B_1B_1'A_1',$$

which is the first identity in Theorem 5.10(a). The second identity can be obtained by noting that

$$A_1' = (AB_1B_1^+)' = (AB_1B_1^+B_1B_1^+)' = (A_1B_1B_1^+)' = B_1B_1^+A_1',$$

and then postmultiplying this identity by A_1B_1. □

Note that in Theorem 5.11, B was transformed to B_1 by the projection matrix of the range space of A^+, while A was transformed to A_1 by the projection matrix of the range space of B_1 and not that of B. Our next result indicates that the range space of B can be used instead of that of B_1, if we do not insist that $AB = A_1B_1$. A proof of this result can be found in Campbell and Meyer (1979).

Theorem 5.12. Let A be an $m\times p$ matrix and B, a $p\times n$ matrix. If we define $B_1 = A^+AB$ and $A_1 = ABB^+$, then $(AB)^+ = B_1^+A_1^+$.

5. THE MOORE–PENROSE INVERSE OF PARTITIONED MATRICES

Suppose that the $m \times n$ matrix A has been partitioned as $A = [U \quad V]$, where U is $m \times n_1$ and V is $m \times n_2$. In some situations, it may be useful to have an expression for A^+ in terms of the submatrices, U and V. We begin with the general case, in which no assumptions can be made regarding U and V.

Theorem 5.13. Let the $m\times n$ matrix A be partitioned as $A = [U \quad V]$, where U is $m \times n_1$, V is $m \times n_2$, and $n = n_1 + n_2$. Then

$$A^+ = \begin{bmatrix} U^+ - U^+V(C^+ + W) \\ C^+ + W \end{bmatrix},$$

where $C = (\mathrm{I}_m - UU^+)V$, $M = \{\mathrm{I}_{n_2} + (\mathrm{I}_{n_2} - C^+C)V'U^{+\prime}U^+V(\mathrm{I}_{n_2} - C^+C)\}^{-1}$, and $W = (\mathrm{I}_{n_2} - C^+C)MV'U^{+\prime}U^+(\mathrm{I}_m - VC^+)$.

The proof of Theorem 5.13, which is rather lengthy, will be omitted. The interested reader should refer to Cline (1964b), Boullion and Odell (1971), or Pringle and Rayner (1971). The proofs of the following consequences of Theorem 5.13 can also be found in these references.

Corollary 5.13.1. Let A and C be defined as in Theorem 5.13, and let $K = (\mathrm{I}_{n_2} + V'U^{+\prime}U^+V)^{-1}$. Then

(a) $A^+ = \begin{bmatrix} U^+ - U^+VKV'U^{+\prime}U^+ \\ C^+ + KV'U^{+\prime}U^+ \end{bmatrix}$

if and only if $C^+CV'U^{+\prime}U^+V = (0)$,

(b) $A^+ = \begin{bmatrix} U^+ - U^+VKV'U^{+\prime}U^+ \\ KV'U^{+\prime}U^+ \end{bmatrix}$

if and only if $C = (0)$,

(c) $A^+ = \begin{bmatrix} U^+ - U^+VC^+ \\ C^+ \end{bmatrix}$

if and only if $C^+CV'U^{+\prime}U^+V = V'U^{+\prime}U^+V$,

(d) $A^+ = \begin{bmatrix} U^+ \\ V^+ \end{bmatrix}$

if and only if $U'V = (0)$.

Our final theorem involves the Moore–Penrose inverse of a partitioned matrix that has the block diagonal form. This result can be easily proven by simply verifying that the conditions of the Moore–Penrose inverse are satisfied.

Theorem 5.14. Let the $m \times n$ marix A be given by

$$A = \begin{bmatrix} A_{11} & (0) & \cdots & (0) \\ (0) & A_{22} & \cdots & (0) \\ \vdots & \vdots & & \vdots \\ (0) & (0) & \cdots & A_{rr} \end{bmatrix},$$

where A_{ii} is $m_i \times n_i$, $m_1 + \cdots + m_r = m$, and $n_1 + \cdots + n_r = n$. Then

$$A^+ = \begin{bmatrix} A_{11}^+ & (0) & \cdots & (0) \\ (0) & A_{22}^+ & \cdots & (0) \\ \vdots & \vdots & & \vdots \\ (0) & (0) & \cdots & A_{rr}^+ \end{bmatrix}$$

6. THE MOORE–PENROSE INVERSE OF A SUM

Theorem 1.7 gave an expression for $(A+CBD)^{-1}$, when the matrices A and B are both square and nonsingular. Although a generalization of this formula to the case of a Moore–Penrose inverse is not available, there are some specialized results for the Moore–Penrose inverse of a sum of matrices. Some of these results are presented in this section. The proofs of our first two results utilize the results of the previous section regarding partitioned matrices. These proofs can be found in Cline (1965) or Boullion and Odell (1971).

Theorem 5.15. Let U be an $m\times n_1$ matrix and V be an $m\times n_2$ matrix. Then

$$(UU' + VV')^+ = (\mathrm{I}_m - C^{+\prime}V')U^{+\prime}KU^+(\mathrm{I}_m - VC^+) + (CC')^+,$$

where $K = \mathrm{I}_{n_1} - U^+V(\mathrm{I}_{n_2} - C^+C)M(U^+V)'$, and C and M are defined as in Theorem 5.13.

Theorem 5.16. Suppose U and V are both $m \times n$ matrices. If $UV' = (0)$, then

$$(U + V)^+ = U^+ + (\mathrm{I}_n - U^+V)(C^+ + W),$$

where C and W are as given in Theorem 5.13.

Theorem 5.16 gives an expression for $(U + V)^+$ that holds when the rows of U are orthogonal to the rows of V. If, in addition, the columns of U are orthogonal to the columns of V, this expression greatly simplifies. This special case is summarized in the following theorem.

Theorem 5.17. If U and V are $m \times n$ matrices satisfying $UV' = (0)$ and $U'V = (0)$, then

$$(U + V)^+ = U^+ + V^+$$

Proof. Using Theorem 5.3(g), we find that

$$U^+V = (U'U)^+U'V = (0)$$

and

$$VU^+ = VU'(UU')^+ = \{(UU')^{+\prime}UV'\}' = (0)$$

Similarly, we have $V^+U = (0)$ and $UV^+ = (0)$. As a result,

$$(U + V)(U^+ + V^+) = UU^+ + VV^+ \tag{5.14}$$

$$(U^+ + V^+)(U + V) = U^+U + V^+V, \tag{5.15}$$

which are both symmetric, so that conditions (5.3) and (5.4) are satisfied. Post-multiplying equation (5.14) by $(U + V)$ and (5.15) by $(U^+ + V^+)$ yields conditions (5.1) and (5.2), so the result follows. □

Theorem 5.17 can be easily generalized to more than two matrices.

Corollary 5.17.1. Let $U_1, \dots, U_k$ be $m \times n$ matrices satisfying $U_iU_j' = (0)$ and $U_i'U_j = (0)$ for all $i \neq j$. Then

$$(U_1 + \cdots + U_k)^+ = U_1^+ + \cdots + U_k^+$$

7. THE CONTINUITY OF THE MOORE–PENROSE INVERSE

It is very useful to establish the continuity of a function since continuous functions enjoy many nice properties. In this section, we will give conditions under which the elements of A^+ are continuous functions of the elements of A. But before doing this, let us first consider the determinant of a square matrix A and the inverse of a nonsingular matrix A. Recall that the determinant of an $m \times m$ matrix A can be expressed as the sum of terms, where each term is +1 or -1 times the product of m of the elements of A. Thus, due to the continuity of sums and the continuity of scalar products, we immediately have the following.

Theorem 5.18. Let A be an $m \times m$ matrix. Then the determinant of A, $|A|$, is a continuous function of the elements of A.

Suppose that A is an $m \times m$ nonsingular matrix so that $|A| \neq 0$. Recall that the inverse of A can be expressed as

$$A^{-1} = |A|^{-1} A_{\#}, \tag{5.16}$$

where $A_{\#}$ is the adjoint matrix of A. If $A_1, A_2, \ldots$ is a sequence of matrices such that $A_i \to A$ as $i \to \infty$, then, due to the continuity of the determinant function, $|A_i| \to |A|$, and so there must exist an N such that $|A_i| \neq 0$ for all $i > N$. Since each element of an adjoint matrix is +1 or -1 times a determinant, it also follows from the continuity of the determinant function that if $A_{i\#}$ is the adjoint of A_i, then $A_{i\#} \to A_{\#}$ as $i \to \infty$. As a result, equation (5.16) has allowed us to establish the following.

Theorem 5.19. Let A be an $m \times m$ nonsingular matrix. Then the inverse of A, A^{-1}, is a continuous function of the elements of A.

The continuity of the Moore–Penrose inverse is not as straightforward as the continuity of the inverse of a nonsingular matrix. If A is an $m \times n$ matrix and $A_1, A_2, \ldots$ is an arbitrary sequence of $m \times n$ matrices satisfying $A_i \to A$ as $i \to \infty$, then we are not assured that $A_i^+ \to A^+$. A simple example will illustrate the potential problem.

Example 5.6. Consider the sequence of 2×2 matrices $A_1, A_2, \ldots$, where

$$A_i = \begin{bmatrix} 1/i & 0 \\ 0 & 1 \end{bmatrix}$$

Clearly, $A_i \rightarrow A$, where

$$A = \begin{bmatrix} 0 & 0 \\ 0 & 1 \end{bmatrix}$$

However, note that rank(A) = 1, while rank(A_i) = 2 for all i. For this reason, we do not have $A_i^+ \rightarrow A^+$. In fact,

$$A_i^+ = \begin{bmatrix} i & 0 \\ 0 & 1 \end{bmatrix}$$

does not converge to anything since its (1, 1)th element, i, converges to ∞. On the other hand

$$A^+ = \begin{bmatrix} 0 & 0 \\ 0 & 1 \end{bmatrix}$$

If we have a sequence of matrices $A_1, A_2, \ldots$ for which rank(A_i) = rank(A) for all i larger than some interger, say N, then we will not encounter the difficulty observed in the example above; that is, as A_i gets closer to A, A_i^+ will get closer to A^+. This continuity property of A^+ is summarized below. A proof of this important result can be found in Penrose (1955) or Campbell and Meyer (1979).

Theorem 5.20. Let A be an $m \times n$ matrix and $A_1, A_2, \ldots$ a sequence of $m \times n$ matrices such that $A_i \rightarrow A$, as $i \rightarrow \infty$. Then

$$A_i^+ \rightarrow A^+, \quad \text{as } i \rightarrow \infty$$

if and only if there exists an integer N such that

$$\text{rank}(A_i) = \text{rank}(A) \quad \text{for all } i > N$$

Example 5.7. The conditions for the continuity of the Moore–Penrose inverse have important implications in estimation and hypothesis testing problems. In particular, in this example, we will discuss a property, referred to as consistency, that some estimators possess. An estimator t, computed from a sample of size n, is said to be a consistent estimator of a parameter θ if t converges in probability to θ; that is, if

$$\lim_{n \rightarrow \infty} \text{P}(|t - \theta| \geq \epsilon) = 0$$

for any $\epsilon > 0$. An important result associated with the property of consistency is that continuous functions of consistent estimators are consistent; that is, if t is a consistent estimator of θ, and $g(t)$ is a continuous function of t, then $g(t)$ is a consistent estimator of $g(\theta)$. We will now apply some of these ideas to a situation involving the estimation of the Moore–Penrose inverse of a matrix of parameters. For instance, let Ω be an $m \times m$ positive semidefinite covariance matrix having rank $r < m$. Suppose that the elements of the matrix Ω are unknown and are, therefore, to be estimated. Suppose, in addition, that our sample estimate of Ω, which we will denote by $\hat{\Omega}$, is positive definite with probability one, so that $\text{rank}(\hat{\Omega}) = m$ with probability one, and $\hat{\Omega}$ is a consistent estimator of Ω; that is, each element of $\hat{\Omega}$ is a consistent estimator of the corresponding element of Ω. However, since $\text{rank}(\Omega) = r < m$, $\hat{\Omega}^+$ is not a consistent estimator of Ω^+. Intuitatively, the problem here is obvious. If $\hat{\Omega} = X\Lambda X'$ is the spectral decomposition of $\hat{\Omega}$ so that $\hat{\Omega}^+ = \hat{\Omega}^{-1} = X\Lambda^{-1}X'$, then the consistency of $\hat{\Omega}$ is implying that as n increases, the $m - r$ smallest diagonal elements of Λ are converging to zero, while the $m - r$ largest diagonal elements of Λ^{-1} are increasing without bound. The difficulty here can be easily avoided if the value of r is known. In this case, $\hat{\Omega}$ can be adjusted to yield an estimator of Ω having rank r. For example, if $\hat{\Omega}$, has eigenvalues $\lambda_1 \geq \lambda_2 \geq \cdots \geq \lambda_m$ and corresponding normalized eigenvectors $\boldsymbol{x}_1, \dots, \boldsymbol{x}_m$ and P_r is the eigenprojection

$$P_r = \sum_{i=1}^{r} \boldsymbol{x}_i \boldsymbol{x}_i',$$

then

$$\hat{\Omega}_* = P_r \hat{\Omega} P_r = \sum_{i=1}^{r} \lambda_i \boldsymbol{x}_i \boldsymbol{x}_i'$$

will be an estimator of Ω having rank of r. It can be shown then that, due to the continuity of eigenprojections, $\hat{\Omega}_*$ is also a consistent estimator of Ω. More importantly, since $\text{rank}(\hat{\Omega}_*) = \text{rank}(\Omega) = r$, Theorem 5.20 guarantees that $\hat{\Omega}_*^+$ is a consistent estimator of Ω^+.

8. SOME OTHER GENERALIZED INVERSES

The Moore–Penrose inverse is just one of many generalized inverses that have been developed in recent years. In this section, we will briefly discuss two other generalized inverses that have applications in statistics. Both of these inverses can be defined by utilizing some of the four conditions (5.1)–(5.4) or, for simplicity, 1–4 of the Moore–Penrose inverse. In fact, we can define a different class of inverses corresponding to each different subset of the conditions 1–4 that the inverse must satisfy.

Definition 5.3. For any $m \times n$ matrix A, let the $n \times m$ matrix denoted $A^{(i_1,\ldots,i_r)}$ be any matrix satisfying conditions $i_1, \ldots, i_r$ from among the four conditions 1–4; $A^{(i_1,\ldots,i_r)}$ will be called a $\{i_1, \ldots, i_r\}$-inverse of A.

Thus, the Moore–Penrose inverse of A is the $\{1, 2, 3, 4\}$-inverse of A; that, is $A^+ = A^{(1,2,3,4)}$. Note that for any proper subset $\{i_1, \ldots, i_r\}$ of $\{1, 2, 3, 4\}$, A^+ will also be a $\{i_1, \ldots, i_r\}$-inverse of A but it may not be the only one. Since in many cases there are many different $\{i_1, \ldots, i_r\}$-inverses of A, it may be easier to compute a $\{i_1, \ldots, i_r\}$-inverse of A than to compute the Moore–Penrose inverse. The rest of this section will be devoted to the $\{1\}$-inverse of A and the $\{1, 3\}$-inverse of A, which have special applications that will be discussed in the next chapter. Discussion of other useful $\{i_1, \ldots, i_r\}$-inverses can be found in Ben-Israel and Greville (1974), Campbell and Meyer (1979), and Rao and Mitra (1971).

In the next chapter, we will see that in solving systems of linear equations, we will only need an inverse matrix satisfying the first condition of the four Moore–Penrose conditions. We will refer to any such $\{1\}$-inverse of A as simply a generalized inverse of A, and we will write it using the fairly common notation A^-; that is, $A^{(1)} = A^-$. One useful way of expressing a generalized inverse of a matrix A makes use of the singular value decomposition of A. The following result, which is stated for a matrix A having less than full rank, can easily be modified for matrices having full row rank or full column rank.

Theorem 5.21. Suppose that the $m \times n$ matrix A has rank $r > 0$ and the singular value decomposition given by

$$A = P \begin{bmatrix} \Delta & (0) \\ (0) & (0) \end{bmatrix} Q',$$

where P and Q are $m \times m$ and $n \times n$ orthogonal matrices, respectively, and Δ is an $r \times r$ nonsingular diagonal matrix. Let

$$B = Q \begin{bmatrix} \Delta^{-1} & E \\ F & G \end{bmatrix} P',$$

where E is $r \times m - r$, F is $n - r \times r$, and G is $(n - r) \times (m - r)$. Then for all choices of E, F, and G, B is a generalized inverse of A, and any generalized inverse of A can be expressed in the form of B for some E, F, and G.

Proof. Note that

$$ABA = P\begin{bmatrix} \Delta & (0) \\ (0) & (0) \end{bmatrix} Q'Q \begin{bmatrix} \Delta^{-1} & E \\ F & G \end{bmatrix} P'P \begin{bmatrix} \Delta & (0) \\ (0) & (0) \end{bmatrix} Q'$$

$$= P\begin{bmatrix} \Delta\Delta^{-1}\Delta & (0) \\ (0) & (0) \end{bmatrix} Q' = P\begin{bmatrix} \Delta & (0) \\ (0) & (0) \end{bmatrix} Q' = A,$$

and so the matrix B above is a generalized inverse of A regardless of the choice of E, F, and G. On the other hand, if we write $Q = [Q_1 \quad Q_2]$, $P = [P_1 \quad P_2]$, where Q_1 is $n \times r$ and P_1 is $m \times r$, then, since $PP' = \mathrm{I}_m$, $QQ' = \mathrm{I}_n$, any generalized inverse B, of A, can be expressed as

$$B = QQ'BPP' = Q\begin{bmatrix} Q_1' \\ Q_2' \end{bmatrix} B[P_1 \quad P_2]P'$$

$$= Q\begin{bmatrix} Q_1'BP_1 & Q_1'BP_2 \\ Q_2'BP_1 & Q_2'BP_2 \end{bmatrix} P',$$

which is in the required form if we can show that $Q_1'BP_1 = \Delta^{-1}$. Since B is a generalized inverse of A, $ABA = A$, or equivalently, $P'ABAQ = P'AQ$. Writing this last identity in partitioned form and equating the (1, 1)th submatrices on both sides, we find that

$$\Delta Q_1'BP_1\Delta = \Delta,$$

from which it immediately follows that $Q_1'BP_1 = \Delta^{-1}$, and so the proof is complete. □

Example 5.8. The 4×3 matrix

$$A = \begin{bmatrix} 1 & 0 & 0.5 \\ 1 & 0 & 0.5 \\ 0 & -1 & -0.5 \\ 0 & -1 & -0.5 \end{bmatrix}$$

has rank $r = 2$ and singular value decomposition with

$$P = \frac{1}{2}\begin{bmatrix} 1 & 1 & 1 & -1 \\ 1 & 1 & -1 & 1 \\ 1 & -1 & 1 & 1 \\ 1 & -1 & -1 & -1 \end{bmatrix}, \qquad Q' = \begin{bmatrix} 1/\sqrt{2} & -1/\sqrt{2} & 0 \\ 1/\sqrt{3} & 1/\sqrt{3} & 1/\sqrt{3} \\ 1/\sqrt{6} & 1/\sqrt{6} & -2/\sqrt{6} \end{bmatrix}$$

and

$$\Delta = \begin{bmatrix} \sqrt{2} & 0 \\ 0 & \sqrt{3} \end{bmatrix}$$

If we take E, F, and G as null matrices, and use the equation for B given in Theorem 5.21, we obtain as a generalized inverse of A the matrix

$$\frac{1}{12}\begin{bmatrix} 5 & 5 & 1 & 1 \\ -1 & -1 & -5 & -5 \\ 2 & 2 & -2 & -2 \end{bmatrix}$$

Actually, from the proof of Theorem 5.1, we know that the matrix above is the Moore–Penrose inverse. Different generalized inverses of A may be constructed through different choices of E, F, and G; for example, if we again take E and F as null matrices but now use

$$G = \begin{bmatrix} 1/\sqrt{6} & 0 \end{bmatrix},$$

then we obtain the generalized inverse

$$\frac{1}{6}\begin{bmatrix} 3 & 2 & 1 & 0 \\ 0 & -1 & -2 & -3 \\ 0 & 2 & -2 & 0 \end{bmatrix}$$

Note that this matrix has rank 3, while the Moore–Penrose inverse has its rank equal to that of A, which is 2.

The following theorem summarizes some of the basic properties of $\{1\}$-inverses.

Theorem 5.22. Let A be an $m \times n$ matrix and let A^- be a generalized inverse of A. Then

(a) $A^{-\prime}$ is a generalized inverse of A',
(b) if α is a nonzero scalar, $\alpha^{-1}A^-$ is a generalized inverse of αA,
(c) if A is square and nonsingular, $A^- = A^{-1}$ uniquely,
(d) if B and C are nonsingular, $C^{-1}A^-B^{-1}$ is a generalized inverse of BAC,
(e) $\text{rank}(A) = \text{rank}(AA^-) = \text{rank}(A^-A) \leq \text{rank}(A^-)$,
(f) $\text{rank}(A) = m$ if and only if $AA^- = I_m$,
(g) $\text{rank}(A) = n$ if and only if $A^-A = I_n$.

Proof. Properties (a)–(d) are easily proven by simply verifying that the one condition of a generalized inverse holds. To prove (e), note that since $A = AA^-A$, we can use Theorem 2.10 to get

$$\text{rank}(A) = \text{rank}(AA^-A) \leq \text{rank}(AA^-) \leq \text{rank}(A)$$

and

$$\text{rank}(A) = \text{rank}(AA^-A) \leq \text{rank}(A^-A) \leq \text{rank}(A),$$

so that $\text{rank}(A) = \text{rank}(AA^-) = \text{rank}(A^-A)$. In addition,

$$\text{rank}(A) = \text{rank}(AA^-A) \leq \text{rank}(A^-A) \leq \text{rank}(A^-),$$

so the result follows. It follows from (e) that $\text{rank}(A) = m$ if and only if AA^- is nonsingular. Premultiplying the equation

$$(AA^-)^2 = (AA^-A)A^- = AA^-$$

by $(AA^-)^{-1}$ yields (f). Similarly, $\text{rank}(A) = n$, if and only if A^-A is nonsingular and so premultiplying

$$(A^-A)^2 = A^-(AA^-A) = A^-A$$

by $(A^-A)^{-1}$ gives (g). □

Example 5.9. Some of the properties possessed by the Moore–Penrose inverse do not carry over to the $\{1\}$-inverse. For instance, we have seen that A is the Moore–Penrose inverse of A^+; that is, $(A^+)^+ = A$. However, in general, we are not guaranteed that A is a generalized inverse of A^-, where A^- is an arbitrary generalized inverse of A. For example, consider the diagonal matrix $A = \text{diag}(0, 2, 4)$. One choice of a generalized inverse of A is $A^- = \text{diag}(1, 0.5, 0.25)$. Here A^- is nonsingular so it has only one generalized inverse, namely, $(A^-)^{-1} = \text{diag}(1, 2, 4)$ and, thus, A is not a generalized inverse of $A^- = \text{diag}(1, 0.5, 0.25)$.

All of the generalized inverses of a matrix A can be expressed in terms of any one particular generalized inverse. This relationship is given below.

Theorem 5.23. Let A^- be any generalized inverse of the $m \times n$ matrix A. Then for any $n \times m$ matrix C,

$$A^- + C - A^-ACAA^-$$

is a generalized inverse of A, and each generalized inverse of A can be expressed in this form for some C.

Proof. Since $AA^-A = A$,

$$\begin{aligned} A(A^- + C - A^-ACAA^-)A &= AA^-A + ACA - AA^-ACAA^-A \\ &= A + ACA - ACA = A, \end{aligned}$$

so $A^- + C - A^-ACAA^-$ is a generalized inverse of A regardless of the choice of A^- and C. Now let B be any generalized inverse of A and define $C = B - A^-$. Then, since $ABA = A$, we have

$$\begin{aligned} A^- + C - A^-ACAA^- &= A^- + (B - A^-) - A^-A(B - A^-)AA^- \\ &= B - A^-ABAA^- + A^-AA^-AA^- \\ &= B - A^-AA^- + A^-AA^- = B, \end{aligned}$$

and so the proof is complete. □

We will find the following result useful in a later chapter.

Theorem 5.24. Let A, B, and C be matrices of sizes $p \times m$, $m \times n$, and $n \times q$, respectively. If $\text{rank}(ABC) = \text{rank}(B)$, then $C(ABC)^-A$ is a generalized inverse of B.

Proof. Our proof follows that of Srivastava and Khatri (1979). Using Theorem 2.10, we have

$$\text{rank}(B) = \text{rank}(ABC) \leq \text{rank}(AB) \leq \text{rank}(B)$$

and

$$\text{rank}(B) = \text{rank}(ABC) \leq \text{rank}(BC) \leq \text{rank}(B),$$

so that evidently

$$\text{rank}(AB) = \text{rank}(BC) = \text{rank}(B) = \text{rank}(ABC) \tag{5.17}$$

Using Theorem 2.12 along with the identity

$$A(BC)\{\mathrm{I}_q - (ABC)^-ABC\} = (0),$$

we find that

$$\text{rank}(ABC) + \text{rank}(BC\{\text{I}_q - (ABC)^- ABC\}) - \text{rank}(BC) \leq \text{rank}\{(0)\} = 0,$$

so that

$$\text{rank}(BC\{\text{I}_q - (ABC)^- ABC\}) \leq \text{rank}(BC) - \text{rank}(ABC) = 0,$$

where the equality follows from (5.17). But this can be true only if

$$BC\{\text{I}_q - (ABC)^- ABC\} = \{\text{I}_q - BC(ABC)^- A\}B(C) = (0)$$

Again applying Theorem 2.12, this time on the middle expression above, we obtain

$$\text{rank}(\{\text{I}_q - BC(ABC)^- A\}B) + \text{rank}(BC) - \text{rank}(B) \leq \text{rank}\{(0)\} = 0,$$

or equivalently,

$$\text{rank}(\{\text{I}_q - BC(ABC)^- A\}B) \leq \text{rank}(B) - \text{rank}(BC) = 0,$$

where, again, the equality follows from (5.17). This implies that

$$\{\text{I}_q - BC(ABC)^- A\}B = B - B\{C(ABC)^- A\}B = (0),$$

and so the result follows. □

We will see in the next chapter that the $\{1, 3\}$-inverse is useful in finding least squares solutions to an inconsistent system of linear equations. Consequently, this inverse is commonly called the least squares inverse. We will denote the $\{1, 3\}$-inverse of A by A^L; that is, $A^{(1,3)} = A^L$. Since a least squares inverse of A is also a $\{1\}$-inverse of A, the properties given in Theorem 5.22 also apply to A^L. Some additional properties of least squares inverses are given below.

Theorem 5.25. Let A be an $m \times n$ matrix. Then

(a) for any least squares inverse, A^L, of A, $AA^L = AA^+$,

(b) $(A'A)^- A'$ is a least squares inverse of A for any generalized inverse, $(A'A)^-$, of $A'A$.

Proof. Since $AA^LA = A$ and $(AA^L)' = AA^L$, we find that

$$\begin{aligned} AA^L &= AA^+AA^L = (AA^+)'(AA^L)' = A^{+\prime}A'A^{L\prime}A' \\ &= A^{+\prime}(AA^LA)' = A^{+\prime}A' = (AA^+)' = AA^+, \end{aligned}$$

and so (a) holds. To prove (b) first note that

$$\begin{aligned} A(A'A)^-A'A &= AA^+A(A'A)^-A'A = (AA^+)'A(A'A)^-A'A = A^{+\prime}A'A(A'A)^-A'A \\ &= A^{+\prime}A'A = (AA^+)'A = AA^+A = A, \end{aligned}$$

so that $(A'A)^-A'$ satisfies condition 1. To verify that condition 3 holds, observe that

$$\begin{aligned} A(A'A)^-A' &= A(A'A)^-A'A^{+\prime}A' = A(A'A)^-A'(AA^+)' \\ &= A(A'A)^-A'AA^+ = AA^+, \end{aligned}$$

where the last equality uses the identity, $A(A'A)^-A'A = A$, just proven. Thus, the symmetry of $A(A'A)^-A'$ follows from the symmetry of AA^+. □

9. COMPUTING GENERALIZED INVERSES

In this section we review some computational formulas for generalized inverses. The emphasis here is not on the development of formulas best suited for the numerical computation of generalized inverses on a computer. For instance, the most common method of computing the Moore–Penrose inverse of a matrix is through the computation of its singular value decomposition; that is, if $A = P_1\Delta Q_1'$ is the singular value decomposition of A as given in Corollary 4.1.1, then A^+ can be easily computed via the formula $A^+ = Q_1\Delta^{-1}P_1'$. The formulas provided here and in the problems are ones that, in some cases, may be useful for the computation of the generalized inverse of matrices of small size but, in most cases, are primarily useful for theoretical purposes.

Greville (1960) obtained an expression for the Moore–Penrose inverse of a matrix partitioned in the form $[B \quad \boldsymbol{c}]$, where, of course, the matrix B and the vector $\boldsymbol{c}$ have the same number of rows. This formula can be then used recursively to compute the Moore–Penrose inverse of an $m \times n$ matrix A. To see this, let $\boldsymbol{a}_j$ denote the jth column of A and define $A_j = (\boldsymbol{a}_1, \dots, \boldsymbol{a}_j)$, so that A_j is the $m \times j$ matrix containing the first j columns of A. Greville has shown that if we write $A_j = [A_{j-1} \quad \boldsymbol{a}_j]$, then

$$A_j^+ = \begin{bmatrix} A_{j-1}^+ - \boldsymbol{d}_j\boldsymbol{b}_j' \\ \boldsymbol{b}_j' \end{bmatrix}, \tag{5.18}$$

where $\boldsymbol{d}_j = A_{j-1}^{+}\boldsymbol{a}_j$,

$$\boldsymbol{b}_j' = \begin{cases} (\boldsymbol{c}_j'\boldsymbol{c}_j)^{-1}\boldsymbol{c}_j', & \text{if } \boldsymbol{c}_j \neq \boldsymbol{0}, \\ (1 + \boldsymbol{d}_j'\boldsymbol{d}_j)^{-1}\boldsymbol{d}_j'A_{j-1}^{+}, & \text{if } \boldsymbol{c}_j = \boldsymbol{0}, \end{cases}$$

and $\boldsymbol{c}_j = \boldsymbol{a}_j - A_{j-1}\boldsymbol{d}_j$. Thus, $A^{+} = A_n^{+}$ can be computed by successively computing $A_2^{+}, A_3^{+}, \dots, A_n^{+}$.

Example 5.10. We will use the procedure above to compute the Moore–Penrose inverse of the matrix

$$A = \begin{bmatrix} 1 & 1 & 2 & 3 \\ 1 & -1 & 0 & 1 \\ 1 & 1 & 2 & 3 \end{bmatrix}$$

We begin by computing the inverse of $A_2 = [\boldsymbol{a}_1 \quad \boldsymbol{a}_2] = [A_1 \quad \boldsymbol{a}_2]$. We find that

$$A_1^{+} = (\boldsymbol{a}_1'\boldsymbol{a}_1)^{-1}\boldsymbol{a}_1' = \left(\frac{1}{3}, \frac{1}{3}, \frac{1}{3}\right),$$

$$d_2 = A_1^{+}\boldsymbol{a}_2 = \frac{1}{3},$$

$$\boldsymbol{c}_2 = \boldsymbol{a}_2 - A_1 d_2 = \boldsymbol{a}_2 - \frac{1}{3}\boldsymbol{a}_1 = \frac{1}{3}\begin{bmatrix} 2 \\ -4 \\ 2 \end{bmatrix}$$

Since $\boldsymbol{c}_2 \neq \boldsymbol{0}$, we get

$$\boldsymbol{b}_2' = \boldsymbol{c}_2^{+} = (\boldsymbol{c}_2'\boldsymbol{c}_2)^{-1}\boldsymbol{c}_2' = \frac{1}{4}[1, \quad -2, \quad 1],$$

and, thus,

$$A_2^{+} = \begin{bmatrix} A_1^{+} - d_2\boldsymbol{b}_2' \\ \boldsymbol{b}_2' \end{bmatrix} = \frac{1}{4}\begin{bmatrix} 1 & 2 & 1 \\ 1 & -2 & 1 \end{bmatrix}$$

The inverse of $A_3 = [A_2 \quad \boldsymbol{a}_3]$ now can be computed by using A_2^{+} and

$$\boldsymbol{d}_3 = A_2^+ \boldsymbol{a}_3 = \begin{bmatrix} 1 \\ 1 \end{bmatrix},$$

$$\boldsymbol{c}_3 = \boldsymbol{a}_3 - A_2 \boldsymbol{d}_3 = \begin{bmatrix} 2 \\ 0 \\ 2 \end{bmatrix} - \begin{bmatrix} 2 \\ 0 \\ 2 \end{bmatrix} = \boldsymbol{0}$$

Since $\boldsymbol{c}_3 = \boldsymbol{0}$, we find that

$$\boldsymbol{b}_3' = (1 + \boldsymbol{d}_3'\boldsymbol{d}_3)^{-1}\boldsymbol{d}_3' A_2^+ = (1+2)^{-1}[1 \quad 1]\,\frac{1}{4}\begin{bmatrix} 1 & 2 & 1 \\ 1 & -2 & 1 \end{bmatrix}$$

$$= \frac{1}{6}[1 \quad 0 \quad 1],$$

and so

$$A_3^+ = \begin{bmatrix} A_2^+ - \boldsymbol{d}_3 \boldsymbol{b}_3' \\ \boldsymbol{b}_3' \end{bmatrix} = \frac{1}{12}\begin{bmatrix} 1 & 6 & 1 \\ 1 & -6 & 1 \\ 2 & 0 & 2 \end{bmatrix}$$

Finally, to obtain the Moore–Penrose inverse of $A = A_4$, we compute

$$\boldsymbol{d}_4 = A_3^+ \boldsymbol{a}_4 = \begin{bmatrix} 1 \\ 0 \\ 1 \end{bmatrix},$$

$$\boldsymbol{c}_4 = \boldsymbol{a}_4 - A_3 \boldsymbol{d}_4 = \begin{bmatrix} 3 \\ 1 \\ 3 \end{bmatrix} - \begin{bmatrix} 3 \\ 1 \\ 3 \end{bmatrix} = \boldsymbol{0},$$

$$\boldsymbol{b}_4' = (1 + \boldsymbol{d}_4'\boldsymbol{d}_4)^{-1}\boldsymbol{d}_4' A_3^+ = \frac{1}{12}[1 \quad 2 \quad 1].$$

Consequently, the Moore–Penrose inverse of A is given by

$$A_4^+ = \begin{bmatrix} A_3^+ - \boldsymbol{d}_4 \boldsymbol{b}_4' \\ \boldsymbol{b}_4' \end{bmatrix} = \frac{1}{12}\begin{bmatrix} 0 & 4 & 0 \\ 1 & -6 & 1 \\ 1 & -2 & 1 \\ 1 & 2 & 1 \end{bmatrix}$$

A common method of computing a generalized inverse, that is, a $\{1\}$-inverse, of a matrix is based on the row reduction of that matrix to Hermite form.

Definition 5.4. An $m \times m$ matrix H is said to be in Hermite form if the following four conditions hold.

(a) H is an upper triangular matrix.
(b) h_{ii} equals 0 or 1 for each i.
(c) If $h_{ii} = 0$, then $h_{ij} = 0$ for all j.
(d) If $h_{ii} = 1$, then $h_{ji} = 0$ for all $j \neq i$.

Before applying this concept of Hermite forms to find a generalized inverse of a matrix, we will need a couple of results regarding matrices in Hermite form. The first of these two results says that any square matrix can be transformed to a matrix in Hermite form through its premultiplication by a nonsingular matrix. Details of the proof are given in Rao (1973).

Theorem 5.26. Let A be an $m \times m$ matrix. Then there exists a nonsingular $m \times m$ matrix C such that $CA = H$, where H is in Hermite form.

The proof of the next result will be left to the reader as an exercise.

Theorem 5.27. Suppose the $m \times m$ matrix H is in Hermite form. Then H is idempotent; that is, $H^2 = H$.

The connection between a generalized inverse of a square matrix A and matrices in Hermite form is established in the following theorem. This result says that any matrix C satisfying the conditions of Theorem 5.26 will be a generalized inverse of A.

Theorem 5.28. Let A be an $m \times m$ matrix and C be an $m \times m$ nonsingular matrix for which $CA = H$, where H is a matrix in Hermite form. Then the matrix C is a generalized inverse of A.

Proof. We need to show that $ACA = A$. Now from Theorem 5.27 we know that H is idempotent and so

$$CACA = H^2 = H = CA$$

The result then follows by premultiplying this equation by C^{-1}. □

The matrix C can be obtained by transforming A, through elementary row transformations, to a matrix in Hermite form. This process is illustrated in the following example.

Example 5.11. We will find a generalized inverse of the 3×3 matrix

$$A = \begin{bmatrix} 2 & 2 & 4 \\ 4 & -2 & 2 \\ 2 & -4 & -2 \end{bmatrix}$$

First, we perform row transformations on A so that the resulting matrix has its first diagonal element equal to one, while the remaining elements in the first column are all equal to zero. This can be achieved via the matrix equation $C_1A = A_1$ where

$$C_1 = \begin{bmatrix} 1/2 & 0 & 0 \\ -2 & 1 & 0 \\ -1 & 0 & 1 \end{bmatrix}, \qquad A_1 = \begin{bmatrix} 1 & 1 & 2 \\ 0 & -6 & -6 \\ 0 & -6 & -6 \end{bmatrix}$$

Next we use row transformations on A_1 so that the resulting matrix has its second diagonal element equal to one, while each of the remaining elements in the second column is zero. This can be written as $C_2A_1 = A_2$, where

$$C_2 = \begin{bmatrix} 1 & 1/6 & 0 \\ 0 & -1/6 & 0 \\ 0 & -1 & 1 \end{bmatrix}, \qquad A_2 = \begin{bmatrix} 1 & 0 & 1 \\ 0 & 1 & 1 \\ 0 & 0 & 0 \end{bmatrix}$$

The matrix A_2 satisfies the conditions of Definition 5.4, and so it is in Hermite form. Thus, we have $C_2A_1 = C_2C_1A = A_2$, so by Theorem 5.28 a generalized inverse of A is given by

$$C = C_2C_1 = \frac{1}{6}\begin{bmatrix} 1 & 1 & 0 \\ 2 & -1 & 0 \\ 6 & -6 & 6 \end{bmatrix}$$

Not only is a generalized inverse not necessarily unique, but this particular method of producing a generalized inverse does not, in general, yield a unique matrix. For instance, in the second transformation given above, $C_2A_1 = A_2$, we could have chosen

$$C_2 = \begin{bmatrix} 1 & 0 & 1/6 \\ 0 & -1/6 & 0 \\ 0 & -2 & 2 \end{bmatrix}$$

In this case, we would have obtained the generalized inverse

$$C = C_2C_1 = \frac{1}{6}\begin{bmatrix} 2 & 0 & 1 \\ 2 & -1 & 0 \\ 12 & -12 & 12 \end{bmatrix}$$

The method of finding a generalized inverse of a matrix by transforming it to a matrix in Hermite form can be easily extended from square matrices to rectangular matrices. The following result indicates how such an extension is possible.

Theorem 5.29. Let A be an $m \times n$ matrix, where $m < n$. Define the matrix A_* as

$$A_* = \begin{bmatrix} A \\ (0) \end{bmatrix},$$

so that A_* is $n \times n$, and let C be any $n \times n$ nonsingular matrix for which CA_* is in Hermite form. If we partition C as $C = [C_1 \quad C_2]$, where C_1 is $n \times m$, then C_1 is a generalized inverse of A.

Proof. We know from Theorem 5.28 that C is a generalized inverse of A_*. Hence, $A_*CA_* = A_*$. Simplifying the left-hand side of this identity, we find that

$$A_*CA_* = \begin{bmatrix} A \\ (0) \end{bmatrix} [C_1 \quad C_2] \begin{bmatrix} A \\ (0) \end{bmatrix}$$

$$= \begin{bmatrix} AC_1 & AC_2 \\ (0) & (0) \end{bmatrix} \begin{bmatrix} A \\ (0) \end{bmatrix} = \begin{bmatrix} AC_1A \\ (0) \end{bmatrix}$$

Equating this to A_*, we get $AC_1A = A$, and so the proof is complete. □

Clearly, an analogous result holds for the case in which $m > n$.

Example 5.12. Suppose that we wish to find a generalized inverse of the matrix

$$A = \begin{bmatrix} 1 & 1 & 2 \\ 1 & 0 & 1 \\ 1 & 1 & 2 \\ 2 & 0 & 2 \end{bmatrix}$$

Consequently, we consider the augmented matrix

$$A_* = [A \quad \mathbf{0}] = \begin{bmatrix} 1 & 1 & 2 & 0 \\ 1 & 0 & 1 & 0 \\ 1 & 1 & 2 & 0 \\ 2 & 0 & 2 & 0 \end{bmatrix}$$

Proceeding as in the previous example, we obtain a nonsingular matrix C so that CA_* is in Hermite form. One such matrix is given by

$$C = \begin{bmatrix} 0 & 1 & 0 & 0 \\ 1 & -1 & 0 & 0 \\ -1 & 0 & 1 & 0 \\ 0 & -2 & 0 & 1 \end{bmatrix}$$

Thus, partitioning this matrix as

$$C = \begin{bmatrix} C_1 \\ C_2 \end{bmatrix},$$

we find that a generalized inverse of A is given by

$$C_1 = \begin{bmatrix} 0 & 1 & 0 & 0 \\ 1 & -1 & 0 & 0 \\ -1 & 0 & 1 & 0 \end{bmatrix}$$

A least squares generalized inverse of a matrix A can be computed by first computing a generalized inverse of $A'A$ and then using the relationship, $A^L = (A'A)^- A'$, established in Theorem 5.25(b).

Example 5.13. To find a least squares inverse of the matrix A from Example 5.12, we first compute

$$A'A = \begin{bmatrix} 7 & 2 & 9 \\ 2 & 2 & 4 \\ 9 & 4 & 13 \end{bmatrix}$$

By transforming this matrix to Hermite form, we find that a generalized inverse of $(A'A)$ is given by

$$(A'A)^- = \frac{1}{10}\begin{bmatrix} 2 & -2 & 0 \\ -2 & 7 & 0 \\ -10 & -10 & 10 \end{bmatrix}$$

Hence, a least squares inverse of A is given by

$$A^L = (A'A)^- A' = \frac{1}{10}\begin{bmatrix} 0 & 2 & 0 & 4 \\ 5 & -2 & 5 & -4 \\ 0 & 0 & 0 & 0 \end{bmatrix}$$

PROBLEMS

1. Prove results (a)–(d) of Theorem 5.3.

2. Use Theorem 5.3(h) to find the Moore–Penrose inverse of

$$A = \begin{bmatrix} 1 & 1 & 1 \\ 0 & 1 & 0 \\ 0 & 1 & 1 \\ 2 & 0 & 1 \end{bmatrix}$$

3. Find the Moore–Penrose inverse of the vector

$$\boldsymbol{a} = \begin{bmatrix} 2 \\ 1 \\ 3 \\ 2 \end{bmatrix}$$

4. Provide the proofs for (f)–(j) of Theorem 5.3.

5. Prove Theorem 5.6.

6. Use the spectral decomposition of the matrix

$$A = \begin{bmatrix} 2 & 0 & 1 \\ 0 & 2 & 3 \\ 1 & 3 & 5 \end{bmatrix}$$

   to find its Moore–Penrose inverse.

7. Consider the matrix

$$A = \begin{bmatrix} 0 & -1 & 2 \\ 0 & -1 & 2 \\ 3 & 2 & -1 \end{bmatrix}$$

(a) Find the Moore–Penrose inverse of AA', and then use Theorem 5.3(g) to find A^+.

(b) Use A^+ to find the projection matrix for the range of A and the projection matrix for the row space of A.

8. Let A be an $m \times n$ matrix with $\text{rank}(A) = 1$. Show that $A^+ = c^{-1}A'$, where $c = \text{tr}(A'A)$.

9. Let $\boldsymbol{x}$ and $\boldsymbol{y}$ be $m \times 1$ vectors, and let $\mathbf{1}$ be the $m \times 1$ vector with each element equal to one. Obtain expressions for the Moore–Penrose inverses of

(a) $\mathbf{1}\mathbf{1}'$ (b) $\text{I}_m - m^{-1}\mathbf{1}\mathbf{1}'$ (c) $\boldsymbol{x}\boldsymbol{x}'$ (d) $\boldsymbol{x}\boldsymbol{y}'$.

10. Let A be an $m \times n$ matrix. Show that each of the matrices, AA^+, A^+A, $(\text{I}_m - AA^+)$, and $(\text{I}_n - A^+A)$ is idempotent.

11. Let A be an $m \times n$ marix. Establish the following identities.

(a) $A'AA^+ = A^+AA' = A'$.

(b) $A'A^{+\prime}A^+ = A^+A^{+\prime}A' = A^+$.

(c) $A(A'A)^+A'A = AA'(AA')^+A = A$.

12. Let A be an $m \times n$ matrix. Show that

(a) $AB = (0)$ if and only if $B^+A^+ = (0)$, where B is an $n \times p$ matrix.

(b) $A^+B = (0)$ if and only if $A'B = (0)$, where B is an $m \times p$ matrix.

13. Let A be an $m \times m$ symmetric matrix having rank r. Show that if A has one nonzero eigenvalue λ of multiplicity r, then $A^+ = \lambda^{-2}A$.

14. Let A be an $m \times n$ matrix and B be an $n \times p$ matrix. Show that if B has full row rank, then

$$AB(AB)^+ = AA^+$$

15. Let A be an $m \times m$ symmetric matrix. Show that

(a) if A is nonnegative definite, then so is A^+,

(b) if $A\boldsymbol{x} = \mathbf{0}$ for some vector $\boldsymbol{x}$, then $A^+\boldsymbol{x} = \mathbf{0}$ also.

16. Let A be an $m \times m$ symmetric matrix with $\text{rank}(A) = r$. Use the spectral decomposition of A to show that if B is any $m \times m$ symmetric matrix with $\text{rank}(B) = m - r$ such that $AB = (0)$, then $A^+A + B^+B = \text{I}_m$.

17. Let A be an $m \times n$ matrix and B be an $n \times m$ matrix. Suppose that $\text{rank}(A) = \text{rank}(B)$ and, further, that the space spanned by the eigenvec-

tors corresponding to the positive eigenvalues of $A'A$ is the same as that spanned by the eigenvectors corresponding to the positive eigenvalues of BB'. Show that $(AB)^+ = B^+A^+$.

18. Prove Theorem 5.8.

19. Prove (b)–(d) of Theorem 5.10.

20. For each case below use Theorem 5.10 to determine whether $(AB)^+ = B^+A^+$.

(a) $A = \begin{bmatrix} 0 & 0 & 0 \\ 1 & 0 & 0 \\ 0 & 1 & 0 \end{bmatrix}, \qquad B = \begin{bmatrix} 1 & 0 & 0 \\ 0 & 0 & 0 \\ 0 & 0 & 2 \end{bmatrix}$

(b) $A = \begin{bmatrix} 1 & 1 & 0 \\ 0 & 1 & 0 \\ 0 & 0 & 0 \end{bmatrix}, \qquad B = \begin{bmatrix} 0 & 0 & 0 \\ 0 & 1 & 1 \\ 0 & 1 & 0 \end{bmatrix}$

21. Let A be an $m \times n$ matrix and B be an $n \times m$ matrix. Show that $(AB)^+ = B^+A^+$ if $A'ABB' = BB'A'A$.

22. Prove Theorem 5.14.

23. Find the Moore–Penrose inverse of the matrix

$$A = \begin{bmatrix} 2 & 1 & 0 & 0 & 0 \\ 1 & 1 & 0 & 0 & 0 \\ 0 & 0 & 1 & 2 & 0 \\ 0 & 0 & 1 & 2 & 0 \\ 0 & 0 & 0 & 0 & 4 \end{bmatrix}$$

24. Use Corollary 5.13.1(d) to find the Moore–Penrose inverse of the matrix $A = [U \quad V]$, where

$$U = \begin{bmatrix} 1 & 1 & 1 \\ 1 & 1 & 1 \\ 1 & 1 & 1 \end{bmatrix}, \qquad V = \begin{bmatrix} 1 & -2 \\ -1 & 1 \\ 0 & 1 \end{bmatrix}$$

25. Use Corollary 5.13.1(c) to find the Moore–Penrose inverse of the matrix $A = [U \quad V]$, where

$$U = \begin{bmatrix} 1 & 1 \\ 1 & -1 \\ 1 & 0 \\ 1 & 0 \\ 0 & 0 \end{bmatrix}, \qquad V = \begin{bmatrix} 2 & 2 \\ 2 & 0 \\ -1 & 0 \\ 1 & -2 \\ 0 & 1 \end{bmatrix}$$

26. Let the vectors $\boldsymbol{w}$, $\boldsymbol{x}$, $\boldsymbol{y}$, and $\boldsymbol{z}$ be given by

$$\boldsymbol{w} = \begin{bmatrix} 1 \\ 1 \\ 1 \\ 1 \end{bmatrix}, \qquad \boldsymbol{x} = \begin{bmatrix} 1 \\ 1 \\ -2 \\ 0 \end{bmatrix}, \qquad \boldsymbol{y} = \begin{bmatrix} 1 \\ -1 \\ 0 \\ 0 \end{bmatrix}, \qquad \boldsymbol{z} = \begin{bmatrix} 1 \\ 1 \\ 1 \\ -3 \end{bmatrix}$$

Use Theorem 5.17 to find the Moore–Penrose inverse of the matrix $A = \boldsymbol{w}\boldsymbol{x}' + \boldsymbol{y}\boldsymbol{z}'$.

27. Find a generalized inverse, different from the Moore–Penrose inverse, of the vector given in Problem 3.

28. Consider the diagonal matrix $A = \text{diag}(0, 2, 3)$.
    (a) Find a generalized inverse of A having rank of 2.
    (b) Find a generalized inverse of A that has rank of 3 and is diagonal.
    (c) Find a generalized inverse of A that is not diagonal.

29. Let A be an $m \times n$ matrix and B be an $n \times p$ matrix. Show that $B^- A^-$ will be a generalized inverse of (AB) for any choice of A^- and B^- if $\text{rank}(B) = n$.

30. Let A be an $m \times n$ matrix and B be an $n \times p$ matrix. Show that for any choice of A^- and B^-, $B^- A^-$ will be a generalized inverse of (AB) if and only if $A^- ABB^-$ is idempotent.

31. Let A, P, and Q by $m \times n$, $p \times m$, and $n \times q$ matrices, respectively. Show that if P has full column rank and Q has full row rank, then $Q^- A^- P^-$ is a generalized inverse of PAQ.

32. Let $(A'A)^-$ be any generalized inverse of the matrix $A'A$, where A is $m \times n$. Establish the following.
    (a) $A(A'A)^- A'$ does not depend on the choice of $(A'A)^-$.
    (b) $A(A'A)^- A'$ is symmetric even if $(A'A)^-$ is not symmetric.

33. Suppose that the $m \times n$ matrix A is partitioned as $A = [A_1 \quad A_2]$, where A_1 is $m \times r$, and $\text{rank}(A) = \text{rank}(A_1) = r$. Show that $A(A'A)^- A' = A_1(A_1'A_1)^- A_1'$.

34. Use the recursive procedure described in Section 9 to obtain the Moore–Penrose inverse of the matrix.

$$A = \begin{bmatrix} 1 & -1 & -1 \\ -1 & 1 & 1 \\ 2 & -1 & 1 \end{bmatrix}$$

35. Find a generalized inverse of the matrix A in the previous exercise by finding a nonsingular matrix that transforms it into a matrix having Hermite form.

36. Find a generalized inverse of the matrix

$$A = \begin{bmatrix} 1 & -1 & -2 & 1 \\ -2 & 4 & 3 & -2 \\ 1 & 1 & -3 & 1 \end{bmatrix}$$

37. Find a least squares inverse for the matrix A given in the previous exercise.

38. It was shown in Theorem 5.28 that a generalized inverse of A can be obtained by finding a nonsingular matrix that row reduces A to Hermite form. Show that there is a similar result for column reduction to Hermite form; that is, show that if C is a nonsingular matrix such that $AC = H$, where H is in Hermite form, then C is a generalized inverse of A.

39. Prove Theorem 5.27.

40. Penrose (1956) obtained the following recursive method for calculating the Moore–Penrose inverse of an $m \times n$ matrix A. Successively calculate $B_2, B_3, \ldots,$ where

$$B_{i+1} = i^{-1}\text{tr}(B_iA'A)\text{I}_n - B_iA'A$$

and B_1 is defined to be the $n \times n$ identity matrix. If $\text{rank}(A) = r$, then $B_{r+1}A'A = (0)$ and

$$A^+ = r\{\text{tr}(B_rA'A)\}^{-1}B_rA'$$

Use this method to compute the Moore–Penrose inverse of the matrix A of Example 5.10.

41. Let λ be the largest eigenvalue of AA', where A is an $m \times n$ matrix. Let α be any constant satisfying $0 < \alpha < 2/\lambda$ and define $X_1 = \alpha A'$. Ben-Israel

(1966) has shown that if we define

$$X_{i+1} = X_i(2\mathrm{I}_m - AX_i)$$

for $i = 1, 2, \ldots$ then $X_i \rightarrow A^+$ as $i \rightarrow \infty$. Use this iterative procedure to compute the Moore–Penrose inverse of the matrix A of Example 5.10 on a computer. Stop the iterative process when

$$\mathrm{tr}\left\{(X_{i+1} - X_i)'(X_{i+1} - X_i)\right\}$$

gets small. Note that λ does not need to be computed since we must have

$$\frac{2}{\mathrm{tr}(AA')} < \frac{2}{\lambda}$$

42. Use the results of Section 5 to obtain the expression given in (5.18) for the Moore–Penrose inverse of the matrix $A_j = [A_{j-1} \quad \boldsymbol{a}_j]$.

CHAPTER SIX

Systems of Linear Equations

1. INTRODUCTION

As mentioned at the beginning of Chapter 5, one of the applications of generalized inverses is in finding solutions to a system of linear equations of the form

$$Ax = c, \tag{6.1}$$

where A is an $m \times n$ matrix of constants, c is an $m \times 1$ vector of constants, and x is an $n \times 1$ vector of variables for which solutions are needed. In this chapter, we discuss such issues as the existence of solutions to (6.1), the form of a general solution, and the number of linearly independent solutions. We conclude the chapter by taking a look at the special application of finding least squares solutions to (6.1), when an exact solution does not exist.

2. CONSISTENCY OF A SYSTEM OF EQUATIONS

In this section, we will obtain necessary and sufficient conditions for the existence of a vector x satisfying equation (6.1). When one or more such vectors exist, the system of equations is said to be consistent; otherwise, the system is referred to as an inconsistent system. Our first necessary and sufficient condition for consistency is that the vector c is in the column space of A or, equivalently, that the rank of the augmented matrix $[A \quad c]$ is the same as the rank of A.

Theorem 6.1. The system of equations, $Ax = c$, is consistent if and only if $\text{rank}([A \quad c]) = \text{rank}(A)$.

Proof. If $a_1, \ldots, a_n$ are the columns of A, then the equation $Ax = c$ can be written as

$$A\boldsymbol{x} = [\boldsymbol{a}_1 \cdots \boldsymbol{a}_n] \begin{bmatrix} x_1 \\ \vdots \\ x_n \end{bmatrix} = \sum_{i=1}^{n} x_i \boldsymbol{a}_i = \boldsymbol{c}$$

Clearly, this holds for some $\boldsymbol{x}$ if and only if $\boldsymbol{c}$ is a linear combination of the columns of A, in which case $\text{rank}[A \quad \boldsymbol{c}] = \text{rank}(A)$. □

Example 6.1. Consider the system of equations which has

$$A = \begin{bmatrix} 1 & 2 \\ 2 & 1 \\ 1 & 0 \end{bmatrix}, \qquad \boldsymbol{c} = \begin{bmatrix} 1 \\ 5 \\ 3 \end{bmatrix}$$

Clearly, the rank of A is 2 while

$$|[A \quad \boldsymbol{c}]| = \begin{vmatrix} 1 & 2 & 1 \\ 2 & 1 & 5 \\ 1 & 0 & 3 \end{vmatrix} = 0,$$

so that the rank of $[A \quad \boldsymbol{c}]$ is also 2. Thus, we know from Theorem 6.1 that the system of equations $A\boldsymbol{x} = \boldsymbol{c}$ is consistent.

Although Theorem 6.1 is useful in determining whether a given system of linear equations is consistent, it does not tell us how to find a solution to the system when it is consistent. Our next result gives an alternative necessary and sufficient condition for consistency utilizing a generalized inverse, A^-, of A. An obvious consequence of this result is that when the system $A\boldsymbol{x} = \boldsymbol{c}$ is consistent, then a solution will be given by $\boldsymbol{x} = A^-\boldsymbol{c}$.

Theorem 6.2. The system of equations $A\boldsymbol{x} = \boldsymbol{c}$ is consistent if and only if for some generalized inverse, A^-, of A, $AA^-\boldsymbol{c} = \boldsymbol{c}$.

Proof. First, suppose that the system is consistent and $\boldsymbol{x}_*$ is a solution, so that $\boldsymbol{c} = A\boldsymbol{x}_*$. Premultiplying this identity by AA^-, where A^- is any generalized inverse of A, yields

$$AA^-\boldsymbol{c} = AA^-A\boldsymbol{x}_* = A\boldsymbol{x}_* = \boldsymbol{c},$$

as is required. Conversely, now suppose that there is a generalized inverse of A satisfying $AA^-\boldsymbol{c} = \boldsymbol{c}$. Define $\boldsymbol{x}_* = A^-\boldsymbol{c}$ and note that

$$A\boldsymbol{x}_* = AA^-\boldsymbol{c} = \boldsymbol{c}$$

Thus, since $\boldsymbol{x}_* = A^-\boldsymbol{c}$ is a solution, the system is consistent and so the proof is complete. □

Suppose that A_1 and A_2 are any two generalized inverses of A so that $AA_1A = AA_2A = A$. In addition, suppose that A_1 satisfies the condition of Theorem 6.2; that is, $AA_1\boldsymbol{c} = \boldsymbol{c}$. Then A_2 satisfies the same condition since

$$AA_2\boldsymbol{c} = AA_2(AA_1\boldsymbol{c}) = (AA_2A)A_1\boldsymbol{c} = AA_1\boldsymbol{c} = \boldsymbol{c}.$$

Thus, in applying Theorem 6.2, one will need to check the given condition for only one generalized inverse of A, and it doesn't matter which generalized inverse is used. In particular, we can use the Moore–Penrose inverse A^+, of A.

The following results involve some special cases regarding the matrix A.

Corollary 6.2.1. If A is an $m \times m$ nonsingular matrix and $\boldsymbol{c}$ is an $m \times 1$ vector of constants, then the system $A\boldsymbol{x} = \boldsymbol{c}$ is consistent.

Corollary 6.2.2. If the $m \times n$ matrix A has rank equal to m, then the system $A\boldsymbol{x} = \boldsymbol{c}$ is consistent.

Proof. Since A has full row rank, it follows from Theorem 5.22(f) that $AA^- = I_m$. As a result, $AA^-\boldsymbol{c} = \boldsymbol{c}$, and so from Theorem 6.2, the system must be consistent. □

Example 6.2. Consider the system of equations $A\boldsymbol{x} = \boldsymbol{c}$, where

$$A = \begin{bmatrix} 1 & 1 & 1 & 2 \\ 1 & 0 & 1 & 0 \\ 2 & 1 & 2 & 2 \end{bmatrix}, \qquad \boldsymbol{c} = \begin{bmatrix} 3 \\ 2 \\ 5 \end{bmatrix}$$

A generalized inverse of the transpose of A was given in Example 5.12. Using this, we find that

$$AA^-\boldsymbol{c} = \begin{bmatrix} 1 & 1 & 1 & 2 \\ 1 & 0 & 1 & 0 \\ 2 & 1 & 2 & 2 \end{bmatrix} \begin{bmatrix} 0 & 1 & -1 \\ 1 & -1 & 0 \\ 0 & 0 & 1 \\ 0 & 0 & 0 \end{bmatrix} \begin{bmatrix} 3 \\ 2 \\ 5 \end{bmatrix}$$

$$= \begin{bmatrix} 1 & 0 & 0 \\ 0 & 1 & 0 \\ 1 & 1 & 0 \end{bmatrix} \begin{bmatrix} 3 \\ 2 \\ 5 \end{bmatrix} = \begin{bmatrix} 3 \\ 2 \\ 5 \end{bmatrix}$$

Since this is $\boldsymbol{c}$, the system of equations is consistent, and a solution is given by

$$A^-c = \begin{bmatrix} 0 & 1 & -1 \\ 1 & -1 & 0 \\ 0 & 0 & 1 \\ 0 & 0 & 0 \end{bmatrix} \begin{bmatrix} 3 \\ 2 \\ 5 \end{bmatrix} = \begin{bmatrix} -3 \\ 1 \\ 5 \\ 0 \end{bmatrix}$$

The system of linear equations $A\boldsymbol{x} = \boldsymbol{c}$ is a special case of the more general system of linear equations given by $AXB = C$, where A is $m \times n$, B is $p \times q$, C is $m \times q$, and X is $n \times p$. A necessary and sufficient condition for the existence of a solution matrix X satisfying this system is given in the following theorem.

Theorem 6.3. Let A, B, and C be matrices of constants, where A is $m \times n$, B is $p \times q$, and C is $m \times q$. Then the system of equations

$$AXB = C,$$

is consistent if and only if for some generalized inverses A^- and B^-,

$$AA^-CB^-B = C \tag{6.2}$$

Proof. Suppose that the system is consistent and the matrix X_* is a solution, so that $C = AX_*B$. Premultiplying by AA^- and postmultiplying by B^-B, where A^- and B^- are any generalized inverses of A and B, we find that

$$AA^-CB^-B = AA^-AX_*BB^-B = AX_*B = C,$$

and so equation (6.2) holds. On the other hand, if A^- and B^- satisfy (6.2), define $X_* = A^-CB^-$, and note that X_* is a solution since

$$AX_*B = AA^-CB^-B = C \qquad \square$$

Using an argument similar to that given after Theorem 6.2, we can verify that if (6.2) is satisfied for any one particular choice of A^- and B^-, then it will hold for all choices of A^- and B^-. Consequently, the application of Theorem 6.3 is not dependent upon the choices of generalized inverses for A and B.

3. SOLUTIONS TO A CONSISTENT SYSTEM OF EQUATIONS

We have seen that if the system of equations $A\boldsymbol{x} = \boldsymbol{c}$ is consistent, then $\boldsymbol{x} = A^-\boldsymbol{c}$ is a solution regardless of the choice of the generalized inverse A^-. Thus, if $A^-\boldsymbol{c}$ is not the same for all choices of A^-, then our system of equations has more than one solution. In fact, we will see that even when $A^-\boldsymbol{c}$ does not depend on the choice of A^-, which is the case if $\boldsymbol{c} = \mathbf{0}$, our system of equations may have many solutions. The following theorem gives a general expression for all solutions to the system.

Theorem 6.4. Suppose that $A\boldsymbol{x} = \boldsymbol{c}$ is a consistent system of equations, and let A^- be any generalized inverse of the $m \times n$ matrix A. Then, for any $n \times 1$ vectory $\boldsymbol{y}$,

$$\boldsymbol{x}_y = A^-\boldsymbol{c} + (\mathrm{I}_n - A^-A)\boldsymbol{y} \tag{6.3}$$

is a solution, and for any solution, $\boldsymbol{x}_*$, there exists a vector $\boldsymbol{y}$ such that $\boldsymbol{x}_* = \boldsymbol{x}_y$.

Proof. Since $A\boldsymbol{x} = \boldsymbol{c}$ is a consistent system of equations, we know from Theorem 6.2 that $AA^-\boldsymbol{c} = \boldsymbol{c}$, and so

$$\begin{aligned} A\boldsymbol{x}_y &= AA^-\boldsymbol{c} + A(\mathrm{I}_n - A^-A)\boldsymbol{y} \\ &= \boldsymbol{c} + (A - AA^-A)\boldsymbol{y} = \boldsymbol{c}, \end{aligned}$$

since $AA^-A = A$. Thus, $\boldsymbol{x}_y$ is a solution regardless of the choice of $\boldsymbol{y}$. On the other hand, if $\boldsymbol{x}_*$ is an arbitrary solution, so that $A\boldsymbol{x}_* = \boldsymbol{c}$, it follows that $A^-A\boldsymbol{x}_* = A^-\boldsymbol{c}$. Consequently,

$$A^-\boldsymbol{c} + (\mathrm{I}_n - A^-A)\boldsymbol{x}_* = A^-\boldsymbol{c} + \boldsymbol{x}_* - A^-A\boldsymbol{x}_* = \boldsymbol{x}_*,$$

so that $\boldsymbol{x}_* = x_{x*}$. This completes the proof. □

The set of solutions given in Theorem 6.4 is expressed in terms of a fixed generalized inverse A^- and an arbitrary $n \times 1$ vector $\boldsymbol{y}$. Alternatively, this set of all solutions can be expressed in terms of an arbitrary generalized inverse of A.

Corollary 6.4.1. Suppose that $A\boldsymbol{x} = \boldsymbol{c}$ is a consistent system of equations, where $\boldsymbol{c} \neq \mathbf{0}$. If B is a generalized inverse of A, then $\boldsymbol{x} = B\boldsymbol{c}$ is a solution, and for any solution $\boldsymbol{x}_*$, there exists a generalized inverse B such that $\boldsymbol{x}_* = B\boldsymbol{c}$.

Proof. Theorem 6.4 was not dependent upon the choice of the generalized inverse, so by choosing $A^- = B$ and $\boldsymbol{y} = \mathbf{0}$ in (6.3), we prove that $\boldsymbol{x} = B\boldsymbol{c}$ is a solution. All that remains to be shown is that for any particular A^- and $\boldsymbol{y}$, we can find a generalized inverse B such that the expression in (6.3) equals $B\boldsymbol{c}$. Now since $\boldsymbol{c} \neq \mathbf{0}$, it has at least one component, say c_i, not equal to 0. Define the $n \times m$ matrix C as $C = c_i^{-1}\boldsymbol{y}\boldsymbol{e}_i'$ so that $C\boldsymbol{c} = \boldsymbol{y}$. Since the system of equations $A\boldsymbol{x} = \boldsymbol{c}$ is consistent, we must have $AA^-\boldsymbol{c} = \boldsymbol{c}$, and so

$$\begin{aligned}\boldsymbol{x}_y &= A^-\boldsymbol{c} + (\mathrm{I}_n - A^-A)\boldsymbol{y} = A^-\boldsymbol{c} + (\mathrm{I}_n - A^-A)C\boldsymbol{c} \\ &= A^-\boldsymbol{c} + C\boldsymbol{c} - A^-AC\boldsymbol{c} = A^-\boldsymbol{c} + C\boldsymbol{c} - A^-ACAA^-\boldsymbol{c} \\ &= (A^- + C - A^-ACAA^-)\boldsymbol{c}\end{aligned}$$

But it follows from Theorem 5.23 that $A^- + C - A^-ACAA^-$ is a generalized inverse of A for any choice of the $n \times m$ matrix C and so the proof is complete. □

Our next theorem gives a result, analogous to Theorem 6.4, for the system of equations $AXB = C$. The proof will be left to the reader as an exercise.

Theorem 6.5. Let $AXB = C$ be a consistent system of equations, where A is $m \times n$, B is $p \times q$, and C is $m \times q$. Then for any generalized inverses, A^- and B^-, and any $n \times p$ matrix, Y,

$$X_Y = A^-CB^- + Y - A^-AYBB^-$$

is a solution, and for any solution, X_*, there exists a matrix Y such that $X_* = X_Y$.

Example 6.3. For the consistent system of equations discussed in Example 6.2, we have

$$A^-A = \begin{bmatrix} 0 & 1 & -1 \\ 1 & -1 & 0 \\ 0 & 0 & 1 \\ 0 & 0 & 0 \end{bmatrix} \begin{bmatrix} 1 & 1 & 1 & 2 \\ 1 & 0 & 1 & 0 \\ 2 & 1 & 2 & 2 \end{bmatrix} = \begin{bmatrix} -1 & -1 & -1 & -2 \\ 0 & 1 & 0 & 2 \\ 2 & 1 & 2 & 2 \\ 0 & 0 & 0 & 0 \end{bmatrix}$$

Consequently, a general solution to this system of equations is given by

$$\begin{aligned}\boldsymbol{x}_y &= A^-\boldsymbol{c} + (\mathrm{I}_n - A^-A)\boldsymbol{y} \\ &= \begin{bmatrix} -3 \\ 1 \\ 5 \\ 0 \end{bmatrix} + \begin{bmatrix} 2 & 1 & 1 & 2 \\ 0 & 0 & 0 & -2 \\ -2 & -1 & -1 & -2 \\ 0 & 0 & 0 & 1 \end{bmatrix} \begin{bmatrix} y_1 \\ y_2 \\ y_3 \\ y_4 \end{bmatrix} \\ &= \begin{bmatrix} -3 + 2y_1 + y_2 + y_3 + 2y_4 \\ 1 - 2y_4 \\ 5 - 2y_1 - y_2 - y_3 - 2y_4 \\ y_4 \end{bmatrix}\end{aligned}$$

where $\boldsymbol{y}$ is an arbitrary 4×1 vector.

In some applications, it may be important to know whether a consistent system of equations yields a unique solution; that is, under what conditions will (6.3) yield the same solution for all choices of $\boldsymbol{y}$?

Theorem 6.6. If $A\boldsymbol{x} = \boldsymbol{c}$ is a consistent system of equations, then the solution $\boldsymbol{x}_* = A^-\boldsymbol{c}$ is a unique solution if and only if $A^-A = \mathrm{I}_n$, where A^- is any generalized inverse of the $m \times n$ matrix A.

Proof. Note that $\boldsymbol{x}_* = A^-\boldsymbol{c}$ is a unique solution if and only if $\boldsymbol{x}_y = \boldsymbol{x}_*$ for all choices of $\boldsymbol{y}$, where $\boldsymbol{x}_y$ is as defined in (6.3). In other words, the solution is unique if and only if

$$(\mathrm{I}_n - A^-A)\boldsymbol{y} = \boldsymbol{0}$$

for all $\boldsymbol{y}$, and clearly this is equivalent to the condition $(\mathrm{I}_n - A^-A) = (0)$ or $A^-A = \mathrm{I}_n$. □

We saw in Theorem 5.22(g) that $\text{rank}(A) = n$ if and only if $A^-A = \mathrm{I}_n$. As a result, we can restate the necessary and sufficient condition of Theorem 6.6 as follows.

Corollary 6.6.1. Suppose that $A\boldsymbol{x} = \boldsymbol{c}$ is a consistent system of equations. Then the solution $\boldsymbol{x}_* = A^-\boldsymbol{c}$ is a unique solution if and only if $\text{rank}(A) = n$.

Example 6.4. We saw in Example 6.1 that the system of equations $A\boldsymbol{x} = \boldsymbol{c}$, where

$$A = \begin{bmatrix} 1 & 2 \\ 2 & 1 \\ 1 & 0 \end{bmatrix}, \qquad \boldsymbol{c} = \begin{bmatrix} 1 \\ 5 \\ 3 \end{bmatrix},$$

is consistent. The Moore–Penrose inverse of the transpose of A was obtained in Example 5.1. Using this, we find that

$$A^+A = \frac{1}{14}\begin{bmatrix} -3 & 6 & 5 \\ 8 & -2 & -4 \end{bmatrix}\begin{bmatrix} 1 & 2 \\ 2 & 1 \\ 1 & 0 \end{bmatrix} = \frac{1}{14}\begin{bmatrix} 14 & 0 \\ 0 & 14 \end{bmatrix} = \mathrm{I}_2$$

Thus, the system of equations $A\boldsymbol{x} = \boldsymbol{c}$ has the unique solution given by

$$A^+\boldsymbol{c} = \frac{1}{14}\begin{bmatrix} -3 & 6 & 5 \\ 8 & -2 & -4 \end{bmatrix}\begin{bmatrix} 1 \\ 5 \\ 3 \end{bmatrix} = \frac{1}{14}\begin{bmatrix} 42 \\ -14 \end{bmatrix} = \begin{bmatrix} 3 \\ -1 \end{bmatrix}$$

Suppose that a system of linear equations has more than one solution, and let $\boldsymbol{x}_1$ and $\boldsymbol{x}_2$ be two different solutions. Then, since $A\boldsymbol{x}_i = \boldsymbol{c}$ for $i = 1$ and 2, it follows that for any scalar α

$$A\{\alpha\boldsymbol{x}_1 + (1-\alpha)\boldsymbol{x}_2\} = \alpha A\boldsymbol{x}_1 + (1-\alpha)A\boldsymbol{x}_2 = \alpha\boldsymbol{c} + (1-\alpha)\boldsymbol{c} = \boldsymbol{c}$$

Thus, $\boldsymbol{x} = \{\alpha\boldsymbol{x}_1 + (1-\alpha)\boldsymbol{x}_2\}$ is also a solution. Since α was arbitrary, we see that if a system has more than one solution, then it has infinitely many solutions. However, the number of linearly independent solutions to a consistent system of equations having $\boldsymbol{c} \neq \boldsymbol{0}$ must be between 1 and n; that is, there exists a set of linearly independent solutions, $\{\boldsymbol{x}_1, \dots, \boldsymbol{x}_r\}$, such that every solution can be expressed as a linear combination of the solutions, $\boldsymbol{x}_1, \dots, \boldsymbol{x}_r$. In other words, any solution $\boldsymbol{x}$ can be written as $\boldsymbol{x} = \alpha_1\boldsymbol{x}_1 + \cdots + \alpha_r\boldsymbol{x}_r$, for some coefficients, $\alpha_1, \dots, \alpha_r$. Note that since $A\boldsymbol{x}_i = \boldsymbol{c}$ for each i, we must have

$$A\boldsymbol{x} = A\left(\sum_{i=1}^{r} \alpha_i\boldsymbol{x}_i\right) = \sum_{i=1}^{r} \alpha_i A\boldsymbol{x}_i = \sum_{i=1}^{r} \alpha_i\boldsymbol{c} = \left(\sum_{i=1}^{r} \alpha_i\right)\boldsymbol{c},$$

and so if $\boldsymbol{x}$ is a solution, the coefficients must satisfy the identity, $\alpha_1 + \cdots + \alpha_r = 1$. Our next result tells us exactly how to determine this number of linearly independent solutions r when $\boldsymbol{c} \neq \boldsymbol{0}$. We will delay the discussion of the situation in which $\boldsymbol{c} = \boldsymbol{0}$ until the next section.

Theorem 6.7. Suppose that the system $A\boldsymbol{x} = \boldsymbol{c}$ is consistent, where A is $m \times n$ and $\boldsymbol{c} \neq \boldsymbol{0}$. Then each solution can be expressed as a linear combination of r linearly independent solutions, where $r = n - \text{rank}(A) + 1$.

Proof. Using (6.3) with the particular generalized inverse A^+, we begin with the $n+1$ solutions, $\boldsymbol{x}_0 = A^+\boldsymbol{c}, \boldsymbol{x}_{e_1} = A^+\boldsymbol{c} + (\text{I}_n - A^+A)\boldsymbol{e}_1, \dots, \boldsymbol{x}_{e_n} = A^+\boldsymbol{c} + (\text{I}_n - A^+A)\boldsymbol{e}_n$, where, as usual, $\boldsymbol{e}_i$ denotes the $n \times 1$ vector whose only nonzero element is 1 in the ith position. Now every solution can be expressed as a linear combination of these solutions since for any $\boldsymbol{y} = (y_1, \dots, y_n)'$,

$$\boldsymbol{x}_y = A^+\boldsymbol{c} + (\text{I}_n - A^+A)\boldsymbol{y} = \left(1 - \sum_{i=1}^{n} y_i\right)\boldsymbol{x}_0 + \sum_{i=1}^{n} y_i\boldsymbol{x}_{e_i}$$

Thus, if we define the $n \times (n+1)$ matrix $X = (\boldsymbol{x}_0, \boldsymbol{x}_{e_1}, \dots, \boldsymbol{x}_{e_n})$, the proof will be complete if we can show that $\text{rank}(X) = n - \text{rank}(A) + 1$. Note that we can write X as $X = BC$, where B and C are the $n \times (n+1)$ and $(n+1) \times (n+1)$

matrices given by $B = (A^+\boldsymbol{c}, \mathrm{I}_n - A^+A)$ and

$$C = \begin{bmatrix} 1 & \mathbf{1}'_n \\ \mathbf{0} & \mathrm{I}_n \end{bmatrix}$$

Clearly, C is nonsingular since it is lower triangular and the product of its diagonal elements is 1. Consequently, from Theorem 1.8, we know that $\mathrm{rank}(X) = \mathrm{rank}(B)$. Note also that

$$\begin{aligned}(\mathrm{I}_n - A^+A)'A^+\boldsymbol{c} &= (\mathrm{I}_n - A^+A)A^+\boldsymbol{c} = (A^+ - A^+AA^+)\boldsymbol{c} \\ &= (A^+ - A^+)\boldsymbol{c} = \mathbf{0},\end{aligned}$$

so that the first column of B is orthogonal to the remaining columns. This implies that

$$\mathrm{rank}(B) = \mathrm{rank}(A^+\boldsymbol{c}) + \mathrm{rank}(\mathrm{I}_n - A^+A) = 1 + \mathrm{rank}(\mathrm{I}_n - A^+A),$$

since the consistency condition, $AA^+\boldsymbol{c} = \boldsymbol{c}$ and $\boldsymbol{c} \neq \mathbf{0}$ guarantee that $A^+\boldsymbol{c} \neq \mathbf{0}$. All that remains is to show that $\mathrm{rank}(\mathrm{I}_n - A^+A) = n - \mathrm{rank}(A)$. Now since A^+A is the projection matrix of $R(A^+) = R(A')$, it follows that $\mathrm{I}_n - A^+A$ is the projection matrix of the orthogonal complement of $R(A')$ or, in other words, the null space of A, $N(A)$. Since $\dim\{N(A)\} = n - \mathrm{rank}(A)$, we must have $\mathrm{rank}(\mathrm{I}_n - A^+A) = n - \mathrm{rank}(A)$. □

Since $\boldsymbol{x}_0 = A^+\boldsymbol{c}$ is orthogonal to the columns of $(\mathrm{I}_n - A^+A)$, when constructing a set of r linearly independent solutions, one of these solutions always will be $\boldsymbol{x}_0$, with the remaining solutions given by $\boldsymbol{x}_y$ for $r-1$ different choices of $\boldsymbol{y} \neq \mathbf{0}$. This statement is not dependent upon the choice of A^+ as the generalized inverse in (6.3), since $A^-\boldsymbol{c}$ and $(\mathrm{I}_n - A^-A)\boldsymbol{y}$ are linearly independent regardless of the choice of A^- if $\boldsymbol{c} \neq \mathbf{0}, \boldsymbol{y} \neq \mathbf{0}$. The proof of this linear independence is left as an exercise.

Example 6.5. We saw that the system of equations $A\boldsymbol{x} = \boldsymbol{c}$ of Examples 6.2 and 6.3 has the set of solutions consisting of all vectors of the form

$$\boldsymbol{x}_y = A^-\boldsymbol{c} + (\mathrm{I}_4 - A^-A)\boldsymbol{y} = \begin{bmatrix} -3 + 2y_1 + y_2 + y_3 + 2y_4 \\ 1 - 2y_4 \\ 5 - 2y_1 - y_2 - y_3 - 2y_4 \\ y_4 \end{bmatrix}$$

Since the last row of the 3×4 matrix

$$A = \begin{bmatrix} 1 & 1 & 1 & 2 \\ 1 & 0 & 1 & 0 \\ 2 & 1 & 2 & 2 \end{bmatrix}$$

is the sum of the first two rows, rank(A) = 2. Thus, the system of equations possesses

$$n - \text{rank}(A) + 1 = 4 - 2 + 1 = 3$$

linearly independent solutions. Three linearly independent solutions can be obtained through appropriate choices of the $\boldsymbol{y}$ vector. For instance, since $A^{-}\boldsymbol{c}$ and $(\text{I}_4 - A^{-}A)\boldsymbol{y}$ are linearly independent, the three solutions

$$A^{-}\boldsymbol{c}, A^{-}\boldsymbol{c} + (\text{I}_4 - A^{-}A)_{\cdot i}, A^{-}\boldsymbol{c} + (\text{I}_4 - A^{-}A)_{\cdot j}$$

will be linearly independent if the ith and jth columns of $(\text{I}_4 - A^{-}A)$ are linearly independent. Looking back at the matrix $(\text{I}_4 - A^{-}A)$ given in Example 6.3, we see that its first and fourth columns are linearly independent. Thus, three linearly independent solutions of $A\boldsymbol{x} = \boldsymbol{c}$ are given by

$$A^{-}\boldsymbol{c} = \begin{bmatrix} -3 \\ 1 \\ 5 \\ 0 \end{bmatrix}, \qquad A^{-}\boldsymbol{c} + (\text{I}_4 - A^{-}A)_{\cdot 1} = \begin{bmatrix} -3 \\ 1 \\ 5 \\ 0 \end{bmatrix} + \begin{bmatrix} 2 \\ 0 \\ -2 \\ 0 \end{bmatrix} = \begin{bmatrix} -1 \\ 1 \\ 3 \\ 0 \end{bmatrix},$$

$$A^{-}\boldsymbol{c} + (\text{I}_4 - A^{-}A)_{\cdot 4} = \begin{bmatrix} -3 \\ 1 \\ 5 \\ 0 \end{bmatrix} + \begin{bmatrix} 2 \\ -2 \\ -2 \\ 1 \end{bmatrix} = \begin{bmatrix} -1 \\ -1 \\ 3 \\ 1 \end{bmatrix}$$

4. HOMOGENEOUS SYSTEMS OF EQUATIONS

The system of equations $A\boldsymbol{x} = \boldsymbol{c}$ is called a nonhomogeneous system of equations when $\boldsymbol{c} \neq \boldsymbol{0}$, while $A\boldsymbol{x} = \boldsymbol{0}$ is referred to as a homogeneous system of equations. In this section, we obtain some results regarding homogeneous systems of equations. One obvious distinction between homogeneous and nonhomogeneous systems is that a homogeneous system of equations must be consistent since it will always have the trivial solution, $\boldsymbol{x} = \boldsymbol{0}$. A homogeneous system will then have a unique solution only when the trivial solution is the only solution. Conditions for the existence of nontrivial solutions, which we state in the next theorem, follow directly from Theorem 6.6 and Corollary 6.6.1.

Theorem 6.8. Suppose that A is an $m \times n$ matrix. The system $A\boldsymbol{x} = \mathbf{0}$ has nontrivial solutions if and only if $A^{-}A \neq \mathrm{I}_n$, or equivalently if and only if $\operatorname{rank}(A) < n$.

If the system $A\boldsymbol{x} = \mathbf{0}$ has more than one solution, and $\{\boldsymbol{x}_1, \ldots, \boldsymbol{x}_r\}$ is a set of r solutions, then $\boldsymbol{x} = \alpha_1\boldsymbol{x}_1 + \cdots + \alpha_r\boldsymbol{x}_r$ is also a solution regardless of the choice of $\alpha_1, \ldots, \alpha_r$, since

$$A\boldsymbol{x} = A\left(\sum_{i=1}^{r} \alpha_i\boldsymbol{x}_i\right) = \sum_{i=1}^{r} \alpha_i A\boldsymbol{x}_i = \sum_{i=1}^{r} \alpha_i\mathbf{0} = \mathbf{0}$$

In fact, we have the following.

Theorem 6.9. If A is an $m \times n$ matrix, then the set of all solutions to the system of equations $A\boldsymbol{x} = \mathbf{0}$ forms a vector subpsace of R^n having dimension $n - \operatorname{rank}(A)$.

Proof. The result follows immediately from the fact that the set of all solutions of $A\boldsymbol{x} = \mathbf{0}$ is the null space of A. □

In contrast to Theorem 6.9, the set of all solutions to a nonhomogeneous system of equations will not form a vector subspace. This is because, as we have seen in the previous section, a linear combination of solutions to a nonhomogeneous system yields another solution only if the coefficients sum to one. Additionally, a nonhomogeneous system cannot have $\mathbf{0}$ as a solution.

The general form of a solution given in Theorem 6.4 applies to both homogeneous and nonhomogeneous systems. Thus, for any $n \times 1$ vector $\boldsymbol{y}$,

$$\boldsymbol{x}_y = (\mathrm{I}_n - A^{-}A)\boldsymbol{y}$$

is a solution to the system $A\boldsymbol{x} = \mathbf{0}$, and for any solution, $\boldsymbol{x}_*$, there exists a vector $\boldsymbol{y}$ such that $\boldsymbol{x}_* = \boldsymbol{x}_y$. The following result shows that the set of solutions of $A\boldsymbol{x} = \boldsymbol{c}$ can be expressed in terms of the set of solutions to $A\boldsymbol{x} = \mathbf{0}$.

Theorem 6.10. Let $\boldsymbol{x}_*$ be any solution to the system of equations $A\boldsymbol{x} = \boldsymbol{c}$. Then

(a) if $\boldsymbol{x}_\#$ is a solution to the system $A\boldsymbol{x}_\# = \mathbf{0}$, $\boldsymbol{x} = \boldsymbol{x}_* + \boldsymbol{x}_\#$ is a solution of $A\boldsymbol{x} = \boldsymbol{c}$, and
(b) for any solution $\boldsymbol{x}$ to the equation $A\boldsymbol{x} = \boldsymbol{c}$, there exists a solution $\boldsymbol{x}_\#$ to the equation $A\boldsymbol{x} = \mathbf{0}$ such that $\boldsymbol{x} = \boldsymbol{x}_* + \boldsymbol{x}_\#$.

Proof. Note that if $\boldsymbol{x}_\#$ is as defined in (a), then

$$A(\boldsymbol{x}_* + \boldsymbol{x}_\#) = A\boldsymbol{x}_* + A\boldsymbol{x}_\# = \boldsymbol{c} + \mathbf{0} = \boldsymbol{c},$$

and so $\boldsymbol{x} = \boldsymbol{x}_* + \boldsymbol{x}_\#$ is a solution to $A\boldsymbol{x} = \boldsymbol{c}$. To prove (b), define $\boldsymbol{x}_\# = \boldsymbol{x} - \boldsymbol{x}_*$, so that $\boldsymbol{x} = \boldsymbol{x}_* + \boldsymbol{x}_\#$. Then since $A\boldsymbol{x} = \boldsymbol{c}$ and $A\boldsymbol{x}_* = \boldsymbol{c}$, it follows that

$$A\boldsymbol{x}_\# = A(\boldsymbol{x} - \boldsymbol{x}_*) = A\boldsymbol{x} - A\boldsymbol{x}_* = \boldsymbol{c} - \boldsymbol{c} = \boldsymbol{0} \qquad \square$$

Our next result, regarding the number of linearly independent solutions possessed by a homogeneous system of equations, follows immediately from Theorem 6.9.

Theorem 6.11. Each solution of the homogeneous system of equations $A\boldsymbol{x} = \boldsymbol{0}$ can be expressed as a linear combination of r linearly independent solutions, where $r = n - \text{rank}(A)$.

Example 6.6. Consider the system of equations $A\boldsymbol{x} = \boldsymbol{0}$, where

$$A = \begin{bmatrix} 1 & 2 \\ 2 & 1 \\ 1 & 0 \end{bmatrix}$$

We saw in Example 6.4 that $A^+A = \text{I}_2$. Thus, the system only has the trivial solution $\boldsymbol{0}$.

Example 6.7. Since the matrix

$$A = \begin{bmatrix} 1 & 1 & 1 & 2 \\ 1 & 0 & 1 & 0 \\ 2 & 1 & 2 & 2 \end{bmatrix}$$

from Example 6.5 has rank of 2, the homogeneous system of equations $A\boldsymbol{x} = \boldsymbol{0}$ has $r = n - \text{rank}(A) = 4 - 2 = 2$ linearly independent solutions. Any set of two linearly independent columns of the matrix $(\text{I}_4 - A^-A)$ will be a set of linearly independent solutions; for example, the first and fourth columns,

$$\begin{bmatrix} 2 \\ 0 \\ -2 \\ 0 \end{bmatrix}, \quad \begin{bmatrix} 2 \\ -2 \\ -2 \\ 1 \end{bmatrix}$$

are linearly independent solutions.

5. LEAST SQUARES SOLUTIONS TO A SYSTEM OF LINEAR EQUATIONS

In some situations in which we have an inconsistent system of equtions $A\boldsymbol{x} = \boldsymbol{c}$, it may be desirable to find the vector or set of vectors which comes "closest" to satisfying the equations. If $\boldsymbol{x}_*$ is one choice for $\boldsymbol{x}$, then $\boldsymbol{x}_*$ will approximately satisfy our system of equations if $A\boldsymbol{x}_* - \boldsymbol{c}$ is close to $\mathbf{0}$. One of the most common ways of measuring the closeness of $A\boldsymbol{x}_* - \boldsymbol{c}$ to $\mathbf{0}$ is through the computation of the sum of squares of the components of the vector $A\boldsymbol{x}_* - \boldsymbol{c}$. Any vector minimizing this sum of squares is referred to as a least squares solution.

Definition 6.1. The $n \times 1$ vector $\boldsymbol{x}_*$ is said to be a least squares solution to the system of equations $A\boldsymbol{x} = \boldsymbol{c}$ if the inequality

$$(A\boldsymbol{x}_* - \boldsymbol{c})'(A\boldsymbol{x}_* - \boldsymbol{c}) \leq (A\boldsymbol{x} - \boldsymbol{c})'(A\boldsymbol{x} - \boldsymbol{c}) \tag{6.4}$$

holds for every $n \times 1$ vector $\boldsymbol{x}$.

Of course, we have already utilized the concept of a least squares solution in many of our examples on regression analysis. In particular, we have seen that if the matrix X has full column rank, then the least squares solution for $\hat{\boldsymbol{\beta}}$ in the fitted regression equation, $\hat{\boldsymbol{y}} = X\hat{\boldsymbol{\beta}}$ is given by $\hat{\boldsymbol{\beta}} = (X'X)^{-1}X'\boldsymbol{y}$. The generalized inverses that we have discussed in this chapter will enable us to obtain a unified treatment of this problem including cases in which X is not of full rank.

In Section 5.8, we briefly discussed the $\{1,3\}$-inverse of a matrix A, that is, any matrix satisfying the first and third conditions of the Moore–Penrose inverse. We referred to this inverse as the least squares inverse of A. The following result motivates this description.

Theorem 6.12. Let A^L by any $\{1,3\}$-inverse of a matrix A. Then the vector $\boldsymbol{x}_* = A^L\boldsymbol{c}$ is a least squares solution to the system of equations $A\boldsymbol{x} = \boldsymbol{c}$.

Proof. We must show that (6.4) holds when $\boldsymbol{x}_* = A^L\boldsymbol{c}$. The right-hand side of (6.4) can be written as

$$\begin{aligned}(A\boldsymbol{x} - \boldsymbol{c})'(A\boldsymbol{x} - \boldsymbol{c}) &= \{(A\boldsymbol{x} - AA^L\boldsymbol{c}) + (AA^L\boldsymbol{c} - \boldsymbol{c})\}'\{(A\boldsymbol{x} - AA^L\boldsymbol{c}) + (AA^L\boldsymbol{c} - \boldsymbol{c})\} \\ &= (A\boldsymbol{x} - AA^L\boldsymbol{c})'(A\boldsymbol{x} - AA^L\boldsymbol{c}) + (AA^L\boldsymbol{c} - \boldsymbol{c})'(AA^L\boldsymbol{c} - \boldsymbol{c}) \\ &\quad + 2(A\boldsymbol{x} - AA^L\boldsymbol{c})'(AA^L\boldsymbol{c} - \boldsymbol{c}) \\ &\geq (AA^L\boldsymbol{c} - \boldsymbol{c})'(AA^L\boldsymbol{c} - \boldsymbol{c}) = (A\boldsymbol{x}_* - \boldsymbol{c})'(A\boldsymbol{x}_* - \boldsymbol{c}),\end{aligned}$$

where the inequality follows from the fact that

$$(A\boldsymbol{x} - AA^L\boldsymbol{c})'(A\boldsymbol{x} - AA^L\boldsymbol{c}) \geq 0,$$

and

$$\begin{aligned}(A\boldsymbol{x} - AA^L\boldsymbol{c})'(AA^L\boldsymbol{c} - \boldsymbol{c}) &= (\boldsymbol{x} - A^L\boldsymbol{c})'A'(AA^L\boldsymbol{c} - \boldsymbol{c}) \\ &= (\boldsymbol{x} - A^L\boldsymbol{c})'A'((AA^L)'\boldsymbol{c} - \boldsymbol{c}) \\ &= (\boldsymbol{x} - A^L\boldsymbol{c})'(A'A^{L\prime}A'\boldsymbol{c} - A'\boldsymbol{c}) \\ &= (\boldsymbol{x} - A^L\boldsymbol{c})'(A'\boldsymbol{c} - A'\boldsymbol{c}) = 0 \end{aligned} \tag{6.5}$$

This completes the proof. □

Corollary 6.12.1. The vector $\boldsymbol{x}_*$ is a least squares solution to the system $A\boldsymbol{x} = \boldsymbol{c}$ if and only if

$$(A\boldsymbol{x}_* - \boldsymbol{c})'(A\boldsymbol{x}_* - \boldsymbol{c}) = \boldsymbol{c}'(\mathrm{I}_m - AA^L)\boldsymbol{c}$$

Proof. From the previous theorem, $A^L\boldsymbol{c}$ is a least squares solution for any choice of A^L, and its sum of squared errors is given by

$$\begin{aligned}(AA^L\boldsymbol{c} - \boldsymbol{c})'(AA^L\boldsymbol{c} - \boldsymbol{c}) &= \boldsymbol{c}'(AA^L - \mathrm{I}_m)'(AA^L - \mathrm{I}_m)\boldsymbol{c} \\ &= \boldsymbol{c}'(AA^L - \mathrm{I}_m)^2\boldsymbol{c} = \boldsymbol{c}'(AA^LAA^L - 2AA^L + \mathrm{I}_m)\boldsymbol{c} \\ &= \boldsymbol{c}'(AA^L - 2AA^L + \mathrm{I}_m)\boldsymbol{c} = \boldsymbol{c}'(\mathrm{I}_m - AA^L)\boldsymbol{c}\end{aligned}$$

The result now follows since, by definition, a least squares solution minimizes the sum of squared errors, and so any other vector $\boldsymbol{x}_*$ will be a least squares solution if and only if its sum of squared errors is equal to this minimum sum of squares, $\boldsymbol{c}'(\mathrm{I}_m - AA^L)\boldsymbol{c}$. □

Example 6.8. Let the system of equations $A\boldsymbol{x} = \boldsymbol{c}$ have A and $\boldsymbol{c}$ given by

$$A = \begin{bmatrix} 1 & 1 & 2 \\ 1 & 0 & 1 \\ 1 & 1 & 2 \\ 2 & 0 & 2 \end{bmatrix}, \qquad \boldsymbol{c} = \begin{bmatrix} 4 \\ 1 \\ 6 \\ 5 \end{bmatrix}$$

In Example 5.13 we computed the least squares inverse

$$A^L = \frac{1}{10}\begin{bmatrix} 0 & 2 & 0 & 4 \\ 5 & -2 & 5 & -4 \\ 0 & 0 & 0 & 0 \end{bmatrix}$$

Since

$$AA^L\boldsymbol{c} = \frac{1}{10}\begin{bmatrix} 5 & 0 & 5 & 0 \\ 0 & 2 & 0 & 4 \\ 5 & 0 & 5 & 0 \\ 0 & 4 & 0 & 8 \end{bmatrix}\begin{bmatrix} 4 \\ 1 \\ 6 \\ 5 \end{bmatrix} = \begin{bmatrix} 5 \\ 2.2 \\ 5 \\ 4.4 \end{bmatrix} \neq \boldsymbol{c},$$

it follows from Theorem 6.2 that the system of equations is inconsistent. A least squares solution is then given by

$$A^L\boldsymbol{c} = \frac{1}{10}\begin{bmatrix} 0 & 2 & 0 & 4 \\ 5 & -2 & 5 & -4 \\ 0 & 0 & 0 & 0 \end{bmatrix}\begin{bmatrix} 4 \\ 1 \\ 6 \\ 5 \end{bmatrix} = \begin{bmatrix} 2.2 \\ 2.8 \\ 0 \end{bmatrix}$$

Since $(AA^L\boldsymbol{c} - \boldsymbol{c})' = (5, 2.2, 5, 4.4) - (4, 1, 6, 5) = (1, 1.2, -1, -0.6)$, the sum of squared errors for the least squares solution is

$$(AA^L\boldsymbol{c} - \boldsymbol{c})'(AA^L\boldsymbol{c} - \boldsymbol{c}) = 3.8.$$

In general, a least squares solution is not unique. For instance, the reader can easily verify that the matrix

$$B = \begin{bmatrix} -2 & -0.8 & -2 & -1.6 \\ -1.5 & -1.2 & -1.5 & -2.4 \\ 2 & 1 & 2 & 2 \end{bmatrix}$$

is also a least squares inverse of A. Consequently,

$$B\boldsymbol{c} = \begin{bmatrix} -2 & -0.8 & -2 & -1.6 \\ -1.5 & -1.2 & -1.5 & -2.4 \\ 2 & 1 & 2 & 2 \end{bmatrix}\begin{bmatrix} 4 \\ 1 \\ 6 \\ 5 \end{bmatrix} = \begin{bmatrix} -28.8 \\ -28.2 \\ 31 \end{bmatrix}$$

is another least squares solution. However, $(AB\boldsymbol{c} - \boldsymbol{c})' = (5, 2.2, 5, 4.4) - (4, 1, 6, 5) = (1, 1.2, -1, -0.6)$, and so the sum of squared errors for this least squares solution is, as it must be, identical to that of the previous solution.

The following result will be useful in establishing the general form of a least squares solution. It indicates that while a least squares solution $\boldsymbol{x}_*$ may not be unique, the vector $A\boldsymbol{x}_*$ will be unique.

Theorem 6.13. The vector $\boldsymbol{x}_*$ is a least squares solution to the system $A\boldsymbol{x} = \boldsymbol{c}$ if and only if

$$A\boldsymbol{x}_* = AA^L\boldsymbol{c} \tag{6.6}$$

Proof. Using Theorem 6.2, we see that the system of equations given in (6.6) is consistent since

$$AA^L(AA^L\boldsymbol{c}) = (AA^LA)A^L\boldsymbol{c} = AA^L\boldsymbol{c}$$

The sum of squared errors for any vector $\boldsymbol{x}_*$ satisfying (6.6) is

$$\begin{aligned}(A\boldsymbol{x}_* - \boldsymbol{c})'(A\boldsymbol{x}_* - \boldsymbol{c}) &= (AA^L\boldsymbol{c} - \boldsymbol{c})'(AA^L\boldsymbol{c} - \boldsymbol{c})\\ &= \boldsymbol{c}'(AA^L - \mathrm{I}_m)^2\boldsymbol{c}\\ &= \boldsymbol{c}'(\mathrm{I}_m - AA^L)\boldsymbol{c},\end{aligned}$$

so by Corollary 6.12.1, $\boldsymbol{x}_*$ is a least squares solution. Conversely, now suppose that $\boldsymbol{x}_*$ is a least squares solution. Then from Corollary 6.12.1 we must have

$$\begin{aligned}(A\boldsymbol{x}_* - \boldsymbol{c})'(A\boldsymbol{x}_* - \boldsymbol{c}) &= \boldsymbol{c}'(\mathrm{I}_m - AA^L)\boldsymbol{c}\\ &= \boldsymbol{c}'(\mathrm{I}_m - AA^L)'(\mathrm{I}_m - AA^L)\boldsymbol{c}\\ &= (AA^L\boldsymbol{c} - \boldsymbol{c})'(AA^L\boldsymbol{c} - \boldsymbol{c}),\end{aligned} \tag{6.7}$$

where we have used the fact that $(\mathrm{I}_m - AA^L)$ is symmetric and idempotent. However, we also have

$$\begin{aligned}(A\boldsymbol{x}_* - \boldsymbol{c})'(A\boldsymbol{x}_* - \boldsymbol{c}) &= \{(A\boldsymbol{x}_* - AA^L\boldsymbol{c}) + (AA^L\boldsymbol{c} - \boldsymbol{c})\}'\\ &\quad \cdot \{(A\boldsymbol{x}_* - AA^L\boldsymbol{c}) + (AA^L\boldsymbol{c} - \boldsymbol{c})\}\\ &= (A\boldsymbol{x}_* - AA^L\boldsymbol{c})'(A\boldsymbol{x}_* - AA^L\boldsymbol{c}) + (AA^L\boldsymbol{c} - \boldsymbol{c})'(AA^L\boldsymbol{c} - \boldsymbol{c}),\end{aligned} \tag{6.8}$$

since $(A\boldsymbol{x}_* - AA^L\boldsymbol{c})'(AA^L\boldsymbol{c} - \boldsymbol{c}) = 0$, as shown in (6.5). Now (6.7) and (6.8) imply that

$$(A\boldsymbol{x}_* - AA^L\boldsymbol{c})'(A\boldsymbol{x}_* - AA^L\boldsymbol{c}) = 0,$$

which can be true only if

$$(A\boldsymbol{x}_* - AA^L\boldsymbol{c}) = 0,$$

and this establishes (6.6). □

We now give an expression for a general least squares solution to a system of equations.

Theorem 6.14. Let A^L be any $\{1, 3\}$-inverse of the $m \times n$ matrix A. Define the vector

$$\boldsymbol{x}_y = A^L\boldsymbol{c} + (\mathrm{I}_n - A^L A)\boldsymbol{y},$$

where $\boldsymbol{y}$ is an arbitrary $n \times 1$ vector. Then, for each $\boldsymbol{y}$, $\boldsymbol{x}_y$ is a least squares solution to the system of equations $A\boldsymbol{x} = \boldsymbol{c}$, and for any least squares solution $\boldsymbol{x}_*$ there exists a vector $\boldsymbol{y}$ such that $\boldsymbol{x}_* = \boldsymbol{x}_y$.

Proof. Since

$$A(\mathrm{I}_n - A^L A)\boldsymbol{y} = (A - AA^L A)\boldsymbol{y} = (A - A)\boldsymbol{y} = \boldsymbol{0},$$

we have $A\boldsymbol{x}_y = AA^L\boldsymbol{c}$, and so by Theorem 6.13 $\boldsymbol{x}_y$ is a least squares solution. Conversely, if $\boldsymbol{x}_*$ is an arbitrary least squares solution, then by using Theorem 6.13 again, we must have

$$A\boldsymbol{x}_* = AA^L\boldsymbol{c},$$

which, when premultiplied by A^L, implies that,

$$\boldsymbol{0} = -A^L A(\boldsymbol{x}_* - A^L\boldsymbol{c})$$

Adding $\boldsymbol{x}_*$ to both sides of this identity, and then rearranging we get

$$\begin{aligned}\boldsymbol{x}_* &= \boldsymbol{x}_* - A^L A(\boldsymbol{x}_* - A^L\boldsymbol{c}) \\ &= A^L\boldsymbol{c} + \boldsymbol{x}_* - A^L\boldsymbol{c} - A^L A(\boldsymbol{x}_* - A^L\boldsymbol{c}) \\ &= A^L\boldsymbol{c} + (\mathrm{I}_n - A^L A)(\boldsymbol{x}_* - A^L\boldsymbol{c})\end{aligned}$$

This completes the proof since we have shown that $\boldsymbol{x}_* = \boldsymbol{x}_y$, where $\boldsymbol{y} = (\boldsymbol{x}_* - A^L\boldsymbol{c})$. □

We saw in Example 6.8 that least squares solutions are not necessarily unique. Theorem 6.14 can be used to obtain a necessary and sufficient condition for the solution to be unique.

Theorem 6.15. If A is an $m \times n$ matrix, then the system of equations $A\boldsymbol{x} = \boldsymbol{c}$ has a unique least squares solution if and only if $\text{rank}(A) = n$.

Proof. It follows immediately from Theorem 6.14 that the least squares solution is unique if and only if $(\text{I} - A^L A) = (0)$, or equivalently, $A^L A = \text{I}_n$. The result now follows from Theorem 5.22(g). □

Even when the least squares solution to a system is not unique, certain linear combinations of the elements of least squares solutions may be unique. This is the subject of our next theorem.

Theorem 6.16. Let $\boldsymbol{x}_*$ be a least squares solution to the system of equations $A\boldsymbol{x} = \boldsymbol{c}$. Then $\boldsymbol{a}'\boldsymbol{x}_*$ is unique if and only if $\boldsymbol{a}$ is in the row space of A.

Proof. Using Theorem 6.14, if $\boldsymbol{a}'\boldsymbol{x}_*$ is unique regardless of the choice of the least squares solution $\boldsymbol{x}_*$, then

$$\boldsymbol{a}'\boldsymbol{x}_y = \boldsymbol{a}'A^L\boldsymbol{c} + \boldsymbol{a}'(\text{I}_n - A^L A)\boldsymbol{y}$$

is the same for all choices of $\boldsymbol{y}$. But this implies that

$$\boldsymbol{a}'(\text{I}_n - A^L A) = \boldsymbol{0}' \tag{6.9}$$

Now if (6.9) holds, then

$$\boldsymbol{a}' = \boldsymbol{b}'A,$$

where $\boldsymbol{b}' = \boldsymbol{a}'A^L$, and so $\boldsymbol{a}$ is in the row space of A. On the other hand, if $\boldsymbol{a}$ is in the row space of A, then there exists some vector $\boldsymbol{b}$ such that $\boldsymbol{a}' = \boldsymbol{b}'A$. This implies that

$$\boldsymbol{a}'(\text{I}_n - A^L A) = \boldsymbol{b}'A(\text{I}_n - A^L A) = \boldsymbol{b}'(A - AA^L A) = \boldsymbol{b}'(A - A) = \boldsymbol{0}',$$

and so the least squares solution must be unique. □

Example 6.9. We will obtain the general least squares solution to the system of equations presented in Example 6.8. First note that

$$A^L A = \begin{bmatrix} 1 & 0 & 1 \\ 0 & 1 & 1 \\ 0 & 0 & 0 \end{bmatrix},$$

so that

$$\begin{aligned} \boldsymbol{x}_y &= A^L \boldsymbol{c} + (\mathrm{I}_3 - A^L A)\boldsymbol{y} \\ &= \frac{1}{10}\begin{bmatrix} 0 & 2 & 0 & 4 \\ 5 & -2 & 5 & -4 \\ 0 & 0 & 0 & 0 \end{bmatrix}\begin{bmatrix} 4 \\ 1 \\ 6 \\ 5 \end{bmatrix} + \begin{bmatrix} 0 & 0 & -1 \\ 0 & 0 & -1 \\ 0 & 0 & 1 \end{bmatrix}\begin{bmatrix} y_1 \\ y_2 \\ y_3 \end{bmatrix} \\ &= \begin{bmatrix} 2.2 - y_3 \\ 2.8 - y_3 \\ y_3 \end{bmatrix} \end{aligned}$$

is a least squares solution for any choice of y_3. The quantity $\boldsymbol{a}'\boldsymbol{x}_y$ does not depend on the choice of y_3 as long as $\boldsymbol{a}$ is in the row space of A; in this case, that corresponds to $\boldsymbol{a}$ being orthogonal to the vector $(-1, -1, 1)'$.

6. LEAST SQUARES ESTIMATION FOR LESS THAN FULL RANK MODELS

In all of our previous examples of least squares estimation for a model of the form

$$\boldsymbol{y} = X\boldsymbol{\beta} + \boldsymbol{\epsilon}, \tag{6.10}$$

where $\boldsymbol{y}$ is $N \times 1$, X is $N \times m$, $\boldsymbol{\beta}$ is $m \times 1$, and $\boldsymbol{\epsilon}$ is $N \times 1$, we have assumed that $\text{rank}(X) = m$. In this case, the normal equations,

$$X'X\hat{\boldsymbol{\beta}} = X'\boldsymbol{y}, \tag{6.11}$$

yield a unique solution, the unique least squares estimator of $\boldsymbol{\beta}$, given by

$$\hat{\boldsymbol{\beta}} = (X'X)^{-1}X'\boldsymbol{y}$$

However, in many applications, the matrix X has less than full rank.

Example 6.10. Consider the univariate one-way classification model, which was written as

$$y_{ij} = \mu_i + \epsilon_{ij},$$

in Example 3.14, where $i = 1, \dots, k$ and $j = 1, \dots, n_i$. This model can be written in the form of (6.10), where $\boldsymbol{\beta} = (\mu_1, \dots, \mu_k)'$ and

$$X = \begin{bmatrix} \mathbf{1}_{n_1} & \mathbf{0} & \cdots & \mathbf{0} \\ \mathbf{0} & \mathbf{1}_{n_2} & \cdots & \mathbf{0} \\ \vdots & \vdots & & \vdots \\ \mathbf{0} & \mathbf{0} & \cdots & \mathbf{1}_{n_k} \end{bmatrix}$$

In this case, X is of full rank, and so $\hat{\boldsymbol{\beta}} = (X'X)^{-1}X'\boldsymbol{y} = \bar{\boldsymbol{y}} = (\sum y_{1j}/n_1, \dots, \sum y_{kj}/n_k)'$. An alternative way of writing this one-way classification model is

$$y_{ij} = \mu + \tau_i + \epsilon_{ij},$$

which has $k+1$ parameters instead of k. Here μ represents an overall effect while τ_i is an effect due to treatment i. In some respects, this form of the model is more natural in that the reduced model, which has all treatment means identical, is simply a submodel with some of the parameters equal to 0, that is, $\tau_1 = \cdots = \tau_k = 0$. If this second form of the model is written as $\boldsymbol{y} = X_* \boldsymbol{\beta}_* + \boldsymbol{\epsilon}$, then $\boldsymbol{\beta}_* = (\mu, \tau_1, \dots, \tau_k)'$ and

$$X_* = \begin{bmatrix} \mathbf{1}_{n_1} & \mathbf{1}_{n_1} & \mathbf{0} & \cdots & \mathbf{0} \\ \mathbf{1}_{n_2} & \mathbf{0} & \mathbf{1}_{n_2} & \cdots & \mathbf{0} \\ \mathbf{1}_{n_3} & \mathbf{0} & \mathbf{0} & \cdots & \mathbf{0} \\ \vdots & \vdots & \vdots & & \vdots \\ \mathbf{1}_{n_k} & \mathbf{0} & \mathbf{0} & \cdots & \mathbf{1}_{n_k} \end{bmatrix}$$

Thus, this second parameterization of the one-way classification model has the design matrix X_* less than full rank since $\text{rank}(X_*) = k$.

In this section, we will apply some of the results of this chapter to the estimation of parameters in the model given by (6.10) when X is less than full rank. First of all, let us consider the task of solving the normal equations given by (6.11); that is, using our usual notation for a system of equations, we want to solve $A\boldsymbol{x} = \boldsymbol{c}$, where $A = X'X, \boldsymbol{x} = \hat{\boldsymbol{\beta}}$, and $\boldsymbol{c} = X'\boldsymbol{y}$. Now from Theorem 6.2, we see that (6.11) is a consistent system of equations since

$$\begin{aligned} X'X(X'X)^+X'\boldsymbol{y} &= X'XX^+X'^+X'\boldsymbol{y} = X'XX^+(XX^+)'\boldsymbol{y} = X'XX^+XX^+\boldsymbol{y} \\ &= X'XX^+\boldsymbol{y} = X'(XX^+)'\boldsymbol{y} = X'X'^+X'\boldsymbol{y} = X'\boldsymbol{y} \end{aligned}$$

Consequently, using Theorem 6.4, we find that the general solution $\hat{\boldsymbol{\beta}}$ can be written as

$$\hat{\boldsymbol{\beta}} = (X'X)^{-}X'\boldsymbol{y} + \{\mathrm{I} - (X'X)^{-}X'X\}\boldsymbol{u}, \tag{6.12}$$

or, if we use the Moore–Penrose generalized inverse, as

$$\begin{aligned}\hat{\boldsymbol{\beta}} &= (X'X)^{+}X'\boldsymbol{y} + \{\mathrm{I} - (X'X)^{+}X'X\}\boldsymbol{u} \\ &= X^{+}\boldsymbol{y} + (\mathrm{I} - X^{+}X)\boldsymbol{u},\end{aligned}$$

where $\boldsymbol{u}$ is an arbitrary $m \times 1$ vector. The same general solution can be obtained by using the least squares results of Section 6.5 on the system of equations

$$\boldsymbol{y} = X\hat{\boldsymbol{\beta}}$$

Thus, using Theorem 6.14 with $A = X$, $\boldsymbol{x} = \hat{\boldsymbol{\beta}}$, and $\boldsymbol{c} = \boldsymbol{y}$, the least squares solution is given by

$$\hat{\boldsymbol{\beta}} = X^{L}\boldsymbol{y} + (\mathrm{I} - X^{L}X)\boldsymbol{u},$$

which is, of course, equivalent to that given by (6.12).

One key difference between the full rank model and the less than full rank model is that the least squares solution is unique only if X has full rank. When X is less than full rank, the model $\boldsymbol{y} = X\boldsymbol{\beta} + \boldsymbol{\epsilon}$ is overparameterized, and so not all of the parameters or linear functions of the parameters are uniquely defined; this is what leads to the infinitely many solutions for $\hat{\boldsymbol{\beta}}$. Thus, when estimating linear functions of the parameters, we must make sure that we are trying to estimate a function of the parameters that is uniquely defined. This leads to the following definition of what is known as an estimable function.

Definition 6.2. The linear function $\boldsymbol{a}'\boldsymbol{\beta}$ of the parameter vector $\boldsymbol{\beta}$ is estimable if and only if there exists some $N \times 1$ vector $\boldsymbol{b}$ such that

$$\boldsymbol{a}'\boldsymbol{\beta} = \mathrm{E}(\boldsymbol{b}'\boldsymbol{y}) = \boldsymbol{b}'\mathrm{E}(\boldsymbol{y}) = \boldsymbol{b}'X\boldsymbol{\beta};$$

that is, if and only if there exists a linear function of the components of $\boldsymbol{y}$, $\boldsymbol{b}'\boldsymbol{y}$, which is an unbiased estimator of $\boldsymbol{a}'\boldsymbol{\beta}$.

The condition that a linear function $\boldsymbol{a}'\boldsymbol{\beta}$ be estimable is equivalent to the condition that the corresponding estimator $\boldsymbol{a}'\hat{\boldsymbol{\beta}}$ be unique. To see this, note that from the definition above, the function $\boldsymbol{a}'\boldsymbol{\beta}$ is estimable if and only if $\boldsymbol{a}$ is in the row space of X, while it follows from Theorem 6.16 that $\boldsymbol{a}'\hat{\boldsymbol{\beta}}$ is unique if and only if $\boldsymbol{a}$ is in the row space of X. In addition, since $X'(XX')^+X$ is the projection matrix for the row space of X, we get the more practical condition for estimability of $\boldsymbol{a}'\boldsymbol{\beta}$ given by

$$X'(XX')^+X\boldsymbol{a} = \boldsymbol{a} \tag{6.13}$$

It follows from Theorems 5.3 and 5.25 that

$$X'(XX')^+X = X'X'^+ = X'X'^L = X'(XX')^-X,$$

and so equation (6.13) is not dependent upon the Moore–Penrose inverse as the choice of the generalized inverse of XX'.

Finally, we will demonstrate the invariance of the vector of fitted values $\hat{\boldsymbol{y}} = X\hat{\boldsymbol{\beta}}$ and its sum of squared errors $(\boldsymbol{y} - \hat{\boldsymbol{y}})'(\boldsymbol{y} - \hat{\boldsymbol{y}})$ to the choice of the least squares solution $\hat{\boldsymbol{\beta}}$. Since $XX^+X = X$

$$\hat{\boldsymbol{y}} = X\hat{\boldsymbol{\beta}} = X\{X^+\boldsymbol{y} + (\mathrm{I} - X^+X)\boldsymbol{u}\} = XX^+\boldsymbol{y} + (X - XX^+X)\boldsymbol{u} = XX^+\boldsymbol{y},$$

which does not depend on the vector $\boldsymbol{u}$. Thus, $\hat{\boldsymbol{y}}$ is unique, while the uniqueness of

$$(\boldsymbol{y} - \hat{\boldsymbol{y}})'(\boldsymbol{y} - \hat{\boldsymbol{y}}) = \boldsymbol{y}'(\mathrm{I} - XX^+)\boldsymbol{y}$$

follows immediately from the uniqueness of $\hat{\boldsymbol{y}}$.

Example 6.11. Let us return to the one-way classification model

$$\boldsymbol{y} = X_*\boldsymbol{\beta}_* + \boldsymbol{\epsilon}$$

of Example 6.10, where $\boldsymbol{\beta}_* = (\mu, \tau_1, \ldots, \tau_k)'$ and

$$X_* = \begin{bmatrix} \mathbf{1}_{n_1} & \mathbf{1}_{n_1} & \mathbf{0} & \cdots & \mathbf{0} \\ \mathbf{1}_{n_2} & \mathbf{0} & \mathbf{1}_{n_2} & \cdots & \mathbf{0} \\ \mathbf{1}_{n_3} & \mathbf{0} & \mathbf{0} & \cdots & \mathbf{0} \\ \vdots & \vdots & \vdots & & \vdots \\ \mathbf{1}_{n_k} & \mathbf{0} & \mathbf{0} & \cdots & \mathbf{1}_{n_k} \end{bmatrix}$$

Since the rank of the $n \times (k+1)$ matrix X_*, where $n = \sum n_i$ is k, the least squares solution for $\boldsymbol{\beta}_*$ is not unique. To find the form of the general solution, note that

$$X_*'X_* = \begin{bmatrix} n & \boldsymbol{n}' \\ \boldsymbol{n} & D_n \end{bmatrix},$$

while a generalized inverse is given by

$$(X_*'X_*)^- = \begin{bmatrix} n^{-1} & \mathbf{0}' \\ \mathbf{0} & D_n^{-1} - n^{-1}\mathbf{1}_k\mathbf{1}_k' \end{bmatrix},$$

where $\boldsymbol{n} = (n_1, \dots, n_k)'$ and $D_n = \text{diag}(n_1, \dots, n_k)$. Thus, using (6.12) we have the general solution

$$\hat{\boldsymbol{\beta}}_* = \begin{bmatrix} n^{-1} & \mathbf{0}' \\ \mathbf{0} & D_n^{-1} - n^{-1}\mathbf{1}_k\mathbf{1}_k' \end{bmatrix} \begin{bmatrix} n\bar{y} \\ D_n\bar{\mathbf{y}} \end{bmatrix} + \left\{ \mathrm{I}_{k+1} - \begin{bmatrix} 1 & n^{-1}\boldsymbol{n}' \\ \mathbf{0} & \mathrm{I}_k - n^{-1}\mathbf{1}_k\boldsymbol{n}' \end{bmatrix} \right\} \boldsymbol{u}$$

$$= \begin{bmatrix} \bar{y} \\ \bar{\mathbf{y}} - \bar{y}\mathbf{1}_k \end{bmatrix} + \begin{bmatrix} 0 & -n^{-1}\boldsymbol{n}' \\ \mathbf{0} & n^{-1}\mathbf{1}_k\boldsymbol{n}' \end{bmatrix} \boldsymbol{u},$$

where $\bar{\mathbf{y}} = (\bar{y}_1, \dots, \bar{y}_k)'$ and $\bar{y} = \sum n_i\bar{y}_i/n$. Choosing $\boldsymbol{u} = \mathbf{0}$, we get the particular least squares solution that has $\hat{\mu} = \bar{y}$ and $\hat{\tau}_i = \bar{y}_i - \bar{y}$ for $i = 1, \dots, k$. Since $\boldsymbol{a}'\boldsymbol{\beta}_*$ is estimable only if $\boldsymbol{a}$ is in the row space of X, we find that the k quantities, $\mu + \tau_i$, $i = 1, \dots, k$, as well as any linear combinations of these quantities, are estimable. In particular, since $\mu + \tau_i = \boldsymbol{a}_i'\boldsymbol{\beta}_*$, where $\boldsymbol{a}_i = (1, \boldsymbol{e}_i')'$, its estimator is given by

$$\boldsymbol{a}_i'\hat{\boldsymbol{\beta}}_* = [1 \quad \boldsymbol{e}_i'] \begin{bmatrix} \bar{y} \\ \bar{\mathbf{y}} - \bar{y} \end{bmatrix} = \bar{y}_i$$

The vector of fitted values is

$$\hat{\boldsymbol{y}} = X_*\hat{\boldsymbol{\beta}}_* = \begin{bmatrix} \mathbf{1}_{n_1} & \mathbf{1}_{n_1} & \mathbf{0} & \cdots & \mathbf{0} \\ \mathbf{1}_{n_2} & \mathbf{0} & \mathbf{1}_{n_2} & \cdots & \mathbf{0} \\ \mathbf{1}_{n_3} & \mathbf{0} & \mathbf{0} & \cdots & \mathbf{0} \\ \vdots & \vdots & \vdots & & \vdots \\ \mathbf{1}_{n_k} & \mathbf{0} & \mathbf{0} & \cdots & \mathbf{1}_{n_k} \end{bmatrix} \begin{bmatrix} \bar{y} \\ \bar{y}_1 - \bar{y} \\ \bar{y}_2 - \bar{y} \\ \vdots \\ \bar{y}_k - \bar{y} \end{bmatrix} = \begin{bmatrix} \bar{y}_1\mathbf{1}_{n_1} \\ \bar{y}_2\mathbf{1}_{n_2} \\ \vdots \\ \bar{y}_k\mathbf{1}_{n_k} \end{bmatrix},$$

while the sum of squared errors is given by

$$(\boldsymbol{y} - \hat{\boldsymbol{y}})'(\boldsymbol{y} - \hat{\boldsymbol{y}}) = \sum_{i=1}^{k} \sum_{j=1}^{n_i} (y_{ij} - \bar{y}_i)^2$$

7. SYSTEMS OF LINEAR EQUATIONS AND THE SINGULAR VALUE DECOMPOSITION

When A is square and nonsingular, then the solution to the system of equations $A\boldsymbol{x} = \boldsymbol{c}$ can be conveniently expressed in terms of the inverse of A, as $\boldsymbol{x} = A^{-1}\boldsymbol{c}$. For this reason, it has seemed somewhat natural to deal with the solutions for the more general case in terms of the generalization of A^{-1}, A^{+}. This is the approach that we have taken throughout this chapter. Alternatively, we can attack this problem by directly using the singular value decomposition, an approach that may offer more insight. In this case, we will always be able to transform our system to a simpler system of equations of the form

$$D\boldsymbol{y} = \boldsymbol{b}, \tag{6.14}$$

where $\boldsymbol{y}$ is an $n \times 1$ vector of variables, $\boldsymbol{b}$ is an $m \times 1$ vector of constants, and D is an $m \times n$ matrix such that $d_{ij} = 0$ if $i \neq j$. In particular, D will have one of the four forms, as given in Theorem 4.1,

$$(a)\ \Delta \qquad (b)\ [\Delta \quad (0)] \qquad (c) \begin{bmatrix} \Delta \\ (0) \end{bmatrix} \qquad (d) \begin{bmatrix} \Delta & (0) \\ (0) & (0) \end{bmatrix},$$

where Δ is an $r \times r$ nonsingular diagonal matrix and $r = \text{rank}(A)$. Now if D has the form given in (a), then the system (6.14) is consistent with the unique solution given by $\boldsymbol{y} = \Delta^{-1}\boldsymbol{b}$. For (b), if we partition $\boldsymbol{y}$ as $\boldsymbol{y} = (\boldsymbol{y}_1', \boldsymbol{y}_2')'$, where $\boldsymbol{y}_1$ is $r \times 1$, then equation (6.14) reduces to

$$\Delta\boldsymbol{y}_1 = \boldsymbol{b}$$

Thus, (6.14) is consistent and has solutions of the form

$$\boldsymbol{y} = \begin{bmatrix} \Delta^{-1}\boldsymbol{b} \\ \boldsymbol{y}_2 \end{bmatrix},$$

where the $(n - r) \times 1$ vector $\boldsymbol{y}_2$ is arbitrary. Since we then have $n - r$ linearly independent choices for $\boldsymbol{y}_2$, the number of linearly independent solutions is $n - r$ if $\boldsymbol{b} = \boldsymbol{0}$ and $n - r + 1$ if $\boldsymbol{b} \neq \boldsymbol{0}$. When D has the form given in (c), the system in (6.14) takes the form

$$\begin{bmatrix} \Delta\boldsymbol{y} \\ \boldsymbol{0} \end{bmatrix} = \begin{bmatrix} \boldsymbol{b}_1 \\ \boldsymbol{b}_2 \end{bmatrix},$$

where $\boldsymbol{b}_1$ is $r \times 1$ and $\boldsymbol{b}_2$ is $(m - r) \times 1$, and so it is consistent only if $\boldsymbol{b}_2 = \boldsymbol{0}$. If this is the case, the system then has a unique solution given by $\boldsymbol{y} = \Delta^{-1}\boldsymbol{b}_1$. For the final form given in (d), the system of equations in (6.14) appears as

$$\begin{bmatrix} \Delta \boldsymbol{y}_1 \\ \boldsymbol{0} \end{bmatrix} = \begin{bmatrix} \boldsymbol{b}_1 \\ \boldsymbol{b}_2 \end{bmatrix},$$

where $\boldsymbol{y}$ and $\boldsymbol{b}$ have been partitioned as before. As in the case of form (c), this system is consistent only if $\boldsymbol{b}_2 = \boldsymbol{0}$, and as in the case of form (b), when consistent, it has $n - r$ linearly independent solutions if $\boldsymbol{b} = \boldsymbol{0}$, and $n - r + 1$ linearly independent solutions if $\boldsymbol{b} \neq \boldsymbol{0}$. The general solution is given by

$$\boldsymbol{y} = \begin{bmatrix} \Delta^{-1}\boldsymbol{b}_1 \\ \boldsymbol{y}_2 \end{bmatrix},$$

where the $(n - r) \times 1$ vector $\boldsymbol{y}_2$ is arbitrary.

All of the above can now be readily applied to the general system of equations,

$$A\boldsymbol{x} = \boldsymbol{c} \tag{6.15}$$

by utilizing the singular value decomposition of A given by $A = PDQ'$ as in Theorem 4.1. Premultiplication of this system of equations by P' produces the system of equations in (6.14), where the vector of variables is given by $\boldsymbol{y} = Q'\boldsymbol{x}$ and the vector of constants is $\boldsymbol{b} = P'\boldsymbol{c}$. Consequently, if $\boldsymbol{y}$ is a solution to (6.14), then $\boldsymbol{x} = Q\boldsymbol{y}$ will be a solution to (6.15). Thus, in the case of forms (a) and (b), (6.15) is consistent with the unique solution given by

$$\boldsymbol{x} = Q\boldsymbol{y} = Q\Delta^{-1}\boldsymbol{b} = Q\Delta^{-1}P'\boldsymbol{c} = A^{-1}\boldsymbol{c},$$

when (a) is the form of D, while for form (b) the general solution is

$$\boldsymbol{x} = Q\boldsymbol{y} = [Q_1 \quad Q_2] \begin{bmatrix} \Delta^{-1}\boldsymbol{b} \\ \boldsymbol{y}_2 \end{bmatrix} = Q_1\Delta^{-1}P'\boldsymbol{c} + Q_2\boldsymbol{y}_2,$$

where Q_1 is $n \times r$ and $\boldsymbol{y}_2$ is an arbitrary $(n - r) \times 1$ vector. The term $Q_2\boldsymbol{y}_2$ has no effect on the value of $A\boldsymbol{x}$ since the columns of the $n \times (n - r)$ matrix Q_2 form a basis for the null space of A. In the case of forms (c) and (d), the system (6.15) is consistent only if $\boldsymbol{c} = P_1\boldsymbol{b}_1$ so that $P_2\boldsymbol{b}_2 = \boldsymbol{0}$, where $P = (P_1, P_2)$ and P_1 is $m \times r$; that is, since the columns of P_1 form a basis for the range of A, the system is consistent if $\boldsymbol{c}$ is in the column space of A. Thus, if we partition $\boldsymbol{c}$ as $\boldsymbol{c} = (\boldsymbol{c}_1', \boldsymbol{c}_2')'$ where $\boldsymbol{c}_1$ is $r \times 1$, then when form (c) holds, the unique solution will be given by

$$x = Qy = Q\Delta^{-1}b_1 = Q\Delta^{-1}P_1'c$$

In the case of form (d), the general solution is

$$x = Qy = [Q_1 \quad Q_2]\begin{bmatrix} \Delta^{-1}b_1 \\ y_2 \end{bmatrix} = Q_1\Delta^{-1}P_1'c + Q_2 y_2$$

8. SPARSE LINEAR SYSTEMS OF EQUATIONS

The typical approach to the numerical computation of solutions to a consistent system of equations $Ax = c$, or least squares solutions when the system is inconsistent, utilizes some factorization of A such as the QR factorization, the singular value decomposition, or the LU decomposition, which factors A into the product of a lower triangular matrix and upper triangular matrix. Any method of this type is referred to as a direct method. One situation in which direct methods may not be appropriate is when our system of equations is large and sparse; that is, m and n are large and a relatively large number of the elements of the $m \times n$ matrix A are equal to zero. Thus, although the size of A may be quite large, its storage will not require an enormous amount of computer memory since we only need store the nonzero values and their location. However, when A is sparse, the factors in its decompositions need not be sparse, so if A is large enough, the computation of these factorizations may easily require more memory than is available.

If there is some particular structure to the sparsity of A, then it may be possible to implement a direct method that exploits this structure. A simple example of such a situation is one in which A is $m \times m$ and tridiagonal; that is, A has the form

$$A = \begin{bmatrix} v_1 & w_1 & 0 & \cdots & 0 & 0 & 0 \\ u_2 & v_2 & w_2 & \cdots & 0 & 0 & 0 \\ \vdots & \vdots & \vdots & & \vdots & \vdots & \vdots \\ 0 & 0 & 0 & \cdots & u_{m-1} & v_{m-1} & w_{m-1} \\ 0 & 0 & 0 & \cdots & 0 & u_m & v_m \end{bmatrix}$$

In this case, if we define

$$L = \begin{bmatrix} r_1 & 0 & \cdots & 0 & 0 \\ u_2 & r_2 & \cdots & 0 & 0 \\ \vdots & \vdots & & \vdots & \vdots \\ 0 & 0 & \cdots & r_{m-1} & 0 \\ 0 & 0 & \cdots & u_m & r_m \end{bmatrix}, \qquad U = \begin{bmatrix} 1 & s_1 & \cdots & 0 & 0 \\ 0 & 1 & \cdots & 0 & 0 \\ \vdots & \vdots & & \vdots & \vdots \\ 0 & 0 & \cdots & 1 & s_{m-1} \\ 0 & 0 & \cdots & 0 & 1 \end{bmatrix},$$

where $r_1 = v_1$, $r_i = v_i - u_i w_{i-1}/r_{i-1}$, and $s_{i-1} = w_{i-1}/r_{i-1}$, for $i = 2, \ldots, m$, then A can be factored as $A = LU$ as long as each $r_i \neq 0$. Thus, the two factors, L and U are also sparse. The system $A\boldsymbol{x} = \boldsymbol{c}$ can easily be solved by first solving the system $L\boldsymbol{y} = \boldsymbol{c}$ and then solving the system $U\boldsymbol{x} = \boldsymbol{y}$. For more details on this, and adaptations of direct methods for other structured matrices such as banded matrices and block tridiagonal matrices see Duff, Erisman, and Reid (1986) and Golub and Van Loan (1989).

A second approach to the solution of sparse systems of equations utilizes iterative methods. In this case, a sequence of vectors, $\boldsymbol{x}_0, \boldsymbol{x}_1, \ldots$ is generated with $\boldsymbol{x}_0$ being some initial vector while $\boldsymbol{x}_j$ for $j = 1, 2, \ldots$ is a vector that is computed using the previous vector $\boldsymbol{x}_{j-1}$, with the property that $\boldsymbol{x}_j \to \boldsymbol{x}$, as $j \to \infty$, where $\boldsymbol{x}$ is the true solution to $A\boldsymbol{x} = \boldsymbol{c}$. Typically, the computation in these methods only involves A through its product with vectors, and this is an operation that will be easy to handle if A is sparse. Two of the oldest and simplest iterative schemes are the Jacobi and Gauss–Seidel methods. If A is $m \times m$ with nonzero diagonal elements, then the system $A\boldsymbol{x} = \boldsymbol{c}$ can be written as

$$(A - D_A)\boldsymbol{x} + D_A\boldsymbol{x} = \boldsymbol{c},$$

which yields the identity

$$\boldsymbol{x} = D_A^{-1}\{\boldsymbol{c} - (A - D_A)\boldsymbol{x}\}$$

This is the motivation for the Jacobi method that computes $\boldsymbol{x}_j$ as

$$\boldsymbol{x}_j = D_A^{-1}\{\boldsymbol{c} - (A - D_A)\boldsymbol{x}_{j-1}\}$$

On the other hand, the Gauss–Seidel method utilizes the splitting of A as $A = A_1 + A_2$, where A_1 is lower triangular and A_2 is upper triangular with each of its diagonal elements equal to zero. In this case, $A\boldsymbol{x} = \boldsymbol{c}$ can be rearranged as

$$A_1\boldsymbol{x} = \boldsymbol{c} - A_2\boldsymbol{x},$$

and this leads to the iterative scheme

$$A_1\boldsymbol{x}_j = \boldsymbol{c} - A_2\boldsymbol{x}_{j-1},$$

which is easily solved for $\boldsymbol{x}_j$ since the system is triangular.

In recent years, some other more sophisticated iterative methods, requiring less computation and having better convergence properties, have been developed. We will briefly discuss a method for solving a system of equations, which utilizes an algorithm known as the Lanczos algorithm [Lanczos (1950)]. For more information on this procedure, including convergence properties, generalizations to a general $m \times n$ matrix, and to the problem of finding least squares solutions, as well as other iterative methods, the reader is referred to Young (1971), Hageman and Young (1981), and Golub and Van Loan (1989).

Consider the function

$$f(\boldsymbol{x}) = \tfrac{1}{2}\boldsymbol{x}'A\boldsymbol{x} - \boldsymbol{x}'\boldsymbol{c},$$

where $\boldsymbol{x}$ is an $m \times 1$ vector and A is an $m \times m$ positive definite matrix. The vector of partial derivatives of $f(\boldsymbol{x})$ given by

$$\nabla f(\boldsymbol{x}) = \left(\frac{\partial f}{\partial x_1}, \dots, \frac{\partial f}{\partial x_m} \right)' = A\boldsymbol{x} - \boldsymbol{c}$$

is sometimes referred to as the gradient of $f(\boldsymbol{x})$. Setting this equal to the zero vector, we find that the vector minimizing f, $\boldsymbol{x} = A^{-1}\boldsymbol{c}$, is also the solution to the system $A\boldsymbol{x} = \boldsymbol{c}$. Thus, a vector which approximately minimizes f will also be an approximate solution to $A\boldsymbol{x} = \boldsymbol{c}$. One iterative method for finding the minimizer $\boldsymbol{x}$ involves successively finding minimizers $\boldsymbol{x}_j$ of f over a j-dimensional subspace of R^m, starting with $j = 1$ and continually increasing j by 1. In particular, for some set of orthonormal $m \times 1$ vectors, $\boldsymbol{q}_1, \dots, \boldsymbol{q}_m$, we will define the jth subspace as the space with the columns of the $m \times j$ matrix, $Q_j = (\boldsymbol{q}_1, \dots, \boldsymbol{q}_j)$, as its basis. Consequently, for some $j \times 1$ vector $\boldsymbol{y}_j$,

$$\boldsymbol{x}_j = Q_j \boldsymbol{y}_j \tag{6.16}$$

and

$$f(\boldsymbol{x}_j) = \min_{\boldsymbol{y} \in R^j} f(Q_j \boldsymbol{y}) = \min_{\boldsymbol{y} \in R^j} g(\boldsymbol{y}) = g(\boldsymbol{y}_j),$$

where

$$g(\boldsymbol{y}) = \tfrac{1}{2}\boldsymbol{y}'(Q_j' A Q_j)\boldsymbol{y} - \boldsymbol{y}' Q_j' \boldsymbol{c}$$

Thus, the gradient of $g(\boldsymbol{y}_j)$ must be equal to the null vector, and so

$$(Q_j' A Q_j)\boldsymbol{y}_j = Q_j' \boldsymbol{c} \tag{6.17}$$

To obtain $\boldsymbol{x}_j$, we can first use (6.17) to calculate $\boldsymbol{y}_j$ and then use this in (6.16) to get $\boldsymbol{x}_j$. The final $\boldsymbol{x}_j$, $\boldsymbol{x}_m$, will be the solution to $A\boldsymbol{x} = \boldsymbol{c}$, but the goal here is to stop the iterative process before $j = m$ with a sufficiently accurate solution $\boldsymbol{x}_j$.

The iterative scheme described above will work with different sets of orthonormal vectors $\boldsymbol{q}_1, \ldots, \boldsymbol{q}_m$ but we will see that by a judicious choice of this set, we may guarantee that the computation involved in computing the $\boldsymbol{x}_j$s will be fairly straightforward even when A is large and sparse. These same vectors are also useful in an iterative procedure for obtaining a few of the largest and smallest eigenvalues of A. We will derive these vectors in the context of this eigenvalue problem and then later return to our discussion of the system of equations $A\boldsymbol{x} = \boldsymbol{c}$. Let λ_1 and λ_m denote the largest and smallest eigenvalues of A, while λ_{1j} and λ_{jj} denote the largest and smallest eigenvalues of the $j \times j$ matrix $Q_j'AQ_j$. Now we have seen in Chapter 3 that $\lambda_{1j} \leq \lambda_1$, $\lambda_{jj} \geq \lambda_m$ and that λ_1 and λ_m are the maximum and minimum values of the Rayleigh quotient,

$$R(\boldsymbol{x}, A) = \frac{\boldsymbol{x}'A\boldsymbol{x}}{\boldsymbol{x}'\boldsymbol{x}}$$

Suppose that we have the j columns of Q_j, and we wish to find an additional vector $\boldsymbol{q}_{j+1}$ so as to form the matrix Q_{j+1} and have $\lambda_{1,j+1}$ and $\lambda_{j+1,j+1}$ as close to λ_1 and λ_m as possible. If $\boldsymbol{u}_j$ is a vector in the space spanned by the columns of Q_j and satisfying $R(\boldsymbol{u}_j, A) = \lambda_{1j}$, then since the gradient

$$\nabla R(\boldsymbol{u}_j, A) = \frac{2}{\boldsymbol{u}_j'\boldsymbol{u}_j} \{A\boldsymbol{u}_j - R(\boldsymbol{u}_j, A)\boldsymbol{u}_j\}$$

gives the direction in which $R(\boldsymbol{u}_j, A)$ is increasing most rapidly, we would want to choose $\boldsymbol{q}_{j+1}$ so that $\nabla R(\boldsymbol{u}_j, A)$ is in the space spanned by the columns of Q_{j+1}. On the other hand, if $\boldsymbol{v}_j$ is a vector in the space spanned by Q_j and satisfying $R(\boldsymbol{v}_j, A) = \lambda_{jj}$, then since $R(\boldsymbol{v}_j, A)$ is decreasing most rapidly in the direction given by $-\nabla R(\boldsymbol{v}_j, A)$, we would want to make sure that $\nabla R(\boldsymbol{v}_j, A)$ is also in the space spanned by the columns of Q_{j+1}. Both of these objectives can be satisfied if the columns of Q_j are spanned by the vectors $\boldsymbol{q}_1, A\boldsymbol{q}_1, \ldots, A^{j-1}\boldsymbol{q}_1$ and we select $\boldsymbol{q}_{j+1}$ so that the columns of Q_{j+1} are spanned by the vectors $\boldsymbol{q}_1, A\boldsymbol{q}_1, \ldots, A^j\boldsymbol{q}_1$, since both $\nabla R(\boldsymbol{u}_j, A)$ and $\nabla R(\boldsymbol{v}_j, A)$ are of the form $aA\boldsymbol{x} + b\boldsymbol{x}$ for some vector $\boldsymbol{x}$ spanned by the columns of Q_j. Thus, we start with an initial unit vector $\boldsymbol{q}_1$, while for $j \geq 2$, $\boldsymbol{q}_j$ is selected as a unit vector orthogonal to $\boldsymbol{q}_1, \ldots, \boldsymbol{q}_{j-1}$ and such that the columns of Q_j are spanned by the vectors $\boldsymbol{q}_1$, $A\boldsymbol{q}_1, \ldots, A^{j-1}\boldsymbol{q}_1$. These particular $\boldsymbol{q}_j$ vectors are known as the Lanczos vectors. The calculation of the $\boldsymbol{q}_j$s can be facilitated by the use of the tridiagonal factorization $A = PTP'$, where P is orthogonal and T has the tridiagonal form

$$T = \begin{bmatrix} \alpha_1 & \beta_1 & 0 & \cdots & 0 & 0 & 0 \\ \beta_1 & \alpha_2 & \beta_2 & \cdots & 0 & 0 & 0 \\ \vdots & \vdots & \vdots & & \vdots & \vdots & \vdots \\ 0 & 0 & 0 & \cdots & \beta_{m-2} & \alpha_{m-1} & \beta_{m-1} \\ 0 & 0 & 0 & \cdots & 0 & \beta_{m-1} & \alpha_m \end{bmatrix}$$

Using this factorization, we find that if choose P and $\boldsymbol{q}_1$ so that $P\boldsymbol{e}_1 = \boldsymbol{q}_1$, then

$$(\boldsymbol{q}_1, A\boldsymbol{q}_1, \ldots, A^{j-1}\boldsymbol{q}_1) = P(\boldsymbol{e}_1, T\boldsymbol{e}_1, \ldots, T^{j-1}\boldsymbol{e}_1).$$

Since $(\boldsymbol{e}_1, T\boldsymbol{e}_1, \ldots, T^{j-1}\boldsymbol{e}_1)$ has upper triangular structure, this means that the first j columns of P span the column space of $(\boldsymbol{q}_1, A\boldsymbol{q}_1, \ldots, A^{j-1}\boldsymbol{q}_1)$; that is, the $\boldsymbol{q}_j$s can be obtained by calculating the factorization $A = PTP'$, or in other words, we can take $Q = (\boldsymbol{q}_1, \ldots, \boldsymbol{q}_m) = P$. Thus, since $AQ = QT$, we have

$$A\boldsymbol{q}_1 = \alpha_1\boldsymbol{q}_1 + \beta_1\boldsymbol{q}_2, \tag{6.18}$$

and

$$A\boldsymbol{q}_j = \beta_{j-1}\boldsymbol{q}_{j-1} + \alpha_j\boldsymbol{q}_j + \beta_j\boldsymbol{q}_{j+1}, \tag{6.19}$$

for $j = 2, \ldots, m-1$. Using these equations and the orthonormality of the $\boldsymbol{q}_j$s, it is easily shown that $\alpha_j = \boldsymbol{q}_j'A\boldsymbol{q}_j$ for all j, and as long as $\boldsymbol{p}_j = (A - \alpha_j \mathrm{I}_m)\boldsymbol{q}_j - \beta_{j-1}\boldsymbol{q}_{j-1} \neq \boldsymbol{0}$, then $\beta_j^2 = \boldsymbol{p}_j'\boldsymbol{p}_j$ and $\boldsymbol{q}_{j+1} = \boldsymbol{p}_j/\beta_j$ for $j = 1, \ldots, m-1$, if we define $\boldsymbol{q}_0 = \boldsymbol{0}$. Thus, we can continue calculating the $\boldsymbol{q}_j$s until we encounter a $\boldsymbol{p}_j = \boldsymbol{0}$. To see the significance of this event, let us suppose that the iterative procedure has proceeded through the first $j-1$ steps with $\boldsymbol{p}_i \neq \boldsymbol{0}$ for each $i = 2, \ldots, j-1$, and so we have obtained the matrix Q_j whose columns form a basis for $(\boldsymbol{q}_1, A\boldsymbol{q}_1, \ldots, A^{j-1}\boldsymbol{q}_1)$. Note that it follows immediately from the relationship $AQ = QT$ that

$$AQ_j = Q_jT_j + \boldsymbol{p}_j\boldsymbol{e}_j',$$

where T_j is the $j \times j$ submatrix of T consisting of its first j rows and j columns. This leads to the equation $Q_j'AQ_j = T_j + Q_j'\boldsymbol{p}_j\boldsymbol{e}_j'$. But $q_i'A\boldsymbol{q}_i = \alpha_i$, while it follows from (6.18) and (6.19) that $\boldsymbol{q}_{i+1}'A\boldsymbol{q}_i = \beta_i$ and $\boldsymbol{q}_k'A\boldsymbol{q}_i = 0$ if $k > i+1$. Thus, $Q_j'AQ_j = T_j$ and so we must have $Q_j'\boldsymbol{p}_j = \boldsymbol{0}$. Now if $\boldsymbol{p}_j \neq \boldsymbol{0}$, then $\boldsymbol{q}_{j+1} = \boldsymbol{p}_j/\beta_j$ is orthogonal to the columns of Q_j. Further, it follows from the fact that $\boldsymbol{q}_{j+1}$ is a linear combination of $A\boldsymbol{q}_j$, $\boldsymbol{q}_j$, and $\boldsymbol{q}_{j-1}$ that the columns of $Q_{j+1} = (Q_j, \boldsymbol{q}_{j+1})$ form a basis for the column space of $(\boldsymbol{q}_1, A\boldsymbol{q}_1, \ldots, A^j\boldsymbol{q}_1)$. If, on the other hand, $\boldsymbol{p}_j = \boldsymbol{0}$, then $AQ_j = Q_jT_j$. From this we see that the vectors $A^j\boldsymbol{q}_1, \ldots, A^{m-1}\boldsymbol{q}_1$ are in the space spanned by the columns of Q_j, that is, the space spanned by the vectors $\boldsymbol{q}_1, A\boldsymbol{q}_1, \ldots, A^{j-1}\boldsymbol{q}_1$. Consequently, the iterative procedure is complete since there are only j $\boldsymbol{q}_i$s.

In the iterative procedure described above, the largest and smallest eigenvalues of T_j serve as approximations to the largest and smallest eigenvalues of A. In practice, the termination of this iterative process is usually not due to the encounter of a $\boldsymbol{p}_j = \boldsymbol{0}$, but due to sufficiently accurate approximations of the eigenvalues of A.

Now let us return to the problem of solving the system of equations $A\boldsymbol{x} = \boldsymbol{c}$ through the iterative procedure based on the calculation of $\boldsymbol{y}_j$ in (6.17) and then $\boldsymbol{x}_j$ in (6.16). We will see that the choice of the Lanczos vectors as the columns of Q_j will simplify the computations involved. For this choice of Q_j, we have already seen that $Q_j'AQ_j = T_j$, so that the system in (6.17) is a special case of the tridiagonal system of equations discussed at the beginning of this section, special in that T_j is symmetric. As a result, the matrix T_j can be factored as $T_j = L_j D_j L_j'$, where $D_j = \mathrm{diag}(d_1, \ldots, d_j)$.

$$L_j = \begin{bmatrix} 1 & 0 & \cdots & 0 & 0 \\ l_1 & 1 & \cdots & 0 & 0 \\ \vdots & \vdots & & \vdots & \vdots \\ 0 & 0 & \cdots & 1 & 0 \\ 0 & 0 & \cdots & l_{j-1} & 1 \end{bmatrix},$$

$d_1 = \alpha_1$, and for $i = 2, \ldots, j$, $l_{i-1} = \beta_{i-1}/d_{i-1}$ and $d_i = \alpha_i - \beta_{i-1}l_{i-1}$. Thus, the solution for $\boldsymbol{y}_j$ in (6.17) can be easily found by first solving $L_j\boldsymbol{w}_j = Q_j'\boldsymbol{c}$, then $D_j\boldsymbol{z}_j = \boldsymbol{w}_j$, and finally $L_j'\boldsymbol{y}_j = \boldsymbol{z}_j$. Even as j increases, the computation required is not extensive since D_{j-1} and L_{j-1} are submatrices of D_j and L_j, and so in the jth iteration we only need to calculate d_j and l_{j-1} to obtain D_j and L_j from D_{j-1} and L_{j-1}.

The next step is to compute $\boldsymbol{x}_j$ from $\boldsymbol{y}_j$ using (6.16). We will see that this also may be done with a small amount of computation. Note that if we define the $m \times j$ matrix $B_j = (\boldsymbol{b}_1, \ldots, \boldsymbol{b}_j)$ so that $B_jL_j' = Q_j$, then by premultiplying the equation $T_j\boldsymbol{y}_j = Q_j'\boldsymbol{c}$ by $Q_jT_j^{-1}$ and using (6.16), we get

$$\boldsymbol{x}_j = Q_jT_j^{-1}Q_j'\boldsymbol{c} = Q_j(L_jD_jL_j')^{-1}Q_j'\boldsymbol{c} = B_j\boldsymbol{z}_j, \tag{6.20}$$

where $\boldsymbol{z}_j$ is as previously defined. It will be easier to compute $\boldsymbol{x}_j$ from (6.20) than from (6.16) since B_j and $\boldsymbol{z}_j$ are simple to compute after B_{j-1} and $\boldsymbol{z}_{j-1}$ have already been calculated. For instance, from the definition of B_j, we see that $\boldsymbol{b}_1 = \boldsymbol{q}_1$ and $\boldsymbol{b}_i = \boldsymbol{q}_i - l_{i-1}\boldsymbol{b}_{i-1}$ for $i > 1$, and consequently, $B_j = (B_{j-1}, \boldsymbol{b}_j)$. Using the defining equations for $\boldsymbol{w}_j$ and $\boldsymbol{z}_j$, we find that

$$L_jD_j\boldsymbol{z}_j = Q_j'\boldsymbol{c} \tag{6.21}$$

If we partition z_j as $z_j = (\boldsymbol{\gamma}_{j-1}', \gamma_j)'$, where $\boldsymbol{\gamma}_{j-1}$ is a $(j-1)\times 1$ vector, then by using the fact that

$$L_j = \begin{bmatrix} L_{j-1} & \mathbf{0} \\ l_{j-1}\boldsymbol{e}_{j-1}' & 1 \end{bmatrix}, \qquad D_j = \begin{bmatrix} D_{j-1} & \mathbf{0} \\ \mathbf{0}' & d_j \end{bmatrix},$$

we see that (6.21) implies that $L_{j-1}D_{j-1}\boldsymbol{\gamma}_{j-1} = Q_{j-1}'\boldsymbol{c}$. But this means that $\boldsymbol{\gamma}_{j-1} = z_{j-1}$, and so to compute z_j, we only need to compute γ_j, which is given by

$$\gamma_j = (\boldsymbol{q}_j'\boldsymbol{c} - l_{j-1}d_{j-1}\gamma_{j-1})/d_j,$$

where γ_{j-1} is the last component of z_{j-1}. Thus, (6.20) becomes

$$\boldsymbol{x}_j = B_j z_j = [B_{j-1}, \boldsymbol{b}_j]\begin{bmatrix} z_{j-1} \\ \gamma_j \end{bmatrix} = B_{j-1}z_{j-1} + \gamma_j\boldsymbol{b}_j = \boldsymbol{x}_{j-1} + \gamma_j\boldsymbol{b}_j,$$

and so we have a simple formula for computing the jth iterative solution from $\boldsymbol{b}_j$, γ_j, and the $(j-1)$th iterative solution $\boldsymbol{x}_{j-1}$.

PROBLEMS

1. Consider the system of equations $A\boldsymbol{x} = \boldsymbol{c}$, where A is the 4×3 matrix given in Problem 5.2 and

$$\boldsymbol{c} = \begin{bmatrix} 1 \\ 3 \\ -1 \\ 0 \end{bmatrix}$$

   (a) Show that the system is consistent.
   (b) Find a solution to this system of equations.
   (c) How many linearly independent solutions are there?

2. The system of equations $A\boldsymbol{x} = \boldsymbol{c}$ has A equal to the 3×4 matrix given in Problem 5.36 and

$$\boldsymbol{c} = \begin{bmatrix} 1 \\ 1 \\ 4 \end{bmatrix}$$

(a) Show that the system of equations is consistent.
(b) Give the general solution.
(c) Find r, the number of linearly independent solutions.
(d) Give a set of r linearly independent solutions.

3. Suppose the system of equations $A\boldsymbol{x} = \boldsymbol{c}$ has

$$A = \begin{bmatrix} 5 & 2 & 1 \\ 3 & 1 & 1 \\ 2 & 1 & 0 \\ 1 & 2 & -3 \end{bmatrix}$$

For each $\boldsymbol{c}$ given below, determine whether or not the system of equations is consistent.

$$(a)\ \boldsymbol{c} = \begin{bmatrix} 1 \\ 1 \\ 1 \\ 1 \end{bmatrix}, \qquad (b)\ \boldsymbol{c} = \begin{bmatrix} 3 \\ 2 \\ 1 \\ -1 \end{bmatrix}, \qquad (c)\ \boldsymbol{c} = \begin{bmatrix} 1 \\ -1 \\ 1 \\ -1 \end{bmatrix}$$

4. Consider the system of equations $A\boldsymbol{x} = \boldsymbol{c}$, where

$$A = \begin{bmatrix} 1 & 1 & -1 & 0 & 2 \\ 2 & 1 & 1 & 1 & 1 \end{bmatrix}, \qquad \boldsymbol{c} = \begin{bmatrix} 3 \\ 1 \end{bmatrix}$$

(a) Show that the system of equations is consistent.
(b) Give the general solution.
(c) Find r, the number of linearly independent solutions.
(d) Give a set of r linearly independent solutions.

5. Prove Theorem 6.5.

6. Consider the system of equations $AXB = C$, where X is a 3×3 matrix of variables and

$$A = \begin{bmatrix} 1 & 3 & 1 \\ 3 & 2 & 1 \end{bmatrix}, \qquad B = \begin{bmatrix} 1 & -1 \\ 1 & 0 \\ 0 & 1 \end{bmatrix}, \qquad C = \begin{bmatrix} 4 & 2 \\ 2 & 1 \end{bmatrix}$$

(a) Show that the system of equations is consistent.
(b) Find the form of the general solution to this system.

7. The general solution of a consistent system of equations was given in Theorem 6.4 as $A^-\boldsymbol{c}+(\mathrm{I}_n-A^-A)\boldsymbol{y}$. Show that the two vectors $A^-\boldsymbol{c}$ and $(\mathrm{I}_n-A^-A)\boldsymbol{y}$ are linearly independent if $\boldsymbol{c} \neq \mathbf{0}$ and $\boldsymbol{y} \neq \mathbf{0}$.

8. Suppose the $m \times n$ matrix A and $m \times 1$ vector $\boldsymbol{c} \neq \mathbf{0}$ are such that $A^-\boldsymbol{c}$ is the same for all choices of A^-. Use Theorem 5.23 to show that, if $A\boldsymbol{x} = \boldsymbol{c}$ is a consistent system of equations, then it has a unique solution.

9. For the homogeneous system of equations $A\boldsymbol{x} = \mathbf{0}$ in which

$$A = \begin{bmatrix} -1 & 3 & -2 & 1 \\ 2 & -3 & 0 & -2 \end{bmatrix},$$

determine r, the number of linearly independent solutions, and find a set of r linearly independent solutions.

10. Show that if the system of equations $AXB = C$ is consistent, then the solution is unique if and only if A has full column rank and B has full row rank.

11. Let

$$A = \begin{bmatrix} 1 & -1 & 1 & 1 \\ 2 & 3 & 1 & -1 \end{bmatrix}, \qquad \boldsymbol{c} = \begin{bmatrix} 1 \\ 2 \end{bmatrix},$$

$$B = \begin{bmatrix} 2 & 1 & 2 & -1 \\ 0 & 1 & 1 & 1 \end{bmatrix}, \qquad \boldsymbol{d} = \begin{bmatrix} 2 \\ 4 \end{bmatrix}.$$

(a) Show that the system $A\boldsymbol{x} = \boldsymbol{c}$ is consistent and has three linearly independent solutions.

(b) Show that the system $B\boldsymbol{x} = \boldsymbol{d}$ is consistent and has three linearly independent solutions.

(c) Show that the systems $A\boldsymbol{x} = \boldsymbol{c}$ and $B\boldsymbol{x} = \boldsymbol{d}$ have a common solution and that this common solution is unique.

12. Consider the systems of equations $AX = C$ and $XB = D$, where A is $m \times n$, B is $p \times q$, C is $m \times p$, and D is $n \times q$.

(a) Show that the two systems of equations have a common solution X if and only if each system is consistent and $AD = CB$.

(b) Show that the general common solution is given by

$$X_* = A^-C + (\mathrm{I} - A^-A)DB^- + (\mathrm{I} - A^-A)Y(\mathrm{I} - BB^-),$$

where Y is an arbitrary $n \times p$ matrix.

13. In Exercise 5.37, a least squares inverse was found for the matrix

$$A = \begin{bmatrix} 1 & -1 & -2 & 1 \\ -2 & 4 & 3 & -2 \\ 1 & 1 & -3 & 1 \end{bmatrix}$$

(a) Use this least squares inverse to show that the system of equations $A\boldsymbol{x} = \boldsymbol{c}$ is inconsistent, where $\boldsymbol{c}' = (2, 1, 5)$.
(b) Find a least squares solution.
(c) Compute the sum of squared errors for a least squares solution to this system of equations.

14. Consider the system of equations $A\boldsymbol{x} = \boldsymbol{c}$, where

$$A = \begin{bmatrix} 1 & 0 & 2 \\ 2 & -1 & 3 \\ -1 & 2 & 0 \\ -2 & 1 & -3 \end{bmatrix}, \qquad \boldsymbol{c} = \begin{bmatrix} 2 \\ 2 \\ 5 \\ 0 \end{bmatrix}$$

(a) Find a least squares inverse of A.
(b) Show that the system of equations is inconsistent.
(c) Find a least squares solution.
(d) Is this solution unique?

15. Show that $\boldsymbol{x}_*$ is a least squares solution to the system of equations $A\boldsymbol{x} = \boldsymbol{c}$ if and only if

$$A'A\boldsymbol{x}_* = A'\boldsymbol{c}$$

16. Let A be an $m \times n$ matrix, and $\boldsymbol{x}_*, \boldsymbol{y}_*$, and $\boldsymbol{c}$, $n \times 1$, $m \times 1$, and $m \times 1$ vectors, respectively. Suppose that $\boldsymbol{x}_*$ and $\boldsymbol{y}_*$ are such that the system of equations

$$\begin{bmatrix} \mathrm{I}_m & A \\ A' & (0) \end{bmatrix} \begin{bmatrix} \boldsymbol{y}_* \\ \boldsymbol{x}_* \end{bmatrix} = \begin{bmatrix} \boldsymbol{c} \\ \boldsymbol{0} \end{bmatrix},$$

holds. Show that $\boldsymbol{x}_*$ then must be a least squares solution to the system $A\boldsymbol{x} = \boldsymbol{c}$.

17. The balanced two-way classification model with interaction is of the form

$$y_{ijk} = \mu + \tau_i + \gamma_j + \eta_{ij} + \epsilon_{ijk},$$

where $i = 1, \ldots, a, j = 1, \ldots, b$, and $k = 1, \ldots, n$. The parameter μ represents an overall effect, τ_i is an effect due to the ith level of factor one, γ_j is an effect due to the jth level of factor two, and η_{ij} is an effect due to the interaction of the ith and jth levels of factors one and two; as usual, the e_{ijk}s represent independent random errors, each distributed as $N(0, \sigma^2)$.

(a) Set up the vectors $\boldsymbol{y}$, $\boldsymbol{\beta}$, and $\boldsymbol{\epsilon}$ and the matrix X so that the two-way model above can be written in the matrix form $\boldsymbol{y} = X\boldsymbol{\beta} + \boldsymbol{\epsilon}$.

(b) Find the rank r, of X. Determine a set of r linear independent estimable functions of the parameters, μ, τ_i, γ_j, and η_{ij}.

(c) Find a least squares solution for the parameter vector $\boldsymbol{\beta}$.

18. Consider the regression model

$$\boldsymbol{y} = X\boldsymbol{\beta} + \boldsymbol{\epsilon},$$

where X is $N \times m$, $\boldsymbol{\epsilon} \sim N_N(\boldsymbol{0}, \sigma^2 C)$, and C is a known positive definite matrix. In Example 4.6, for the case in which X is full column rank, we obtained the generalized least squares estimator $\hat{\boldsymbol{\beta}} = (X'C^{-1}X)^{-1}X'C^{-1}\boldsymbol{y}$, which minimizes

$$(\boldsymbol{y} - X\hat{\boldsymbol{\beta}})'C^{-1}(\boldsymbol{y} - X\hat{\boldsymbol{\beta}}) \tag{6.22}$$

Show that if X is less than full column rank, then the generalized least squares estimator of $\boldsymbol{\beta}$ which minimizes (6.22) is given by

$$\hat{\boldsymbol{\beta}} = (X'C^{-1}X)^{-}X'C^{-1}\boldsymbol{y} + \{\mathrm{I} - (X'C^{-1}X)^{-}X'C^{-1}X\}\boldsymbol{u},$$

where $\boldsymbol{u}$ is an arbitrary $m \times 1$ vector.

19. Restricted least squares obtains the vectors $\hat{\boldsymbol{\beta}}$ that minimize

$$(\boldsymbol{y} - X\hat{\boldsymbol{\beta}})'(\boldsymbol{y} - X\hat{\boldsymbol{\beta}}),$$

subject to the restriction that $\hat{\boldsymbol{\beta}}$ satisfies $B\hat{\boldsymbol{\beta}} = \boldsymbol{b}$, where B is $p \times m$ and $\boldsymbol{b}$ is $p \times 1$ such that $BB^{-}\boldsymbol{b} = \boldsymbol{b}$. Use Theorem 6.4 to find the general solution $\hat{\boldsymbol{\beta}}_u$ to the consistent system of equations $B\hat{\boldsymbol{\beta}} = \boldsymbol{b}$, where $\hat{\boldsymbol{\beta}}_u$ depends on an arbitrary vector $\boldsymbol{u}$. Substitute this expression for $\hat{\boldsymbol{\beta}}$ into $(\boldsymbol{y} - X\hat{\boldsymbol{\beta}})'(\boldsymbol{y} - X\hat{\boldsymbol{\beta}})$, and then use Theorem 6.14 to obtain the general least squares solution $\boldsymbol{u}_w$, for $\boldsymbol{u}$, where $\boldsymbol{u}_w$ depends on an arbitrary vector $\boldsymbol{w}$. Substitute $\boldsymbol{u}_w$ for $\boldsymbol{u}$ in $\hat{\boldsymbol{\beta}}_u$ to show that the general restricted least squares solution for $\boldsymbol{\beta}$ is given by

$$\begin{aligned}\hat{\boldsymbol{\beta}}_w = B^{-}\boldsymbol{b} + (\mathrm{I} - B^{-}B)\{&[X(\mathrm{I} - B^{-}B)]^{L}(\boldsymbol{y} - XB^{-}\boldsymbol{b}) \\ &+ (\mathrm{I} - [X(\mathrm{I} - B^{-}B)]^{L}X(\mathrm{I} - B^{-}B))\boldsymbol{w}\}.\end{aligned}$$

20. In the previous exercise, show that if we use the Moore–Penrose inverse as the least squares inverse of $[X(\mathrm{I} - B^- B)]$ in the expression given for $\hat{\boldsymbol{\beta}}_w$, then it simplifies to

$$\hat{\boldsymbol{\beta}}_w = B^- \boldsymbol{b} + [X(\mathrm{I} - B^- B)]^+ (\boldsymbol{y} - XB^- \boldsymbol{b}) \\ + (\mathrm{I} - B^- B)\{\mathrm{I} - [X(\mathrm{I} - B^- B)]^+ X(\mathrm{I} - B^- B)\}\boldsymbol{w}.$$

21. Consider the iterative procedure, based on the Lanczos vectors, for solving the system of equations $A\boldsymbol{x} = \boldsymbol{c}$. Suppose that for the initial Lanczos vector $\boldsymbol{q}_1$ we use $(\boldsymbol{c}'\boldsymbol{c})^{-1/2}\boldsymbol{c}$.
    (a) Show that if for some j, $\boldsymbol{p}_j = (A - \alpha_j \mathrm{I}_m)\boldsymbol{q}_j - \beta_{j-1}\boldsymbol{q}_{j-1} = \boldsymbol{0}$, then $A\boldsymbol{x}_j = \boldsymbol{c}$.
    (b) Show that for any j, the procedure easily yields a measure of the adequacy of the jth iterative solution since

$$(A\boldsymbol{x}_j - \boldsymbol{c})'(A\boldsymbol{x}_j - \boldsymbol{c}) = \beta_j^2 y_{jj}^2,$$

where y_{jj} is the jth component of the vector $\boldsymbol{y}_j$ in (6.16).

CHAPTER SEVEN

Special Matrices and Matrix Operators

1. INTRODUCTION

The concept of partitioning matrices was first introduced in Chapter 1, and we have subsequently used partitioned matrices throughout this text. In this chapter we develop some specialized formulas for the determinant and inverse of partitioned matrices. In addition to partitioned matrices, we will look at some other special types of structured matrices that we have not previously discussed. In this chapter, we will also introduce and develop properties of some special matrix operators. In many situations, a seemingly complicated matrix expression can be written in a fairly simple form by making use of one or more of these matrix operators.

2. PARTITIONED MATRICES

Up to this point, most of our applications involving partitioned matrices have utilized only the simple operations of matrix addition and multiplication. In this section, we will obtain expressions for the inverse and determinant of an $m \times m$ matrix A that is partitioned into the 2×2 block form given by

$$A = \begin{bmatrix} A_{11} & A_{12} \\ A_{21} & A_{22} \end{bmatrix}, \tag{7.1}$$

where A_{11} is $m_1 \times m_1$, A_{12} is $m_1 \times m_2$, A_{21} is $m_2 \times m_1$, and A_{22} is $m_2 \times m_2$. We wish to obtain expressions for the inverse and determinant of A in terms of its submatrices. We begin with the inverse of A.

Theorem 7.1. Let the $m \times m$ matrix A be partitioned as in (7.1), and suppose that A, A_{11}, and A_{22} are nonsingular matrices. For notational convenience write $B = A^{-1}$ and partition B as

$$B = \begin{bmatrix} B_{11} & B_{12} \\ B_{21} & B_{22} \end{bmatrix},$$

where the submatrices of B are of the same sizes as the corresponding submatrices of A. Then we have

(a) $B_{11} = (A_{11} - A_{12}A_{22}^{-1}A_{21})^{-1} = A_{11}^{-1} + A_{11}^{-1}A_{12}B_{22}A_{21}A_{11}^{-1}$,
(b) $B_{22} = (A_{22} - A_{21}A_{11}^{-1}A_{12})^{-1} = A_{22}^{-1} + A_{22}^{-1}A_{21}B_{11}A_{12}A_{22}^{-1}$,
(c) $B_{12} = -A_{11}^{-1}A_{12}B_{22}$,
(d) $B_{21} = -A_{22}^{-1}A_{21}B_{11}$.

Proof. The matrix equation

$$AB = \begin{bmatrix} A_{11} & A_{12} \\ A_{21} & A_{22} \end{bmatrix} \begin{bmatrix} B_{11} & B_{12} \\ B_{21} & B_{22} \end{bmatrix} = \begin{bmatrix} \mathrm{I}_{m_1} & (0) \\ (0) & \mathrm{I}_{m_2} \end{bmatrix} = \mathrm{I}_m$$

yields the four equations

$$A_{11}B_{11} + A_{12}B_{21} = \mathrm{I}_{m_1}, \tag{7.2}$$
$$A_{21}B_{12} + A_{22}B_{22} = \mathrm{I}_{m_2}, \tag{7.3}$$
$$A_{11}B_{12} + A_{12}B_{22} = (0), \tag{7.4}$$
$$A_{21}B_{11} + A_{22}B_{21} = (0) \tag{7.5}$$

Solving (7.4) and (7.5) for B_{12} and B_{21}, respectively, immediately leads to the expressions given in (c) and (d). Substituting these solutions for B_{12} and B_{21} into (7.2) and (7.3) and solving for B_{11} and B_{22} yields the first expressions given for B_{11} and B_{22} in (a) and (b). The second expressions in (a) and (b) follow immediately from the first after using Theorem 1.7. □

Example 7.1. Consider the regression model

$$\boldsymbol{y} = X\boldsymbol{\beta} + \boldsymbol{\epsilon},$$

where $\boldsymbol{y}$ is $N \times 1$, X is $N \times (k+1)$, $\boldsymbol{\beta}$ is $(k+1) \times 1$, and $\boldsymbol{\epsilon}$ is $N \times 1$. Suppose that $\boldsymbol{\beta}$ and X are partitioned as $\boldsymbol{\beta} = (\boldsymbol{\beta}_1', \boldsymbol{\beta}_2')'$ and $X = (X_1, X_2)$ so that the product $X_1\boldsymbol{\beta}_1$ is defined, and we are interested in comparing the complete regression model given above to the reduced regression model

$$\boldsymbol{y} = X_1\boldsymbol{\beta}_1 + \boldsymbol{\epsilon}$$

If X has full column rank, then the least squares estimators for the two models are $\hat{\boldsymbol{\beta}} = (X'X)^{-1}X'\boldsymbol{y}$ and $\hat{\boldsymbol{\beta}}_1 = (X_1'X_1)^{-1}X_1'\boldsymbol{y}$, respectively, and the difference in the sums of squared errors for the two models

$$\begin{aligned}(\boldsymbol{y} - X_1\hat{\boldsymbol{\beta}}_1)'(\boldsymbol{y} - X_1\hat{\boldsymbol{\beta}}_1) &- (\boldsymbol{y} - X\hat{\boldsymbol{\beta}})'(\boldsymbol{y} - X\hat{\boldsymbol{\beta}}) \\ &= \boldsymbol{y}'(\mathrm{I} - X_1(X_1'X_1)^{-1}X_1')\boldsymbol{y} - \boldsymbol{y}'(\mathrm{I} - X(X'X)^{-1}X')\boldsymbol{y} \\ &= \boldsymbol{y}'X(X'X)^{-1}X'\boldsymbol{y} - \boldsymbol{y}'X_1(X_1'X_1)^{-1}X_1'\boldsymbol{y} \qquad (7.6)\end{aligned}$$

gives the reduction in the sum of squared errors attributable to the inclusion of the term $X_2\boldsymbol{\beta}_2$ in the complete model. By using the geometrical properties of least squares regression in Example 2.10, we showed that this reduction in the sum of squared errors simplifies to

$$\boldsymbol{y}'X_{2*}(X_{2*}'X_{2*})^{-1}X_{2*}'\boldsymbol{y},$$

where $X_{2*} = (\mathrm{I} - X_1(X_1'X_1)^{-1}X_1')X_2$. An alternative way of showing this, which we illustrate here, uses Theorem 7.1. Now $X'X$ can be partitioned as

$$X'X = \begin{bmatrix} X_1'X_1 & X_1'X_2 \\ X_2'X_1 & X_2'X_2 \end{bmatrix},$$

and so if we let $C = (X_2'X_2 - X_2'X_1(X_1'X_1)^{-1}X_1'X_2)^{-1} = (X_{2*}'X_{2*})^{-1}$, we find from a direct application of Theorem 7.1 that

$$(X'X)^{-1} = \begin{bmatrix} (X_1'X_1)^{-1} + (X_1'X_1)^{-1}X_1'X_2CX_2'X_1(X_1'X_1)^{-1} & -(X_1'X_1)^{-1}X_1'X_2C \\ -CX_2'X_1(X_1'X_1)^{-1} & C \end{bmatrix}$$

Substituting this into (7.6) and then simplifying, we get $\boldsymbol{y}'X_{2*}(X_{2*}'X_{2*})^{-1}X_{2*}'\boldsymbol{y}$, as required.

Before obtaining expressions for the determinant of A, we will first consider some special cases.

Theorem 7.2. Let the $m \times m$ matrix A be partitioned as in (7.1). If $A_{22} = \mathrm{I}_{m_2}$ and $A_{12} = (0)$ or $A_{21} = (0)$, then $|A| = |A_{11}|$.

Proof. To find the determinant

$$|A| = \begin{vmatrix} A_{11} & (0) \\ A_{21} & \mathrm{I}_{m_2} \end{vmatrix},$$

first use the cofactor expansion formula for a determinant on the last column of A to obtain

$$|A| = \begin{vmatrix} A_{11} & (0) \\ B & \mathrm{I}_{m_2-1} \end{vmatrix},$$

where B is the $(m_2 - 1) \times m_1$ matrix obtained by deleting the last row from A_{21}. Repeating this process another $(m_2 - 1)$ times yields $|A| = |A_{11}|$. In a similar fashion, we obtain $|A| = |A_{11}|$, when $A_{21} = (0)$, by repeatedly expanding along the last row. □

Clearly we have a result analogous to Theorem 7.2 when $A_{11} = \mathrm{I}_{m_1}$ and $A_{12} = (0)$ or $A_{21} = (0)$. Also, Theorem 7.2 can be generalized to the following.

Theorem 7.3. Let the $m \times m$ matrix A be partitioned as in (7.1). If $A_{12} = (0)$ or $A_{21} = (0)$, then $|A| = |A_{11}||A_{22}|$.

Proof. Observe that

$$|A| = \begin{vmatrix} A_{11} & (0) \\ A_{21} & A_{22} \end{vmatrix} = \begin{vmatrix} A_{11} & (0) \\ A_{21} & \mathrm{I}_{m_2} \end{vmatrix} \begin{vmatrix} \mathrm{I}_{m_1} & (0) \\ (0) & A_{22} \end{vmatrix} = |A_{11}||A_{22}|,$$

where the last equality follows from Theorem 7.2. A similar proof yields $|A| = |A_{11}||A_{22}|$ when $A_{21} = (0)$. □

We are now ready to find an expression for the determinant of A in the general case.

Theorem 7.4. Let the $m \times m$ matrix A be partitioned as in (7.1). Then

(a) $|A| = |A_{22}||A_{11} - A_{12}A_{22}^{-1}A_{21}|$, if A_{22} is nonsingular, and
(b) $|A| = |A_{11}||A_{22} - A_{21}A_{11}^{-1}A_{12}|$, if A_{11} is nonsingular.

Proof. Suppose that A_{22} is nonsingular. Note that in this case the identity

$$\begin{bmatrix} \mathrm{I}_{m_1} & -A_{12}A_{22}^{-1} \\ (0) & \mathrm{I}_{m_2} \end{bmatrix} \begin{bmatrix} A_{11} & A_{12} \\ A_{21} & A_{22} \end{bmatrix} \begin{bmatrix} \mathrm{I}_{m_1} & (0) \\ -A_{22}^{-1}A_{21} & \mathrm{I}_{m_2} \end{bmatrix}$$
$$= \begin{bmatrix} A_{11} - A_{12}A_{22}^{-1}A_{21} & (0) \\ (0) & A_{22} \end{bmatrix}$$

holds. After taking the determinant of both sides of this identity and using the previous theorem, we immediately get (a). The proof of (b) is similar. □

Example 7.2. We will find the determinant and inverse of the $2m \times 2m$ matrix A given by

$$A = \begin{bmatrix} a\mathrm{I}_m & \mathbf{1}_m\mathbf{1}'_m \\ \mathbf{1}_m\mathbf{1}'_m & b\mathrm{I}_m \end{bmatrix},$$

where a and b are nonzero scalars. Using (a) of Theorem 7.4, we find that

$$\begin{aligned} |A| &= |b\mathrm{I}_m|\ |a\mathrm{I}_m - \mathbf{1}_m\mathbf{1}'_m(b\mathrm{I}_m)^{-1}\mathbf{1}_m\mathbf{1}'_m| \\ &= b^m \left| a\mathrm{I}_m - \frac{m}{b}\ \mathbf{1}_m\mathbf{1}'_m \right| \\ &= b^m a^{m-1}\left(a - \frac{m^2}{b} \right), \end{aligned}$$

where we have used the result of Problem 3.18(e) in the last step. The matrix A will be nonsingular if $|A| \neq 0$ or, equivalently, if

$$a \neq \frac{m^2}{b}$$

In this case, using Theorem 7.1, we find that

$$\begin{aligned} B_{11} &= (a\mathrm{I}_m - \mathbf{1}_m\mathbf{1}'_m(b\mathrm{I}_m)^{-1}\mathbf{1}_m\mathbf{1}'_m)^{-1} \\ &= \left(a\mathrm{I}_m - \frac{m}{b}\ \mathbf{1}_m\mathbf{1}'_m \right)^{-1} \\ &= a^{-1}\mathrm{I}_m + \left\{ \frac{m}{a(ab - m^2)} \right\} \mathbf{1}_m\mathbf{1}'_m, \end{aligned}$$

where this last expression follows from Problem 3.18(d). In a similar fashion, we find that

$$\begin{aligned} B_{22} &= (b\mathrm{I}_m - \mathbf{1}_m\mathbf{1}'_m(a\mathrm{I}_m)^{-1}\mathbf{1}_m\mathbf{1}'_m)^{-1} \\ &= \left(b\mathrm{I}_m - \frac{m}{a}\ \mathbf{1}_m\mathbf{1}'_m \right)^{-1} \\ &= b^{-1}\mathrm{I}_m + \left\{ \frac{m}{b(ab - m^2)} \right\} \mathbf{1}_m\mathbf{1}'_m \end{aligned}$$

The remaining submatrices of $B = A^{-1}$ are given by

$$B_{12} = -(a\mathrm{I}_m)^{-1}\mathbf{1}_m\mathbf{1}'_m\left(b^{-1}\mathrm{I}_m + \left\{\frac{m}{b(ab-m^2)}\right\}\mathbf{1}_m\mathbf{1}'_m\right)$$
$$= -(ab-m^2)^{-1}\mathbf{1}_m\mathbf{1}'_m,$$

and, since A is symmetric, $B_{21} = B'_{12} = B_{12}$. Putting this all together, we have

$$A^{-1} = B = \begin{bmatrix} a^{-1}(\mathrm{I}_m + mc\mathbf{1}_m\mathbf{1}'_m) & -c\mathbf{1}_m\mathbf{1}'_m \\ -c\mathbf{1}_m\mathbf{1}'_m & b^{-1}(\mathrm{I}_m + mc\mathbf{1}_m\mathbf{1}'_m) \end{bmatrix},$$

where $c = (ab - m^2)^{-1}$.

We will use Theorem 7.4 to establish the following useful result.

Theorem 7.5. Let A and B be $m \times n$ and $n \times m$ matrices, respectively. Then

$$|\mathrm{I}_m + AB| = |\mathrm{I}_n + BA|$$

Proof. Note that

$$\begin{bmatrix} \mathrm{I}_m & A \\ -B & \mathrm{I}_n \end{bmatrix}\begin{bmatrix} \mathrm{I}_m & (0) \\ B & \mathrm{I}_n \end{bmatrix} = \begin{bmatrix} \mathrm{I}_m + AB & A \\ (0) & \mathrm{I}_n \end{bmatrix},$$

so that by taking the determinant of both sides and using Theorem 7.4, we obtain the identity

$$\begin{vmatrix} \mathrm{I}_m & A \\ -B & \mathrm{I}_n \end{vmatrix} = |\mathrm{I}_m + AB| \tag{7.7}$$

Similarly, observe that

$$\begin{bmatrix} \mathrm{I}_m & (0) \\ B & \mathrm{I}_n \end{bmatrix}\begin{bmatrix} \mathrm{I}_m & A \\ -B & \mathrm{I}_n \end{bmatrix} = \begin{bmatrix} \mathrm{I}_m & A \\ (0) & \mathrm{I}_n + BA \end{bmatrix},$$

so that

$$\begin{vmatrix} \mathrm{I}_m & A \\ -B & \mathrm{I}_n \end{vmatrix} = |\mathrm{I}_n + BA| \tag{7.8}$$

The result now follows by equating (7.7) and (7.8). □

Our final result follows directly from Theorem 7.5 if we replace A by $-\lambda A$.

Corollary 7.5.1. Let A and B be $m \times n$ and $n \times m$ matrices. Then the nonzero eigenvalues of AB are the same as the nonzero eigenvalues of BA.

3. THE KRONECKER PRODUCT

Some matrices possess a special type of structure that permits them to be expressed as a product, commonly referred to as the Kronecker product, of two other matrices. If A is an $m \times n$ matrix and B is a $p \times q$ matrix, then the Kronecker product of A and B, denoted by $A \otimes B$, is the $mp \times nq$ matrix

$$\begin{bmatrix} a_{11}B & a_{12}B & \cdots & a_{1n}B \\ a_{21}B & a_{22}B & \cdots & a_{2n}B \\ \vdots & \vdots & & \vdots \\ a_{m1}B & a_{m2}B & \cdots & a_{mn}B \end{bmatrix} \tag{7.9}$$

The Kronecker product defined above is more precisely known as the right Kronecker product, and it is the most common definition of Kronecker product appearing in the literature. However, some authors [for example, Graybill (1983)] define the Kronecker product as the left Kronecker product, which has $B \otimes A$ as the matrix given in (7.9). Throughout this book, any reference to the Kronecker product will be referring to the right Kronecker product. The special structure of the matrix given in (7.9) leads to simplified formulas for the computation of such things as its inverse, determinant, and eigenvalues. In this section, we will develop some of these formulas as well as some of the more basic properties of the Kronecker product.

Unlike ordinary matrix multiplication, the Kronecker product $A \otimes B$ is defined regardless of the sizes of A and B. However, as with ordinary matrix multiplication, the Kronecker product is not, in general, commutative as is demonstrated in the following example.

Example 7.3. Let A and B be the 1×3 and 2×2 matrices given by

$$A = [0 \quad 1 \quad 2], \qquad B = \begin{bmatrix} 1 & 2 \\ 3 & 4 \end{bmatrix}$$

Then we find that

$$A \otimes B = [0B \quad 1B \quad 2B] = \begin{bmatrix} 0 & 0 & 1 & 2 & 2 & 4 \\ 0 & 0 & 3 & 4 & 6 & 8 \end{bmatrix},$$

while

$$B \otimes A = \begin{bmatrix} 1A & 2A \\ 3A & 4A \end{bmatrix} = \begin{bmatrix} 0 & 1 & 2 & 0 & 2 & 4 \\ 0 & 3 & 6 & 0 & 4 & 8 \end{bmatrix}$$

Some of the basic properties of the Kronecker product, which are easily proven from its definition, are summarized below. The proofs are left to the reader as an exercise.

Theorem 7.6. Let A, B, and C be any matrices and $\boldsymbol{a}$ and $\boldsymbol{b}$ be any two vectors. Then

(a) $\alpha \otimes A = A \otimes \alpha = \alpha A$, for any scalar α,
(b) $(\alpha A) \otimes (\beta B) = \alpha\beta(A \otimes B)$, for any scalars α and β,
(c) $(A \otimes B) \otimes C = A \otimes (B \otimes C)$,
(d) $(A + B) \otimes C = (A \otimes C) + (B \otimes C)$, if A and B are of the same size,
(e) $A \otimes (B + C) = (A \otimes B) + (A \otimes C)$, if B and C are of the same size,
(f) $(A \otimes B)' = A' \otimes B'$,
(g) $\boldsymbol{a}\boldsymbol{b}' = \boldsymbol{a} \otimes \boldsymbol{b}' = \boldsymbol{b}' \otimes \boldsymbol{a}$.

We have the following very useful property involving the Kronecker product and ordinary matrix multiplication.

Theorem 7.7. Let A, B, C, and D be matrices of sizes $m \times h$, $p \times k$, $h \times n$, and $k \times q$, respectively. Then

$$(A \otimes B)(C \otimes D) = AC \otimes BD \tag{7.10}$$

Proof. The left-hand side of (7.10) is

$$\begin{bmatrix} a_{11}B & \cdots & a_{1h}B \\ \vdots & & \vdots \\ a_{m1}B & \cdots & a_{mh}B \end{bmatrix} \begin{bmatrix} c_{11}D & \cdots & c_{1n}D \\ \vdots & & \vdots \\ c_{h1}D & \cdots & c_{hn}D \end{bmatrix} = \begin{bmatrix} F_{11} & \cdots & F_{1n} \\ \vdots & & \vdots \\ F_{m1} & \cdots & F_{mn} \end{bmatrix},$$

where

$$F_{ij} = \sum_{l=1}^{h} a_{il}c_{lj}BD = (A)_{i\cdot}(C)_{\cdot j}BD = (AC)_{ij}BD$$

The result now follows since

$$AC \otimes BD = \begin{bmatrix} (AC)_{11}BD & \cdots & (AC)_{1n}BD \\ \vdots & & \vdots \\ (AC)_{m1}BD & \cdots & (AC)_{mn}BD \end{bmatrix}$$ □

Our next result demonstrates that the trace of the Kronecker product $A \otimes B$ can be expressed in terms of the trace of A and the trace of B when A and B are square matrices.

Theorem 7.8. Let A be an $m \times m$ matrix and B be a $p \times p$ matrix. Then

$$\text{tr}(A \otimes B) = \text{tr}(A)\text{tr}(B)$$

Proof. Using (7.9) when $n = m$, we see that

$$\text{tr}(A \otimes B) = \sum_{i=1}^{m} a_{ii}\, \text{tr}(B) = \left(\sum_{i=1}^{m} a_{ii} \right) \text{tr}(B) = \text{tr}(A)\, \text{tr}(B),$$

so that the result holds. □

Theorem 7.8 gives a simplified expression for the trace of a Kronecker product. There is an analogous result for the determinant of a Kronecker product. But before we get to that, let us first consider the inverse of $A \otimes B$ and the eigenvalues of $A \otimes B$ when A and B are square matrices.

Theorem 7.9. Let A be an $m \times n$ matrix and B be a $p \times q$ matrix. Then

(a) $(A \otimes B)^{-1} = A^{-1} \otimes B^{-1}$, if $m = n$, $p = q$ and $A \otimes B$ is nonsingular,
(b) $(A \otimes B)^{+} = A^{+} \otimes B^{+}$,
(c) $(A \otimes B)^{-} = A^{-} \otimes B^{-}$, for any generalized inverses, A^{-} and B^{-}, of A and B.

Proof. Using Theorem 7.7, we find that

$$(A^{-1} \otimes B^{-1})(A \otimes B) = (A^{-1}A \otimes B^{-1}B) = \text{I}_m \otimes \text{I}_p = \text{I}_{mp},$$

so (a) holds. We will leave the verification of (b) and (c) as an exercise for the reader. □

Theorem 7.10. Let $\lambda_1, \ldots, \lambda_m$ be the eigenvalues of the $m \times m$ matrix A, and let $\theta_1, \ldots, \theta_p$ be the eigenvalues of the $p \times p$ matrix B. Then the set of mp eigenvalues of $A \otimes B$ is given by $\{\lambda_i \theta_j \colon i = 1, \ldots, m; j = 1, \ldots, p\}$.

Proof. It follows from Theorem 4.12 that there exist nonsingular matrices P and Q such that

$$P^{-1}AP = T_1, \qquad Q^{-1}BQ = T_2,$$

where T_1 and T_2 are upper triangular matrices with the eigenvalues of A and B, respectively, as diagonal elements. The eigenvalues of $A \otimes B$ are the same as those of

$$\begin{aligned}(P \otimes Q)^{-1}(A \otimes B)(P \otimes Q) &= (P^{-1} \otimes Q^{-1})(A \otimes B)(P \otimes Q)\\ &= P^{-1}AP \otimes Q^{-1}BQ = T_1 \otimes T_2,\end{aligned}$$

which must be upper triangular since T_1 and T_2 are upper triangular. The result now follows since the eigenvalues of $T_1 \otimes T_2$ are its diagonal elements, and these are clearly given by $\{\lambda_i \theta_j:\ i = 1, \ldots, m; j = 1, \ldots, p\}$. □

A simplified expression for the determinant of $A \otimes B$, when A and B are square matrices, is most easily obtained by using the fact that the determinant of a matrix is given by the product of its eigenvalues.

Theorem 7.11. Let A be an $m \times m$ matrix and B be a $p \times p$ matrix. Then

$$|A \otimes B| = |A|^p |B|^m$$

Proof. Let $\lambda_1, \ldots, \lambda_m$ be the eigenvalues of A, and let $\theta_1, \ldots, \theta_p$ be the eigenvalues of B. Then we have

$$|A| = \prod_{i=1}^{m} \lambda_i, \qquad |B| = \prod_{j=1}^{p} \theta_j,$$

and from the previous theorem

$$\begin{aligned}|A \otimes B| &= \prod_{j=1}^{p} \prod_{i=1}^{m} \lambda_i \theta_j = \prod_{j=1}^{p} \theta_j^m \left(\prod_{i=1}^{m} \lambda_i \right) = \prod_{j=1}^{p} \theta_j^m |A|\\ &= |A|^p \left(\prod_{j=1}^{p} \theta_j \right)^m = |A|^p |B|^m\end{aligned}$$

□

Our final result on Kronecker products identifies a relationship between $\text{rank}(A \otimes B)$, and $\text{rank}(A)$ and $\text{rank}(B)$.

Theorem 7.12. Let A be an $m \times n$ matrix and B be a $p \times q$ matrix. Then

$$\text{rank}(A \otimes B) = \text{rank}(A)\text{rank}(B)$$

Proof. Our proof utilizes Theorem 3.11, which states that the rank of a symmetric matrix equals the number of its nonzero eigenvalues. Although $A \otimes B$ as given is not necessarily symmetric, the matrix $(A \otimes B)(A \otimes B)'$, as well as AA' and BB', is symmetric. Now from Theorem 2.10 we have

$$\text{rank}(A \otimes B) = \text{rank}\{(A \otimes B)(A \otimes B)'\} = \text{rank}(AA' \otimes BB')$$

Since $AA' \otimes BB'$ is symmetric, its rank is given by the number of its nonzero eigenvalues. Now if $\lambda_1, \ldots, \lambda_m$ are the eigenvalues of AA', and $\theta_1, \ldots, \theta_p$ are the eigenvalues of BB' then, by Theorem 7.10, the eigenvalues of $AA' \otimes BB'$ are given by $\{\lambda_i \theta_j : i = 1, \ldots, m; j = 1, \ldots, p\}$. Clearly, the number of nonzero values in this set is the number of nonzero λ_is times the number of nonzero θ_js. But, since AA' and BB' are symmetric, the number of nonzero λ_is is given by $\text{rank}(AA') = \text{rank}(A)$, and the number of nonzero θ_js is given by $\text{rank}(BB') = \text{rank}(B)$. The proof is now complete. □

Example 7.4. The computations involved in an analysis of variance are sometimes particularly well suited for the use of the Kronecker product. For example, consider the univariate one-way classification model

$$y_{ij} = \mu + \tau_i + \epsilon_{ij},$$

which was discussed in Examples 3.14, 6.10, and 6.11. Suppose that we have the same number of observations available from each of the k treatments, so that $j = 1, \ldots, n$ for each i. In this case, the model may be written as

$$\boldsymbol{y} = X\boldsymbol{\beta} + \boldsymbol{\epsilon},$$

where $X = (\mathbf{1}_k \otimes \mathbf{1}_n, \mathrm{I}_k \otimes \mathbf{1}_n)$, $\boldsymbol{\beta} = (\mu, \tau_1, \ldots, \tau_k)'$, $\boldsymbol{y} = (\boldsymbol{y}_1', \ldots, \boldsymbol{y}_k')'$, and $\boldsymbol{y}_i = (y_{i1}, \ldots, y_{in})'$. Consequently, a least squares solution for $\boldsymbol{\beta}$ is easily computed as

$$\begin{aligned}\hat{\boldsymbol{\beta}} = (X'X)^- X'\boldsymbol{y} &= \left\{\begin{bmatrix} \mathbf{1}_k' \otimes \mathbf{1}_n' \\ \mathrm{I}_k \otimes \mathbf{1}_n' \end{bmatrix} [\mathbf{1}_k \otimes \mathbf{1}_n \quad \mathrm{I}_k \otimes \mathbf{1}_n]\right\}^- \begin{bmatrix} \mathbf{1}_k' \otimes \mathbf{1}_n' \\ \mathrm{I}_k \otimes \mathbf{1}_n' \end{bmatrix} \boldsymbol{y} \\ &= \begin{bmatrix} nk & n\mathbf{1}_k' \\ n\mathbf{1}_k & n\mathrm{I}_k \end{bmatrix}^- \begin{bmatrix} \mathbf{1}_k' \otimes \mathbf{1}_n' \\ \mathrm{I}_k \otimes \mathbf{1}_n' \end{bmatrix} \boldsymbol{y}\end{aligned}$$

$$= \begin{bmatrix} (nk)^{-1} & \mathbf{0}' \\ \mathbf{0} & n^{-1}(\mathrm{I}_k - k^{-1}\mathbf{1}_k\mathbf{1}_k') \end{bmatrix} \begin{bmatrix} \mathbf{1}_k' \otimes \mathbf{1}_n' \\ \mathrm{I}_k \otimes \mathbf{1}_n' \end{bmatrix} \boldsymbol{y}$$

$$= \begin{bmatrix} (nk)^{-1}(\mathbf{1}_k' \otimes \mathbf{1}_n') \\ n^{-1}(\mathrm{I}_k \otimes \mathbf{1}_n') - (nk)^{-1}(\mathbf{1}_k\mathbf{1}_k' \otimes \mathbf{1}_n') \end{bmatrix} \boldsymbol{y}$$

This yields $\hat{\mu} = \bar{y}$ and $\hat{\tau}_i = \bar{y}_i - \bar{y}$, where

$$\bar{y} = \frac{1}{nk} \sum_{i=1}^{k} \sum_{j=1}^{n} y_{ij}, \qquad \bar{y}_i = \frac{1}{n} \sum_{j=1}^{n} y_{ij}$$

Note that this solution is not unique since X is not full rank, and hence the solution depends on the choice of the generalized inverse of $X'X$. However, for each i, $\mu + \tau_i$ is estimable and its estimate is given by $\hat{\mu} + \hat{\tau}_i = \bar{y}_i$. In addition, the sum of squared errors for the model is always unique and is given by

$$(\boldsymbol{y} - X\hat{\boldsymbol{\beta}})'(\boldsymbol{y} - X\hat{\boldsymbol{\beta}}) = \boldsymbol{y}'(\mathrm{I}_{nk} - X(X'X)^{-}X')\boldsymbol{y} = \boldsymbol{y}'(\mathrm{I}_{nk} - n^{-1}(\mathrm{I}_k \otimes \mathbf{1}_n\mathbf{1}_n'))\boldsymbol{y}$$
$$= \sum_{i=1}^{k} \boldsymbol{y}_i'(\mathrm{I}_n - n^{-1}\mathbf{1}_n\mathbf{1}_n')\boldsymbol{y}_i = \sum_{i=1}^{k} \sum_{j=1}^{n} (y_{ij} - \bar{y}_i)^2$$

Since $\{(\mathbf{1}_k \otimes \mathbf{1}_n)'(\mathbf{1}_k \otimes \mathbf{1}_n)\}^{-1}(\mathbf{1}_k \otimes \mathbf{1}_n)'\boldsymbol{y} = \bar{y}$, the reduced model

$$y_{ij} = \mu + \epsilon_{ij}$$

has the least squares estimate $\hat{\mu} = \bar{y}$, while its sum of squared errors is

$$\{\boldsymbol{y} - \bar{y}(\mathbf{1}_k \otimes \mathbf{1}_n)\}'\{\boldsymbol{y} - \bar{y}(\mathbf{1}_k \otimes \mathbf{1}_n)\} = \sum_{i=1}^{k} \sum_{j=1}^{n} (y_{ij} - \bar{y})^2$$

The difference in the sums of squared errors for these two models, the so-called sum of squares for treatments (SST), is then

$$\mathrm{SST} = \sum_{i=1}^{k} \sum_{j=1}^{n} (y_{ij} - \bar{y})^2 - \sum_{i=1}^{k} \sum_{j=1}^{n} (y_{ij} - \bar{y}_i)^2$$
$$= \sum_{i=1}^{k} n(\bar{y}_i - \bar{y})^2$$

Example 7.5. In this example, we will illustrate some of the computations involved in the analysis of the two-way classification model with interaction, which is of the form

$$y_{ijk} = \mu + \tau_i + \gamma_j + \eta_{ij} + \epsilon_{ijk},$$

where $i = 1, \dots, a$, $j = 1, \dots, b$, and $k = 1, \dots, n$ (see Problem 6.17). Here μ can be described as an overall effect, while τ_i is an effect due to the ith level of factor A, γ_j is an effect due to the jth level of factor B, and η_{ij} is an effect due to the interaction of the ith and jth levels of factors A and B. If we define the parameter vector, $\boldsymbol{\beta} = (\mu, \tau_1, \dots, \tau_a, \gamma_1, \dots, \gamma_b, \eta_{11}, \eta_{12}, \dots, \eta_{ab-1}, \eta_{ab})'$ and the response vector, $\boldsymbol{y} = (y_{111}, \dots, y_{11n}, y_{121}, \dots, y_{1bn}, y_{211}, \dots, y_{abn})'$, then the model above can be written as

$$\boldsymbol{y} = X\boldsymbol{\beta} + \boldsymbol{\epsilon},$$

where

$$X = (\mathbf{1}_a \otimes \mathbf{1}_b \otimes \mathbf{1}_n, \mathrm{I}_a \otimes \mathbf{1}_b \otimes \mathbf{1}_n, \mathbf{1}_a \otimes \mathrm{I}_b \otimes \mathbf{1}_n, \mathrm{I}_a \otimes \mathrm{I}_b \otimes \mathbf{1}_n).$$

Now it is easily verified that the matrix

$$X'X = \begin{bmatrix} abn & bn\mathbf{1}_a' & an\mathbf{1}_b' & n\mathbf{1}_a' \otimes \mathbf{1}_b' \\ bn\mathbf{1}_a & bn\mathrm{I}_a & n\mathbf{1}_a \otimes \mathbf{1}_b' & n\mathrm{I}_a \otimes \mathbf{1}_b' \\ an\mathbf{1}_b & n\mathbf{1}_a' \otimes \mathbf{1}_b & an\mathrm{I}_b & n\mathbf{1}_a' \otimes \mathrm{I}_b \\ n\mathbf{1}_a \otimes \mathbf{1}_b & n\mathrm{I}_a \otimes \mathbf{1}_b & n\mathbf{1}_a \otimes \mathrm{I}_b & n\mathrm{I}_a \otimes \mathrm{I}_b \end{bmatrix}$$

has as a generalized inverse the matrix

$$\mathrm{diag}((abn)^{-1}, (bn)^{-1}(\mathrm{I}_a - a^{-1}\mathbf{1}_a\mathbf{1}_a'), (an)^{-1}(\mathrm{I}_b - b^{-1}\mathbf{1}_b\mathbf{1}_b'), C),$$

where

$$C = n^{-1}\mathrm{I}_a \otimes \mathrm{I}_b - (bn)^{-1}\mathrm{I}_a \otimes \mathbf{1}_b\mathbf{1}_b' - (an)^{-1}\mathbf{1}_a\mathbf{1}_a' \otimes \mathrm{I}_b \\ + (abn)^{-1}\mathbf{1}_a\mathbf{1}_a' \otimes \mathbf{1}_b\mathbf{1}_b'$$

Using this generalized inverse, we find that a least squares solution for $\boldsymbol{\beta}$ is given by

$$\hat{\boldsymbol{\beta}} = (X'X)^{-}X'\boldsymbol{y} = \begin{bmatrix} \bar{y}_{\cdot\cdot} \\ \bar{y}_{1\cdot} - \bar{y}_{\cdot\cdot} \\ \vdots \\ \bar{y}_{a\cdot} - \bar{y}_{\cdot\cdot} \\ \bar{y}_{\cdot 1} - \bar{y}_{\cdot\cdot} \\ \vdots \\ \bar{y}_{\cdot b} - \bar{y}_{\cdot\cdot} \\ \bar{y}_{11} - \bar{y}_{1\cdot} - \bar{y}_{\cdot 1} + \bar{y}_{\cdot\cdot} \\ \vdots \\ \bar{y}_{ab} - \bar{y}_{a\cdot} - \bar{y}_{\cdot b} + \bar{y}_{\cdot\cdot} \end{bmatrix},$$

where

$$\bar{y}_{\cdot\cdot} = (abn)^{-1} \sum_{i=1}^{a} \sum_{j=1}^{b} \sum_{k=1}^{n} y_{ijk}, \qquad \bar{y}_{i\cdot} = (bn)^{-1} \sum_{j=1}^{b} \sum_{k=1}^{n} y_{ijk}$$

$$\bar{y}_{\cdot j} = (an)^{-1} \sum_{i=1}^{a} \sum_{k=1}^{n} y_{ijk}, \qquad \bar{y}_{ij} = n^{-1} \sum_{k=1}^{n} y_{ijk}$$

Clearly, $\mu + \tau_i + \gamma_j + \eta_{ij}$ is estimable, and its estimate, which is the fitted value for y_{ijk}, is $\hat{\mu} + \hat{\tau}_i + \hat{\gamma}_j + \hat{\eta}_{ij} = \bar{y}_{ij}$. We will leave the computation of some of the sums of squares associated with the analysis of this model for the reader as an exercise.

4. THE DIRECT SUM

The direct sum is a matrix operator that transforms several square matrices into one block diagonal matrix with these matrices appearing as the submatrices along the diagonal. Recall that a block diagonal matrix is of the form

$$\operatorname{diag}(A_1, \ldots, A_r) = \begin{bmatrix} A_1 & (0) & \cdots & (0) \\ (0) & A_2 & \cdots & (0) \\ \vdots & \vdots & & \vdots \\ (0) & (0) & \cdots & A_r \end{bmatrix},$$

where A_i is an $m_i \times m_i$ matrix. This block diagonal matrix is said to be the direct sum of the matrices $A_1, \ldots, A_r$ and is sometimes written as

$$\text{diag}(A_1, \dots, A_r) = A_1 \oplus \cdots \oplus A_r$$

Clearly, the commutative property does not hold for the direct sum since, for instance,

$$A_1 \oplus A_2 = \begin{bmatrix} A_1 & (0) \\ (0) & A_2 \end{bmatrix} \neq \begin{bmatrix} A_2 & (0) \\ (0) & A_1 \end{bmatrix} = A_2 \oplus A_1,$$

unless $A_1 = A_2$. Direct sums of a matrix with itself can be expressed as Kronecker products; that is, if $A_1 = \cdots = A_r = A$, then

$$A_1 \oplus \cdots \oplus A_r = A \oplus \cdots \oplus A = \mathrm{I}_r \otimes A$$

Some of the basic properties of the direct sum are summarized in the following theorem. The proofs, which are fairly straightforward, are left to the reader.

Theorem 7.13. Let $A_1, \dots, A_r$ be matrices, where A_i is $m_i \times m_i$. Then

(a) $\text{tr}(A_1 \oplus \cdots \oplus A_r) = \text{tr}(A_1) + \cdots + \text{tr}(A_r)$,
(b) $|A_1 \oplus \cdots \oplus A_r| = |A_1| \cdots |A_r|$,
(c) if each A_i is nonsingular, $A = A_1 \oplus \cdots \oplus A_r$ is also nonsingular and $A^{-1} = A_1^{-1} \oplus \cdots \oplus A_r^{-1}$,
(d) $\text{rank}(A_1 \oplus \cdots \oplus A_r) = \text{rank}(A_1) + \cdots + \text{rank}(A_r)$,
(e) if the eigenvalues of A_i are denoted by $\lambda_{i,1}, \dots, \lambda_{i,m_i}$, the eigenvalues of $A_1 \oplus \cdots \oplus A_r$ are given by $\{\lambda_{i,j}: i = 1, \dots, r; j = 1, \dots, m_i\}$.

5. THE VEC OPERATOR

There are situations in which it is useful to transform a matrix to a vector that has as its elements the elements of the matrix. One such situation in statistics involves the study of the distribution of the sample covariance matrix S. It is usually more convenient mathematically in distribution theory to express density functions and moments of jointly distributed random variables in terms of the vector with these random variables as its components. Thus, the distribution of the random matrix S is usually given in terms of the vector formed by stacking columns of S, one underneath the other.

The operator that transforms a matrix to a vector is known as the vec operator. If the $m \times n$ matrix A has $\boldsymbol{a}_i$ as its ith column, then $\text{vec}(A)$ is the $mn \times 1$ vector given by

$$\text{vec}(A) = \begin{bmatrix} \boldsymbol{a}_1 \\ \boldsymbol{a}_2 \\ \vdots \\ \boldsymbol{a}_n \end{bmatrix}$$

Example 7.6. If A is the 2×3 matrix given by

$$A = \begin{bmatrix} 2 & 0 & 5 \\ 8 & 1 & 3 \end{bmatrix},$$

then $\text{vec}(A)$ is the 6×1 vector given by

$$\text{vec}(A) = \begin{bmatrix} 2 \\ 8 \\ 0 \\ 1 \\ 5 \\ 3 \end{bmatrix}$$

In this section, we develop some of the basic algebra associated with this operator. For instance, if $\boldsymbol{a}$ is $m \times 1$ and $\boldsymbol{b}$ is $n \times 1$, then $\boldsymbol{ab}'$ is $m \times n$ and

$$\text{vec}(\boldsymbol{ab}') = \text{vec}([b_1\boldsymbol{a}, b_2\boldsymbol{a}, \ldots, b_n\boldsymbol{a}]) = \begin{bmatrix} b_1\boldsymbol{a} \\ b_2\boldsymbol{a} \\ \vdots \\ b_n\boldsymbol{a} \end{bmatrix} = \boldsymbol{b} \otimes \boldsymbol{a}$$

Our first theorem gives this result and some others that follow directly from the definition of the vec operator.

Theorem 7.14. Let $\boldsymbol{a}$ and $\boldsymbol{b}$ be any two vectors, while A and B are two matrices of the same size. Then

(a) $\text{vec}(\boldsymbol{a}) = \text{vec}(\boldsymbol{a}') = \boldsymbol{a}$,
(b) $\text{vec}(\boldsymbol{ab}') = \boldsymbol{b} \otimes \boldsymbol{a}$,
(c) $\text{vec}(\alpha A + \beta B) = \alpha\, \text{vec}(A) + \beta\, \text{vec}(B)$, where α and β are scalars.

The trace of a product of two matrices can be expressed in terms of the vecs of those two matrices. This result is given next.

Theorem 7.15. Let A and B both be $m \times n$ matrices. Then

$$\operatorname{tr}(A'B) = \{\operatorname{vec}(A)\}' \operatorname{vec}(B)$$

Proof. As usual, let $\boldsymbol{a}_1, \ldots, \boldsymbol{a}_n$ denote the columns of A and $\boldsymbol{b}_1, \ldots, \boldsymbol{b}_n$ denote the columns of B. Then

$$\operatorname{tr}(A'B) = \sum_{i=1}^{n} (A'B)_{ii} = \sum_{i=1}^{n} \boldsymbol{a}_i'\boldsymbol{b}_i = [\boldsymbol{a}_1', \ldots, \boldsymbol{a}_n'] \begin{bmatrix} \boldsymbol{b}_1 \\ \vdots \\ \boldsymbol{b}_n \end{bmatrix}$$
$$= \{\operatorname{vec}(A)\}' \operatorname{vec}(B)$$
□

A generalization of Theorem 7.14(b) to the situation involving the vec of the product of three matrices is our next result.

Theorem 7.16. Let A, B, and C be matrices of sizes $m \times n$, $n \times p$, and $p \times q$, respectively. Then

$$\operatorname{vec}(ABC) = (C' \otimes A)\operatorname{vec}(B)$$

Proof. Note that if $\boldsymbol{b}_1, \ldots, \boldsymbol{b}_p$ are the columns of B, then B can be written as

$$B = \sum_{i=1}^{p} \boldsymbol{b}_i \boldsymbol{e}_i',$$

where $\boldsymbol{e}_i$ is the ith column of I_p. Thus,

$$\operatorname{vec}(ABC) = \operatorname{vec}\left\{A\left(\sum_{i=1}^{p} \boldsymbol{b}_i \boldsymbol{e}_i'\right)C\right\} = \sum_{i=1}^{p} \operatorname{vec}(A\boldsymbol{b}_i \boldsymbol{e}_i' C)$$
$$= \sum_{i=1}^{p} \operatorname{vec}\{(A\boldsymbol{b}_i)(C'\boldsymbol{e}_i)'\} = \sum_{i=1}^{p} C'\boldsymbol{e}_i \otimes A\boldsymbol{b}_i$$
$$= (C' \otimes A) \sum_{i=1}^{p} (\boldsymbol{e}_i \otimes \boldsymbol{b}_i),$$

where the second last equality follows from Theorem 7.14(b). The result now follows since, by again using Theorem 7.14(b), we find that

$$\sum_{i=1}^{p} (\boldsymbol{e}_i \otimes \boldsymbol{b}_i) = \sum_{i=1}^{p} \text{vec}(\boldsymbol{b}_i \boldsymbol{e}_i') = \text{vec}\left(\sum_{i=1}^{p} \boldsymbol{b}_i \boldsymbol{e}_i' \right) = \text{vec}(B) \qquad \square$$

Example 7.7. In Chapter 6, we discussed systems of linear equations of the form $A\boldsymbol{x} = \boldsymbol{c}$, as well as systems of equations of the form $AXB = C$. Using the vec operator and Theorem 7.16, this second system of equations can be equivalently expressed as

$$\text{vec}(AXB) = (B' \otimes A)\text{vec}(X) = \text{vec}(C);$$

that is, this is an equation of the form $A\boldsymbol{x} = \boldsymbol{c}$, where in place of A, $\boldsymbol{x}$, and $\boldsymbol{c}$, we have $(B' \otimes A)$, $\text{vec}(X)$, and $\text{vec}(C)$. As a result, Theorem 6.4, which gives the general form of a solution to $A\boldsymbol{x} = \boldsymbol{c}$, can be used to prove Theorem 6.5, which gives the general form of a solution to $AXB = C$. The details of this proof are left to the reader.

Theorem 7.15 also can be generalized to a result involving the product of more than two matrices.

Theorem 7.17. Let A, B, C, and D be matrices of sizes $m \times n$, $n \times p$, $p \times q$, and $q \times m$, respectively. Then

$$\text{tr}(ABCD) = \{\text{vec}(A')\}'(D' \otimes B)\text{vec}(C)$$

Proof. Using Theorem 7.15, it follows that

$$\text{tr}(ABCD) = \text{tr}\{A(BCD)\} = \{\text{vec}(A')\}' \,\text{vec}(BCD)$$

But from the previous theorem, we know that $\text{vec}(BCD) = (D' \otimes B)\text{vec}(C)$, and so the proof is complete. $\square$

The proofs of the following consequences of Theorem 7.17 are left to the reader as an exercise.

Corollary 7.17.1. Let A and C be matrices of sizes $m \times n$ and $n \times m$, respectively, while B and D are $n \times n$. Then

(a) $\text{tr}(ABC) = \{\text{vec}(A')\}'(\text{I}_m \otimes B)\text{vec}(C)$,
(b) $\text{tr}(AD'BDC) = \{\text{vec}(D)\}'(A'C' \otimes B)\text{vec}(D)$.

Other transformations of a matrix, A, to a vector may be useful when the matrix A has some special structure. One such transformation for an $m \times m$

matrix, denoted by v(A), is defined so as to produce the $m(m+1)/2 \times 1$ vector obtained from vec(A) by deleting from it all of the elements that are above the diagonal of A. Thus, if A is a lower triangular matrix, v(A) contains all of the elements of A except for the zeros in the upper triangular portion of A. Yet another transformation of the $m \times m$ matrix A to a vector will be denoted by $\tilde{\text{v}}(A)$ and yields the $m(m-1)/2 \times 1$ vector formed from v(A) by deleting from it all of the diagonal elements of A; that is, $\tilde{\text{v}}(A)$ is the vector obtained by stacking only the portion of the columns of A that are below its diagonal. If A is a skew-symmetric matrix, then A can be reconstructed from $\tilde{\text{v}}(A)$ since the diagonal elements of A must be zero, while $a_{ji} = -a_{ij}$ if $i \neq j$. The notation we use here, that is, v(A) and $\tilde{\text{v}}(A)$, corresponds to that used by Magnus (1988). Others [see, for example, Henderson and Searle (1979)] use the notation vech(A) and veck(A). In Section 8, we will discuss some transformations which relate the v and $\tilde{\text{v}}$ operators to the vec operator.

Example 7.8. The v and $\tilde{\text{v}}$ operators are particularly useful when dealing with covariance and correlation matrices. For instance, suppose that we are interested in the distribution of the sample covariance matrix or the distribution of the sample correlation matrix computed from a sample of observations on three different variables. The resulting sample covariance and correlation matrices would be of the form

$$S = \begin{bmatrix} s_{11} & s_{12} & s_{13} \\ s_{12} & s_{22} & s_{23} \\ s_{13} & s_{23} & s_{33} \end{bmatrix}, \qquad R = \begin{bmatrix} 1 & r_{12} & r_{13} \\ r_{12} & 1 & r_{23} \\ r_{13} & r_{23} & 1 \end{bmatrix},$$

so that

$$\text{vec}(S) = (s_{11}, s_{12}, s_{13}, s_{12}, s_{22}, s_{23}, s_{13}, s_{23}, s_{33})',$$
$$\text{vec}(R) = (1, r_{12}, r_{13}, r_{12}, 1, r_{23}, r_{13}, r_{23}, 1)'$$

Since both S and R are symmetric, there are redundant elements in vec(S) and vec(R). The elimination of these results in v(S) and v(R) given by

$$\text{v}(S) = (s_{11}, s_{12}, s_{13}, s_{22}, s_{23}, s_{33})',$$
$$\text{v}(R) = (1, r_{12}, r_{13}, 1, r_{23}, 1)'$$

Finally, by eliminating the nonrandom 1s from v(R), we obtain

$$\tilde{\text{v}}(R) = (r_{12}, r_{13}, r_{23})',$$

which contains all of the random variables in R.

6. THE HADAMARD PRODUCT

A matrix operator that is a little more obscure than our other matrix operators, but one which is finding increasing applications in statistics, is known as the Hadamard product. This operator, which we will denote by the symbol $\odot$, simply performs the elementwise multiplication of two matrices; that is, if A and B are each $m \times n$, then

$$A \odot B = \begin{bmatrix} a_{11}b_{11} & \cdots & a_{1n}b_{1n} \\ \vdots & & \vdots \\ a_{m1}b_{m1} & \cdots & a_{mn}b_{mn} \end{bmatrix}$$

Clearly, this operation is only defined if the two matrices involved are of the same size.

Example 7.9. If A and B are the 2×3 matrices given by

$$A = \begin{bmatrix} 1 & 4 & 2 \\ 0 & 2 & 3 \end{bmatrix}, \qquad B = \begin{bmatrix} 3 & 1 & 3 \\ 6 & 5 & 1 \end{bmatrix},$$

then

$$A \odot B = \begin{bmatrix} 3 & 4 & 6 \\ 0 & 10 & 3 \end{bmatrix}$$

One of the situations in which the Hadamard product finds application in statistics is in expressions for the covariance structure of certain functions of the sample covariance and sample correlation matrices. We will see examples of this later in Section 9.7. In this section, we will investigate some of the properties of this operator. For a more complete treatment, along with some other examples of applications of the operator in statistics, the reader is referred to Styan (1973) and Horn and Johnson (1991). We begin with some elementary properties that follow directly from the definition of the Hadamard product.

Theorem 7.18. Let A, B, and C be $m \times n$ matrices. Then

(a) $A \odot B = B \odot A$,

(b) $(A \odot B) \odot C = A \odot (B \odot C)$,

(c) $(A + B) \odot C = A \odot C + B \odot C$,

(d) $(A \odot B)' = A' \odot B'$,

(e) $A \odot (0) = (0)$,

(f) $A \odot \mathbf{1}_m \mathbf{1}_n' = A$,

(g) $A \odot \mathrm{I}_m = D_A = \mathrm{diag}(a_{11}, \ldots, a_{mm})$, if $m = n$,

(h) $C(A \odot B) = (CA) \odot B = A \odot (CB)$ and $(A \odot B)C = (AC) \odot B = A \odot (BC)$, if $m = n$ and C is diagonal,

(i) $\boldsymbol{ab}' \odot \boldsymbol{cd}' = (\boldsymbol{a} \odot \boldsymbol{c})(\boldsymbol{b} \odot \boldsymbol{d})'$, where $\boldsymbol{a}$ and $\boldsymbol{c}$ are $m \times 1$ vectors and $\boldsymbol{b}$ and $\boldsymbol{d}$ are $n \times 1$ vectors.

We will now show how $A \odot B$ is related to the Kronecker product $A \otimes B$; specifically, $A \odot B$ is a submatrix of $A \otimes B$. To see this, define the $m \times m^2$ matrix Ψ_m as

$$\Psi_m = \sum_{i=1}^{m} \boldsymbol{e}_{i,m}(\boldsymbol{e}_{i,m} \otimes \boldsymbol{e}_{i,m})',$$

where $\boldsymbol{e}_{i,m}$ is the ith column of the identity matrix I_m. Note that if A and B are $m \times n$, then $\Psi_m(A \otimes B)\Psi_n'$ forms the $m \times n$ submatrix of the $m^2 \times n^2$ matrix $A \otimes B$, containing rows $1, m+2, 2m+3, \ldots, m^2$ and columns $1, n+2, 2n+3, \ldots, n^2$. Taking a closer look at this submatrix, we find that

$$\begin{aligned}
\Psi_m(A \otimes B)\Psi_n' &= \sum_{i=1}^{m} \sum_{j=1}^{n} \boldsymbol{e}_{i,m}(\boldsymbol{e}_{i,m} \otimes \boldsymbol{e}_{i,m})'(A \otimes B)(\boldsymbol{e}_{j,n} \otimes \boldsymbol{e}_{j,n})\boldsymbol{e}_{j,n}' \\
&= \sum_{i=1}^{m} \sum_{j=1}^{n} \boldsymbol{e}_{i,m}(\boldsymbol{e}_{i,m}' A \boldsymbol{e}_{j,n} \otimes \boldsymbol{e}_{i,m}' B \boldsymbol{e}_{j,n})\boldsymbol{e}_{j,n}' \\
&= \sum_{i=1}^{m} \sum_{j=1}^{n} a_{ij} b_{ij} \boldsymbol{e}_{i,m}\boldsymbol{e}_{j,n}' = A \odot B
\end{aligned}$$

Although the rank of $A \odot B$ is not determined, in general, by the rank of A and the rank of B, we do have the following bound.

Theorem 7.19. Let A and B be $m \times n$ matrices. Then

$$\mathrm{rank}(A \odot B) \le \mathrm{rank}(A)\,\mathrm{rank}(B)$$

Proof. Let $r_A = \mathrm{rank}(A)$ and $r_B = \mathrm{rank}(B)$. It follows from the singular value decomposition theorem (Theorem 4.1 and Corollary 4.1.1) that there exist $m \times r_A$ and $n \times r_A$ matrices $U = (\boldsymbol{u}_1, \ldots, \boldsymbol{u}_{r_A})$ and $V = (\boldsymbol{v}_1, \ldots, \boldsymbol{v}_{r_A})$, and $m \times r_B$ and $n \times r_B$ matrices $W = (\boldsymbol{w}_1, \ldots, \boldsymbol{w}_{r_B})$ and $X = (\boldsymbol{x}_1, \ldots, \boldsymbol{x}_{r_B})$, such that $A = UV'$ and $B = WX'$. Then

$$A \odot B = UV' \odot WX' = \left(\sum_{i=1}^{r_A} \boldsymbol{u}_i \boldsymbol{v}_i' \right) \odot \left(\sum_{j=1}^{r_B} \boldsymbol{w}_j \boldsymbol{x}_j' \right)$$

$$= \sum_{i=1}^{r_A} \sum_{j=1}^{r_B} (\boldsymbol{u}_i \boldsymbol{v}_i' \odot \boldsymbol{w}_j \boldsymbol{x}_j') = \sum_{i=1}^{r_A} \sum_{j=1}^{r_B} (\boldsymbol{u}_i \odot \boldsymbol{w}_j)(\boldsymbol{v}_i \odot \boldsymbol{x}_j)',$$

where we have used Theorem 7.18(c) and (i). The result follows since we have now expressed $A \odot B$ as the sum of $r_A r_B$ matrices, each having rank of at most one. □

Example 7.10. While Theorem 7.19 gives an upper bound for rank($A \odot B$) in terms of rank(A) and rank(B), there is no corresponding lower bound. In other words, it is possible that both A and B have full rank while $A \odot B$ has rank equal to 0. For instance, each of the matrices

$$A = \begin{bmatrix} 0 & 1 & 0 \\ 1 & 0 & 0 \\ 0 & 0 & 1 \end{bmatrix}, \qquad B = \begin{bmatrix} 1 & 0 & 0 \\ 0 & 0 & 1 \\ 0 & 1 & 0 \end{bmatrix}$$

clearly has rank 3, yet $A \odot B$ has rank 0 since $A \odot B = (0)$.

The following result shows that a bilinear form in a Hadamard product of two matrices may be written as a trace.

Theorem 7.20. Let A and B be $m \times n$ matrices, and let $\boldsymbol{x}$ and $\boldsymbol{y}$ be $m \times 1$ and $n \times 1$ vectors, respectively. Then

(a) $\mathbf{1}_m'(A \odot B)\mathbf{1}_n = \text{tr}(AB')$
(b) $\boldsymbol{x}'(A \odot B)\boldsymbol{y} = \text{tr}(D_{\boldsymbol{x}} A D_{\boldsymbol{y}} B')$,

where $D_{\boldsymbol{x}} = \text{diag}(x_1, \ldots, x_m)$ and similarly for $D_{\boldsymbol{y}}$.

Proof. (a) follows since

$$\mathbf{1}_m'(A \odot B)\mathbf{1}_n = \sum_{i=1}^{m} \sum_{j=1}^{n} (A \odot B)_{ij} = \sum_{i=1}^{m} \sum_{j=1}^{n} a_{ij} b_{ij}$$

$$= \sum_{i=1}^{m} (A)_{i\cdot}(B')_{\cdot i} = \sum_{i=1}^{m} (AB')_{ii} = \text{tr}(AB')$$

To prove (b), note that $\boldsymbol{x} = D_x \mathbf{1}_m$ and $\boldsymbol{y} = D_y \mathbf{1}_n$, so that by using Theorem 7.20(a) and Theorem 7.18(h), we find that

$$\boldsymbol{x}'(A \odot B)\boldsymbol{y} = \mathbf{1}_m' D_x (A \odot B) D_y \mathbf{1}_n = \mathbf{1}_m' (D_x A \odot B D_y) \mathbf{1}_n = \operatorname{tr}(D_x A D_y B') \qquad \square$$

The following result can be helpful in determining whether the Hadamard product of two symmetric matrices is nonnegative definite or positive definite.

Theorem 7.21. Let A and B each be an $m \times m$ symmetric matrix. Then

(a) $A \odot B$ is nonnegative definite if A and B are nonnegative definite,

(b) $A \odot B$ is positive definite if A and B are positive definite.

Proof. Clearly, if A and B are symmetric, then so also is $A \odot B$. Let $B = X \Lambda X'$ be the spectral decomposition of B so that $b_{ij} = \sum \lambda_k x_{ik} x_{jk}$, where $\lambda_k \geq 0$ for all k since B is nonnegative definite. Then we find that for any $m \times 1$ vector $\boldsymbol{y}$,

$$\begin{aligned} \boldsymbol{y}'(A \odot B)\boldsymbol{y} &= \sum_{i=1}^{m} \sum_{j=1}^{m} a_{ij} b_{ij} y_i y_j = \sum_{k=1}^{m} \left(\sum_{i=1}^{m} \sum_{j=1}^{m} \lambda_k (y_i x_{ik}) a_{ij} (y_j x_{jk}) \right) \\ &= \sum_{k=1}^{m} \lambda_k (\boldsymbol{y} \odot \boldsymbol{x}_k)' A (\boldsymbol{y} \odot \boldsymbol{x}_k), \end{aligned} \tag{7.11}$$

where $\boldsymbol{x}_k$ represents the kth column of X. Since A is nonnegative definite, the sum in (7.11) must be nonnegative, and so $A \odot B$ is also nonnegative definite. This proves (a). Now if A is positive definite, then (7.11) will be positive for any $\boldsymbol{y} \neq \mathbf{0}$ that satisfies $\boldsymbol{y} \odot \boldsymbol{x}_k \neq \mathbf{0}$ for at least one k for which $\lambda_k > 0$. But if B is also positive definite, then $\lambda_k > 0$ for all k and if $\boldsymbol{y}$ has its hth component $y_h \neq 0$, then $\boldsymbol{y} \odot \boldsymbol{x}_k = \mathbf{0}$ for all k only if the hth row of X has all zeros. This is not possible since X is nonsingular. Consequently, there is no $\boldsymbol{y} \neq \mathbf{0}$ for which (7.11) equals zero and so (b) follows. $\square$

Theorem 7.21(b) gives a sufficient condition for the matrix $A \odot B$ to be positive definite. The following example demonstrates that this condition is not necessary.

Example 7.11. Consider the 2×2 matrices

$$A = \begin{bmatrix} 1 & 1 \\ 1 & 1 \end{bmatrix}, \qquad B = \begin{bmatrix} 4 & 2 \\ 2 & 2 \end{bmatrix}$$

The matrix B is positive definite since, for instance, $B = VV'$, where

$$V = \begin{bmatrix} 2 & 0 \\ 1 & 1 \end{bmatrix}$$

and $\text{rank}(V) = 2$. Clearly, $A \odot B$ is also positive definite since $A \odot B = B$. However, A is not positive definite since $\text{rank}(A) = 1$.

A sufficient condition for the positive definiteness of $A \odot B$, weaker than that given in Theorem 7.21(b), is given in our next theorem.

Theorem 7.22. Let A and B each be an $m \times m$ symmetric matrix. If B is positive definite and A is nonnegative definite with positive diagonal elements, then $A \odot B$ is positive definite.

Proof. We need to show that for any $\boldsymbol{x} \neq \mathbf{0}, \boldsymbol{x}'(A \odot B)\boldsymbol{x} > 0$. Since B is positive definite, there exists a nonsingular matrix T such that $B = TT'$. It follows then from Theorem 7.20(b) that

$$\boldsymbol{x}'(A \odot B)\boldsymbol{x} = \text{tr}(D_x A D_x B') = \text{tr}(D_x A D_x TT') = \text{tr}(T' D_x A D_x T) \tag{7.12}$$

Since A is nonnegative definite, so is $D_x A D_x$. In addition, if $\boldsymbol{x} \neq \mathbf{0}$, and A has no diagonal elements equal to zero, then $D_x A D_x \neq (0)$; that is, $D_x A D_x$ has rank of at least one, and so it has at least one positive eigenvalue. Since T is nonsingular, $\text{rank}(D_x A D_x) = \text{rank}(T' D_x A D_x T)$, and so $T' D_x A D_x T$ is also nonnegative definite with at least one positive eigenvalue. The result now follows since (7.12) implies that $\boldsymbol{x}'(A \odot B)\boldsymbol{x}$ is the sum of the eigenvalues of $T' D_x A D_x T$. □

The following result, which gives a relationship between the determinant of a positive definite matrix and its diagonal elements, is commonly known as the Hadamard inequality.

Theorem 7.23. If A is an $m \times m$ positive definite matrix, then

$$|A| \leq \prod_{i=1}^{m} a_{ii},$$

with equality if and only if A is a diagonal matrix.

Proof. Our proof is by induction. If $m = 2$, then

$$|A| = a_{11}a_{22} - a_{12}^2 \leq a_{11}a_{22},$$

with equality if and only if $a_{12} = 0$, and so the result clearly holds when $m = 2$. For general m, use the cofactor expansion formula for the determinant of A to obtain

$$|A| = a_{11} \begin{vmatrix} a_{22} & a_{23} & \cdots & a_{2m} \\ a_{32} & a_{33} & \cdots & a_{3m} \\ \vdots & \vdots & & \vdots \\ a_{m2} & a_{m3} & \cdots & a_{mm} \end{vmatrix} + \begin{vmatrix} 0 & a_{12} & \cdots & a_{1m} \\ a_{21} & a_{22} & \cdots & a_{2m} \\ \vdots & \vdots & & \vdots \\ a_{m1} & a_{m2} & \cdots & a_{mm} \end{vmatrix}$$

$$= a_{11}|A_1| + \begin{vmatrix} 0 & \boldsymbol{a}' \\ \boldsymbol{a} & A_1 \end{vmatrix}, \tag{7.13}$$

where A_1 is the $(m - 1) \times (m - 1)$ submatrix of A formed by deleting the first row and column of A and $\boldsymbol{a}' = (a_{12}, \ldots, a_{1m})$. Since A is positive definite, A_1 also must be positive definite. Consequently, we can use Theorem 7.4(a) to simplify the second term in the right-hand side of (7.13), leading to the equation

$$|A| = a_{11}|A_1| - \boldsymbol{a}'A_1^{-1}\boldsymbol{a}|A_1|$$

Since A_1 and A_1^{-1} are positive definite, it follows that

$$|A| \leq a_{11}|A_1|,$$

with equality if and only if $\boldsymbol{a} = \boldsymbol{0}$. Thus, the result holds for the $m \times m$ matrix A if the result holds for the $(m - 1) \times (m - 1)$ matrix A_1, and so our induction proof is complete. □

Corollary 7.23.1. Let B be an $m \times m$ nonsingular matrix. Then

$$|B|^2 \leq \prod_{i=1}^{m} \left(\sum_{j=1}^{m} b_{ij}^2 \right),$$

with equality if and only if the rows of B are orthogonal.

Proof. Since B is nonsingular, the matrix $A = BB'$ is positive definite. Note that

$$|A| = |BB'| = |B||B'| = |B|^2$$

and

$$a_{ii} = (BB')_{ii} = (B)_{i\cdot}(B')_{\cdot i} = (B)_{i\cdot}(B)'_{i\cdot} = \sum_{j=1}^{m} b_{ij}^2,$$

and so the result follows immediately from Theorem 7.23. □

Theorem 7.23 also holds for positive semidefinite matrices except that in this case A need not be diagonal for equality since one or more of its diagonal elements may equal zero. Likewise, Corollary 7.23.1 holds for singular matrices except for the statement concerning equality.

Hadamard's inequality given in Theorem 7.23 can be expressed, using the Hadamard product, as

$$|A|\left(\prod_{i=1}^{m} 1\right) \leq |A \odot \mathrm{I}_m|, \tag{7.14}$$

where the term $(\prod 1)$ corresponds to the product of the diagonal elements of I_m. Theorem 7.25 will show that the inequality (7.14) holds for other matrices besides the identity. But first we will need the following result.

Theorem 7.24. Let A be an $m \times m$ positive definite matrix and define

$$A_\alpha = A - \alpha \boldsymbol{e}_1 \boldsymbol{e}_1',$$

where $\alpha = |A|/|A_1|$ and A_1 is the $(m-1) \times (m-1)$ submatrix of A formed by deleting its first row and column. Then A_α is nonnegative definite.

Proof. Let A be partitioned as

$$A = \begin{bmatrix} a_{11} & \boldsymbol{a}' \\ \boldsymbol{a} & A_1 \end{bmatrix},$$

and note that since A is positive definite, so is A_1. Thus, using Theorem 7.4, we find that

$$|A| = \begin{vmatrix} a_{11} & \boldsymbol{a}' \\ \boldsymbol{a} & A_1 \end{vmatrix} = |A_1|(a_{11} - \boldsymbol{a}'A_1^{-1}\boldsymbol{a}),$$

and so $\alpha = |A|/|A_1| = (a_{11} - \boldsymbol{a}'A_1^{-1}\boldsymbol{a})$. Consequently, A_α may be written as

$$A_\alpha = \begin{bmatrix} a_{11} & \boldsymbol{a}' \\ \boldsymbol{a} & A_1 \end{bmatrix} - \begin{bmatrix} (a_{11} - \boldsymbol{a}'A_1^{-1}\boldsymbol{a}) & \boldsymbol{0}' \\ \boldsymbol{0} & (0) \end{bmatrix}$$

$$= \begin{bmatrix} \boldsymbol{a}'A_1^{-1}\boldsymbol{a} & \boldsymbol{a}' \\ \boldsymbol{a} & A_1 \end{bmatrix} = \begin{bmatrix} \boldsymbol{a}'A_1^{-1} \\ \mathrm{I}_{m-1} \end{bmatrix} A_1[A_1^{-1}\boldsymbol{a} \quad \mathrm{I}_{m-1}]$$

Since A_1 is positive definite, there exists an $(m-1)\times(m-1)$ matrix T such that $A_1 = TT'$. If we let $U' = T'[A_1^{-1}\boldsymbol{a} \quad \mathrm{I}_{m-1}]$, then $A_\alpha = UU'$, and so A_α is nonnegative definite. □

Theorem 7.25. Let A and B be $m\times m$ nonnegative definite matrices. Then

$$|A| \prod_{i=1}^{m} b_{ii} \leq |A \odot B|$$

Proof. The result follows immediately if A is singular since $|A| = 0$, while $|A\odot B| \geq 0$ is guaranteed by Theorem 7.21. For the case in which A is positive definite, we will prove the result by induction. The result holds when $m = 2$, since in this case

$$|A \odot B| = \begin{vmatrix} a_{11}b_{11} & a_{12}b_{12} \\ a_{12}b_{12} & a_{22}b_{22} \end{vmatrix} = a_{11}a_{22}b_{11}b_{22} - (a_{12}b_{12})^2$$
$$= (a_{11}a_{22} - a_{12}^2)b_{11}b_{22} + a_{12}^2(b_{11}b_{22} - b_{12}^2)$$
$$= |A|b_{11}b_{22} + a_{12}^2|B| \geq |A|b_{11}b_{22}$$

To prove the result for general m, assume that it holds for $m-1$, so that

$$|A_1| \prod_{i=2}^{m} b_{ii} \leq |A_1 \odot B_1|, \tag{7.15}$$

where A_1 and B_1 are the submatrices of A and B formed by deleting their first row and first column. From Theorem 7.24 we know that $(A - \alpha\boldsymbol{e}_1\boldsymbol{e}_1')$ is nonnegative definite, where $\alpha = |A|/|A_1|$. Thus, by using Theorem 7.21(a), Theorem 7.18(c), and the expansion formula for determinants, we find that

$$0 \leq |(A - \alpha\boldsymbol{e}_1\boldsymbol{e}_1')\odot B| = |A\odot B - \alpha\boldsymbol{e}_1\boldsymbol{e}_1'\odot B| = |A\odot B - \alpha b_{11}\boldsymbol{e}_1\boldsymbol{e}_1'|$$
$$= |A\odot B| - \alpha b_{11}|(A\odot B)_1|,$$

where $(A\odot B)_1$ denotes the $(m-1)\times(m-1)$ submatrix of $A\odot B$ formed by deleting its first row and column. But $(A \odot B)_1 = A_1 \odot B_1$ so that the inequality above, along with (7.15) and the identity $\alpha|A_1| = |A|$, implies that

$$|A \odot B| \geq \alpha b_{11}|A_1 \odot B_1| \geq \alpha b_{11}\left(|A_1| \prod_{i=2}^{m} b_{ii}\right) = |A| \prod_{i=1}^{m} b_{ii}$$

The proof is now complete. □

Our final results on Hadamard products involve their eigenvalues. First we obtain bounds for each eigenvalue of the matrix $A \odot B$ when A and B are symmetric.

Theorem 7.26. Let A and B be $m \times m$ symmetric matrices. If B is nonnegative definite, then the ith largest eigenvalue of $A \odot B$ satisfies

$$\lambda_m(A)\left\{\min_{1 \leq i \leq m} b_{ii}\right\} \leq \lambda_i(A \odot B) \leq \lambda_1(A)\left\{\max_{1 \leq i \leq m} b_{ii}\right\}$$

Proof. Since B is nonnegative definite there exists an $m \times m$ matrix T such that $B = TT'$. Let $\boldsymbol{t}_j$ be the jth column of T, while t_{ij} denotes the (i, j)th element of T. For any $m \times 1$ vector, $\boldsymbol{x} \neq \boldsymbol{0}$, we find that

$$\begin{aligned}
\boldsymbol{x}'(A \odot B)\boldsymbol{x} &= \sum_{i=1}^{m} \sum_{j=1}^{m} a_{ij} b_{ij} x_i x_j = \sum_{i=1}^{m} \sum_{j=1}^{m} a_{ij} \left(\sum_{h=1}^{m} t_{ih} t_{jh}\right) x_i x_j \\
&= \sum_{h=1}^{m} \left(\sum_{i=1}^{m} \sum_{j=1}^{m} (x_i t_{ih}) a_{ij} (x_j t_{jh})\right) = \sum_{h=1}^{m} (\boldsymbol{x} \odot \boldsymbol{t}_h)' A (\boldsymbol{x} \odot \boldsymbol{t}_h) \\
&\leq \lambda_1(A) \sum_{h=1}^{m} (\boldsymbol{x} \odot \boldsymbol{t}_h)'(\boldsymbol{x} \odot \boldsymbol{t}_h) = \lambda_1(A) \sum_{h=1}^{m} \sum_{j=1}^{m} x_j^2 t_{jh}^2 \\
&= \lambda_1(A) \sum_{j=1}^{m} x_j^2 \left(\sum_{h=1}^{m} t_{jh}^2\right) = \lambda_1(A) \sum_{j=1}^{m} x_j^2 b_{jj} \\
&\leq \lambda_1(A)\{\max_{1 \leq i \leq m} b_{ii}\}\boldsymbol{x}'\boldsymbol{x},
\end{aligned} \tag{7.16}$$

where the first inequality arises from the relation

$$\lambda_1(A) = \max_{y'y \neq 0} \frac{y'Ay}{y'y}$$

given in Theorem 3.15. Using this same relationship for $A \odot B$, along with (7.16), we find that for any i, $1 \leq i \leq m$,

$$\lambda_i(A \odot B) \leq \lambda_1(A \odot B) = \max_{x'x \neq 0} \frac{x'(A \odot B)x}{x'x} \leq \lambda_1(A)\left\{\max_{1 \leq i \leq m} b_{ii}\right\},$$

which is the required upper bound on $\lambda_i(A \odot B)$. The lower bound is obtained in a similar fashion by using the identity

$$\lambda_m(A) = \min_{y'y \neq 0} \frac{y'Ay}{y'y} \qquad \square$$

Our final result provides an alternative lower bound for the eigenvalues of $(A \odot B)$. The derivation of this bound will make use of the following result.

Theorem 7.27. Let A be an $m \times m$ positive definite matrix. Then the matrix $(A \odot A^{-1}) - \text{I}_m$ is nonnegative definite.

Proof. Let $\sum_{i=1}^m \lambda_i x_i x_i'$ be the spectral decomposition of A so that $A^{-1} = \sum_{i=1}^m \lambda_i^{-1} x_i x_i'$. Then

$$(A \odot A^{-1}) - \text{I}_m = (A \odot A^{-1}) - \text{I}_m \odot \text{I}_m$$

$$= \left(\sum_{i=1}^m \lambda_i x_i x_i' \odot \sum_{j=1}^m \lambda_j^{-1} x_j x_j'\right) - \left(\sum_{i=1}^m x_i x_i' \odot \sum_{j=1}^m x_j x_j'\right)$$

$$= \sum_{i=1}^m \sum_{j=1}^m (\lambda_i \lambda_j^{-1} - 1)(x_i x_i' \odot x_j x_j')$$

$$= \sum_{i \neq j} (\lambda_i \lambda_j^{-1} - 1)(x_i x_i' \odot x_j x_j')$$

$$= \sum_{i<j} (\lambda_i \lambda_j^{-1} + \lambda_j \lambda_i^{-1} - 2)(x_i \odot x_j)(x_i \odot x_j)' = XDX',$$

where X is the $m \times m(m-1)/2$ matrix having $(x_i \odot x_j)$, $i < j$ as its columns, while D is the diagonal matrix with its corresponding diagonal elements given by $(\lambda_i \lambda_j^{-1} + \lambda_j \lambda_i^{-1} - 2)$, $i < j$. Since A is positive definite, $\lambda_i > 0$ for all i, and so

$$(\lambda_i \lambda_j^{-1} + \lambda_j \lambda_i^{-1} - 2) = \lambda_i^{-1} \lambda_j^{-1} (\lambda_i - \lambda_j)^2 \geq 0$$

Thus, D is nonnegative definite and consequently so is XDX'. □

Theorem 7.28. Let A and B be $m \times m$ nonnegative definite matrices. Then

$$\lambda_m(A \odot B) \geq \lambda_m(AB)$$

Proof. Due to Theorem 7.21, $A \odot B$ is nonnegative definite, and so the inequality is obvious if either A or B is singular since in this case AB will have a zero eigenvalue. Suppose that A and B are positive definite, and let T be any matrix such that $TT' = B$. Note that $TAT' - \lambda_m(AB)\mathrm{I}_m$ is nonnegative definite since its ith largest eigenvalues is $\lambda_i(TAT') - \lambda_m(AB)$, and $\lambda_m(AB) = \lambda_m(TAT')$. As a result,

$$T'^{-1}(T'AT - \lambda_m(AB)\mathrm{I}_m)T^{-1} = A - \lambda_m(AB)B^{-1}$$

is also nonnegative definite. Thus, $(A - \lambda_m(AB)B^{-1}) \odot B$ is nonnegative definite due to Theorem 7.21, while $\lambda_m(AB)\{(B^{-1} \odot B) - \mathrm{I}_m\}$ is nonnegative definite due to Theorem 7.27, and so the sum of these two matrices, which is given by

$$\begin{aligned}
\{(A - \lambda_m(AB)B^{-1}) \odot B\} &+ \lambda_m(AB)\{(B^{-1} \odot B) - \mathrm{I}_m\} \\
&= A \odot B - \lambda_m(AB)(B^{-1} \odot B) + \lambda_m(AB)(B^{-1} \odot B) - \lambda_m(AB)\mathrm{I}_m \\
&= A \odot B - \lambda_m(AB)\mathrm{I}_m
\end{aligned}$$

is also nonnegative definite. Consequently, for any $\boldsymbol{x}$,

$$\boldsymbol{x}'(A \odot B)\boldsymbol{x} \geq \lambda_m(AB)\boldsymbol{x}'\boldsymbol{x},$$

and so the result follows from Theorem 3.15. □

7. THE COMMUTATION MATRIX

An $m \times m$ permutation matrix was defined in Section 1.10 to be any matrix that can be obtained from I_m by permuting its columns. In this section, we discuss a special class of permutation matrices, known as commutation matrices, which are very useful when computing the moments of the multivariate normal and related distributions. We will establish some of the basic properties of commutation matrices. A more complete treatment of this subject can be found in Magnus and Neudecker (1979) and Magnus (1988).

Definition 7.1. Let H_{ij} be the $m \times n$ matrix that has its only nonzero element, a one, in the (i,j)th position. Then the $mn \times mn$ commutation matrix, denoted by K_{mn}, is given by

$$K_{mn} = \sum_{i=1}^{m} \sum_{j=1}^{n} (H_{ij} \otimes H'_{ij}) \tag{7.17}$$

The matrix H_{ij} can be conveniently expressed in terms of columns from the identity matrices I_m and I_n. If $\boldsymbol{e}_{i,m}$ is the ith column of I_m and $\boldsymbol{e}_{j,n}$ is the jth column of I_n, then $H_{ij} = \boldsymbol{e}_{i,m}\boldsymbol{e}'_{j,n}$.

Note that, in general, there is more than one commutation matrix of order mn. For example, for $mn = 6$, we have the four commutation matrices, K_{16}, K_{23}, K_{32}, and K_{61}. Using (7.17), it is easy to verify that $K_{16} = K_{61} = \mathrm{I}_6$, while

$$K_{23} = \begin{bmatrix} 1 & 0 & 0 & 0 & 0 & 0 \\ 0 & 0 & 1 & 0 & 0 & 0 \\ 0 & 0 & 0 & 0 & 1 & 0 \\ 0 & 1 & 0 & 0 & 0 & 0 \\ 0 & 0 & 0 & 1 & 0 & 0 \\ 0 & 0 & 0 & 0 & 0 & 1 \end{bmatrix},$$

$$K_{32} = \begin{bmatrix} 1 & 0 & 0 & 0 & 0 & 0 \\ 0 & 0 & 0 & 1 & 0 & 0 \\ 0 & 1 & 0 & 0 & 0 & 0 \\ 0 & 0 & 0 & 0 & 1 & 0 \\ 0 & 0 & 1 & 0 & 0 & 0 \\ 0 & 0 & 0 & 0 & 0 & 1 \end{bmatrix}$$

The fact that $K_{32} = K'_{23}$ is not a coincidence, since this is a general property that follows from the definition of K_{mn}.

Theorem 7.29. The commutation matrix satisfies the properties

(a) $K_{m1} = K_{1m} = \mathrm{I}_m$,
(b) $K'_{mn} = K_{nm}$,
(c) $K_{mn}^{-1} = K_{nm}$.

Proof. When H_{ij} is $m \times 1$, then $H_{ij} = \boldsymbol{e}_{i,m}$ and so

$$K_{m1} = \sum_{i=1}^{m} (\boldsymbol{e}_{i,m} \otimes \boldsymbol{e}'_{i,m}) = \mathrm{I}_m = \sum_{i=1}^{m} (\boldsymbol{e}'_{i,m} \otimes \boldsymbol{e}_{i,m}) = K_{1m},$$

proving (a). To prove (b), note that

$$K'_{mn} = \sum_{i=1}^{m} \sum_{j=1}^{n} (H_{ij} \otimes H'_{ij})' = \sum_{i=1}^{m} \sum_{j=1}^{n} (H'_{ij} \otimes H_{ij}) = K_{nm}$$

Finally, (c) follows since

$$H_{ij}H'_{kl} = \boldsymbol{e}_{i,m}\boldsymbol{e}'_{j,n}\boldsymbol{e}_{l,n}\boldsymbol{e}'_{k,m} = \begin{cases} \boldsymbol{e}_{i,m}\boldsymbol{e}'_{k,m}, & \text{if } j = l, \\ (0), & \text{if } j \neq l, \end{cases}$$

$$H'_{ij}H_{kl} = \boldsymbol{e}_{j,n}\boldsymbol{e}'_{i,m}\boldsymbol{e}_{k,m}\boldsymbol{e}'_{l,n} = \begin{cases} \boldsymbol{e}_{j,n}\boldsymbol{e}'_{l,n}, & \text{if } i = k, \\ (0), & \text{if } i \neq k, \end{cases}$$

and so

$$\begin{aligned} K_{mn}K_{nm} = K_{mn}K'_{mn} &= \left\{ \sum_{i=1}^{m} \sum_{j=1}^{n} (H_{ij} \otimes H'_{ij}) \right\} \left\{ \sum_{k=1}^{m} \sum_{l=1}^{n} (H_{kl} \otimes H'_{kl})' \right\} \\ &= \sum_{i=1}^{m} \sum_{j=1}^{n} \sum_{k=1}^{m} \sum_{l=1}^{n} (H_{ij}H'_{kl} \otimes H'_{ij}H_{kl}) \\ &= \sum_{i=1}^{m} \sum_{j=1}^{n} (\boldsymbol{e}_{i,m}\boldsymbol{e}'_{i,m} \otimes \boldsymbol{e}_{j,n}\boldsymbol{e}'_{j,n}) \\ &= \left(\sum_{i=1}^{m} \boldsymbol{e}_{i,m}\boldsymbol{e}'_{i,m} \right) \otimes \left(\sum_{j=1}^{n} \boldsymbol{e}_{j,n}\boldsymbol{e}'_{j,n} \right) = \mathrm{I}_m \otimes \mathrm{I}_n = \mathrm{I}_{mn} \qquad \square \end{aligned}$$

Commutation matrices have important relationships with the vec operator and the Kronecker product. For an $m \times n$ matrix A, the two vectors vec(A) and vec(A') are related since they contain the same elements arranged in a different order; that is, an appropriate reordering of the elements of vec(A) will produce vec(A'). The commutation matrix K_{mn} is the matrix multiplier which transforms vec(A) to vec(A').

Theorem 7.30. For any $m \times n$ matrix A,

$$K_{mn}\,\text{vec}(A) = \text{vec}(A')$$

Proof. Clearly, since $a_{ij}H'_{ij}$ is the $n \times m$ matrix whose only nonzero element, a_{ij}, is in the (j, i)th position, we have

$$A' = \sum_{i=1}^{m} \sum_{j=1}^{n} a_{ij}H'_{ij} = \sum_{i=1}^{m} \sum_{j=1}^{n} (\boldsymbol{e}'_{i,m}A\boldsymbol{e}_{j,n})\boldsymbol{e}_{j,n}\boldsymbol{e}'_{i,m}$$

$$= \sum_{i=1}^{m} \sum_{j=1}^{n} \boldsymbol{e}_{j,n}(\boldsymbol{e}'_{i,m}A\boldsymbol{e}_{j,n})\boldsymbol{e}'_{i,m} = \sum_{i=1}^{m} \sum_{j=1}^{n} (\boldsymbol{e}_{j,n}\boldsymbol{e}'_{i,m})A(\boldsymbol{e}_{j,n}\boldsymbol{e}'_{i,m})$$

$$= \sum_{i=1}^{m} \sum_{j=1}^{n} H'_{ij}AH'_{ij}$$

The result now follows by taking the vec of both sides and using Theorem 7.16, since

$$\text{vec}(A') = \text{vec}\left(\sum_{i=1}^{m} \sum_{j=1}^{n} H'_{ij}AH'_{ij}\right) = \sum_{i=1}^{m} \sum_{j=1}^{n} \text{vec}(H'_{ij}AH'_{ij})$$

$$= \sum_{i=1}^{m} \sum_{j=1}^{n} (H_{ij} \otimes H'_{ij})\text{vec}(A) = K_{mn}\text{vec}(A) \qquad \square$$

The term *commutation* arises from the fact that commutation matrices provide the factors that allow a Kronecker product to commute. This property is summarized in Theorem 7.31.

Theorem 7.31. Let A be an $m \times n$ matrix, B be a $p \times q$ matrix, $\boldsymbol{x}$ be an $m \times 1$ vector, and $\boldsymbol{y}$ be a $p \times 1$ vector. Then

(a) $K_{pm}(A \otimes B) = (B \otimes A)K_{qn}$,
(b) $K_{pm}(A \otimes B)K_{nq} = B \otimes A$,
(c) $K_{pm}(A \otimes \boldsymbol{y}) = \boldsymbol{y} \otimes A$,
(d) $K_{mp}(\boldsymbol{y} \otimes A) = A \otimes \boldsymbol{y}$,
(e) $K_{pm}(\boldsymbol{x} \otimes \boldsymbol{y}) = \boldsymbol{y} \otimes \boldsymbol{x}$,
(f) $\text{tr}\{(B \otimes A)K_{mn}\} = \text{tr}(BA)$, if $p = n$ and $q = m$.

Proof. If X is a $q \times n$ matrix, then by using Theorems 7.16 and 7.30, we find that

$$\begin{aligned} K_{pm}(A \otimes B)\text{vec}(X) &= K_{pm}\,\text{vec}(BXA') = \text{vec}\{(BXA')'\} \\ &= \text{vec}(AX'B') = (B \otimes A)\text{vec}(X') \\ &= (B \otimes A)K_{qn}\,\text{vec}(X) \end{aligned}$$

Thus, if X is chosen so that vec(X) equals the ith column of I_{qn}, we observe that the ith column of $K_{pm}(A \otimes B)$ must be the same as the ith column of $(B \otimes A)K_{qn}$, so (a) follows. Postmultiplying (a) by K_{nq} and then using Theorem 7.29(c) yields (b). Properties (c)–(e) follow from (a) and Theorem 7.29(a) since

$$\begin{aligned} K_{pm}(A \otimes \boldsymbol{y}) &= (\boldsymbol{y} \otimes A)K_{1n} = \boldsymbol{y} \otimes A, \\ K_{mp}(\boldsymbol{y} \otimes A) &= (A \otimes \boldsymbol{y})K_{n1} = A \otimes \boldsymbol{y}, \\ K_{pm}(\boldsymbol{x} \otimes \boldsymbol{y}) &= (\boldsymbol{y} \otimes \boldsymbol{x})K_{11} = \boldsymbol{y} \otimes \boldsymbol{x} \end{aligned}$$

To prove (f), use the definition of the commutation matrix to get

$$\begin{aligned} \text{tr}\{(B \otimes A)K_{mn}\} &= \sum_{i=1}^{m} \sum_{j=1}^{n} \text{tr}\{(B \otimes A)(H_{ij} \otimes H'_{ij})\} \\ &= \sum_{i=1}^{m} \sum_{j=1}^{n} \{\text{tr}(BH_{ij})\}\{\text{tr}(AH'_{ij})\} \\ &= \sum_{i=1}^{m} \sum_{j=1}^{n} (\boldsymbol{e}'_{j,n} B\boldsymbol{e}_{i,m})(\boldsymbol{e}'_{i,m} A\boldsymbol{e}_{j,n}) = \sum_{i=1}^{m} \sum_{j=1}^{n} b_{ji}a_{ij} \\ &= \sum_{j=1}^{n} (B)_{j\cdot}(A)_{\cdot j} = \sum_{j=1}^{n} (BA)_{jj} = \text{tr}(BA) \end{aligned}$$

□

The commutation matrix also can be utilized to obtain a relationship between the vec of a Kronecker product and the Kronecker product of the corresponding vecs.

Theorem 7.32. Let A be an $m \times n$ matrix and B be a $p \times q$ matrix. Then

$$\text{vec}(A \otimes B) = (\mathrm{I}_n \otimes K_{qm} \otimes \mathrm{I}_p)\{\text{vec}(A) \otimes \text{vec}(B)\}$$

Proof. Our proof follows that given by Magnus (1988). Let $\boldsymbol{a}_1, \ldots, \boldsymbol{a}_n$ be the columns of A and $\boldsymbol{b}_1, \ldots, \boldsymbol{b}_q$ the columns of B. Then since A and B can be written as

$$A = \sum_{i=1}^{n} \boldsymbol{a}_i \boldsymbol{e}'_{i,n}, \qquad B = \sum_{j=1}^{q} \boldsymbol{b}_j \boldsymbol{e}'_{j,q},$$

we have

$$\begin{aligned}
\text{vec}(A \otimes B) &= \sum_{i=1}^{n} \sum_{j=1}^{q} \text{vec}(\boldsymbol{a}_i \boldsymbol{e}'_{i,n} \otimes \boldsymbol{b}_j \boldsymbol{e}'_{j,q}) \\
&= \sum_{i=1}^{n} \sum_{j=1}^{q} \text{vec}\{(\boldsymbol{a}_i \otimes \boldsymbol{b}_j)(\boldsymbol{e}'_{i,n} \otimes \boldsymbol{e}'_{j,q})\} \\
&= \sum_{i=1}^{n} \sum_{j=1}^{q} \{(\boldsymbol{e}_{i,n} \otimes \boldsymbol{e}_{j,q}) \otimes (\boldsymbol{a}_i \otimes \boldsymbol{b}_j)\} \\
&= \sum_{i=1}^{n} \sum_{j=1}^{q} \{\boldsymbol{e}_{i,n} \otimes K_{qm}(\boldsymbol{a}_i \otimes \boldsymbol{e}_{j,q}) \otimes \boldsymbol{b}_j\} \\
&= \sum_{i=1}^{n} \sum_{j=1}^{q} (\text{I}_n \otimes K_{qm} \otimes \text{I}_p)(\boldsymbol{e}_{i,n} \otimes \boldsymbol{a}_i \otimes \boldsymbol{e}_{j,q} \otimes \boldsymbol{b}_j) \\
&= (\text{I}_n \otimes K_{qm} \otimes \text{I}_p)\left\{\sum_{i=1}^{n} (\boldsymbol{e}_{i,n} \otimes \boldsymbol{a}_i)\right\} \otimes \left\{\sum_{j=1}^{q} (\boldsymbol{e}_{j,q} \otimes \boldsymbol{b}_j)\right\} \\
&= (\text{I}_n \otimes K_{qm} \otimes \text{I}_p)\left\{\sum_{i=1}^{n} \text{vec}(\boldsymbol{a}_i \boldsymbol{e}'_{i,n})\right\} \otimes \left\{\sum_{j=1}^{q} \text{vec}(\boldsymbol{b}_j \boldsymbol{e}'_{j,q})\right\} \\
&= (\text{I}_n \otimes K_{qm} \otimes \text{I}_p)\{\text{vec}(A) \otimes \text{vec}(B)\} \qquad \square
\end{aligned}$$

Our next theorem establishes some results for the special commutation matrix K_{mm}. Corresponding results for the general commutation matrix K_{mn} can be found in Magnus and Neudecker (1979) or Magnus (1988).

Theorem 7.33. The commutation matrix K_{mm} has the eigenvalue $+1$ repeated $\frac{1}{2}m(m + 1)$ times and the eigenvalue -1 repeated $\frac{1}{2}m(m - 1)$ times. In addition,

$$\text{tr}(K_{mm}) = m \quad \text{and} \quad |K_{mm}| = (-1)^{m(m-1)/2}$$

Proof. Since K_{mm} is real and symmetric, we know from Theorem 3.8 that its eigenvalues are also real. Further, since K_{mm} is orthogonal, the square of each eigenvalue must be 1, so it has eigenvalues +1 and -1 only. Let p be the number of eigenvalues equal to -1, implying that $m^2 - p$ is the number of eigenvalues equal to +1. Since the trace equals the sum of the eigenvalues, we must have $\text{tr}(K_{mm}) = p(-1) + (m^2 - p)(1) = m^2 - 2p$. But by using basic properties of the trace, we also find that

$$\text{tr}(K_{mm}) = \text{tr}\left\{\sum_{i=1}^{m}\sum_{j=1}^{m}(\boldsymbol{e}_i\boldsymbol{e}_j' \otimes \boldsymbol{e}_j\boldsymbol{e}_i')\right\} = \sum_{i=1}^{m}\sum_{j=1}^{m} \text{tr}(\boldsymbol{e}_i\boldsymbol{e}_j' \otimes \boldsymbol{e}_j\boldsymbol{e}_i')$$

$$= \sum_{i=1}^{m}\sum_{j=1}^{m} \{\text{tr}(\boldsymbol{e}_i\boldsymbol{e}_j')\}\{\text{tr}(\boldsymbol{e}_j\boldsymbol{e}_i')\} = \sum_{i=1}^{m}\sum_{j=1}^{m} (\boldsymbol{e}_i'\boldsymbol{e}_j)^2$$

$$= \sum_{i=1}^{m} 1 = m$$

Evidently, $m^2 - 2p = m$, so that $p = \frac{1}{2}m(m-1)$ as claimed. Finally, the formula given for the determinant follows directly from the fact that the determinant equals the product of the eigenvalues. □

We will see later that the commutation matrix K_{mm} appears in some important matrix moment formulas through the term $N_m = \frac{1}{2}(\text{I}_{m^2} + K_{mm})$. Consequently, we will establish some basic properties of N_m.

Theorem 7.34. Let $N_m = \frac{1}{2}(\text{I}_{m^2} + K_{mm})$, and let A and B be $m \times m$ matrices. Then

(a) $N_m = N_m' = N_m^2$,
(b) $N_m K_{mm} = N_m = K_{mm} N_m$,
(c) $N_m \text{vec}(A) = \frac{1}{2}\text{vec}(A + A')$,
(d) $N_m(A \otimes B)N_m = N_m(B \otimes A)N_m$.

Proof. The symmetry of N_m follows from the symmetry of I_{m^2} and K_{mm}, while

$$N_m^2 = \frac{1}{4}(\text{I}_{m^2} + K_{mm})^2 = \frac{1}{4}(\text{I}_{m^2} + 2K_{mm} + K_{mm}^2) = \frac{1}{2}(\text{I}_{m^2} + K_{mm}) = N_m,$$

since $K_{mm}^2 = \text{I}$ follows from the fact that $K_{mm}^{-1} = K_{mm}$. Similarly, (b) follows from the fact that $K_{mm}^2 = \text{I}_{m^2}$. Part (c) is an immediate consequence of

$$\mathrm{I}_{m^2}\,\mathrm{vec}(A) = \mathrm{vec}(A), \qquad K_{mm}\,\mathrm{vec}(A) = \mathrm{vec}(A')$$

Finally, to prove (d), note that, by using Theorem 7.31(b), and Theorem 7.34(b),

$$N_m(A \otimes B)N_m = N_m K_{mm}(B \otimes A)K_{mm}N_m = N_m(B \otimes A)N_m \qquad \square$$

The proof of our final result will be left to the reader as an exercise.

Theorem 7.35. Let A and B be $m \times m$ matrices such that $A = BB'$. Then

(a) $N_m(B \otimes B)N_m = (B \otimes B)N_m = N_m(B \otimes B)$,
(b) $(B \otimes B)N_m(B' \otimes B') = N_m(A \otimes A)$.

8. SOME OTHER MATRICES ASSOCIATED WITH THE VEC OPERATOR

In this section, we introduce several other matrices that, like the commutation matrix, have important relationships with the vec operator. However, each of the matrices we discuss here is useful in working with vec(A) when the matrix A is square and has some particular structure. A more thorough discussion of this and other related material can be found in Magnus (1988).

When the $m \times m$ matrix A is symmetric, then vec(A) contains redundant elements since $a_{ij} = a_{ji}$ for $i \neq j$. For this reason, we previously defined v(A) to be the $m(m+1)/2 \times 1$ vector formed by stacking the columns of the lower triangular portion of A. The matrix that transforms v(A) into vec(A) is called the duplication matrix; that is, if we denote this duplication matrix by D_m, then for any $m \times m$ symmetric matrix A,

$$D_m \mathrm{v}(A) = \mathrm{vec}(A) \tag{7.18}$$

For instance, the duplication matrix D_3 is given by

$$D_3 = \begin{bmatrix} 1 & 0 & 0 & 0 & 0 & 0 \\ 0 & 1 & 0 & 0 & 0 & 0 \\ 0 & 0 & 1 & 0 & 0 & 0 \\ 0 & 1 & 0 & 0 & 0 & 0 \\ 0 & 0 & 0 & 1 & 0 & 0 \\ 0 & 0 & 0 & 0 & 1 & 0 \\ 0 & 0 & 1 & 0 & 0 & 0 \\ 0 & 0 & 0 & 0 & 1 & 0 \\ 0 & 0 & 0 & 0 & 0 & 1 \end{bmatrix}$$

For an explicit expression for the $m^2 \times m(m+1)/2$ duplication matrix D_m, refer to Magnus (1988) or Problem 7.54.

Some properties of the duplication matrix and its Moore–Penrose inverse are summarized in Theorem 7.36.

Theorem 7.36. Let D_m be the $m^2 \times m(m+1)/2$ duplication matrix and D_m^+ its Moore–Penrose inverse. Then

(a) $\text{rank}(D_m) = m(m+1)/2$,
(b) $D_m^+ = (D_m' D_m)^{-1} D_m'$,
(c) $D_m^+ D_m = \text{I}_{m(m+1)/2}$,
(d) $D_m^+ \text{vec}(A) = \text{v}(A)$ for every $m \times m$ symmetric matrix A.

Proof. Clearly, for every $m(m+1)/2 \times 1$ vector $\boldsymbol{x}$ there exists an $m \times m$ symmetric matrix A such that $\boldsymbol{x} = \text{v}(A)$. But if for some symmetric A, $D_m \text{v}(A) = \mathbf{0}$, then from the definition of D_m, $\text{vec}(A) = \mathbf{0}$, which then implies that $\text{v}(A) = \mathbf{0}$. Thus, $D_m \boldsymbol{x} = \mathbf{0}$ only if $\boldsymbol{x} = \mathbf{0}$, and so D_m has full column rank. Parts (b) and (c) follow immediately from (a) and Theorem 5.3, while (d) is obtained by premultiplying (7.18) by D_m^+ and then using (c). □

The duplication matrix and its Moore–Penrose inverse have some important relationships with K_{mm} and N_m.

Theorem 7.37. Let D_m be the $m^2 \times m(m+1)/2$ duplication matrix and D_m^+ its Moore–Penrose inverse. Then

(a) $K_{mm} D_m = N_m D_m = D_m$,
(b) $D_m^+ K_{mm} = D_m^+ N_m = D_m^+$,
(c) $D_m D_m^+ = N_m$.

Proof. For any $m \times m$ symmetric matrix A, it follows that

$$K_{mm} D_m \, \text{v}(A) = K_{mm} \, \text{vec}(A) = \text{vec}(A') = \text{vec}(A) = D_m \, \text{v}(A) \tag{7.19}$$

Similarly, we have

$$N_m D_m \text{v}(A) = N_m \, \text{vec}(A) = \frac{1}{2} \, \text{vec}(A + A') = \text{vec}(A) = D_m \, \text{v}(A) \tag{7.20}$$

Since $\{\text{v}(A)\colon A \; m \times m \text{ and } A' = A\}$ is all of $m(m+1)/2$-dimensional space, (7.19) and (7.20) establish (a). To prove (b), take the transpose of (a), premultiply all three sides by $(D_m' D_m)^{-1}$, and then use Theorem 7.36(b). We will prove (c) by showing that for any $m \times m$ matrix A,

$$D_m D_m^+ \operatorname{vec}(A) = N_m \operatorname{vec}(A)$$

If we define $A_* = \frac{1}{2}(A + A')$, then A_* is symmetric and

$$\begin{aligned} N_m \operatorname{vec}(A) &= \frac{1}{2}(\mathrm{I}_{m^2} + K_{mm})\operatorname{vec}(A) = \frac{1}{2}\{\operatorname{vec}(A) + \operatorname{vec}(A')\} \\ &= \operatorname{vec}(A_*) \end{aligned}$$

Using this and (b), we find that

$$\begin{aligned} D_m D_m^+ \operatorname{vec}(A) &= D_m D_m^+ N_m \operatorname{vec}(A) = D_m D_m^+ \operatorname{vec}(A_*) \\ &= D_m \operatorname{v}(A_*) = \operatorname{vec}(A_*) = N_m \operatorname{vec}(A), \end{aligned}$$

and so the proof is complete. □

We will need the following result in the next chapter

Theorem 7.38. If A is an $m \times m$ nonsingular matrix, then $D_m'(A \otimes A)D_m$ is nonsingular and its inverse is given by $D_m^+(A^{-1} \otimes A^{-1})D_m^{+\prime}$.

Proof. To prove the result we simply show that the product of the two matrices given above yields $\mathrm{I}_{m(m+1)/2}$. Using Theorem 7.35(a), Theorem 7.36(c), and Theorem 7.37(a) and (c), we have

$$\begin{aligned} &D_m'(A \otimes A)D_m D_m^+(A^{-1} \otimes A^{-1})D_m^{+\prime} \\ &\quad = D_m'(A \otimes A)N_m(A^{-1} \otimes A^{-1})D_m^{+\prime} \\ &\quad = D_m' N_m (A \otimes A)(A^{-1} \otimes A^{-1})D_m^{+\prime} = D_m' N_m D_m^{+\prime} \\ &\quad = (N_m D_m)' D_m^{+\prime} = D_m' D_m^{+\prime} = (D_m^+ D_m)' = \mathrm{I}_{m(m+1)/2} \end{aligned}$$ □

We next consider the situation in which the $m \times m$ matrix A is lower triangular. In this case, the elements of vec(A) are identical to those of v(A) except that vec(A) has some additional zeros. We will denote by L_m' the $m^2 \times m(m+1)/2$ matrix which transforms v(A) into vec(A); that is, L_m' satisfies

$$L_m' \operatorname{v}(A) = \operatorname{vec}(A) \tag{7.21}$$

Thus, for instance, for $m = 3$,

$$L_3' = \begin{bmatrix} 1 & 0 & 0 & 0 & 0 & 0 \\ 0 & 1 & 0 & 0 & 0 & 0 \\ 0 & 0 & 1 & 0 & 0 & 0 \\ 0 & 0 & 0 & 0 & 0 & 0 \\ 0 & 0 & 0 & 1 & 0 & 0 \\ 0 & 0 & 0 & 0 & 1 & 0 \\ 0 & 0 & 0 & 0 & 0 & 0 \\ 0 & 0 & 0 & 0 & 0 & 0 \\ 0 & 0 & 0 & 0 & 0 & 1 \end{bmatrix}$$

Note that L_m' can be obtained from D_m by replacing $m(m-1)/2$ of the rows of D_m by rows of zeros. The following properties of the matrix L_m can be proven directly from its definition given in (7.21).

Theorem 7.39. The $m(m+1)/2 \times m^2$ matrix L_m satisfies

(a) $\text{rank}(L_m) = m(m+1)/2$,
(b) $L_m L_m' = I_{m(m+1)/2}$,
(c) $L_m^+ = L_m'$,
(d) $L_m \text{vec}(A) = \text{v}(A)$, for every $m \times m$ matrix A.

Proof. Note that if A is lower triangular, then $\text{vec}(A)'\text{vec}(A) = \text{v}(A)'\text{v}(A)$ and so (7.21) implies

$$\text{v}(A)' L_m L_m' \text{v}(A) - \text{v}(A)'\text{v}(A) = \text{v}(A)'(L_m L_m' - I_{m(m+1)/2})\text{v}(A) = 0$$

for all lower triangular matrices A. But this can be true only if (b) holds since $\{\text{v}(A)\text{: } A\ m \times m \text{ and lower triangular}\} = R^{m(m+1)/2}$. Part (a) follows immediately from (b), as does (c) since $L_m^+ = (L_m' L_m)^{-1} L_m'$. To prove (d), note that every matrix A can be written $A = A_L + A_U$, where A_L is lower triangular, and A_U is upper triangular with each diagonal element equal to zero. Clearly,

$$0 = \text{vec}(A_L)' \text{vec}(A_U) = \text{v}(A_L)' L_m \text{vec}(A_U),$$

and since, for fixed A_U, this must hold for all choices of the lower triangular matrix A_L, it follows that

$$L_m \text{vec}(A_U) = \mathbf{0}$$

Thus, using this along with (7.21), (b), and the fact that $\text{v}(A_L) = \text{v}(A)$, we have

$$L_m \operatorname{vec}(A) = L_m \operatorname{vec}(A_L + A_U) = L_m \operatorname{vec}(A_L) = L_m L_m' \,\text{v}(A_L) = \text{v}(A_L) = \text{v}(A) \quad \square$$

We see from property (d) in Theorem 7.39 that L_m is the matrix that eliminates the zeros in vec(A) coming from the upper triangular portion of A so as to yield v(A). For this reason, L_m is sometimes referred to as the elimination matrix. Our next result gives some relationships between L_m and the matrices D_m and N_m. We will leave the proofs of these results as an exercise for the reader.

Theorem 7.40. The elimination matrix L_m satisfies

(a) $L_m D_m = \text{I}_{m(m+1)/2}$,
(b) $D_m L_m N_m = N_m$,
(c) $D_m^+ = L_m N_m$.

The last matrix related to vec(A) that we will discuss is another sort of elimination matrix. Suppose now that the $m \times m$ matrix A is a strictly lower triangular matrix; that is, it is lower triangular and all of its diagonal elements are zero. In this case, $\tilde{\text{v}}(A)$ contains all of the relevant elements of A. We denote by $\tilde{L}_m'$ the $m^2 \times m(m-1)/2$ matrix that transforms $\tilde{\text{v}}(A)$ into vec(A); that is,

$$\tilde{L}_m' \,\tilde{\text{v}}(A) = \operatorname{vec}(A)$$

Thus, for $m = 3$ we have

$$\tilde{L}_3 = \begin{bmatrix} 0 & 1 & 0 & 0 & 0 & 0 & 0 & 0 & 0 \\ 0 & 0 & 1 & 0 & 0 & 0 & 0 & 0 & 0 \\ 0 & 0 & 0 & 0 & 0 & 1 & 0 & 0 & 0 \end{bmatrix}$$

Since $\tilde{L}_m$ is very similar to L_m, some of its basic properties parallel those of L_m. For instance, the following results are analogous to those in Theorem 7.39. The proofs, which we omit, are similar to those of Theorem 7.39.

Theorem 7.41. The $m(m-1)/2 \times m^2$ matrix $\tilde{L}_m$ satisfies

(a) $\operatorname{rank}(\tilde{L}_m) = m(m-1)/2$,
(b) $\tilde{L}_m \tilde{L}_m' = \text{I}_{m(m-1)/2}$,
(c) $\tilde{L}_m^+ = \tilde{L}_m'$,
(d) $\tilde{L}_m \operatorname{vec}(A) = \tilde{\text{v}}(A)$, for every $m \times m$ matrix A.

Our final theorem gives some relationships between $\tilde{L}_m$, L_m, D_m, K_{mm} and N_m. The proof is left to the reader as an exercise.

Theorem 7.42. The $m(m-1)/2 \times m^2$ matrix $\tilde{L}_m$ satisfies

(a) $\tilde{L}_m K_{mm} \tilde{L}'_m = (0)$,
(b) $\tilde{L}_m K_{mm} L'_m = (0)$,
(c) $\tilde{L}_m D_m = \tilde{L}_m L'_m$,
(d) $L'_m L_m \tilde{L}'_m = \tilde{L}'_m$,
(e) $D_m L_m \tilde{L}'_m = 2N_m \tilde{L}'_m$,
(f) $\tilde{L}_m L'_m L_m \tilde{L}'_m = \mathrm{I}_{m(m-1)/2}$.

9. NONNEGATIVE MATRICES

The topic of this section, nonnegative and positive matrices, should not be confused with nonnegative definite and positive definite matrices, which we have discussed earlier on several occasions. An $m \times n$ matrix A is a nonnegative matrix, indicated by $A \geq (0)$, if each element of A is nonnegative. Similarly, A is a positive matrix, indicated by $A > (0)$, if each element of A is positive. We will write $A \geq B$ and $A > B$ to mean that $A - B \geq (0)$ and $A - B > (0)$, respectively. Any matrix A can be transformed to a nonnegative matrix by replacing each of its elements by its absolute value. This will be denoted by abs(A); that is, if A is an $m \times n$ matrix, then abs(A) is also an $m \times n$ matrix with (i, j)th element given by $|a_{ij}|$. We will investigate some of the properties of nonnegative square matrices as well as indicate some of their applications in stochastic processes. For a more exhaustive coverage of this topic the reader is referred to the texts on nonnegative matrices by Berman and Plemmons (1994), Minc (1988), and Seneta (1973), as well as the books by Gantmacher (1959) and Horn and Johnson (1985). Most of the proofs that we present here follow along the lines of the derivations, based on matrix norms, given in Horn and Johnson (1985).

We begin with some results regarding the spectral radius of nonnegative and positive matrices.

Theorem 7.43. Let A be an $m \times m$ matrix and $\boldsymbol{x}$ be an $m \times 1$ vector. If $A \geq (0)$ and $\boldsymbol{x} > \boldsymbol{0}$, then

$$\min_{1 \leq i \leq m} \sum_{j=1}^{m} a_{ij} \leq \rho(A) \leq \max_{1 \leq i \leq m} \sum_{j=1}^{m} a_{ij}, \tag{7.22}$$

$$\min_{1 \leq i \leq m} x_i^{-1} \sum_{j=1}^{m} a_{ij} x_j \leq \rho(A) \leq \max_{1 \leq i \leq m} x_i^{-1} \sum_{j=1}^{m} a_{ij} x_j \tag{7.23}$$

with similar inequalities holding when minimizing and maximizing over columns instead of rows.

Proof. Let

$$\alpha = \min_{1 \leq i \leq m} \sum_{j=1}^{m} a_{ij}$$

and define the $m \times m$ matrix B to have (i, h)th element

$$b_{ih} = \alpha a_{ih} \left(\sum_{j=1}^{m} a_{ij} \right)^{-1}$$

if $\alpha > 0$ and $b_{ih} = 0$ if $\alpha = 0$. Note that $\|B\|_\infty = \alpha$ and $b_{ih} \leq a_{ih}$, so that $A \geq B$. Clearly, it follows that for any positive integer k, $A^k \geq B^k$ and this then implies that $\|A^k\|_\infty \geq \|B^k\|_\infty$ or, equivalently,

$$\{\|A^k\|_\infty\}^{1/k} \geq \{\|B^k\|_\infty\}^{1/k}$$

Taking the limit as $k \rightarrow \infty$, it follows from Theorem 4.24 that $\rho(A) \geq \rho(B)$. But this proves the lower bound in (7.22) since $\rho(B) = \alpha$ follows from the fact that $\rho(B) \geq \alpha$ since

$$B\mathbf{1}_m = \alpha \mathbf{1}_m$$

and $\rho(B) \leq \|B\|_\infty = \alpha$ due to Theorem 4.19. The upper bound is proven in a similar fashion using

$$\alpha = \max_{1 \leq i \leq m} \sum_{j=1}^{m} a_{ij}$$

The bounds in (7.23) follow directly from those in (7.22) since if we define the matrix $C = D_x^{-1} A D_x$, then $C \geq (0)$, $\rho(C) = \rho(A)$, and $c_{ij} = a_{ij} x_i^{-1} x_j$. □

Theorem 7.44. Let A be an $m \times m$ positive matrix. Then $\rho(A)$ is positive and is an eigenvalue of A. In addition, there exists a positive eigenvector of A corresponding to the eigenvalue $\rho(A)$.

Proof. $\rho(A) > 0$ follows immediately from Theorem 7.43 since A is positive. By the definition of $\rho(A)$, there exists an eigenvalue of A, λ, such that

$|\lambda| = \rho(A)$. Let $\boldsymbol{x}$ be an eigenvector of A corresponding to λ so that $A\boldsymbol{x} = \lambda\boldsymbol{x}$. Note that

$$\rho(A)\text{abs}(\boldsymbol{x}) = |\lambda|\,\text{abs}(\boldsymbol{x}) = \text{abs}(\lambda\boldsymbol{x}) = \text{abs}(A\boldsymbol{x}) \leq \text{abs}(A)\text{abs}(\boldsymbol{x}) = A\,\text{abs}(\boldsymbol{x}),$$

where the inequality clearly follows from the fact that

$$\left|\sum_{j=1}^{m} a_{ij}x_j\right| \leq \sum_{j=1}^{m} |a_{ij}||x_j|$$

for each i. Thus, the vector $\boldsymbol{y} = A\,\text{abs}(\boldsymbol{x}) - \rho(A)\text{abs}(\boldsymbol{x})$ is nonnegative. The vector $\boldsymbol{z} = A\,\text{abs}(\boldsymbol{x})$ is positive since A is positive and the eigenvector $\boldsymbol{x}$ must be a nonnull vector. Now if we assume that $\boldsymbol{y}$ is positive, then again since A is positive, we have

$$0 < A\boldsymbol{y} = A\boldsymbol{z} - \rho(A)\boldsymbol{z},$$

or simply $A\boldsymbol{z} > \rho(A)\boldsymbol{z}$. Premultiplying this inequality by D_z^{-1}, we get

$$D_z^{-1}A\boldsymbol{z} > \rho(A)\mathbf{1}_m$$

or in other words,

$$z_i^{-1}\sum_{j=1}^{m} a_{ij}z_j > \rho(A)$$

holds for each i. But using Theorem 7.43 implies that $\rho(A) > \rho(A)$. Thus, $\boldsymbol{y}$ cannot be positive and, since we have already shown that it is nonnegative, we must have $\boldsymbol{y} = \mathbf{0}$. This yields $A\,\text{abs}(\boldsymbol{x}) = \rho(A)\text{abs}(\boldsymbol{x})$, so that $\text{abs}(\boldsymbol{x})$ is an eigenvector corresponding to $\rho(A)$, and from this we get $\text{abs}(\boldsymbol{x}) = \rho(A)^{-1}A\,\text{abs}(\boldsymbol{x})$, which shows that $\text{abs}(\boldsymbol{x})$ is positive since $\rho(A) > 0$ and $A\,\text{abs}(\boldsymbol{x}) > \mathbf{0}$. This completes the proof. □

An immediate consequence of the proof of Theorem 7.44 is the following.

Corollary 7.44.1. Let A be an $m \times m$ positive matrix and suppose that λ is an eigenvalue of A satisfying $|\lambda| = \rho(A)$. If $\boldsymbol{x}$ is any eigenvector corresponding to λ, then

$$A\,\text{abs}(\boldsymbol{x}) = \rho(A)\text{abs}(\boldsymbol{x})$$

Before determining the dimensionality of the eigenspace associated with the eigenvalue $\rho(A)$, we need the following result.

Theorem 7.45. Let $\boldsymbol{x}$ be an eigenvector corresponding to the eigenvalue λ of the $m \times m$ positive matrix A. If $|\lambda| = \rho(A)$, then there exists some angle θ such that $e^{-i\theta}\boldsymbol{x} > \boldsymbol{0}$.

Proof. Note that

$$\text{abs}(A\boldsymbol{x}) = \text{abs}(\lambda\boldsymbol{x}) = \rho(A)\text{abs}(\boldsymbol{x}), \tag{7.24}$$

while it follows from Corollary 7.44.1 that

$$A\,\text{abs}(\boldsymbol{x}) = \rho(A)\text{abs}(\boldsymbol{x}) \tag{7.25}$$

Now by using (7.24) and (7.25), we find that

$$\begin{aligned}\rho(A)|x_j| = |\lambda||x_j| = |\lambda x_j| &= \left|\sum_{k=1}^{m} a_{jk}x_k\right| \leq \sum_{k=1}^{m} |a_{jk}||x_k| \\ &= \sum_{k=1}^{m} a_{jk}|x_k| = \rho(A)|x_j|\end{aligned}$$

holds for each j. Evidently

$$\left|\sum_{k=1}^{m} a_{jk}x_k\right| = \sum_{k=1}^{m} |a_{jk}||x_k|,$$

and this can happen only if the, possibly complex, numbers $a_{jk}x_k = r_k e^{i\theta_k} = r_k(\cos\theta_k + i\sin\theta_k)$, for $k = 1, \dots, m$, have identical angles; that is, there exists some angle θ such that each $a_{jk}x_k$, for $k = 1, \dots, m$ can be written in the form $a_{jk}x_k = r_k e^{i\theta} = r_k(\cos\theta + i\sin\theta)$. In this case, $e^{-i\theta}a_{jk}x_k = r_k > 0$, which implies that $e^{-i\theta}x_k > 0$ since $a_{jk} > 0$. □

The following result not only indicates that the eigenspace corresponding to $\rho(A)$ has dimension one, but also that $\rho(A)$ is the only eigenvalue of A having modulus equal to $\rho(A)$.

Theorem 7.46. If A is an $m \times m$ positive matrix, then the dimension of the eigenspace corresponding to the eigenvalue $\rho(A)$ is one. Further, if λ is an eigenvalue of A and $\lambda \neq \rho(A)$, then $|\lambda| < \rho(A)$.

Proof. The first statement will be proven by showing that if $\boldsymbol{u}$ and $\boldsymbol{v}$ are nonnull vectors satisfying $A\boldsymbol{u} = \rho(A)\boldsymbol{u}$ and $A\boldsymbol{v} = \rho(A)\boldsymbol{v}$, then there exists some scalar c such that $\boldsymbol{v} = c\boldsymbol{u}$. Now from Theorem 7.45, we know there exist angles θ_1 and θ_2 such that $\boldsymbol{s} = e^{-i\theta_1}\boldsymbol{u} > \boldsymbol{0}$ and $\boldsymbol{t} = e^{-i\theta_2}\boldsymbol{v} > \boldsymbol{0}$. Define $\boldsymbol{w} = \boldsymbol{t} - d\boldsymbol{s}$, where

$$d = \min_{1 \leq j \leq m} s_j^{-1} t_j,$$

so that $\boldsymbol{w}$ is nonnegative with at least one component equal to 0. If $\boldsymbol{w} \neq \boldsymbol{0}$, then clearly $A\boldsymbol{w} > \boldsymbol{0}$ since A is positive. This leads to a contradiction since

$$A\boldsymbol{w} = A\boldsymbol{t} - dA\boldsymbol{s} = \rho(A)\boldsymbol{t} - \rho(A)d\boldsymbol{s} = \rho(A)\boldsymbol{w}$$

then implies that $\boldsymbol{w} > \boldsymbol{0}$. Thus, we must have $\boldsymbol{w} = \boldsymbol{0}$, so $\boldsymbol{t} = d\boldsymbol{s}$ and $\boldsymbol{v} = c\boldsymbol{u}$, where $c = de^{i(\theta_2 - \theta_1)}$. To prove the second statement of the theorem, first note that from the definition of the spectral radius, $|\lambda| \leq \rho(A)$ for any eigenvalue λ, of A. Now if $\boldsymbol{x}$ is an eigenvector corresponding to λ and $|\lambda| = \rho(A)$, then it follows from Theorem 7.45 that there exists an angle θ such that $\boldsymbol{u} = e^{-i\theta}\boldsymbol{x} > \boldsymbol{0}$. Clearly, $A\boldsymbol{u} = \lambda\boldsymbol{u}$. Premultiplying this identity by D_u^{-1}, we get

$$D_{\boldsymbol{u}}^{-1} A\boldsymbol{u} = \lambda \mathbf{1}_m,$$

so that

$$u_i^{-1} \sum_{j=1}^{m} a_{ij} u_j = \lambda$$

holds for each i. Now applying Theorem 7.43, we get $\lambda = \rho(A)$. □

We will see that the first statement in the previous theorem actually can be replaced by the stronger condition that $\rho(A)$ must be a simple eigenvalue of A. But first we have the following results, the last of which is a very useful limiting result for A.

Theorem 7.47. Suppose that A is an $m \times m$ positive matrix, and $\boldsymbol{x}$ and $\boldsymbol{y}$ are positive vectors satisfying $A\boldsymbol{x} = \rho(A)\boldsymbol{x}$, $A'\boldsymbol{y} = \rho(A)\boldsymbol{y}$, and $\boldsymbol{x}'\boldsymbol{y} = 1$. Then the following hold.

(a) $(A - \rho(A)\boldsymbol{x}\boldsymbol{y}')^k = A^k - \rho(A)^k\boldsymbol{x}\boldsymbol{y}'$, for $k = 1, 2, \ldots$.
(b) Each nonzero eigenvalue of $A - \rho(A)\boldsymbol{x}\boldsymbol{y}'$ is an eigenvalue of A.
(c) $\rho(A)$ is not an eigenvalue of $A - \rho(A)\boldsymbol{x}\boldsymbol{y}'$.
(d) $\rho(A - \rho(A)\boldsymbol{x}\boldsymbol{y}') < \rho(A)$.
(e) $\lim_{k\to\infty}\{\rho(A)^{-1}A\}^k = \boldsymbol{x}\boldsymbol{y}'$.

Proof. (a) is easily established by induction, since it clearly holds for $k = 1$, and if it holds for $k = j - 1$, then

$$\begin{aligned}(A - \rho(A)\boldsymbol{x}\boldsymbol{y}')^j &= (A - \rho(A)\boldsymbol{x}\boldsymbol{y}')^{j-1}(A - \rho(A)\boldsymbol{x}\boldsymbol{y}')\\ &= (A^{j-1} - \rho(A)^{j-1}\boldsymbol{x}\boldsymbol{y}')(A - \rho(A)\boldsymbol{x}\boldsymbol{y}')\\ &= A^j - \rho(A)A^{j-1}\boldsymbol{x}\boldsymbol{y}' - \rho(A)^{j-1}\boldsymbol{x}\boldsymbol{y}'A + \rho(A)^j\boldsymbol{x}\boldsymbol{y}'\boldsymbol{x}\boldsymbol{y}'\\ &= A^j - \rho(A)^j\boldsymbol{x}\boldsymbol{y}' - \rho(A)^j\boldsymbol{x}\boldsymbol{y}' + \rho(A)^j\boldsymbol{x}\boldsymbol{y}' = A^j - \rho(A)^j\boldsymbol{x}\boldsymbol{y}'\end{aligned}$$

Next, suppose that $\lambda \neq 0$ and $\boldsymbol{u}$ are an eigenvalue and eigenvector of $(A - \rho(A)\boldsymbol{x}\boldsymbol{y}')$, so that

$$(A - \rho(A)\boldsymbol{x}\boldsymbol{y}')\boldsymbol{u} = \lambda\boldsymbol{u}$$

Premultiplying this equation by $\boldsymbol{x}\boldsymbol{y}'$ and observing that $\boldsymbol{x}\boldsymbol{y}'(A - \rho(A)\boldsymbol{x}\boldsymbol{y}') = \boldsymbol{0}$, we see that we must have $\boldsymbol{x}\boldsymbol{y}'\boldsymbol{u} = \boldsymbol{0}$. Consequently,

$$A\boldsymbol{u} = (A - \rho(A)\boldsymbol{x}\boldsymbol{y}')\boldsymbol{u} = \lambda\boldsymbol{u},$$

and so λ is also an eigenvalue of A, as is required for (b). To prove (c), suppose that $\lambda = \rho(A)$ is an eigenvalue of $A - \rho(A)\boldsymbol{x}\boldsymbol{y}'$ with $\boldsymbol{u}$ a corresponding eigenvector. But we have just seen that this would imply that $\boldsymbol{u}$ is also an eigenvector of A corresponding to the eigenvalue $\rho(A)$. Thus, from Theorem 7.46, $\boldsymbol{u} = c\boldsymbol{x}$ for some scalar c and

$$\rho(A)\boldsymbol{u} = (A - \rho(A)\boldsymbol{x}\boldsymbol{y}')\boldsymbol{u} = (A - \rho(A)\boldsymbol{x}\boldsymbol{y}')c\boldsymbol{x} = \rho(A)c\boldsymbol{x} - \rho(A)c\boldsymbol{x} = \boldsymbol{0}$$

But this is impossible since $\rho(A) > 0$ and $\boldsymbol{u} \neq \boldsymbol{0}$, and so (c) holds. Now (d) follows directly from (b), (c), and Theorem 7.46. Finally, to prove (e), note that by dividing both sides of the equation given in (a) by $\rho(A)^k$ and rearranging, we get

$$\{\rho(A)^{-1}A\}^k = \boldsymbol{x}\boldsymbol{y}' + \{\rho(A)^{-1}A - \boldsymbol{x}\boldsymbol{y}'\}^k$$

Take the limit, as $k \to \infty$, of both sides of this equation and observe that from (d),

$$\rho\{\rho(A)^{-1}A - \boldsymbol{x}\boldsymbol{y}'\} = \frac{\rho\{A - \rho(A)\boldsymbol{x}\boldsymbol{y}'\}}{\rho(A)} < 1,$$

and so

$$\lim_{k\to\infty} \{\rho(A)^{-1}A - \boldsymbol{x}\boldsymbol{y}'\}^k = (0)$$

follows from Theorem 4.23. □

Theorem 7.48. Let A be an $m \times m$ positive matrix. Then the eigenvalue $\rho(A)$ is a simple eigenvalue of A.

Proof. Let $A = XTX^*$ be the Schur decomposition of A, so that X is a unitary matrix and T is an upper triangular matrix with the eigenvalues of A as its diagonal elements. Write $T = T_1 + T_2$, where T_1 is diagonal and T_2 is upper triangular with each diagonal element equal to 0. Suppose that we have chosen X so that the diagonal elements of T_1 are ordered as $T_1 = \text{diag}(\rho(A), \ldots, \rho(A), \lambda_{r+1}, \ldots, \lambda_m)$, where r is the multiplicity of the eigenvalue $\rho(A)$ and $|\lambda_j| < \rho(A)$ for $j = r+1, \ldots, m$, due to Theorem 7.46. We need to show that $r = 1$. Note that, for any upper triangular matrix U with ith diagonal element u_{ii}, U^k is also upper triangular with its ith diagonal element given by u_{ii}^k. Using this, we find that

$$\begin{aligned}
\lim_{k\to\infty} \{\rho(A)^{-1}A\}^k &= X\{\lim_{k\to\infty} \{\rho(A)^{-1}(T_1 + T_2)\}^k\}X^* \\
&= X\left\{\lim_{k\to\infty} \text{diag}\left(1, \ldots, 1, \left\{\frac{\lambda_{r+1}}{\rho(A)}\right\}^k, \ldots,\right.\right. \\
&\qquad \left.\left.\left\{\frac{\lambda_m}{\rho(A)}\right\}^k\right) + T_3\right\}X^* \\
&= X\{\text{diag}(1, \ldots, 1, 0, \ldots, 0) + T_3\}X^*,
\end{aligned}$$

where this last diagonal matrix has r 1s and T_3 is an upper triangular matrix with each diagonal element equal to 0. Clearly, this limiting matrix has rank at least r. But from Theorem 7.47(e), we see that the limiting matrix must have rank 1. This proves the result. □

To this point, we have concentrated on positive matrices. Our next step is to extend some of the results above to nonnegative matrices. We will see that many of these results generalize to the class of irreducible nonnegative matrices.

Definition 7.2. An $m \times m$ matrix A, with $m \geq 2$, is called a reducible matrix if there exist some integer r and $m \times m$ permutation matrix P such that

$$PAP' = \begin{bmatrix} B & C \\ (0) & D \end{bmatrix},$$

where B is $r \times r$, C is $r \times (m - r)$, and D is $(m - r) \times (m - r)$. If A is not reducible, then it is said to be irreducible.

We will need the following result regarding irreducible nonnegative matrices.

Theorem 7.49. An $m \times m$ nonnegative matrix A is irreducible if and only if $(\mathrm{I}_m + A)^{m-1} > (0)$.

Proof. First suppose that A is irreducible. We will show that if $\boldsymbol{x}$ is an $m \times 1$ nonnegative vector with r positive components $1 \leq r \leq m - 1$, then $(\mathrm{I}_m + A)\boldsymbol{x}$ has at least $r + 1$ positive components. Repeated use of this result verifies that $(\mathrm{I}_m + A)^{m-1} > (0)$ since each column of $\mathrm{I}_m + A$ has at least one positive component. Since $A \geq (0)$, $(\mathrm{I}_m + A)\boldsymbol{x} = \boldsymbol{x} + A\boldsymbol{x}$ must have at least r positive components. If it has exactly r positive components, then the jth component of $A\boldsymbol{x}$ must be 0 for every j for which $x_j = 0$. Equivalently, for any permutation matrix P, the jth component of $PA\boldsymbol{x}$ must be 0 for every j for which the jth component of $P\boldsymbol{x}$ is 0. If we choose a permutation matrix for which $\boldsymbol{y} = P\boldsymbol{x}$ has its $m-r$ 0s in the last $m-r$ positions, then we find that the jth component of $PA\boldsymbol{x} = PAP'\boldsymbol{y}$ must be 0 for $j = r + 1, \ldots, m$. Since $PAP' \geq (0)$ and the first r components of $\boldsymbol{y}$ are positive, PAP' would have to be of the form

$$PAP' = \begin{bmatrix} B & C \\ (0) & D \end{bmatrix}$$

Since this contradicts the fact that A is irreducible, the number of positive components in the vector $(\mathrm{I}_m + A)\boldsymbol{x}$ must exceed r. Conversely, now suppose that $(\mathrm{I}_m + A)^{m-1} > (0)$ so that, clearly, $(\mathrm{I}_m + A)^{m-1}$ is irreducible. Now A cannot be reducible since, if for some permutation matrix P,

$$PAP' = \begin{bmatrix} B & C \\ (0) & D \end{bmatrix},$$

then

$$P(\mathrm{I}_m + A)^{m-1}P' = \begin{bmatrix} \mathrm{I}_r + B & C \\ (0) & \mathrm{I}_{m-r} + D \end{bmatrix}^{m-1},$$

and the matrix on the right-hand side of this last equation has the upper triangular form given in Definition 7.2. □

We will generalize the result of Theorem 7.44 by showing that $\rho(A)$ is positive, is an eigenvalue of A, and has a positive eigenvector when A is an irreducible nonnegative matrix. But first we need the following result.

Theorem 7.50. Let A be an $m \times m$ irreducible nonnegative matrix, $\boldsymbol{x}$ be an $m \times 1$ nonnegative vector, and define the function

$$f(\boldsymbol{x}) = \min_{x_i \neq 0} x_i^{-1}(A)_{i\cdot}\boldsymbol{x} = \min_{x_i \neq 0} x_i^{-1} \sum_{j=1}^{m} a_{ij}x_j$$

Then there exists an $m \times 1$ nonnegative vector $\boldsymbol{b}$ such that $\boldsymbol{b}'\mathbf{1}_m = 1$ and $f(\boldsymbol{b}) \geq f(\boldsymbol{x})$ holds for any nonnegative $\boldsymbol{x}$.

Proof. Define the set

$$S = \{\boldsymbol{y}\colon\ \boldsymbol{y} = (\mathrm{I}_m + A)^{m-1}\boldsymbol{x}_*, \boldsymbol{x}_* \in R^m, \boldsymbol{x}_* \geq \mathbf{0}, \boldsymbol{x}_*'\mathbf{1}_m = 1\}$$

Since S is a closed and bounded set, and f is a continuous function on S due to the fact that $\boldsymbol{y} > \mathbf{0}$ if $\boldsymbol{y} \in S$, there exists a $\boldsymbol{c} \in S$ such that $f(\boldsymbol{c}) \geq f(\boldsymbol{y})$ for all $\boldsymbol{y} \in S$. Define $\boldsymbol{b} = \boldsymbol{c}/(\boldsymbol{c}'\mathbf{1}_m)$, and note that f is unaffected by scale changes, so $f(\boldsymbol{b}) = f(\boldsymbol{c})$. Let $\boldsymbol{x}$ be an arbitrary nonnegative vector and define $\boldsymbol{x}_* = \boldsymbol{x}/(\boldsymbol{x}'\mathbf{1}_m)$ and $\boldsymbol{y} = (\mathrm{I}_m + A)^{m-1}\boldsymbol{x}_*$. Now it follows from the definition of f that

$$A\boldsymbol{x}_* - f(\boldsymbol{x}_*)\boldsymbol{x}_* \geq \mathbf{0}$$

Premultiplying this equation by $(\mathrm{I}_m + A)^{m-1}$ and using the fact that $(\mathrm{I}_m + A)^{m-1}A = A(\mathrm{I}_m + A)^{m-1}$, we find that

$$A\boldsymbol{y} - f(\boldsymbol{x}_*)\boldsymbol{y} \geq \mathbf{0}$$

But $\alpha = f(\boldsymbol{y})$ is the largest value for which $A\boldsymbol{y} - \alpha\boldsymbol{y} \geq \mathbf{0}$ since at least one component of $A\boldsymbol{y} - f(\boldsymbol{y})\boldsymbol{y}$ is 0; that is, for some k, $f(\boldsymbol{y}) = y_k^{-1}(A)_{k\cdot}\boldsymbol{y}$ and, consequently, the kth component of $A\boldsymbol{y} - f(\boldsymbol{y})\boldsymbol{y}$ will be 0. Thus, we have shown that $f(\boldsymbol{y}) \geq f(\boldsymbol{x}_*) = f(\boldsymbol{x})$. The result then follows from the fact that $f(\boldsymbol{y}) \leq f(\boldsymbol{c}) = f(\boldsymbol{b})$. □

Theorem 7.51. Let A be an $m \times m$ irreducible nonnegative matrix. Then A has the positive eigenvalue $\rho(A)$ and associated with it a positive eigenvector $\boldsymbol{x}$.

Proof. We first show that $f(\boldsymbol{b})$ is a positive eigenvalue of A, where $f(\boldsymbol{b})$ is defined as in Theorem 7.50, and $\boldsymbol{b}$ is a nonnegative vector satisfying $\boldsymbol{b}'\mathbf{1}_m = 1$ and maximizing f. Since $\boldsymbol{b}$ maximizes $f(\boldsymbol{x})$ over all nonnegative $\boldsymbol{x}$, we have

$$f(\boldsymbol{b}) \geq f(m^{-1}\mathbf{1}_m) = \min_{1 \leq i \leq m} (1/m)^{-1}(A)_{i\cdot}(m^{-1}\mathbf{1}_m)$$
$$= \min_{1 \leq i \leq m} \sum_{j=1}^{m} a_{ij} > 0,$$

since A is nonnegative and irreducible. To prove that $f(\boldsymbol{b})$ is an eigenvalue of A, recall that from the definition of f it follows that $A\boldsymbol{b} - f(\boldsymbol{b})\boldsymbol{b} \geq \mathbf{0}$. If $A\boldsymbol{b} - f(\boldsymbol{b})\boldsymbol{b}$ has at least one positive component, then since $(\mathrm{I}_m + A)^{m-1} > (0)$, we must have

$$(\mathrm{I}_m + A)^{m-1}(A\boldsymbol{b} - f(\boldsymbol{b})\boldsymbol{b}) = A\boldsymbol{y} - f(\boldsymbol{b})\boldsymbol{y} > \mathbf{0},$$

where $\boldsymbol{y} = (\mathrm{I}_m + A)^{m-1}\boldsymbol{b}$. But $\alpha = f(\boldsymbol{y})$ is the largest value for which $A\boldsymbol{y} - \alpha\boldsymbol{y} \geq \mathbf{0}$, so we would have $f(\boldsymbol{y}) > f(\boldsymbol{b})$ which cannot be true since $\boldsymbol{b}$ maximizes $f(\boldsymbol{y})$ over all $\boldsymbol{y} \geq \mathbf{0}$. Thus, $A\boldsymbol{b} - f(\boldsymbol{b})\boldsymbol{b} = \mathbf{0}$ and so $f(\boldsymbol{b})$ is an eigenvalue of A and $\boldsymbol{b}$ is a corresponding eigenvector. Our next step is to show that $f(\boldsymbol{b}) = \rho(A)$ by showing that $f(\boldsymbol{b}) \geq |\lambda_i|$, where λ_i is an arbitrary eigenvalue of A. Now if $\boldsymbol{u}$ is an eigenvector of A corresponding to λ_i, then $A\boldsymbol{u} = \lambda_i\boldsymbol{u}$ or

$$\lambda_i u_h = \sum_{j=1}^{m} a_{hj} u_j$$

for $h = 1, \ldots, m$. Consequently,

$$|\lambda_i||u_h| \leq \sum_{j=1}^{m} a_{hj} |u_j|,$$

for $h = 1, \ldots, m$ or simply

$$A\,\mathrm{abs}(\boldsymbol{u}) - |\lambda_i|\,\mathrm{abs}(\boldsymbol{u}) \geq \mathbf{0},$$

and this implies that $|\lambda_i| \leq f(\mathrm{abs}(\boldsymbol{u})) \leq f(\boldsymbol{b})$. Finally, we must find a positive eigenvector associated with the eigenvalue $\rho(A) = f(\boldsymbol{b})$. We have already found a nonnegative eigenvector, $\boldsymbol{b}$. Note that $A\boldsymbol{b} = f(\boldsymbol{b})\boldsymbol{b}$ implies that $(\mathrm{I}_m + A)^{m-1}\boldsymbol{b} = \{1 + f(\boldsymbol{b})\}^{m-1}\boldsymbol{b}$, and so

$$\boldsymbol{b} = \frac{(\mathrm{I}_m + A)^{m-1}\boldsymbol{b}}{\{1 + f(\boldsymbol{b})\}^{m-1}}$$

Thus, using Theorem 7.49, we find that $\boldsymbol{b}$ is actually positive. □

The proof of the following result will be left to the reader as an exercise.

Theorem 7.52. If A is an $m \times m$ irreducible nonnegative matrix, then $\rho(A)$ is a simple eigenvalue of A.

Although $\rho(A)$ is a simple eigenvalue of an irreducible nonnegative matrix A, there may be other eigenvalues of A that have absolute value $\rho(A)$. Consequently, Theorem 7.47(e) does not immediately extend to irreducible nonnegative matrices. This leads us to the following definition.

Definition 7.3. An $m \times m$ nonnegative matrix A is said to be primitive if it is irreducible and has only one eigenvalue satisfying $|\lambda_i| = \rho(A)$.

Clearly, the result of Theorem 7.47(e) does extend to primitive matrices and this is summarized below.

Theorem 7.53. Let A be an $m \times m$ primitive nonnegative matrix and suppose that the $m \times 1$ vectors $\boldsymbol{x}$ and $\boldsymbol{y}$ satisfy $A\boldsymbol{x} = \rho(A)\boldsymbol{x}$, $A'\boldsymbol{y} = \rho(A)\boldsymbol{y}$, $\boldsymbol{x} > \boldsymbol{0}$, $\boldsymbol{y} > \boldsymbol{0}$, and $\boldsymbol{x}'\boldsymbol{y} = 1$. Then

$$\lim_{k \to \infty} \{\rho(A)^{-1}A\}^k = \boldsymbol{x}\boldsymbol{y}'$$

Our final theorem of this section gives a general limit result that holds for all irreducible nonnegative matrices. A proof of this result can be found in Horn and Johnson (1985).

Theorem 7.54. Let A be an $m \times m$ irreducible nonnegative matrix and suppose that the $m \times 1$ vectors $\boldsymbol{x}$ and $\boldsymbol{y}$ satisfy $A\boldsymbol{x} = \rho(A)\boldsymbol{x}$, $A'\boldsymbol{y} = \rho(A)\boldsymbol{y}$, and $\boldsymbol{x}'\boldsymbol{y} = \boldsymbol{1}$. Then

$$\lim_{N \to \infty} \left(N^{-1} \sum_{k=1}^{N} \{\rho(A)^{-1}A\}^k \right) = \boldsymbol{x}\boldsymbol{y}'$$

Nonnegative matrices play an important role in the study of stochastic processes. We will illustrate some of their applications to a particular type of stochastic process known as a Markov chain. Additional information on Markov chains, and stochastic processes in general, can be found in texts such as Bhattacharya and Waymire (1990), Medhi (1994), and Taylor and Karlin (1984).

Example 7.12. Suppose that we are observing some random phenomenon over time, and at any one point in time our observation can take on any one of the m values, sometimes referred to as states, $1, \dots, m$. In other words, we have a sequence of random variables X_t, for time periods $t = 0, 1, \dots$, where each random variable can be equal to any one of the numbers, $1, \dots, m$. If the probability that X_t is in state i depends only on the state that X_{t-1} is in and not on the states of prior time periods, then this process is said to be a Markov chain. If this probability also does not depend on the value of t, then the Markov chain is said to be homogeneous. In this case, the state probabilities for any time period can be computed from the initial state probabilities and what are known as the transition probabilities. We will write the initial state probability vector $\boldsymbol{p}^{(0)} = (p_1^{(0)}, \dots, p_m^{(0)})'$, where $p_i^{(0)}$ gives the probability that the process starts out at time 0 in state i. The matrix of transition probabilities is the $m \times m$ matrix P whose (i,j)th element, p_{ij}, gives the probability of X_t being in state i given that X_{t-1} is in state j. Thus, if $\boldsymbol{p}^{(t)} = (p_1^{(t)}, \dots, p_m^{(t)})'$ and $p_i^{(t)}$ is the probability that the system is in state i at time t, then, clearly,

$$\boldsymbol{p}^{(1)} = P\boldsymbol{p}^{(0)}, \qquad \boldsymbol{p}^{(2)} = P\boldsymbol{p}^{(1)} = PP\boldsymbol{p}^{(0)} = P^2\boldsymbol{p}^{(0)},$$

or for general t,

$$\boldsymbol{p}^{(t)} = P^t\boldsymbol{p}^{(0)}$$

If we have a large population of individuals subject to the random process discussed above, then $p_i^{(t)}$ could be described as the proportion of individuals in state i at time t, while $p_i^{(0)}$ would be the proportion of individuals starting out in state i. A natural question then is what is happening to these proportions as t increases? That is, can we determine the limiting behavior of $\boldsymbol{p}^{(t)}$? Note that this depends on the limiting behavior of P^t, and P is a nonnegative matrix since each of its elements is a probability. Thus, if P is a primitive matrix, we can apply Theorem 7.53. Now, since the jth column of P gives the probabilities of the various states for time period t when we are in state j at time period $t - 1$, the column sum must be 1; that is, $\mathbf{1}_m' P = \mathbf{1}_m'$ or $P'\mathbf{1}_m = \mathbf{1}_m$, so P has an eigenvalue equal to 1. Further, a simple application of Theorem 7.43 assures us that $\rho(P) \leq 1$, so we must have $\rho(P) = 1$. Consequently, if P is primitive and $\boldsymbol{\pi}$ is the $m \times 1$ positive vector satisfying $P\boldsymbol{\pi} = \boldsymbol{\pi}$ and $\boldsymbol{\pi}'\mathbf{1}_m = 1$, then

$$\lim_{t \to \infty} \{\rho(P)^{-1}P\}^t = \lim_{t \to \infty} P^t = \boldsymbol{\pi}\mathbf{1}_m'$$

Using this, we see that

$$\lim_{t \to \infty} \boldsymbol{p}^{(t)} = \lim_{t \to \infty} P^t\boldsymbol{p}^{(0)} = \boldsymbol{\pi}\mathbf{1}_m'\boldsymbol{p}^{(0)} = \boldsymbol{\pi},$$

where the last step follows from the fact that $\mathbf{1}_m'\boldsymbol{p}^{(0)} = 1$. Thus, the system approaches a point of equilibrium in which the proportions for the various states are given by the components of $\boldsymbol{\pi}$, and these proportions do not change from time period to time period. Further, this limiting behavior is not dependent upon the initial proportions in $\boldsymbol{p}^{(0)}$.

As a specific example, let us consider the problem of social mobility that involves the transition between social classes over successive generations in a family. Suppose that each individual is classified according to his occupation as being upper, middle, or lower class, and these have been labelled as states 1, 2, and 3, respectively. Suppose that the transition matrix relating a son's class to his father's class is given by

$$P = \begin{bmatrix} 0.45 & 0.05 & 0.05 \\ 0.45 & 0.70 & 0.50 \\ 0.10 & 0.25 & 0.45 \end{bmatrix},$$

so that, for instance, the probabilities that a son will have an upper, middle, or lower class occupation when his father has an upper class occupation are given by the entries in the first column of P. Since P is positive, the limiting result just discussed applies. A simple eigenanalysis of the matrix P reveals that the positive vector $\boldsymbol{\pi}$, satisfying $P\boldsymbol{\pi} = \boldsymbol{\pi}$ and $\boldsymbol{\pi}'\mathbf{1}_m = 1$, is given by $\boldsymbol{\pi} = (0.083, 0.620, 0.297)'$. Thus, if this random process satisfies the conditions of a homogeneous Markov chain, then after many generations, the male population would consist of 8.3% in the upper class, 62% in the middle class, and 29.7% in the lower class.

10. CIRCULANT AND TOEPLITZ MATRICES

In this section, we briefly discuss some structured matrices that have applications in stochastic processes and time series analysis. For a more comprehensive treatment of the first of these classes of matrices, the reader is referred to Davis (1979).

An $m \times m$ matrix A is said to be a circulant matrix if each row of A can be obtained from the previous row by a circular rotation of elements; that is, if we shift each element in the ith row over one column, with the element in the last column being shifted back to the first column, we get the $(i+1)$th row, unless $i = m$, in which case we get the first row. Thus, if the elements of the first row of A are $a_1, a_2, \dots, a_m$, then to be a circulant matrix, A must have the form

$$A = \begin{bmatrix} a_1 & a_2 & a_3 & \cdots & a_{m-1} & a_m \\ a_m & a_1 & a_2 & \cdots & a_{m-2} & a_{m-1} \\ a_{m-1} & a_m & a_1 & \cdots & a_{m-3} & a_{m-2} \\ \vdots & \vdots & \vdots & & \vdots & \vdots \\ a_3 & a_4 & a_5 & \cdots & a_1 & a_2 \\ a_2 & a_3 & a_4 & \cdots & a_m & a_1 \end{bmatrix} \tag{7.26}$$

We will sometimes use the notation $A = \mathrm{circ}(a_1, a_2, \dots, a_m)$ to refer to the circulant matrix in (7.26). One special circulant matrix, which we will denote by Π_m, is $\mathrm{circ}(0, 1, 0, \dots, 0)$. This matrix, which also can be written as

$$\Pi_m = (\boldsymbol{e}_m, \boldsymbol{e}_1, \ldots, \boldsymbol{e}_{m-1}) = \begin{bmatrix} \boldsymbol{e}_2' \\ \boldsymbol{e}_3' \\ \vdots \\ \boldsymbol{e}_m' \\ \boldsymbol{e}_1' \end{bmatrix},$$

is a permutation matrix, so $\Pi_m^{-1} = \Pi_m'$. Note that if we use $\boldsymbol{a}_1, \ldots, \boldsymbol{a}_m$ to denote the columns of an arbitrary $m \times m$ matrix A and $\boldsymbol{b}_1', \ldots, \boldsymbol{b}_m'$ to denote the rows, then

$$A\Pi_m = (\boldsymbol{a}_1, \boldsymbol{a}_2, \ldots, \boldsymbol{a}_m)(\boldsymbol{e}_m, \boldsymbol{e}_1, \ldots, \boldsymbol{e}_{m-1}) = (\boldsymbol{a}_m, \boldsymbol{a}_1, \ldots, \boldsymbol{a}_{m-1}), \tag{7.27}$$

$$\Pi_m A = \begin{bmatrix} \boldsymbol{e}_2' \\ \boldsymbol{e}_3' \\ \vdots \\ \boldsymbol{e}_m' \\ \boldsymbol{e}_1' \end{bmatrix} \begin{bmatrix} \boldsymbol{b}_1' \\ \boldsymbol{b}_2' \\ \vdots \\ \boldsymbol{b}_{m-1}' \\ \boldsymbol{b}_m \end{bmatrix} = \begin{bmatrix} \boldsymbol{b}_2' \\ \boldsymbol{b}_3' \\ \vdots \\ \boldsymbol{b}_m' \\ \boldsymbol{b}_1' \end{bmatrix}, \tag{7.28}$$

and (7.27) equals (7.28) if and only if A is of the form given in (7.26). Thus we have the following result.

Theorem 7.55. The $m \times m$ matrix A is a circulant matrix if and only if

$$A = \Pi_m A \Pi_m'$$

Our next theorem gives an expression for an $m \times m$ circulant matrix in terms of a sum of m matrices.

Theorem 7.56. The circulant matrix $A = \text{circ}(a_1, \ldots, a_m)$ can be expressed as

$$A = a_1 \mathrm{I}_m + a_2 \Pi_m + a_3 \Pi_m^2 + \cdots + a_m \Pi_m^{m-1}$$

Proof. Using (7.26), we see that

$$\begin{aligned} A = a_1 \mathrm{I}_m &+ a_2(\boldsymbol{e}_m, \boldsymbol{e}_1, \ldots, \boldsymbol{e}_{m-1}) + a_3(\boldsymbol{e}_{m-1}, \boldsymbol{e}_m, \boldsymbol{e}_1, \ldots, \boldsymbol{e}_{m-2}) + \cdots \\ &+ a_m(\boldsymbol{e}_2, \boldsymbol{e}_3, \ldots, \boldsymbol{e}_m, \boldsymbol{e}_1) \end{aligned}$$

Since the postmultiplication of any $m \times m$ matrix by Π_m shifts the columns of that matrix one place to the right, we find that

$$\Pi_m^2 = (\boldsymbol{e}_{m-1}, \boldsymbol{e}_m, \ldots, \boldsymbol{e}_{m-2})$$

$$\vdots$$

$$\Pi_m^{m-1} = (\boldsymbol{e}_2, \boldsymbol{e}_3, \ldots, \boldsymbol{e}_m, \boldsymbol{e}_1),$$

and so the result follows. □

Certain operations on circulant matrices produce another circulant matrix. Some of these are given in the following theorem.

Theorem 7.57. Let A and B be $m \times m$ circulant matrices. Then

(a) A' is circulant,
(b) for any scalars α and β, $\alpha A + \beta B$ is circulant,
(c) for any positive integer r, A^r is circulant,
(d) A^{-1} is circulant, if A is nonsingular,
(e) AB is circulant.

Proof. If $A = \text{circ}(a_1, \ldots, a_m)$ and $B = \text{circ}(b_1, \ldots, b_m)$, it follows directly from (7.26) that $A' = \text{circ}(a_1, a_m, a_{m-1}, \ldots, a_2)$ and

$$\alpha A + \beta B = \text{circ}(\alpha a_1 + \beta b_1, \ldots, \alpha a_m + \beta b_m)$$

Since A is circulant, we must have $A = \Pi_m A \Pi_m'$. But Π_m is an orthogonal matrix, so

$$A^r = (\Pi_m A \Pi_m')^r = \Pi_m A^r \Pi_m',$$

and so by Theorem 7.55, A^r is also a circulant matrix. In a similar fashion, we find that if A is nonsingular, then

$$A^{-1} = (\Pi_m A \Pi_m')^{-1} = \Pi_m'^{-1} A^{-1} \Pi_m^{-1} = \Pi_m A^{-1} \Pi_m',$$

and so A^{-1} is circulant. Finally, to prove (e), note that we must have both $A = \Pi_m A \Pi_m'$ and $B = \Pi_m B \Pi_m'$, implying that

$$AB = (\Pi_m A \Pi_m')(\Pi_m B \Pi_m') = \Pi_m AB \Pi_m',$$

and so the proof is complete. □

The representation of a circulant matrix given in Theorem 7.56 provides a simple way of proving the following result.

Theorem 7.58. Suppose that A and B are $m \times m$ circulant matrices. Then their product commutes; that is, $AB = BA$.

Proof. If $A = \text{circ}(a_1, \dots, a_m)$ and $B = \text{circ}(b_1, \dots, b_m)$, then it follows from Theorem 7.56 that

$$A = \sum_{i=1}^{m} a_i \Pi_m^{i-1}, \qquad B = \sum_{j=1}^{m} b_j \Pi_m^{j-1},$$

where $\Pi_m^0 = \text{I}_m$. Consequently,

$$\begin{aligned} AB &= \left(\sum_{i=1}^{m} a_i \Pi_m^{i-1} \right) \left(\sum_{j=1}^{m} b_j \Pi_m^{j-1} \right) = \sum_{i=1}^{m} \sum_{j=1}^{m} (a_i \Pi_m^{i-1})(b_j \Pi_m^{j-1}) \\ &= \sum_{i=1}^{m} \sum_{j=1}^{m} a_i b_j \Pi_m^{i+j-2} = \sum_{i=1}^{m} \sum_{j=1}^{m} (b_j \Pi_m^{j-1})(a_i \Pi_m^{i-1}) \\ &= \left(\sum_{j=1}^{m} b_j \Pi_m^{j-1} \right) \left(\sum_{i=1}^{m} a_i \Pi_m^{i-1} \right) = BA \end{aligned}$$

□

All circulant matrices are diagonalizable. We will show this by determining the eigenvalues and eigenvectors of a circulant matrix. But first let us find the eigenvalues and eigenvectors of the special circulant matrix Π_m.

Theorem 7.59. Let $\lambda_1, \dots, \lambda_m$ be the m solutions to the polynomial equation $\lambda^m - 1 = 0$; that is, $\lambda_j = \theta^{j-1}$, where $\theta = \exp(2\pi i/m) = \cos(2\pi/m) + i\sin(2\pi/m)$ and $i = \sqrt{-1}$. Define Λ to be the diagonal matrix $\text{diag}(1, \theta, \dots, \theta^{m-1})$ and let

$$F = \frac{1}{\sqrt{m}} \begin{bmatrix} 1 & 1 & 1 & \cdots & 1 \\ 1 & \theta & \theta^2 & \cdots & \theta^{m-1} \\ 1 & \theta^2 & \theta^4 & \cdots & \theta^{2(m-1)} \\ \vdots & \vdots & \vdots & & \vdots \\ 1 & \theta^{m-1} & \theta^{2(m-1)} & \cdots & \theta^{(m-1)(m-1)} \end{bmatrix}$$

Then the diagonalization of Π_m is given by $\Pi_m = F\Lambda F^*$, where F^* is the conjugate transpose of F; that is, the diagonal elements of Λ are the eigenvalues of Π_m, while the columns of F are corresponding eigenvectors.

Proof. The eigenvalue–eigenvector equation, $\Pi_m \boldsymbol{x} = \lambda \boldsymbol{x}$, yields the equations

$$x_{j+1} = \lambda x_j,$$

for $j = 1, \dots, m-1$, and

$$x_1 = \lambda x_m$$

After repeated substitution, we obtain for any j, $x_j = \lambda^m x_j$. Thus, $\lambda^m = 1$, and so the eigenvalues of Π_m are $1, \theta, \dots, \theta^{m-1}$. Substituting the eigenvalue θ^{j-1} and $x_1 = m^{-1/2}$ into the equations above, we find that an eigenvector corresponding to the eigenvalue θ^{j-1} is given by $\boldsymbol{x} = m^{-1/2}(1, \theta^{j-1}, \dots, \theta^{(m-1)(j-1)})'$. Thus, we have shown that the diagonal elements of Λ are the eigenvalues of Π_m and the columns of F are corresponding eigenvectors. The remainder of the proof, which simply involves the verification that $F^{-1} = F^*$, is left to the reader as an exercise. □

The matrix F given in Theorem 7.59 is sometimes referred to as the Fourier matrix of order m. The diagonalization of an arbitrary circulant matrix, which follows directly from Theorems 7.56 and 7.59, is given in our next theorem.

Theorem 7.60. Let A be the $m \times m$ circulant matrix circ$(a_1, \dots, a_m)$. Then

$$A = F\Delta F^*,$$

where $\Delta = \text{diag}(\delta_1, \dots, \delta_m)$, $\delta_j = a_1 + a_2\lambda_j^1 + \cdots + a_m\lambda_j^{m-1}$, and λ_j and F are defined as in Theorem 7.59.

Proof. Since $\Pi_m = F\Lambda F^*$ and $FF^* = \text{I}_m$, we have $\Pi_m^j = F\Lambda^j F^*$, for $j = 2, \dots, m-1$, and so by using Theorem 7.56, we find that

$$\begin{aligned} A &= a_1\text{I}_m + a_2\Pi_m + a_3\Pi_m^2 + \cdots + a_m\Pi_m^{m-1} \\ &= a_1FF^* + a_2F\Lambda^1F^* + a_3F\Lambda^2F^* + \cdots + a_mF\Lambda^{m-1}F^* \\ &= F(a_1\text{I}_m + a_2\Lambda^1 + a_3\Lambda^2 + \cdots + a_m\Lambda^{m-1})F^* = F\Delta F^* \end{aligned}$$ □

The class of circulant matrices is a subclass of a larger class of matrices known as Toeplitz matrices. The elements of an $m \times m$ Toeplitz matrix A satisfy $a_{ij} = a_{j-i}$ for scalars $a_{-m+1}, a_{-m+2}, \dots, a_{m-1}$; that is, A has the form

$$A = \begin{bmatrix} a_0 & a_1 & a_2 & \cdots & a_{m-2} & a_{m-1} \\ a_{-1} & a_0 & a_1 & \cdots & a_{m-3} & a_{m-2} \\ a_{-2} & a_{-1} & a_0 & \cdots & a_{m-4} & a_{m-3} \\ \vdots & \vdots & \vdots & & \vdots & \vdots \\ a_{-m+2} & a_{-m+3} & a_{-m+4} & \cdots & a_0 & a_1 \\ a_{-m+1} & a_{-m+2} & a_{-m+3} & \cdots & a_{-1} & a_0 \end{bmatrix}$$

If $a_j = a_{-j}$ for $j = 1, \ldots, m-1$, then the matrix A is a symmetric Toeplitz matrix. One important and fairly simple symmetric Toeplitz matrix is one that has $a_j = a_{-j} = 0$ for $j = 2, \ldots, m-1$, so that

$$A = \begin{bmatrix} a_0 & a_1 & 0 & \cdots & 0 & 0 \\ a_1 & a_0 & a_1 & \cdots & 0 & 0 \\ 0 & a_1 & a_0 & \cdots & 0 & 0 \\ \vdots & \vdots & \vdots & & \vdots & \vdots \\ 0 & 0 & 0 & \cdots & a_0 & a_1 \\ 0 & 0 & 0 & \cdots & a_1 & a_0 \end{bmatrix} \tag{7.29}$$

Some specialized results, such as formulas for eigenvalues and formulas for the computation of the inverse of a Toeplitz matrix, can be found in Grenander and Szego (1984) and Heinig and Rost (1984).

11. HADAMARD AND VANDERMONDE MATRICES

In this section, we discuss some matrices that have applications in the areas of design of experiments and response surface methodology. We begin with a class of matrices known as Hadamard matrices. An $m \times m$ matrix H is said to be a Hadamard matrix if first, each element of H is either +1 or -1, and second, H satisfies

$$H'H = HH' = m\mathrm{I}_m; \tag{7.30}$$

that is, the columns of H form an orthogonal set of vectors, and the rows form an orthogonal set as well. For instance, a 2×2 Hadamard matrix is given by

$$H = \begin{bmatrix} 1 & 1 \\ 1 & -1 \end{bmatrix},$$

while a 4×4 Hadamard matrix is given by

$$H = \begin{bmatrix} 1 & 1 & 1 & 1 \\ -1 & -1 & 1 & 1 \\ 1 & -1 & 1 & -1 \\ 1 & -1 & -1 & 1 \end{bmatrix}$$

Some of the basic properties of Hadamard matrices are given in the following theorem.

Theorem 7.61. Let H_m denote any $m \times m$ Hadamard matrix. Then

(a) $m^{-1/2}H_m$ is an $m \times m$ orthogonal matrix,
(b) $|H_m| = \pm m^{m/2}$,
(c) $H_m \otimes H_n$ is an $mn \times mn$ Hadamard matrix.

Proof. (a) follows directly from (7.30). Also using (7.30), we find that

$$|H'_m H_m| = |m\mathrm{I}_m| = m^m$$

But

$$|H'_m H_m| = |H'_m||H_m| = |H_m|^2,$$

and so (b) follows. To prove (c), note that each element of $H_m \otimes H_n$ is +1 or -1 since each element is the product of an element from H_m and an element from H_n, and

$$(H_m \otimes H_n)'(H_m \otimes H_n) = H'_m H_m \otimes H'_n H_n = m\mathrm{I}_m \otimes n\mathrm{I}_n = mn\mathrm{I}_{mn} \qquad \square$$

Hadamard matrices which have all of the elements of the first row equal to +1 are called normalized Hadamard matrices. Our next result addresses the existence of normalized Hadamard matrices.

Theorem 7.62. If there exists an $m \times m$ Hadamard matrix, then there exists an $m \times m$ normalized Hadamard matrix.

Proof. Suppose that H is an $m \times m$ Hadamard matrix. Let D be the diagonal matrix with the elements of the first row of H as its diagonal elements; that is, $D = \mathrm{diag}(h_{11}, \dots, h_{1m})$. Note that $D^2 = \mathrm{I}_m$ since each diagonal element of D is +1 or -1. Consider the $m \times m$ matrix $H_* = HD$. Each column of H_* is the corresponding column of H multiplied by either +1 or -1, so clearly each element of H_* is +1 or -1. The jth element in the first row of H_* is $h_{1j}^2 = 1$, so H_* has all of its elements of the first row equal to +1. In addition,

$$H_*' H_* = (HD)' HD = D' H' HD = D(m\mathrm{I}_m)D = mD^2 = m\mathrm{I}_m$$

Thus, H_* is an $m \times m$ normalized Hadamard matrix and so the proof is complete. □

Hadamard matrices of size $m \times m$ do not exist for every choice of m. We have already given an example of a 2×2 Hadamard matrix, and this matrix can be used repeatedly in Theorem 7.61(c) to obtain a $2^n \times 2^n$ Hadamard matrix for any $n \geq 2$. However, $m \times m$ Hadamard matrices do exist for some values of $m \neq 2^n$. Our next result gives a necessary condition on the order m so that Hadamard matrices of order m exist.

Theorem 7.63. If H is an $m \times m$ Hadamard matrix, where $m > 2$, then m is a multiple of 4.

Proof. The result can be proven by using the fact that any three rows of H are orthogonal to one another. Consequently, we will refer to the first three rows of H, and, due to Theorem 7.62, we may assume that H is a normalized Hadamard matrix, so that all of the elements in the first row are +1. Since the second and third rows are orthogonal to the first row, they must each have r +1s and r −1s, where $r = n/2$; thus clearly,

$$n = 2r, \tag{7.31}$$

or in other words, n is a multiple of 2. Let n_{+-} be the number of columns in which row 2 has a +1 and row 3 has a −1. Similarly, define n_{-+}, n_{++}, and n_{--}. Note that the value of any one of these ns determines the others since $n_{++} + n_{+-} = r$, $n_{++} + n_{-+} = r$, and $n_{--} + n_{+-} = r$. For instance, if $n_{++} = s$, then $n_{+-} = (r - s), n_{-+} = (r - s)$, and $n_{--} = s$. But the orthogonality of rows 2 and 3 guarantee that $n_{++} + n_{--} = n_{-+} + n_{+-}$, which yields the relationship

$$2s = 2(r - s)$$

Thus, $r = 2s$, and so using (7.31) we get $n = 4s$, which completes the proof. □

Some additional results on Hadamard matrices can be found in Hedayat and Wallis (1978) and Agaian (1985).

An $m \times m$ matrix A is said to be a Vandermonde matrix if it has the form

$$A = \begin{bmatrix} 1 & 1 & 1 & \cdots & 1 \\ a_1 & a_2 & a_3 & \cdots & a_m \\ a_1^2 & a_2^2 & a_3^2 & \cdots & a_m^2 \\ \vdots & \vdots & \vdots & & \vdots \\ a_1^{m-1} & a_2^{m-1} & a_3^{m-1} & \cdots & a_m^{m-1} \end{bmatrix} \tag{7.32}$$

For instance, if F is the $m \times m$ Fourier matrix discussed in Section 7.10 then $A = m^{1/2}F$ is a Vandermonde matrix with $a_i = \theta^{i-1}$, for $i = 1, \dots, m$. Our final result of this chapter gives an expression for the determinant of a Vandermonde matrix.

Theorem 7.64. Let A be the $m \times m$ Vandermonde matrix given in (7.31). Then its determinant is given by

$$|A| = \prod_{1 \le i < j \le m} (a_j - a_i) \tag{7.33}$$

Proof. Our proof is by induction. For $m = 2$, we find that

$$|A| = \begin{vmatrix} 1 & 1 \\ a_1 & a_2 \end{vmatrix} = a_2 - a_1,$$

and so (7.33) holds when A is 2×2. Next we assume that (7.33) holds for Vandermonde matrices of order $m - 1$ and show that then it must also hold for order m. Thus, if B is the $(m-1) \times (m-1)$ matrix obtained from A by deleting its last row and first column, then, since B is a Vandermonde matrix of order $m - 1$, we must have

$$|B| = \prod_{2 \le i < j \le m} (a_j - a_i)$$

Define the $m \times m$ matrix

$$C = \begin{bmatrix} 1 & 0 & 0 & \cdots & 0 & 0 \\ -a_1 & 1 & 0 & \cdots & 0 & 0 \\ 0 & -a_1 & 1 & \cdots & 0 & 0 \\ \vdots & \vdots & \vdots & & \vdots & \vdots \\ 0 & 0 & 0 & \cdots & 1 & 0 \\ 0 & 0 & 0 & \cdots & -a_1 & 1 \end{bmatrix},$$

and note that by repeatedly using the cofactor expansion formula for a determinant on the first row, we find that $|C| = 1$. Thus, $|A| = |CA|$. But it is easily verified that $CA = E$, where

$$E = \begin{bmatrix} 1 & \mathbf{1}'_{m-1} \\ \mathbf{0} & BD \end{bmatrix},$$

and $D = \text{diag}((a_2 - a_1), (a_3 - a_1), \ldots, (a_m - a_1))$. Consequently,

$$|A| = |E| = |BD| = |B||D| = \left\{ \prod_{2 \le i < j \le m} (a_j - a_i) \right\} \left\{ \prod_{2 \le j \le m} (a_j - a_1) \right\}$$

$$= \prod_{1 \le i < j \le m} (a_j - a_i),$$

where the second equality was obtained by using the cofactor expansion formula on the first column of E. This completes the proof. □

PROBLEMS

1. Consider the $2m \times 2m$ matrix

$$A = \begin{bmatrix} a\mathrm{I}_m & b\mathrm{I}_m \\ c\mathrm{I}_m & d\mathrm{I}_m \end{bmatrix},$$

where a, b, c, and d are nonzero scalars.
(a) Give an expression for the determinant of A.
(b) For what values of a, b, c, and d will A be nonsingular?
(c) Find an expression for A^{-1}.

2. Let A be of the form

$$A = \begin{bmatrix} A_{11} & A_{12} \\ A_{21} & (0) \end{bmatrix},$$

where each submatrix is $m \times m$ and the matrices A_{12} and A_{21} are nonsingular. Find an expression for the inverse of A in terms of A_{11}, A_{12}, and A_{21} by utilizing equations (7.2)–(7.5).

3. Generalize Example 7.2 by obtaining the determinant, conditions for nonsingularity, and the inverse of the $2m \times 2m$ matrix

$$A = \begin{bmatrix} a\mathrm{I}_m & c\mathbf{1}_m\mathbf{1}_m' \\ d\mathbf{1}_m\mathbf{1}_m' & b\mathrm{I}_m \end{bmatrix},$$

where a, b, c, and d are nonzero scalars.

4. Let the matrix G be given by

$$G = \begin{bmatrix} A & B & C \\ (0) & D & E \\ (0) & (0) & F \end{bmatrix},$$

where each of the matrices A, D, and F is square and nonsingular. Find the inverse of G.

5. Use Theorems 7.1 and 7.4 to find the determinant and inverse of the matrix

$$A = \begin{bmatrix} 4 & 0 & 0 & 1 & 2 \\ 0 & 3 & 0 & 1 & 2 \\ 0 & 0 & 2 & 2 & 3 \\ 0 & 0 & 1 & 2 & 3 \\ 1 & 1 & 0 & 1 & 2 \end{bmatrix}$$

6. Let A be an $m \times n$ matrix partitioned as

$$A = \begin{bmatrix} A_{11} & A_{12} \\ A_{21} & A_{22} \end{bmatrix},$$

where A_{11} is $r \times r$ and $\text{rank}(A) = \text{rank}(A_{11}) = r$.

(a) Show that $A_{22} = A_{21}A_{11}^{-1}A_{12}$.

(b) Use the result of part (a) to show that

$$B = \begin{bmatrix} A_{11}^{-1} & (0) \\ (0) & (0) \end{bmatrix}$$

is a generalized inverse of A.

(c) Show that the Moore–Penrose inverse of A is given by

$$A^+ = \begin{bmatrix} A'_{11} \\ A'_{12} \end{bmatrix} C[A'_{11} \quad A'_{21}],$$

where $C = (A_{11}A'_{11} + A_{12}A'_{12})^{-1}A_{11}(A'_{11}A_{11} + A'_{21}A_{21})^{-1}$.

7. Use Theorem 7.4 to show that if

$$A = \begin{bmatrix} A_{11} & A_{12} \\ A_{21} & A_{22} \end{bmatrix}$$

is nonsingular and A_{11} is nonsingular, then $A_{22} - A_{21}A_{11}^{-1}A_{12}$ is nonsingular.

8. Let A be an $m \times m$ positive definite matrix and let B be its inverse. Partition A and B as

$$A = \begin{bmatrix} A_{11} & A_{12} \\ A'_{12} & A_{22} \end{bmatrix}, \qquad B = \begin{bmatrix} B_{11} & B_{12} \\ B'_{12} & B_{22} \end{bmatrix},$$

where A_{11} and B_{11} are $r \times r$ matrices. Show that the matrix

$$\begin{bmatrix} A_{11} - B_{11}^{-1} & A_{12} \\ A'_{12} & A_{22} \end{bmatrix}$$

is positive semidefinite with rank of $m - r$.

9. Consider the $m \times m$ matrix

$$A = \begin{bmatrix} A_{11} & \boldsymbol{a} \\ \boldsymbol{a}' & a_{mm} \end{bmatrix},$$

where the $(m-1) \times (m-1)$ matrix A_{11} is positive definite.
(a) Prove that $|A| \leq a_{mm}|A_{11}|$ with equality if and only if $\boldsymbol{a} = \mathbf{0}$.
(b) Use part (a) to obtain an alternative proof of Theorem 7.23; that is, generalize the result of part (a) by proving that if $a_{11}, \ldots, a_{mm}$ are the diagonal elements of a positive definite matrix A, then $|A| \leq a_{11} \cdots a_{mm}$ with equality if and only if A is a diagonal matrix.

10. Let A be an $m \times m$ matrix and define A_i to be the $i \times i$ matrix obtained by deleting the last $m - i$ rows and columns of A. The leading principal minors of A are given by the determinants, $|A_1|, \ldots, |A_m|$, where $A_m = A$.

Show that if A is a symmetric matrix, then it is positive definite if and only if all of its leading principal minors are positive.

11. Let the 2×2 matrices A and B be given by

$$A = \begin{bmatrix} 2 & 3 \\ 1 & 2 \end{bmatrix}, \qquad B = \begin{bmatrix} 5 & 3 \\ 3 & 2 \end{bmatrix}$$

   (a) Compute $A \otimes B$ and $B \otimes A$.
   (b) Find $\text{tr}(A \otimes B)$.
   (c) Compute $|A \otimes B|$.
   (d) Give the eigenvalues of $A \otimes B$.
   (e) Find $(A \otimes B)^{-1}$.

12. Give a simplified expression for $\mathrm{I}_m \otimes \mathrm{I}_n$.

13. Prove the properties given in Theorem 7.6.

14. Prove results (b) and (c) of Theorem 7.9.

15. Show that if A and B are symmetric matrices, then $A \otimes B$ is also symmetric.

16. Find the rank of $A \otimes B$, where

$$A = \begin{bmatrix} 2 & 6 \\ 1 & 4 \\ 3 & 1 \end{bmatrix}, \qquad B = \begin{bmatrix} 5 & 2 & 4 \\ 2 & 1 & 1 \\ 1 & 0 & 2 \end{bmatrix}$$

17. For matrices A and B of any size, show that $A \otimes B = (0)$ if and only if $A = (0)$ or $B = (0)$.

18. Let $\boldsymbol{x}_i$ be an eigenvector of the $m \times m$ matrix A corresponding to the eigenvalue λ_i. Let $\boldsymbol{y}_j$ be an eigenvector of the $p \times p$ matrix B corresponding to the eigenvalue θ_j.
   (a) Show that $\boldsymbol{x}_i \otimes \boldsymbol{y}_j$ is an eigenvector of $A \otimes B$.
   (b) Give an example of matrices A and B such that $A \otimes B$ has an eigenvector that is not the Kronecker product of an eigenvector of A and an eigenvector of B.

19. Show that if A and B are positive definite matrices, then $A \otimes B$ is also positive definite.

20. Let $\boldsymbol{x}$ be an $m \times 1$ vector and $\boldsymbol{y}$ be an $n \times 1$ vector. Verify that the three matrices $\boldsymbol{x}\boldsymbol{y}'$, $\boldsymbol{y}' \otimes \boldsymbol{x}$, and $\boldsymbol{x} \otimes \boldsymbol{y}'$ are identical.

21. Compute the sum of squared errors $\text{SSE} = (\boldsymbol{y} - \hat{\boldsymbol{y}})'(\boldsymbol{y} - \hat{\boldsymbol{y}})$ for the two-way classification model with interaction discussed in Example 7.5.

22. Consider the two-way classification model without interaction given by

$$y_{ijk} = \mu + \tau_i + \gamma_j + \epsilon_{ijk},$$

where $i = 1, \ldots, a$, $j = 1, \ldots, b$, and $k = 1, \ldots, n$.

(a) Find a least squares solution for $\boldsymbol{\beta} = (\mu, \tau_1, \ldots, \tau_a, \gamma_1, \ldots, \gamma_b)'$, and use this to obtain the vector of fitted values and the sum of squared errors for this model.

(b) Compute the sum of squared errors for the reduced model $y_{ijk} = \mu + \gamma_j + \epsilon_{ijk}$ and use this along with the SSE computed in (a) to show that the sum of squares for factor A is

$$\text{SSA} = nb \sum_{i=1}^{a} (\bar{y}_{i\cdot} - \bar{y}_{\cdot\cdot})^2$$

(c) In a similar fashion, show that the sum of squares for factor B is

$$\text{SSB} = na \sum_{j=1}^{b} (\bar{y}_{\cdot j} - \bar{y}_{\cdot\cdot})^2$$

(d) Find a set of as many linearly independent estimable functions of μ, τ_i, and γ_j as possible.

(e) Use the sum of squared errors computed in (a) and the sum of squared errors computed in Problem 21 to show that the sum of squares for interaction in the model of Problem 21 is given by

$$\text{SSAB} = n \sum_{i=1}^{a} \sum_{j=1}^{b} (\bar{y}_{ij} - \bar{y}_{i\cdot} - \bar{y}_{\cdot j} + \bar{y}_{\cdot\cdot})^2$$

23. Prove Theorem 7.13.

24. Let A_1, A_2, A_3, and A_4 be square matrices. Show that, when the sizes of these matrices are such that the appropriate operations are defined,

(a) $(A_1 \oplus A_2) + (A_3 \oplus A_4) = (A_1 + A_3) \oplus (A_2 + A_4)$,

(b) $(A_1 \oplus A_2)(A_3 \oplus A_4) = A_1A_3 \oplus A_2A_4$,
(c) $(A_1 \oplus A_2) \otimes A_3 = (A_1 \otimes A_3) \oplus (A_2 \otimes A_3)$.

25. Give an example to show that, in general,

$$A_1 \otimes (A_2 \oplus A_3) \neq (A_1 \otimes A_2) \oplus (A_1 \otimes A_3)$$

26. Complete the details of Example 7.7; that is, use Theorem 6.4 and Theorem 7.16 to prove Theorem 6.5.

27. Prove the results of Corollary 7.17.1.

28. Let A and B be $m \times n$ and $n \times p$ matrices, respectively, while $\boldsymbol{c}$ and $\boldsymbol{d}$ are $p \times 1$ and $n \times 1$ vectors. Show that
(a) $AB\boldsymbol{c} = (\boldsymbol{c}' \otimes A)\,\text{vec}(B) = (A \otimes \boldsymbol{c}')\,\text{vec}(B')$,
(b) $\boldsymbol{d}'B\boldsymbol{c} = (\boldsymbol{c}' \otimes \boldsymbol{d}')\,\text{vec}(B)$.

29. For any matrix A and any vector $\boldsymbol{b}$, show that

$$\text{vec}(A \otimes \boldsymbol{b}) = \text{vec}(A) \otimes \boldsymbol{b}$$

30. Let A be an $m \times m$ matrix, B be an $n \times n$ matrix, and C be an $m \times n$ matrix. Prove that

$$\text{vec}(AC + CB) = \{(\text{I}_n \otimes A) + (B' \otimes \text{I}_m)\}\text{vec}(C)$$

31. If $\boldsymbol{e}_i$ is the ith column of the identity matrix I_m, verify that

$$\text{vec}(\text{I}_m) = \sum_{i=1}^{m} (\boldsymbol{e}_i \otimes \boldsymbol{e}_i)$$

32. Prove property (h) of Theorem 7.18.

33. Let the 2×2 matrices A and B be given by

$$A = \begin{bmatrix} 1 & 2 \\ 2 & 4 \end{bmatrix}, \qquad B = \begin{bmatrix} 4 & 1 \\ 1 & 3 \end{bmatrix}$$

(a) Compute $A \odot B$.
(b) Which of the matrices, A, B, and $A \odot B$, are positive definite or positive semidefinite? How does this relate to Theorem 7.22?

34. Give an example of matrices A and B such that neither is nonnegative definite, yet $A \odot B$ is positive definite.

35. Let A, B, and C be $m \times n$ matrices. Show that

$$\mathrm{tr}\{(A' \odot B')C\} = \mathrm{tr}\{A'(B \odot C)\}$$

36. Suppose that the $m \times m$ matrix A is diagonalizable; that is, there exist a nonsingular matrix X and a diagonal matrix $\Lambda = \mathrm{diag}(\lambda_1, \dots, \lambda_m)$ such that $A = X\Lambda X^{-1}$. Show that if we define the vector of diagonal elements of A, $\boldsymbol{a} = (a_{11}, \dots, a_{mm})'$ and the vector of eigenvalues of A, $\boldsymbol{\lambda} = (\lambda_1, \dots, \lambda_m)'$, then

$$(X \odot X'^{-1})\boldsymbol{\lambda} = \boldsymbol{a},$$

and

$$(X \odot X'^{-1})\mathbf{1}_m = (X \odot X'^{-1})'\mathbf{1}_m = \mathbf{1}_m$$

37. Let A and B be $m \times m$ nonnegative definite matrices. Show that
(a) $|A \odot B| \geq |A||B|$,
(b) $|A \odot A^{-1}| \geq 1$, if A is positive definite.

38. For each of the following pairs of 2×2 matrices, compute the smaller eigenvalue $\lambda_2(A \odot B)$ and the lower bounds for this eigenvalue given by Theorem 7.26 and Theorem 7.28. Which bound is closer to the actual value?

$$(a)\ A = \begin{bmatrix} 4 & 0 \\ 0 & 1 \end{bmatrix}, \qquad B = \begin{bmatrix} 2 & 0 \\ 0 & 3 \end{bmatrix}$$

$$(b)\ A = \begin{bmatrix} 1 & 0 \\ 0 & 1 \end{bmatrix}, \qquad B = \begin{bmatrix} 2 & \sqrt{2} \\ \sqrt{2} & 3 \end{bmatrix}$$

39. Let A be an $m \times m$ positive definite matrix. Use Theorem 7.24 to show that, if $B = A^{-1}$ then $a_{11}b_{11} \geq 1$. Show how this generalizes to $a_{ii}b_{ii} \geq 1$ for $i = 1, \dots, m$.

40. Let A and B be $m \times m$ positive definite matrices and consider the inequality

$$|A \odot B| + |A||B| \geq |A| \prod_{i=1}^{m} b_{ii} + |B| \prod_{i=1}^{m} a_{ii}$$

(a) Show that this inequality is equivalent to

$$|R_A \odot R_B| + |R_A||R_B| \geq |R_A| + |R_B|,$$

where R_A and R_B represent the correlation matrices computed from A and B.

(b) Use Theorem 7.25 on $|R_A \odot C|$, where $C = R_B - (\boldsymbol{e}_1' R_B^{-1} \boldsymbol{e}_1)^{-1} \boldsymbol{e}_1 \boldsymbol{e}_1'$, to establish the inequality given in (a).

41. Suppose that A and B are $m \times m$ positive definite matrices. Show that $A \odot B = AB$ if and only if both A and B are diagonal matrices.

42. Let A be an $m \times m$ positive definite matrix and B be an $m \times m$ positive semidefinite matrix with exactly r positive diagonal elements. Show that $\text{rank}(A \odot B) = r$.

43. Show that if A and B are singular 2×2 matrices then $A \odot B$ is also singular.

44. Let R be an $m \times m$ positive definite correlation matrix having λ as its smallest eigenvalue. Show that if τ is the smallest eigenvalue of $R \odot R$ and $R \neq \text{I}_m$, then $\tau > \lambda$.

45. Consider the matrix

$$\Psi_m = \sum_{i=1}^{m} \boldsymbol{e}_{i,m}(\boldsymbol{e}_{i,m} \otimes \boldsymbol{e}_{i,m})',$$

which we have seen satisfies $\Psi_m(A \otimes B)\Psi_m' = A \odot B$ for any $m \times m$ matrices A and B. Define $\text{w}(A)$ to be the $m \times 1$ vector containing the diagonal elements of A; that is, $\text{w}(A) = (a_{11}, \ldots, a_{mm})'$. Also let Λ_m be the $m^2 \times m^2$ matrix given by

$$\Lambda_m = \sum_{i=1}^{m} (E_{ii} \otimes E_{ii}) = \sum_{i=1}^{m} (\boldsymbol{e}_{i,m} \boldsymbol{e}_{i,m}' \otimes \boldsymbol{e}_{i,m} \boldsymbol{e}_{i,m}')$$

Show that

(a) $\Psi_m' \text{w}(A) = \text{vec}(A)$ for every diagonal matrix A,

(b) $\Psi_m \text{vec}(A) = \text{w}(A)$ for every matrix A,

(c) $\Psi_m \Psi_m' = \text{I}_m$ so that $\Psi_m^+ = \Psi_m'$,

(d) $\Psi_m' \Psi_m = \Lambda_m$,

(e) $\Lambda_m N_m = N_m \Lambda_m = \Lambda_m$,

(f) $\{\text{vec}(A)\}'\Lambda_m(B \otimes B)\Lambda_m \,\text{vec}(A) = \{\text{w}(A)\}'(B \odot B)\text{w}(A)$.

Additional properties of Ψ_m can be found in Magnus (1988).

46. Verify that the commutation matrix K_{mn} is a permutation matrix; that is, show that each column of K_{mn} is a column of I_{mn} and each column of I_{mn} is a column of K_{mn}.

47. Write out the commutation matrices K_{22} and K_{24}.

48. The eigenvalues of K_{mm} were given in Theorem 7.33. Show that corresponding eigenvectors are given by the vectors of the form $\boldsymbol{e}_l \otimes \boldsymbol{e}_l$, $(\boldsymbol{e}_l \otimes \boldsymbol{e}_k) + (\boldsymbol{e}_k \otimes \boldsymbol{e}_l)$, and $(\boldsymbol{e}_l \otimes \boldsymbol{e}_k) - (\boldsymbol{e}_k \otimes \boldsymbol{e}_l)$.

49. Show that the commutation matrix K_{mn} can be expressed as

$$K_{mn} = \sum_{i=1}^{m} (\boldsymbol{e}_i \otimes \text{I}_n \otimes \boldsymbol{e}_i'),$$

where $\boldsymbol{e}_i$ is the ith column of I_m. Use this to show that if A is $n \times m$, $\boldsymbol{x}$ is $m \times 1$, $\boldsymbol{y}$ is an arbitrary vector, then

$$K_{mn}'(\boldsymbol{x} \otimes A \otimes \boldsymbol{y}') = A \otimes \boldsymbol{x}\boldsymbol{y}'$$

50. Let A be an $m \times n$ matrix with rank r and let $\lambda_1, \ldots, \lambda_r$ be the nonzero eigenvalues of $A'A$. If we define

$$P = K_{mn}(A' \otimes A),$$

show that

(a) P is symmetric,

(b) $\text{rank}(P) = r^2$,

(c) $\text{tr}(P) = \text{tr}(A'A)$,

(d) $P^2 = (AA') \otimes (A'A)$,

(e) the nonzero eigenvalues of P are $\lambda_1, \ldots, \lambda_r$ and $\pm(\lambda_i\lambda_j)^{1/2}$ for all $i < j$.

51. Prove the results of Theorem 7.35.

52. Show that if A and B are $m \times m$ matrices, then

$$N_m(A \otimes B + B \otimes A)N_m = (A \otimes B + B \otimes A)N_m = N_m(A \otimes B + B \otimes A)$$
$$= 2N_m(A \otimes B)N_m$$

53. Write out the matrices N_2 and N_3.

54. For $i = 1, \dots, m$, $j = 1, \dots, i$, define the $m(m+1)/2 \times 1$ vector $\boldsymbol{u}_{ij}$ to be the vector with one in its $\{(j-1)m + i - j(j-1)/2\}$th position and zeros elsewhere. It can be easily verified that these vectors are the columns of the identity matrix of order $m(m+1)/2$; that is,

$$\mathrm{I}_{m(m+1)/2} = (\boldsymbol{u}_{11}, \boldsymbol{u}_{21}, \dots, \boldsymbol{u}_{m1}, \boldsymbol{u}_{22}, \dots, \boldsymbol{u}_{m2}, \boldsymbol{u}_{33}, \dots, \boldsymbol{u}_{mm})$$

Let E_{ij} be the $m \times m$ matrix whose only nonzero element is a one in the (i,j)th position, and define

$$T_{ij} = \begin{cases} E_{ij} + E_{ji}, & \text{if } i \neq j, \\ E_{ii}, & \text{if } i = j \end{cases}$$

Show that $D_m = \sum_{i \geq j} \{\mathrm{vec}(T_{ij})\}\boldsymbol{u}'_{ij}$; that is, verify that

$$\sum_{i \geq j} \{\mathrm{vec}(T_{ij})\}\boldsymbol{u}'_{ij}\, \mathrm{v}(A) = \mathrm{vec}(A),$$

where A is an arbitrary $m \times m$ symmetric matrix.

55. Prove the results of Theorem 7.40.

56. If A is an $m \times m$ matrix show that
(a) $D_m D_m^+(A \otimes A)D_m = (A \otimes A)D_m$,
(b) $\{D_m^+(A \otimes A)D_m\}^i = D_m^+(A^i \otimes A^i)D_m$, where i is any positive integer.

57. If $\boldsymbol{u}_{ij}$ and E_{ij} are defined as in Problem 54, show that $L'_m = \sum_{i \geq j}\{\mathrm{vec}(E_{ij})\}\boldsymbol{u}'_{ij}$; that is, verify that

$$\sum_{i \geq j} \{\mathrm{vec}(E_{ij})\}\boldsymbol{u}'_{ij}\, \mathrm{v}(A) = \mathrm{vec}(A),$$

where A is an arbitrary $m \times m$ lower triangular matrix.

58. Prove Theorem 7.41.

59. For $i = 2, \dots, m$, $j = 1, \dots, i-1$, define the $m(m-1)/2 \times 1$ vector $\tilde{\boldsymbol{u}}_{ij}$ to be the vector with one in its $\{(j-1)m + i - j(j+1)/2\}$th position and zeros elsewhere. It can be easily verified that these vectors are the columns of the identity matrix of order $m(m-1)/2$; that is,

$$\mathrm{I}_{m(m-1)/2} = (\tilde{\boldsymbol{u}}_{21}, \dots, \tilde{\boldsymbol{u}}_{m1}, \tilde{\boldsymbol{u}}_{32}, \dots, \tilde{\boldsymbol{u}}_{m2}, \tilde{\boldsymbol{u}}_{43}, \dots, \tilde{\boldsymbol{u}}_{mm-1})$$

Show that $\tilde{L}_m' = \sum_{i>j}\{\mathrm{vec}(E_{ij})\}\tilde{\boldsymbol{u}}_{ij}'$; that is, verify that

$$\sum_{i>j} \{\mathrm{vec}(E_{ij})\}\tilde{\boldsymbol{u}}_{ij}' \, \tilde{\mathrm{v}}(A) = \mathrm{vec}(A),$$

where A is an arbitrary $m \times m$ strictly lower triangular matrix.

60. Prove the results of Theorem 7.42.

61. Find a 2×2 nonnegative matrix A which has its spectral radius equal to 1, yet A^k does not converge to anything as $k \to \infty$.

62. Show that if A is a nonnegative matrix and, for some positive integer k, A^k is a positive matrix, then $\rho(A) > 0$.

63. It can be shown [see, for example, Horn and Johnson (1985)] that if A is an $m \times m$ nonnegative matrix, then $\rho(A)$ is an eigenvalue of A and there exists a nonnegative eigenvector $\boldsymbol{x}$ corresponding to the eigenvalue $\rho(A)$. This result is weaker than the result for irreducible nonnegative matrices. For each of the following, find a 2×2 nonnull reducible matrix A such that the stated condition holds.
    (a) $\rho(A) = 0$.
    (b) $\boldsymbol{x}$ is not positive for any $\boldsymbol{x}$ satisfying $A\boldsymbol{x} = \rho(A)\boldsymbol{x}$.
    (c) $\rho(A)$ is a multiple eigenvalue.

64. Verify that the absolute value of each of the eigenvalues of the 2×2 irreducible matrix

$$A = \begin{bmatrix} 0 & 1 \\ 1 & 0 \end{bmatrix}$$

is equal to $\rho(A)$.

65. Let A be an $m \times m$ irreducible nonnegative matrix.
    (a) Show that $\rho(\mathrm{I}_m + A) = 1 + \rho(A)$.

(b) Show that if $A^k > (0)$ for some positive integer k, then $\rho(A)$ is a simple eigenvalue of A.

(c) Apply part (b) on the matrix $(\mathrm{I}_m + A)$ to prove Theorem 7.52; that is, prove that for any irreducible nonnegative matrix A, $\rho(A)$ must be a simple eigenvalue.

66. Consider the homogeneous Markov chain that has three states and the matrix of transition probabilities given by

$$P = \begin{bmatrix} 0.50 & 0.25 & 0 \\ 0.50 & 0.50 & 0.25 \\ 0 & 0.25 & 0.75 \end{bmatrix}$$

(a) Show that P is primitive.

(b) Determine the equilibrium distribution; that is, find $\boldsymbol{\pi}$ such that $\lim_{t \to \infty} \boldsymbol{p}^{(t)} = \boldsymbol{\pi}$.

67. Let A be the $m \times m$ circulant matrix $\mathrm{circ}(a_1, \ldots, a_m)$.

(a) Find the trace of A.

(b) Find the determinant of A.

68. Show that the conjugate transpose of the matrix F given in Theorem 7.59 is

$$F^* = \frac{1}{\sqrt{m}} \begin{bmatrix} 1 & 1 & 1 & \cdots & 1 \\ 1 & \theta^{-1} & \theta^{-2} & \cdots & \theta^{-(m-1)} \\ 1 & \theta^{-2} & \theta^{-4} & \cdots & \theta^{-2(m-1)} \\ \vdots & \vdots & \vdots & & \vdots \\ 1 & \theta^{-(m-1)} & \theta^{-2(m-1)} & \cdots & \theta^{-(m-1)(m-1)} \end{bmatrix}$$

Then use the geometric series partial sum formula

$$\sum_{j=0}^{n} r^j = \frac{1 - r^{n+1}}{1 - r}$$

to prove that $F^{-1} = F^*$.

69. Let F be defined as in Theorem 7.59 and let $\Gamma = (\boldsymbol{e}_1, \boldsymbol{e}_m, \boldsymbol{e}_{m-1}, \ldots, \boldsymbol{e}_2)$. Show that

(a) $F^2 = \Gamma$,

(b) $F^4 = \mathrm{I}_m$,

(c) $F^3 = F^*$.

70. Let Π_m be the circulant matrix defined in Section 7.10. Show that
    (a) $\Pi_m^{m-1} = \Pi_m^{-1}$,
    (b) $\Pi_m^m = I_m$,
    (c) $\Pi_m^{mn+r} = \Pi_m^r$, for any integers n and r.

71. If $A = \text{circ}(a_1, \ldots, a_m)$ and $B = \text{circ}(b_1, \ldots, b_m)$, find the eigenvalues of $A + B$ and AB.

72. Use Theorem 7.60 to find the eigenvalues of the circulant matrix $A = \text{circ}(1, \ldots, 1)$.

73. Show that if A is a singular circulant matrix, then its Moore–Penrose inverse, A^+, is also a circulant matrix.

74. Find square matrices A and B of the same order such that A and B are not circulant matrices yet their product AB is a circulant matrix.

75. Let B be the $m \times m$ Jordan block matrix $J_m(0)$. Show that an $m \times m$ matrix A is a Toeplitz matrix if and only if it can be written in the form

$$A = a_0 I_m + \sum_{j=1}^{m-1} (a_j B^j + a_{-j} B'^j)$$

76. Consider the $m \times m$ Toeplitz matrix

$$A = \begin{bmatrix} 1 & b & b^2 & \cdots & b^{m-1} \\ a & 1 & b & \cdots & b^{m-2} \\ a^2 & a & 1 & \cdots & b^{m-3} \\ \vdots & \vdots & \vdots & & \vdots \\ a^{m-1} & a^{m-2} & a^{m-3} & \cdots & 1 \end{bmatrix},$$

where $ab \neq 1$. Verify by multiplication that the inverse of A is given by

$$A^{-1} = \begin{bmatrix} c & -bc & 0 & \cdots & 0 & 0 \\ -ac & (ab+1)c & -bc & \cdots & 0 & 0 \\ 0 & -ac & (ab+1)c & \cdots & 0 & 0 \\ \vdots & \vdots & \vdots & & \vdots & \vdots \\ 0 & 0 & 0 & \cdots & (ab+1)c & -bc \\ 0 & 0 & 0 & \cdots & -ac & c \end{bmatrix},$$

where $c = (1 - ab)^{-1}$. Show that A is singular if $ab = 1$.

77. Suppose that $z_1, \dots, z_{m+1}$ are independent random variables each having mean 0 and variance 1. Let $\boldsymbol{x}$ be the $m \times 1$ random vector that has as its ith component

$$x_i = z_{i+1} - \rho z_i,$$

where ρ is a constant. Show that the covariance matrix of $\boldsymbol{x}$ is a Toeplitz matrix of the form given in (7.29), and find the values of a_0 and a_1.

78. Find a Hadamard matrix of order 8.

79. Give a Hadamard matrix of order 12, thereby illustrating the existence of a Hadamard matrix of order m, where $m \neq 2^n$ for any positive integer n.

80. Show that the determinant of a Hadamard matrix attains the upper bound of the Hadamard inequality given in Corollary 7.23.1.

81. Let A, B, C, and D be $m \times m$ matrices with all of their elements equal to $+1$ and -1, and define H as

$$H = \begin{bmatrix} A & B & C & D \\ -B & A & -D & C \\ -C & D & A & -B \\ -D & -C & B & A \end{bmatrix}$$

Show that if

$$AA' + BB' + CC' + DD' = 4m\mathrm{I}_m$$

and

$$XY' = YX'$$

for every pair of matrices X and Y, chosen from A, B, C, and D, then H is a Hadamard matrix of order $4m$.

82. Show that the Vandermonde matrix A given in (7.32) is nonsingular if and only if the m elements of the second row are distinct.

83. Let A be the $m \times m$ Vandermonde matrix given in (7.32). Prove that if there are r distinct values in the set $\{a_1, \dots, a_m\}$, then $\text{rank}(A) = r$.

84. Let P be the $m \times m$ orthogonal matrix $(\boldsymbol{e}_m, \boldsymbol{e}_{m-1}, \dots, \boldsymbol{e}_1)$. Show that if A is an $m \times m$ Vandermonde matrix, then PAA' and $AA'P$ are Toeplitz matrices.

CHAPTER EIGHT

Matrix Derivatives and Related Topics

1. INTRODUCTION

Differential calculus has widespread applications in statistics. For example, estimation procedures such as the maximum likelihood method and the method of least squares utilize the optimization properties of derivatives, whereas the so-called delta method for obtaining the asymptotic distribution of a function of random variables uses the first derivative to obtain a first-order Taylor series approximation. These and other applications of differential calculus often involve vectors or matrices. In this chapter, we obtain some of the most commonly encountered matrix derivatives.

2. MULTIVARIABLE DIFFERENTIAL CALCULUS

We will begin with a brief review of some of the basic notation, concepts, and results of elementary and multivariable differential calculus. Throughout this section, we will assume differentiability or multiple differentiability of the functions we discuss. For more details on the conditions for differentiability see Magnus and Neudecker (1988). If f is a real-valued function of one variable, x, then its derivative at x, if it exists, is given by

$$f^{(1)}(x) = f'(x) = \frac{\mathrm{d}}{\mathrm{d}x} f(x) = \lim_{u \to 0} \frac{f(x+u) - f(x)}{u}$$

Equivalently, $f'(x)$ is the quantity that gives the first-order Taylor formula for $f(x+u)$. In other words,

$$f(x+u) = f(x) + uf'(x) + r_1(u,x), \tag{8.1}$$

where the remainder $r_1(u,x)$ is a function of u and x satisfying

$$\lim_{u \to 0} \frac{r_1(u, x)}{u} = 0$$

The quantity

$$d_u f(x) = uf'(x) \tag{8.2}$$

appearing in (8.1) is called the first differential of f at x with increment u. This increment u is the differential of x. Later we will use dx in place of u, that is, write $f(x + dx)$ instead of $f(x + u)$, to emphasize the fact that u is the differential of x. For notational convenience, we will often denote the differential given in (8.2) simply by df. Generalizations of (8.1) can be obtained by taking higher-ordered derivatives; that is, with the ith derivative of f at x defined as

$$f^{(i)}(x) = \frac{d^i}{dx^i} f(x) = \lim_{u \to 0} \frac{f^{(i-1)}(x + u) - f^{(i-1)}(x)}{u},$$

we have the kth-order Taylor formula

$$\begin{aligned} f(x + u) &= f(x) + \sum_{i=1}^{k} \frac{u^i f^{(i)}(x)}{i!} + r_k(u, x) \\ &= f(x) + \sum_{i=1}^{k} \frac{d_u^i f(x)}{i!} + r_k(u, x), \end{aligned}$$

where $r_k(u, x)$ is a function of u and x satisfying

$$\lim_{u \to 0} \frac{r_k(u, x)}{u^k} = 0,$$

and

$$d_u^i f(x) = u^i f^{(i)}(x),$$

or simply $d^i f$, is the ith differential of f at x with increment u.

The chain rule is a useful formula for calculating the derivative of a composite function. If y, g, and f are functions such that $y(x) = g(f(x))$, then

$$y'(x) = g'(f(x))f'(x) \tag{8.3}$$

If f is a real-valued function of the $n \times 1$ vector $\boldsymbol{x} = (x_1, \ldots, x_n)'$, then its derivative at $\boldsymbol{x}$, if it exists, is given by the $1 \times n$ row vector

$$\frac{\partial}{\partial \boldsymbol{x}'} f(\boldsymbol{x}) = \left[\frac{\partial}{\partial x_1} f(\boldsymbol{x}) \ \cdots \ \frac{\partial}{\partial x_n} f(\boldsymbol{x}) \right],$$

where

$$\frac{\partial}{\partial x_i} f(\boldsymbol{x}) = \lim_{u_i \to 0} \frac{f(\boldsymbol{x} + u_i \boldsymbol{e}_i) - f(\boldsymbol{x})}{u_i}$$

is the partial derivative of f with respect to x_i, and $\boldsymbol{e}_i$ is the ith column of I_n. The first-order Taylor formula analogous to (8.1) is given by

$$f(\boldsymbol{x} + \boldsymbol{u}) = f(\boldsymbol{x}) + \left(\frac{\partial}{\partial \boldsymbol{x}'} f(\boldsymbol{x}) \right) \boldsymbol{u} + r_1(\boldsymbol{u}, \boldsymbol{x}), \tag{8.4}$$

where the remainder, $r_1(\boldsymbol{u}, \boldsymbol{x})$, satisfies

$$\lim_{\boldsymbol{u} \to \boldsymbol{0}} \frac{r_1(\boldsymbol{u}, \boldsymbol{x})}{(\boldsymbol{u}'\boldsymbol{u})^{1/2}} = 0$$

The second term on the right-hand side of (8.4) is the first differential of f at $\boldsymbol{x}$ with incremental vector $\boldsymbol{u}$; that is,

$$\mathrm{d}f = \mathrm{d}_{\boldsymbol{u}} f(\boldsymbol{x}) = \left(\frac{\partial}{\partial \boldsymbol{x}'} f(\boldsymbol{x}) \right) \boldsymbol{u} = \sum_{i=1}^{n} u_i \frac{\partial}{\partial x_i} f(\boldsymbol{x})$$

It is important to note the relationship between the first differential and the first derivative; the first differential of f at $\boldsymbol{x}$ in $\boldsymbol{u}$ is the first derivative of f at $\boldsymbol{x}$ times $\boldsymbol{u}$. The higher-order differentials of f at $\boldsymbol{x}$ in the vector $\boldsymbol{u}$ are given by

$$\mathrm{d}^i f = \mathrm{d}^i_{\boldsymbol{u}} f(\boldsymbol{x}) = \sum_{j_1=1}^{n} \cdots \sum_{j_i=1}^{n} u_{j_1} \cdots u_{j_i} \frac{\partial^i}{\partial x_{j_1} \cdots \partial x_{j_i}} f(\boldsymbol{x}),$$

and these appear in the kth-order Taylor formula,

$$f(\boldsymbol{x} + \boldsymbol{u}) = f(\boldsymbol{x}) + \sum_{i=1}^{k} \frac{\mathrm{d}^i f}{i!} + r_k(\boldsymbol{u}, \boldsymbol{x}),$$

where the remainder $r_k(\boldsymbol{u}, \boldsymbol{x})$ satisfies

$$\lim_{\boldsymbol{u} \to \boldsymbol{0}} \frac{r_k(\boldsymbol{u}, \boldsymbol{x})}{(\boldsymbol{u}'\boldsymbol{u})^{k/2}} = 0$$

The second differential, $\mathrm{d}^2 f$, can be written as a quadratic form in the vector $\boldsymbol{u}$; that is,

$$\mathrm{d}^2 f = \boldsymbol{u}' H_f \boldsymbol{u},$$

where H_f, called the Hessian matrix, is the matrix of second-order partial derivatives given by

$$H_f = \begin{bmatrix} \dfrac{\partial^2}{\partial x_1^2} f(\boldsymbol{x}) & \dfrac{\partial^2}{\partial x_1 \partial x_2} f(\boldsymbol{x}) & \cdots & \dfrac{\partial^2}{\partial x_1 \partial x_n} f(\boldsymbol{x}) \\ \dfrac{\partial^2}{\partial x_2 \partial x_1} f(\boldsymbol{x}) & \dfrac{\partial^2}{\partial x_2^2} f(\boldsymbol{x}) & \cdots & \dfrac{\partial^2}{\partial x_2 \partial x_n} f(\boldsymbol{x}) \\ \vdots & \vdots & & \vdots \\ \dfrac{\partial^2}{\partial x_n \partial x_1} f(\boldsymbol{x}) & \dfrac{\partial^2}{\partial x_n \partial x_2} f(\boldsymbol{x}) & \cdots & \dfrac{\partial^2}{\partial x_n^2} f(\boldsymbol{x}) \end{bmatrix}$$

3. VECTOR AND MATRIX FUNCTIONS

Suppose now that $f_1, \ldots, f_m$ each is a function of the same $n \times 1$ vector $\boldsymbol{x} = (x_1, \ldots, x_n)'$. These m functions can be conveniently expressed as components of the vector function

$$\boldsymbol{f}(\boldsymbol{x}) = \begin{bmatrix} f_1(\boldsymbol{x}) \\ \vdots \\ f_m(\boldsymbol{x}) \end{bmatrix}$$

The function $\boldsymbol{f}$ is differentiable at $\boldsymbol{x}$ if and only if each component function f_i is differentiable at $\boldsymbol{x}$. The Taylor formulas from the previous section can be applied componentwise to $\boldsymbol{f}$. For instance, the first-order Taylor formula is given by

$$\boldsymbol{f}(\boldsymbol{x} + \boldsymbol{u}) = \boldsymbol{f}(\boldsymbol{x}) + \left(\frac{\partial}{\partial \boldsymbol{x}'} \boldsymbol{f}(\boldsymbol{x}) \right) \boldsymbol{u} + \boldsymbol{r}_1(\boldsymbol{u}, \boldsymbol{x}) = \boldsymbol{f}(\boldsymbol{x}) + \mathrm{d}\boldsymbol{f}(\boldsymbol{x}) + \boldsymbol{r}_1(\boldsymbol{u}, \boldsymbol{x}),$$

where the vector remainder, $\boldsymbol{r}_1(\boldsymbol{u},\boldsymbol{x})$, satisfies

$$\lim_{\boldsymbol{u}\to\boldsymbol{0}} \frac{\boldsymbol{r}_1(\boldsymbol{u},\boldsymbol{x})}{(\boldsymbol{u}'\boldsymbol{u})^{1/2}} = \boldsymbol{0}$$

and the first derivative of $\boldsymbol{f}$ at $\boldsymbol{x}$ is given by

$$\frac{\partial}{\partial \boldsymbol{x}'}\boldsymbol{f}(\boldsymbol{x}) = \begin{bmatrix} \frac{\partial}{\partial x_1} f_1(\boldsymbol{x}) & \frac{\partial}{\partial x_2} f_1(\boldsymbol{x}) & \cdots & \frac{\partial}{\partial x_n} f_1(\boldsymbol{x}) \\ \frac{\partial}{\partial x_1} f_2(\boldsymbol{x}) & \frac{\partial}{\partial x_2} f_2(\boldsymbol{x}) & \cdots & \frac{\partial}{\partial x_n} f_2(\boldsymbol{x}) \\ \vdots & \vdots & & \vdots \\ \frac{\partial}{\partial x_1} f_m(\boldsymbol{x}) & \frac{\partial}{\partial x_2} f_m(\boldsymbol{x}) & \cdots & \frac{\partial}{\partial x_n} f_m(\boldsymbol{x}) \end{bmatrix}$$

This matrix of partial derivatives is sometimes referred to as the Jacobian matrix of $\boldsymbol{f}$ at $\boldsymbol{x}$. Again, it is crucial to understand the relationship between the first differential and the first derivative. If we obtain the first differential of $\boldsymbol{f}$ at $\boldsymbol{x}$ in $\boldsymbol{u}$ and write it in the form

$$\mathrm{d}\boldsymbol{f} = B\boldsymbol{u},$$

then the $m \times n$ matrix B must be the derivative of $\boldsymbol{f}$ at $\boldsymbol{x}$.

If y and g are real-valued functions satisfying $y(\boldsymbol{x}) = g(\boldsymbol{f}(\boldsymbol{x}))$, then the generalization of the chain rule given in (8.3) is

$$\frac{\partial}{\partial x_i} y(\boldsymbol{x}) = \sum_{j=1}^{m} \left(\frac{\partial}{\partial f_j} g(\boldsymbol{f}) \right) \left(\frac{\partial}{\partial x_i} f_j(\boldsymbol{x}) \right) = \left(\frac{\partial}{\partial \boldsymbol{f}'} g(\boldsymbol{f}) \right) \left(\frac{\partial}{\partial x_i} \boldsymbol{f}(\boldsymbol{x}) \right)$$

for $i = 1, \ldots, n$, or simply

$$\frac{\partial}{\partial \boldsymbol{x}'} y(\boldsymbol{x}) = \left(\frac{\partial}{\partial \boldsymbol{f}'} g(\boldsymbol{f}) \right) \left(\frac{\partial}{\partial \boldsymbol{x}'} \boldsymbol{f}(\boldsymbol{x}) \right)$$

In some applications the f_js or the x_is are arranged in a matrix instead of a vector. Thus, the most general case involves the $p \times q$ matrix function

$$F(X) = \begin{bmatrix} f_{11}(X) & f_{12}(X) & \cdots & f_{1q}(X) \\ f_{21}(X) & f_{22}(X) & \cdots & f_{2q}(X) \\ \vdots & \vdots & & \vdots \\ f_{p1}(X) & f_{p2}(X) & \cdots & f_{pq}(X) \end{bmatrix}$$

of the $m \times n$ matrix X. Results for the vector function $\boldsymbol{f}(\boldsymbol{x})$ can be easily extended to the matrix function $F(X)$ by utilizing the vec operator; that is, let $\boldsymbol{f}$ be the $pq \times 1$ vector function such that $\boldsymbol{f}(\text{vec}(X)) = \text{vec}(F(X))$. Then, for instance, the Jacobian matrix of F at X is given by the $pq \times mn$ matrix

$$\frac{\partial}{\partial\, \text{vec}(X)'} \boldsymbol{f}(\text{vec}(X)) = \frac{\partial}{\partial\, \text{vec}(X)'} \text{vec}(F(X)),$$

which has as its (i, j)th element, the partial derivative of the ith element of $\text{vec}(F(X))$ with respect to the jth element of $\text{vec}(X)$. This could then be used to obtain the first-order Taylor formula for $\text{vec}(F(X + U))$. The differentials of the matrix $F(X)$ are defined by the equations

$$\text{vec}(\mathrm{d}^i F) = \text{vec}(\mathrm{d}^i_U F(X)) = \mathrm{d}^i \boldsymbol{f} = \mathrm{d}^i_{\text{vec}(U)} \boldsymbol{f}(\text{vec}(X));$$

that is, $\mathrm{d}^i F$, the ith order differential of F at X in the incremental matrix U, is defined to be the $p \times q$ matrix obtained by unstacking the ith-order differential of $\boldsymbol{f}$ at $\text{vec}(X)$ in the incremental vector $\text{vec}(U)$.

Basic properties of vector and matrix differentials follow in a fairly straightforward fashion from the corresponding properties of scalar differentials. We will summarize some of these properties here. If x and y are functions and α is a constant, then the differential operator, d, satisfies

(a) $\mathrm{d}\alpha = 0$,
(b) $\mathrm{d}(\alpha x) = \alpha\, \mathrm{d}x$,
(c) $\mathrm{d}(x + y) = \mathrm{d}x + \mathrm{d}y$,
(d) $\mathrm{d}(xy) = (\mathrm{d}x)y + x(\mathrm{d}y)$,
(e) $\mathrm{d}x^{\alpha} = \alpha x^{\alpha - 1}\, \mathrm{d}x$,
(f) $\mathrm{d}e^x = e^x\, \mathrm{d}x$,
(g) $\mathrm{d}\log(x) = x^{-1}\, \mathrm{d}x$.

For instance, to illustrate property (d), note that

$$(x + \mathrm{d}x)(y + \mathrm{d}y) = xy + x(\mathrm{d}y) + (\mathrm{d}x)y + (\mathrm{d}x)(\mathrm{d}y),$$

and $\mathrm{d}(xy)$ will be given by the first-degree term in $\mathrm{d}x$ and $\mathrm{d}y$, which is $(\mathrm{d}x)y + x(\mathrm{d}y)$ as required. Using the properties above and the definition of a matrix differential, it is easily shown that if X and Y are matrix functions and A is a matrix of constants, then

(h) $\text{d}A = (0)$,
(i) $\text{d}(\alpha X) = \alpha\, \text{d}X$,
(j) $\text{d}(X') = (\text{d}X)'$,
(k) $\text{d}(X + Y) = \text{d}X + \text{d}Y$,
(l) $\text{d}(XY) = (\text{d}X)Y + X(\text{d}Y)$.

We will verify property (l). Thus, we must show that the (i,j)th element of the matrix on the left-hand side of the equation, $(\text{d}(XY))_{ij}$, is the same as the (i,j)th element on the right-hand side, $(\text{d}X)_{i\cdot}(Y)_{\cdot j} + (X)_{i\cdot}(\text{d}Y)_{\cdot j}$, where X is $m \times n$ and Y is $n \times m$. Using properties (c) and (d), we find that

$$\begin{aligned}
(\text{d}(XY))_{ij} &= \text{d}\{(X)_{i\cdot}(Y)_{\cdot j}\} = \text{d}\left\{\sum_{k=1}^{n} x_{ik}y_{kj}\right\} \\
&= \sum_{k=1}^{n} \text{d}(x_{ik}y_{kj}) = \sum_{k=1}^{n} \{(\text{d}x_{ik})y_{kj} + x_{ik}\, \text{d}y_{kj}\} \\
&= \sum_{k=1}^{n} (\text{d}x_{ik})y_{kj} + \sum_{k=1}^{n} x_{ik}\, \text{d}y_{kj} = (\text{d}X)_{i\cdot}(Y)_{\cdot j} + (X)_{i\cdot}(\text{d}Y)_{\cdot j},
\end{aligned}$$

and so (l) is proven.

We illustrate the use of some of these properties first by finding the derivatives of some simple scalar functions of a vector $\boldsymbol{x}$, and then by finding the derivatives of some simple matrix functions of a matrix X.

Example 8.1. Let $\boldsymbol{x}$ be an $m \times 1$ vector of unrelated variables and define the functions

$$f(\boldsymbol{x}) = \boldsymbol{a}'\boldsymbol{x},$$

where $\boldsymbol{a}$ is an $m \times 1$ vector of constants, and

$$g(\boldsymbol{x}) = \boldsymbol{x}'A\boldsymbol{x},$$

where A is an $m \times m$ symmetric matrix of constants. The differential of the first function is

$$\mathrm{d}f = \mathrm{d}(\boldsymbol{a}'\boldsymbol{x}) = \boldsymbol{a}'\,\mathrm{d}\boldsymbol{x}$$

Since this differential and the derivative are related through the equation

$$\mathrm{d}f = \left(\frac{\partial}{\partial \boldsymbol{x}'} f\right)\,\mathrm{d}\boldsymbol{x},$$

we immediately observe that the derivative is given by

$$\frac{\partial}{\partial \boldsymbol{x}'} f = \boldsymbol{a}'$$

The differential and derivative of our second function are given by

$$\begin{aligned}\mathrm{d}g &= \mathrm{d}(\boldsymbol{x}'A\boldsymbol{x}) = \mathrm{d}(\boldsymbol{x}')A\boldsymbol{x} + \boldsymbol{x}'\,\mathrm{d}(A\boldsymbol{x}) = (\mathrm{d}\boldsymbol{x})'A\boldsymbol{x} + \boldsymbol{x}'A\,\mathrm{d}\boldsymbol{x}\\ &= \{(\mathrm{d}\boldsymbol{x})'A\boldsymbol{x}\}' + \boldsymbol{x}'A\,\mathrm{d}\boldsymbol{x} = \boldsymbol{x}'A'\,\mathrm{d}\boldsymbol{x} + \boldsymbol{x}'A\,\mathrm{d}\boldsymbol{x} = 2\boldsymbol{x}'A\,\mathrm{d}\boldsymbol{x},\end{aligned}$$

and

$$\frac{\partial}{\partial \boldsymbol{x}'} g = 2\boldsymbol{x}'A$$

Example 8.2. Let X be an $m \times n$ matrix of unrelated variables and define the functions

$$F(X) = AX,$$

where A is a $p \times m$ matrix of constants, and

$$G(X) = (X - C)'B(X - C),$$

where B is an $m \times m$ symmetric matrix of constants and C is an $m \times n$ matrix of constants. We will find the Jacobian matrices by first obtaining the differentials of these functions. For our first function, we find that

$$\mathrm{d}F = \mathrm{d}(AX) = A\,\mathrm{d}X,$$

so that

$$\mathrm{d}\,\mathrm{vec}(F) = \mathrm{vec}(\mathrm{d}F) = \mathrm{vec}(A\,\mathrm{d}X) = (\mathrm{I}_n \otimes A)\mathrm{vec}(\mathrm{d}X) = (\mathrm{I}_n \otimes A)\mathrm{d}\,\mathrm{vec}(X)$$

Thus, we must have

$$\frac{\partial}{\partial\, \mathrm{vec}(X)'}\, \mathrm{vec}(F) = \mathrm{I}_n \otimes A$$

The differential of our second function is

$$\begin{aligned} \mathrm{d}G &= \mathrm{d}\{(X - C)'B(X - C)\} \\ &= \{\mathrm{d}(X' - C')\}B(X - C) + (X - C)'B\{\mathrm{d}(X - C)\} \\ &= (\mathrm{d}X)'B(X - C) + (X - C)'B\,\mathrm{d}X \end{aligned}$$

From this we obtain

$$\begin{aligned} \mathrm{d}\,\mathrm{vec}(G) &= \{(X - C)'B \otimes \mathrm{I}_n\}\mathrm{vec}(\mathrm{d}X') + \{\mathrm{I}_n \otimes (X - C)'B\}\mathrm{vec}(\mathrm{d}X) \\ &= \{(X - C)'B \otimes \mathrm{I}_n\}K_{mn}\,\mathrm{vec}(\mathrm{d}X) + \{\mathrm{I}_n \otimes (X - C)'B\}\mathrm{vec}(\mathrm{d}X) \\ &= K_{nn}\{\mathrm{I}_n \otimes (X - C)'B\}\mathrm{vec}(\mathrm{d}X) + \{\mathrm{I}_n \otimes (X - C)'B\}\mathrm{vec}(\mathrm{d}X) \\ &= (\mathrm{I}_{n^2} + K_{nn})\{\mathrm{I}_n \otimes (X - C)'B\}\mathrm{vec}(\mathrm{d}X) \\ &= 2N_n\{\mathrm{I}_n \otimes (X - C)'B\}\,\mathrm{d}\,\mathrm{vec}(X), \end{aligned}$$

where we have used properties of the vec operator and the commutation matrix. Consequently, we have

$$\frac{\partial}{\partial\, \mathrm{vec}(X)'}\, \mathrm{vec}(G) = 2N_n\{\mathrm{I}_n \otimes (X - C)'B\}$$

In our next example, we show how the Jacobian matrix of the simple transformation $\boldsymbol{z} = \boldsymbol{c} + A\boldsymbol{x}$ can be used to obtain the multivariate normal density function given in (1.13).

Example 8.3. Suppose that $\boldsymbol{z}$ is an $m \times 1$ random vector with density function $f_1(\boldsymbol{z})$ that is positive for all $\boldsymbol{z} \in S_1 \subseteq R^m$. Let the $m \times 1$ vector $\boldsymbol{x} = \boldsymbol{x}(\boldsymbol{z})$ represent a one-to-one transformation of S_1 onto $S_2 \subseteq R^m$, so that the inverse transformation $\boldsymbol{z} = \boldsymbol{z}(\boldsymbol{x})$, $\boldsymbol{x} \in S_2$ is unique. Denote the Jacobian matrix of $\boldsymbol{z}$ at $\boldsymbol{x}$ as

$$J = \frac{\partial}{\partial \boldsymbol{x}'}\, \boldsymbol{z}(\boldsymbol{x})$$

If the partial derivatives in J exist and are continuous functions on the set S_2, then the density of $\boldsymbol{x}$ is given by

$$f_2(\boldsymbol{x}) = f_1(\boldsymbol{z}(\boldsymbol{x}))|J|$$

We will use the formula above to obtain the multivariate normal density, given in (1.13), from the standard normal density. Now recall that by definition,

$\boldsymbol{x} \sim \mathrm{N}_m(\boldsymbol{\mu}, \Omega)$ if $\boldsymbol{x}$ can be expressed as $\boldsymbol{x} = \boldsymbol{\mu} + T\boldsymbol{z}$, where $TT' = \Omega$ and the components of $\boldsymbol{z}, z_1, \dots, z_m$ are independently distributed each as N(0, 1). Thus, the density function of $\boldsymbol{z}$ is given by

$$f_1(\boldsymbol{z}) = \prod_{i=1}^{m} \frac{1}{\sqrt{2\pi}} \exp\left(-\frac{1}{2} z_i^2 \right) = \frac{1}{(2\pi)^{m/2}} \exp\left(-\frac{1}{2} \boldsymbol{z}'\boldsymbol{z} \right)$$

The differential of the inverse transformation $\boldsymbol{z} = T^{-1}(\boldsymbol{x} - \boldsymbol{\mu})$ is $\mathrm{d}\boldsymbol{z} = T^{-1}\,\mathrm{d}\boldsymbol{x}$, and so the necessary Jacobian matrix is $J = T^{-1}$. Consequently, we find that the density of $\boldsymbol{x}$ is given by

$$\begin{aligned} f_2(\boldsymbol{x}) &= \frac{1}{(2\pi)^{m/2}} \exp\left(-\frac{1}{2} \{T^{-1}(\boldsymbol{x} - \boldsymbol{\mu})\}'T^{-1}(\boldsymbol{x} - \boldsymbol{\mu}) \right) |T^{-1}| \\ &= \frac{1}{(2\pi)^{m/2}|T|} \exp\left(-\frac{1}{2} (\boldsymbol{x} - \boldsymbol{\mu})'T'^{-1}T^{-1}(\boldsymbol{x} - \boldsymbol{\mu}) \right) \\ &= \frac{1}{(2\pi)^{m/2}|\Omega|^{1/2}} \exp\left(-\frac{1}{2} (\boldsymbol{x} - \boldsymbol{\mu})'\Omega^{-1}(\boldsymbol{x} - \boldsymbol{\mu}) \right) \end{aligned}$$

4. SOME USEFUL MATRIX DERIVATIVES

In this section we will obtain the differentials and the corresponding derivatives of some important scalar functions and matrix functions of matrices. Throughout this section, when dealing with functions of the form $f(X)$ or $F(X)$ we will assume that the $m \times n$ matrix X is composed of mn unrelated variables; that is, X is assumed not to have any particular structure such as symmetry, triangularity, and so on. We begin with some scalar functions of X.

Theorem 8.1. Let X be an $m \times m$ matrix. Then

(a) $\mathrm{d}\{\mathrm{tr}(X)\} = \mathrm{vec}(\mathrm{I}_m)'\mathrm{d}\,\mathrm{vec}(X); \quad \dfrac{\partial}{\partial\,\mathrm{vec}(X)'}\,\mathrm{tr}(X) = \mathrm{vec}(\mathrm{I}_m)',$

(b) $\mathrm{d}|X| = \mathrm{tr}(X_\#\,\mathrm{d}X) = |X|\,\mathrm{tr}(X^{-1}\,\mathrm{d}X); \quad \dfrac{\partial}{\partial\,\mathrm{vec}(X)'}\,|X| = \mathrm{vec}(X_\#')',$

where $X_\#$ is the adjoint matrix of X.

Proof. Part (a) follows directly from the fact that

$$\mathrm{d}\,\mathrm{tr}(X) = \mathrm{tr}(\mathrm{d}X) = \mathrm{tr}(\mathrm{I}_m\,\mathrm{d}X) = \mathrm{vec}(\mathrm{I}_m)'\,\mathrm{vec}(\mathrm{d}X) = \mathrm{vec}(\mathrm{I}_m)'\,\mathrm{d}\,\mathrm{vec}(X),$$

with the third equality following from Theorem 7.15. Since $X_\#$ is the transpose of the matrix of cofactors of X, to obtain the derivative in (b), we simply need to show that

$$\frac{\partial}{\partial x_{ij}} |X| = X_{ij},$$

where X_{ij} is the cofactor of x_{ij}. By using the cofactor expansion formula on the ith row of X, we can write the determinant of X as

$$|X| = \sum_{k=1}^{m} x_{ik} X_{ik}$$

Note that for each k, X_{ik} is a determinant computed after deleting the ith row so that each X_{ik} does not involve the element x_{ij}. Consequently, we have

$$\frac{\partial}{\partial x_{ij}} |X| = \frac{\partial}{\partial x_{ij}} \sum_{k=1}^{m} x_{ik} X_{ik} = \sum_{k=1}^{m} \left(\frac{\partial}{\partial x_{ij}} x_{ik} \right) X_{ik} = X_{ij}$$

Using the relationship between the first differential and derivative and the fact that $X^{-1} = |X|^{-1} X_\#$, we also get

$$\mathrm{d}|X| = \{\mathrm{vec}(X'_\#)\}' \,\mathrm{vec}(\mathrm{d}X) = \mathrm{tr}(X_\# \mathrm{d}X) = |X| \,\mathrm{tr}(X^{-1}\, \mathrm{d}X). \qquad \square$$

An immediate consequence of Theorem 8.1(b) is the following.

Corollary 8.1.1. Let X be an $m \times m$ nonsingular matrix. Then

$$\mathrm{d}\{\log(|X|)\} = \mathrm{tr}(X^{-1}\, \mathrm{d}X); \qquad \frac{\partial}{\partial\, \mathrm{vec}(X)'} \log(|X|) = \mathrm{vec}(X^{-1\prime})'$$

Our next result gives the differential and derivative of the inverse of a nonsingular matrix.

Theorem 8.2. If X is a nonsingular $m \times m$ matrix, then

$$\mathrm{d}X^{-1} = -X^{-1}(\mathrm{d}X)X^{-1}; \qquad \frac{\partial}{\partial\, \mathrm{vec}(X)'} \mathrm{vec}(X^{-1}) = -(X^{-1\prime} \otimes X^{-1})$$

Proof. Computing the differential of both sides of the equation $\mathrm{I}_m = XX^{-1}$, we find that

$$(0) = \mathrm{dI}_m = \mathrm{d}(XX^{-1}) = (\mathrm{d}X)X^{-1} + X(\mathrm{d}X^{-1})$$

Premultiplying this equation by X^{-1}, and then solving for $\mathrm{d}X^{-1}$ yields

$$\mathrm{d}X^{-1} = -X^{-1}(\mathrm{d}X)X^{-1}$$

The expression given for the derivative now follows since

$$\begin{aligned}\mathrm{d}\,\mathrm{vec}(X^{-1}) &= \mathrm{vec}(\mathrm{d}X^{-1}) = -\mathrm{vec}(X^{-1}(\mathrm{d}X)X^{-1}) \\ &= -(X^{-1\prime} \otimes X^{-1})\mathrm{vec}(\mathrm{d}X) = -(X^{-1\prime} \otimes X^{-1})\,\mathrm{d}\,\mathrm{vec}(X)\end{aligned}$$ □

A natural generalization of Theorem 8.2 is one that gives the differential and derivative of the Moore–Penrose inverse of a matrix. The following theorem gives the form of these when they exist at a matrix X.

Theorem 8.3. If X is an $m \times n$ matrix and X^+ is its Moore–Penrose inverse, then

$$\mathrm{d}X^+ = (\mathrm{I} - X^+X)(\mathrm{d}X')X^{+\prime}X^+ + X^+X^{+\prime}(\mathrm{d}X')(\mathrm{I} - XX^+) - X^+(\mathrm{d}X)X^+$$

and

$$\begin{aligned}\frac{\partial}{\partial\,\mathrm{vec}(X)'}\,\mathrm{vec}(X^+) = \{&X^{+\prime}X^+ \otimes (\mathrm{I} - X^+X) + (\mathrm{I} - XX^+) \otimes X^+X^{+\prime}\} \\ &\cdot K_{mn} - (X^{+\prime} \otimes X^+)\end{aligned}$$

Proof. Note that

$$\mathrm{d}(XX^+) = (\mathrm{d}X)X^+ + X\,\mathrm{d}X^+,$$

from which we get

$$X\mathrm{d}X^+ = \mathrm{d}(XX^+) - (\mathrm{d}X)X^+ \tag{8.5}$$

Since $X^+ = X^+XX^+$, we also have

$$\begin{aligned}\mathrm{d}X^+ &= \mathrm{d}(X^+XX^+) = \mathrm{d}(X^+X)X^+ + X^+X\,\mathrm{d}X^+ \\ &= \mathrm{d}(X^+X)X^+ + X^+\,\mathrm{d}(XX^+) - X^+(\mathrm{d}X)X^+,\end{aligned} \tag{8.6}$$

where we have used (8.5) in the last step. Thus, if we obtain expressions for $\mathrm{d}(X^+X)$ and $\mathrm{d}(XX^+)$ in terms of $\mathrm{d}X$, we can then find $\mathrm{d}X^+$. To find $\mathrm{d}(XX^+)$, use the fact that XX^+ is symmetric and idempotent to get

$$\begin{aligned} d(XX^+) &= d(XX^+XX^+) = d(XX^+)XX^+ + XX^+d(XX^+) \\ &= d(XX^+)XX^+ + (d(XX^+)XX^+)', \end{aligned} \tag{8.7}$$

since $d(XX^+)' = d((XX^+)') = d(XX^+)$. But

$$d(XX^+)X = dX - XX^+\,dX = (I - XX^+)\,dX, \tag{8.8}$$

since $X = XX^+X$ implies that

$$dX = d(XX^+X) = d(XX^+)X + XX^+\,dX$$

Now substituting (8.8) in (8.7), we find that

$$\begin{aligned} dXX^+ &= (I - XX^+)(dX)X^+ + \{(I - XX^+)(dX)X^+\}' \\ &= (I - XX^+)(dX)X^+ + X^{+\prime}(dX')(I - XX^+) \end{aligned} \tag{8.9}$$

By using the fact that X^+X is symmetric and idempotent, we can show in a similar fashion that

$$dX^+X = X^+(dX)(I - X^+X) + (I - X^+X)(dX')X^{+\prime} \tag{8.10}$$

Substituting (8.9) and (8.10) into (8.6) and noting that $(I - X^+X)X^+ = (0)$ and $X^+(I - XX^+) = (0)$, we get

$$dX^+ = (I - X^+X)(dX')X^{+\prime}X^+ + X^+X^{+\prime}(dX')(I - XX^+) - X^+(dX)X^+,$$

as is required. The expression given for the derivative follows since, when we take the vec of both sides of the equation above, we get

$$\begin{aligned} d\,\mathrm{vec}(X^+) &= \{X^{+\prime}X^+ \otimes (I - X^+X)\}\mathrm{vec}(dX') + \{(I - XX^+) \otimes X^+X^{+\prime}\}\mathrm{vec}(dX') \\ &\quad - (X^{+\prime} \otimes X^+)\mathrm{vec}(dX) \\ &= \{X^{+\prime}X^+ \otimes (I - X^+X) + (I - XX^+) \otimes X^+X^{+\prime}\}K_{mn}\,d\,\mathrm{vec}(X) \\ &\quad - (X^{+\prime} \otimes X^+)\,d\,\mathrm{vec}(X) \end{aligned}$$

□

5. DERIVATIVES OF FUNCTIONS OF PATTERNED MATRICES

In this section, we consider the computation of the derivative of a function of an $m \times n$ matrix X when some of the variables of X are related to one another.

In particular, we will focus on the situation in which X is square and symmetric. For a more general treatment of the topic of derivatives of functions of patterned matrices see Nel (1980).

If X is an $m \times m$ symmetric matrix of variables, then due to the symmetry it only contains $m(m+1)/2$ mathematically independent variables. These variables are precisely the variables comprising the vector $\mathrm{v}(X)$. If $f(X)$ is some vector function of the matrix X, then the derivative of f will be given by the matrix

$$\frac{\partial}{\partial\, \mathrm{v}(X)'} f(X)$$

We can compute derivatives of this form by utilizing the derivative

$$\frac{\partial}{\partial\, \mathrm{vec}(X)'} f(X),$$

for a general nonsymmetric matrix X, along with the chain rule. Specifically, from the chain rule we have

$$\frac{\partial}{\partial\, \mathrm{v}(X)'} f(X) = \left(\frac{\partial}{\partial\, \mathrm{vec}(X)'} f(X) \right) \left(\frac{\partial}{\partial\, \mathrm{v}(X)'} \mathrm{vec}(X) \right)$$

It must be emphasized here that the first of the two derivatives on the right-hand side of this equation is computed ignoring the symmetry of X. The second of these two derivatives can be conveniently expressed by making use of the duplication matrix D_m. Since $D_m\, \mathrm{v}(X) = \mathrm{vec}(X)$, we immediately get $D_m\, \mathrm{d}\, \mathrm{v}(X) = \mathrm{d}\, \mathrm{vec}(X)$, and so

$$\frac{\partial}{\partial\, \mathrm{v}(X)'} f(X) = \left(\frac{\partial}{\partial\, \mathrm{vec}(X)'} f(X) \right) D_m$$

Consequently, the following results follow directly from Theorems 8.1–8.3.

Theorem 8.4. Let X be an $m \times m$ symmetric matrix of variables. Then

(a) $\dfrac{\partial}{\partial\, \mathrm{v}(X)'} |X| = \mathrm{vec}(X'_{\#})' D_m,$

(b) $\dfrac{\partial}{\partial\, \mathrm{v}(X)'} \mathrm{vec}(X^{-1}) = -(X^{-1} \otimes X^{-1}) D_m,$

(c) $\dfrac{\partial}{\partial\, \mathrm{v}(X)'} \mathrm{vec}(X^{+}) = (\{X^{+}X^{+} \otimes (\mathrm{I} - X^{+}X) + (\mathrm{I} - XX^{+}) \otimes X^{+}X^{+}\}$

$\cdot K_{mm} - (X^{+} \otimes X^{+})) D_m$

The derivatives given in (b) and (c) of Theorem 8.4 still have some redundant elements due to the symmetry of X^{-1} and X^{+}. In general, if X is an $m \times m$ symmetric matrix of variables and the $m \times m$ matrix function $F(X)$ is also symmetric, then all derivatives of elements of $F(X)$ with respect to elements of X will be contained in the matrix derivative

$$\frac{\partial}{\partial\, \mathrm{v}(X)'}\, \mathrm{v}\{F(X)\}$$

This matrix derivative can be easily computed from the derivative

$$A = \frac{\partial}{\partial\, \mathrm{v}(X)'}\, \mathrm{vec}\{F(X)\}, \tag{8.11}$$

by again using the relationship $\mathrm{vec}(F) = D_m \mathrm{v}(F)$. Thus, since (8.11) implies that $\mathrm{d\,vec}(F) = A\,\mathrm{d\,v}(X)$, we have

$$D_m\, \mathrm{d\,v}(F) = A\, \mathrm{d\,v}(X),$$

or

$$D_m^+ D_m\, \mathrm{d\,v}(F) = \mathrm{d\,v}(F) = D_m^+ A\, \mathrm{d\,v}(X),$$

since $D_m^+ D_m = \mathrm{I}$ by Theorem 7.36. Using this we obtain the following derivatives.

Corollary 8.4.1. Let X be an $m \times m$ symmetric matrix of variables. Then

(a) $\dfrac{\partial}{\partial\, \mathrm{v}(X)'}\, \mathrm{v}(X^{-1}) = -D_m^+(X^{-1} \otimes X^{-1})D_m,$

(b) $\dfrac{\partial}{\partial\, \mathrm{v}(X)'}\, \mathrm{v}(X^{+}) = D_m^+(\{X^+X^+ \otimes (\mathrm{I} - X^+X) + (\mathrm{I} - XX^+) \otimes X^+X^+\}$

$\cdot K_{mm} - (X^+ \otimes X^+))D_m.$

6. THE PERTURBATION METHOD

The perturbation method is a technique, closely related to the method utilizing the differential operator, for finding successive terms in a Taylor expansion formula. In this section, we will use this method to obtain Taylor formulas for some important matrix functions. A more rigorous treatment of this subject can be found in texts such as Hinch (1991), Kato (1982), or Nayfeh (1981).

Suppose that the elements of dX are small, which we can emphasize by writing d$X = \epsilon Y$, where ϵ is a small scalar and Y is an $m \times n$ matrix. Then $X + \epsilon Y$ represents a small perturbation of the $m \times n$ matrix X. The Taylor formula for the vector function $\boldsymbol{f}$ of X would then be of the form

$$\boldsymbol{f}(X + \epsilon Y) = \boldsymbol{f}(X) + \sum_{i=1}^{\infty} \epsilon^i \boldsymbol{g}_i(X, Y),$$

where $\boldsymbol{g}_i(X, Y)$ represents some vector function of the two matrices X and Y. Similarly, if we have a matrix function F then the expansion would be of the form

$$F(X + \epsilon Y) = F(X) + \sum_{i=1}^{\infty} \epsilon^i G_i(X, Y) \tag{8.12}$$

Our goal is to determine the first few terms in the summations given above. These then can be used in an approximation of $\boldsymbol{f}(X + \epsilon Y)$ or $F(X + \epsilon Y)$ when ϵ is small. For instance, suppose that $m = n$ and our function is the matrix inverse function; that is, $F(X) = X^{-1}$. For notational simplicity write $G_i(X, Y) = G_i$ and suppose that the $m \times m$ matrices X and $(X + \epsilon Y)$ are nonsingular. Then (8.12) can be written

$$(X + \epsilon Y)^{-1} = X^{-1} + \epsilon G_1 + \epsilon^2 G_2 + \epsilon^3 G_3 + \cdots$$

But we must have

$$\begin{aligned} \mathrm{I}_m &= (X + \epsilon Y)(X + \epsilon Y)^{-1} \\ &= (X + \epsilon Y)(X^{-1} + \epsilon G_1 + \epsilon^2 G_2 + \epsilon^3 G_3 + \cdots) \\ &= \mathrm{I}_m + \epsilon(YX^{-1} + XG_1) + \epsilon^2(YG_1 + XG_2) + \epsilon^3(YG_2 + XG_3) + \cdots \end{aligned}$$

If this is to hold for all ϵ, then we must have $(YX^{-1} + XG_1) = (0)$ or, equivalently,

$$G_1 = -X^{-1}YX^{-1}$$

Similarly, we must have $(YG_1 + XG_2) = (0)$ so that

$$G_2 = -X^{-1}YG_1 = X^{-1}YX^{-1}YX^{-1},$$

and, in fact, it should be apparent that we have the recursive relationship

$$G_h = -X^{-1}YG_{h-1}$$

As a result, we have

$$(X+\epsilon Y)^{-1} = X^{-1} - \epsilon X^{-1}YX^{-1} + \epsilon^2 X^{-1}YX^{-1}YX^{-1} - \epsilon^3 X^{-1}YX^{-1}YX^{-1}YX^{-1} + \cdots,$$

or, if we return to the notation $\mathrm{d}X = \epsilon Y$,

$$(X+\mathrm{d}X)^{-1} = X^{-1} - X^{-1}(\mathrm{d}X)X^{-1} + X^{-1}(\mathrm{d}X)X^{-1}(\mathrm{d}X)X^{-1} - X^{-1}(\mathrm{d}X)X^{-1}(\mathrm{d}X)X^{-1}(\mathrm{d}X)X^{-1} + \cdots$$

Next we will use this perturbation method to determine the first few terms in the Taylor series expansion for an eigenvalue of a symmetric matrix. Such an expansion will be possible only if the corresponding eigenvalue of the unperturbed matrix X is distinct. We will first consider the special case in which X is a diagonal matrix.

Theorem 8.5. Suppose $X = \mathrm{diag}(x_1, \ldots, x_m)$, where $x_1 \geq \cdots \geq x_{l-1} > x_l > x_{l+1} \geq \cdots \geq x_m$, so that the lth diagonal element x_l differs from the other diagonal elements of X. Let U be an $m \times m$ symmetric matrix and denote the lth largest eigenvalue and corresponding normalized eigenvector of $X + U$ by $\lambda_l(X+U)$ and $\boldsymbol{\gamma}_l(X+U)$, respectively. Then

$$\lambda_l(X+U) \simeq x_l + u_{ll} - \sum_{i \neq l} \frac{u_{il}^2}{(x_i - x_l)} - \sum_{i \neq l} \frac{u_{ll}u_{il}^2}{(x_i - x_l)^2} + \sum_{i \neq l} \sum_{j \neq l} \frac{u_{il}u_{jl}u_{ij}}{(x_i - x_l)(x_j - x_l)},$$

$$\gamma_{ll}(X+U) \simeq 1 - \frac{1}{2} \sum_{i \neq l} \frac{u_{il}^2}{(x_i - x_l)^2} - \sum_{i \neq l} \frac{u_{ll}u_{il}^2}{(x_i - x_l)^3} + \sum_{i \neq l} \sum_{j \neq l} \frac{u_{il}u_{jl}u_{ij}}{(x_i - x_l)^2(x_j - x_l)},$$

and for $h \neq l$

$$\gamma_{hl}(X+U) \simeq -\frac{u_{hl}}{(x_h - x_l)} - \frac{u_{ll}u_{hl}}{(x_h - x_l)^2} + \sum_{i \neq l} \frac{u_{il}u_{hi}}{(x_h - x_l)(x_i - x_l)} - \frac{u_{ll}^2 u_{hl}}{(x_h - x_l)^3}$$
$$+ \sum_{i \neq l} \frac{u_{ll}u_{hi}u_{il}}{(x_h - x_l)^2(x_i - x_l)} + \sum_{i \neq l} \frac{u_{hl}u_{il}^2}{(x_h - x_l)^2(x_i - x_l)}$$
$$+ \sum_{i \neq l} \frac{u_{ll}u_{hi}u_{il}}{(x_h - x_l)(x_i - x_l)^2}$$
$$- \sum_{i \neq l} \sum_{j \neq l} \frac{u_{hi}u_{ij}u_{jl}}{(x_h - x_l)(x_i - x_l)(x_j - x_l)}$$
$$+ \frac{1}{2} \sum_{i \neq l} \frac{u_{hl}u_{il}^2}{(x_h - x_l)(x_i - x_l)^2},$$

where $\gamma_{hl}(X+U)$ denotes the hth element of $\boldsymbol{\gamma}_l(X+U)$, and the approximations above are accurate up through third-order terms in the us.

Proof. Here U is the perturbation matrix, and we wish to write $\lambda_l = \lambda_l(X + U)$ and $\boldsymbol{\gamma}_l = \boldsymbol{\gamma}_l(X + U)$ in the form

$$\lambda_l = x_l + a_1 + a_2 + a_3 + \cdots, \tag{8.13}$$
$$\boldsymbol{\gamma}_l = \boldsymbol{e}_l + \boldsymbol{b}_1 + \boldsymbol{b}_2 + \boldsymbol{b}_3 + \cdots, \tag{8.14}$$

where a_i and $\boldsymbol{b}_i$ only involve ith degree terms in the elements of U. Substituting these expressions in the defining equation $(X + U)\boldsymbol{\gamma}_l = \lambda_l \boldsymbol{\gamma}_l$ and then equating ith degree terms in the elements of U on the left-hand side of this equation to those on the right-hand side, we obtain

$$X\boldsymbol{e}_l = x_l \boldsymbol{e}_l, \tag{8.15}$$
$$X\boldsymbol{b}_1 + U\boldsymbol{e}_l = x_l \boldsymbol{b}_1 + a_1 \boldsymbol{e}_l, \tag{8.16}$$
$$X\boldsymbol{b}_2 + U\boldsymbol{b}_1 = x_l \boldsymbol{b}_2 + a_1 \boldsymbol{b}_1 + a_2 \boldsymbol{e}_l, \tag{8.17}$$
$$X\boldsymbol{b}_3 + U\boldsymbol{b}_2 = x_l \boldsymbol{b}_3 + a_1 \boldsymbol{b}_2 + a_2 \boldsymbol{b}_1 + a_3 \boldsymbol{e}_l \tag{8.18}$$

In a similar fashion, the normalizing equation $\boldsymbol{\gamma}_l' \boldsymbol{\gamma}_l = 1$ yields the identities

$$\boldsymbol{e}_l' \boldsymbol{e}_l = 1, \tag{8.19}$$
$$\boldsymbol{e}_l' \boldsymbol{b}_1 + \boldsymbol{b}_1' \boldsymbol{e}_l = 0, \tag{8.20}$$
$$\boldsymbol{e}_l' \boldsymbol{b}_2 + \boldsymbol{b}_1' \boldsymbol{b}_1 + \boldsymbol{b}_2' \boldsymbol{e}_l = 0, \tag{8.21}$$
$$\boldsymbol{e}_l' \boldsymbol{b}_3 + \boldsymbol{b}_1' \boldsymbol{b}_2 + \boldsymbol{b}_2' \boldsymbol{b}_1 + \boldsymbol{b}_3' \boldsymbol{e}_l = 0. \tag{8.22}$$

Equations (8.15) and (8.19) are trivially true, while equations (8.16) and (8.20) can be used to find a_1 and $\boldsymbol{b}_1$. Premultiplying (8.16) by $\boldsymbol{e}_l'$ and then solving for a_1, we find that

$$a_1 = \boldsymbol{e}_l' U \boldsymbol{e}_l = u_{ll} \tag{8.23}$$

We can then rewrite (8.16) as the system of linear equations

$$(X - x_l \mathrm{I}_m)\boldsymbol{b}_1 = -(U - u_{ll}\mathrm{I}_m)\boldsymbol{e}_l,$$

with the general solution for $\boldsymbol{b}_1$ given by

$$\boldsymbol{b}_1 = -(X - x_l \mathrm{I}_m)^+ (U - u_{ll}\mathrm{I}_m)\boldsymbol{e}_l + c_1 \boldsymbol{e}_l,$$

where c_1 is an arbitrary constant. Since $(X - x_l \mathrm{I}_m)^+ \boldsymbol{e}_l = \mathbf{0}$ and (8.20) implies that $\boldsymbol{e}_l' \boldsymbol{b}_1 = \mathbf{0}$, it follows that $c_1 = 0$ and thus,

$$\boldsymbol{b}_1 = -(X - x_l \mathrm{I}_m)^+ U \boldsymbol{e}_l \tag{8.24}$$

Next, we will use (8.17) and (8.21) to find a_2 and $\boldsymbol{b}_2$. Premultiplying (8.17) by $\boldsymbol{e}_l'$ and then solving for a_2, we find, after again using the fact that $\boldsymbol{e}_l' \boldsymbol{b}_1 = 0$, that

$$a_2 = \boldsymbol{e}_l' U \boldsymbol{b}_1 = -\boldsymbol{e}_l' U (X - x_l \mathrm{I}_m)^+ U \boldsymbol{e}_l \tag{8.25}$$

Rewriting (8.17) as the system of equations in $\boldsymbol{b}_2$,

$$(X - x_l \mathrm{I}_m)\boldsymbol{b}_2 = a_2 \boldsymbol{e}_l - (U - a_1 \mathrm{I}_m)\boldsymbol{b}_1,$$

which for any scalar c_2 has as a solution

$$\boldsymbol{b}_2 = (X - x_l \mathrm{I}_m)^+ \{a_2 \boldsymbol{e}_l - (U - a_1 \mathrm{I}_m)\boldsymbol{b}_1\} + c_2 \boldsymbol{e}_l$$

Now since $(X - x_l \mathrm{I}_m)^+ \boldsymbol{e}_l = \mathbf{0}$ and (8.21) implies that $\boldsymbol{e}_l' \boldsymbol{b}_2 = -\frac{1}{2}\boldsymbol{b}_1' \boldsymbol{b}_1$, we find that

$$c_2 = -\frac{1}{2}\,\boldsymbol{b}_1' \boldsymbol{b}_1 = -\frac{1}{2}\,\boldsymbol{e}_l' U \{(X - x_l \mathrm{I}_m)^+\}^2 U \boldsymbol{e}_l = -\frac{1}{2} \sum_{i \neq l} \frac{u_{il}^2}{(x_i - x_l)^2},$$

and so with this value for c_2, the solution for $\boldsymbol{b}_2$ is given by

$$\boldsymbol{b}_2 = (X - x_l \mathrm{I}_m)^+ (U - u_{ll}\mathrm{I}_m)(X - x_l \mathrm{I}_m)^+ U \boldsymbol{e}_l + c_2 \boldsymbol{e}_l \tag{8.26}$$

To find a_3, premultiply (8.18) by $\boldsymbol{e}_l'$ and solve for a_3, after using $\boldsymbol{e}_l'\boldsymbol{b}_1 = 0$, to get

$$\begin{aligned} a_3 &= \boldsymbol{e}_l'(U - a_1 \mathrm{I}_m)\boldsymbol{b}_2 \\ &= \boldsymbol{e}_l'(U - u_{ll}\mathrm{I}_m)\{(X - x_l\mathrm{I}_m)^+(U - u_{ll}\mathrm{I}_m)(X - x_l\mathrm{I}_m)^+ U\boldsymbol{e}_l + c_2\boldsymbol{e}_l\} \\ &= \boldsymbol{e}_l' U(X - x_l\mathrm{I}_m)^+(U - u_{ll}\mathrm{I}_m)(X - x_l\mathrm{I}_m)^+ U\boldsymbol{e}_l \end{aligned} \tag{8.27}$$

Equation (8.18) can be expressed as

$$(X - x_l\mathrm{I}_m)\boldsymbol{b}_3 = a_3\boldsymbol{e}_l + a_2\boldsymbol{b}_1 - (U - a_1\mathrm{I}_m)\boldsymbol{b}_2,$$

so that the solution for $\boldsymbol{b}_3$ will be given by

$$\begin{aligned} \boldsymbol{b}_3 &= (X - x_l\mathrm{I}_m)^+\{a_3\boldsymbol{e}_l + a_2\boldsymbol{b}_1 - (U - a_1\mathrm{I}_m)\boldsymbol{b}_2\} + c_3\boldsymbol{e}_l \\ &= (X - x_l\mathrm{I}_m)^+\{a_2\boldsymbol{b}_1 - (U - a_1\mathrm{I}_m)\boldsymbol{b}_2\} + c_3\boldsymbol{e}_l, \end{aligned} \tag{8.28}$$

where c_3 is an arbitrary constant. By premultiplying this equation by $\boldsymbol{e}_l'$ and using $\boldsymbol{e}_l'\boldsymbol{b}_3 = -\boldsymbol{b}_1'\boldsymbol{b}_2$, which follows from (8.22), we find that

$$\begin{aligned} c_3 = -\boldsymbol{b}_1'\boldsymbol{b}_2 &= \boldsymbol{e}_l' U\{(X - x_l\mathrm{I}_m)^+\}^2(U - u_{ll}\mathrm{I}_m)(X - x_l\mathrm{I}_m)^+ U\boldsymbol{e}_l \\ &= -\sum_{i \neq l} \frac{u_{ll}u_{il}^2}{(x_i - x_l)^3} + \sum_{i \neq l}\sum_{j \neq l} \frac{u_{il}u_{jl}u_{ij}}{(x_i - x_l)^2(x_j - x_l)} \end{aligned}$$

The results now follow by substituting (8.23), (8.25), and (8.27) in (8.13) and (8.24), (8.26), and (8.28) in (8.14). □

Theorem 8.5 can be used to obtain expansion formulas for a general symmetric matrix; that is, if Z is an $m \times m$ symmetric matrix and W is its associated symmetric perturbation matrix, then we can obtain expansion formulas for $\lambda_l(Z+W)$ and $\boldsymbol{\gamma}_l(Z+W)$. Let $Z = QXQ'$ be the spectral decomposition of Z, so that $X = \mathrm{diag}(x_1, \ldots, x_m)$ with x_l being an eigenvalue of Z corresponding to the eigenvector $\boldsymbol{q}_l$, which is the lth column of Q. As in Theorem 8.5, we assume that x_l is a distinct eigenvalue. If we let $U = Q'WQ$, then the eigenvalue–eigenvector equation

$$(Z + W)\{\boldsymbol{\gamma}_l(Z + W)\} = \{\lambda_l(Z + W)\}\{\boldsymbol{\gamma}_l(Z + W)\}$$

can be equivalently expressed as

$$(X + U)Q'\{\boldsymbol{\gamma}_l(Z + W)\} = \{\lambda_l(Z + W)\}Q'\{\boldsymbol{\gamma}_l(Z + W)\};$$

that is, U is the perturbation matrix of X, and $\lambda_l(Z+W)$ is an eigenvalue of $(X+U)$ corresponding to the eigenvector $Q'\boldsymbol{\gamma}_l(Z+W)$. Thus, if we use the elements of $U = QWQ'$ in place of those of U in the formulas in Theorem 8.5, we will obtain expansions for $\lambda_l(Z+W)$ and $Q'\boldsymbol{\gamma}_l(Z+W)$. For instance, first-order approximations of $\lambda_l(Z+W)$ and $\boldsymbol{\gamma}_l(Z+W)$ are given by

$$\begin{aligned}\lambda_l(Z+W) &\simeq x_l + \boldsymbol{q}_l' W \boldsymbol{q}_l,\\ \boldsymbol{\gamma}_l(Z+W) &\simeq Q\{\boldsymbol{e}_l - (X - x_l \mathrm{I}_m)^+ (Q'WQ)\boldsymbol{e}_l\}\\ &= \boldsymbol{q}_l - (Z - x_l\mathrm{I}_m)^+ W\boldsymbol{q}_l\end{aligned}$$

The following is an immediate consequence of the first-order Taylor expansion formulas given above.

Theorem 8.6. Let $\lambda_l(Z)$ be the eigenvalue defined on $m \times m$ symmetric matrices Z, and let $\boldsymbol{\gamma}_l(Z)$ be a corresponding normalized eigenvector. If the matrix Z is such that the eigenvalue $\lambda_l(Z)$ is distinct, then differentials and derivatives at that matrix Z are given by

$$\mathrm{d}\lambda_l = \boldsymbol{\gamma}_l'(\mathrm{d}Z)\boldsymbol{\gamma}_l, \qquad \frac{\partial}{\partial\, \mathrm{v}(Z)'}\lambda_l(Z) = (\boldsymbol{\gamma}_l' \otimes \boldsymbol{\gamma}_l')D_m,$$

$$\mathrm{d}\boldsymbol{\gamma}_l = -(Z - \lambda_l \mathrm{I}_m)^+(\mathrm{d}Z)\boldsymbol{\gamma}_l, \qquad \frac{\partial}{\partial\, \mathrm{v}(Z)'}\boldsymbol{\gamma}_l(Z) = -\{\boldsymbol{\gamma}_l' \otimes (Z - \lambda_l\mathrm{I}_m)^+\}D_m$$

The expansions given in and immediately following Theorem 8.5 do not hold when the eigenvalue x_l is not distinct. Suppose, for instance, that again $x_1 \geq \cdots \geq x_m$, but now $x_l = x_{l+1} = \cdots = x_{l+r-1}$, so that the value x_l is repeated as an eigenvalue of $Z = QXQ'$, r times. In this case, we can get expansions for $\overline{\lambda}_{l,l+r-1}(Z+W)$, the average of the perturbed eigenvalues $\lambda_l(Z+W), \ldots, \lambda_{l+r-1}(Z+W)$, and the total eigenprojection Φ_l associated with this collection of eigenvalues; if $P_{Z+W}\{\lambda_{l+i-1}(Z+W)\}$ represents the eigenprojection of $Z+W$ associated with the eigenvalue $\lambda_{l+i-1}(Z+W)$, then this total eigenprojection is given by

$$\begin{aligned}\Phi_l &= \sum_{i=1}^{r} P_{Z+W}\{\lambda_{l+i-1}(Z+W)\}\\ &= \sum_{i=1}^{r} \boldsymbol{\gamma}_{l+i-1}(Z+W)(\boldsymbol{\gamma}_{l+i-1}(Z+W))'\end{aligned}$$

These expansions are summarized below. The proof, which is similar to that of Theorem 8.5, is left to the reader.

Theorem 8.7. Let Z be an $m \times m$ symmetric matrix with eigenvalues $x_1 \geq \cdots \geq x_{l-1} > x_l = x_{l+1} = \cdots = x_{l+r-1} > x_{l+r} \geq \cdots \geq x_m$, so that x_l is an eigenvalue with multiplicity r. Suppose that W is an $m \times m$ symmetric matrix and let $\lambda_1 \geq \lambda_2 \geq \cdots \geq \lambda_m$ be the eigenvalues of $Z+W$, while $\overline{\lambda}_{l,l+r-1} = r^{-1}(\lambda_l + \cdots + \lambda_{l+r-1})$. Denote the eigenprojection of Z corresponding to the repeated eigenvalue x_l by P_l and denote the total eigenprojection of $Z+W$ corresponding to the collection of eigenvalues $\lambda_l, \ldots, \lambda_{l+r-1}$ by Φ_l. Define $Y = (Z - x_l \mathrm{I}_m)^+$. Then the third-order Taylor approximations

$$\overline{\lambda}_{l,l+r-1} \simeq x_l + a_1 + a_2 + a_3,$$
$$\Phi_l \simeq P_l + B_1 + B_2 + B_3,$$

have

$$
\begin{aligned}
a_1 &= \frac{1}{r}\,\mathrm{tr}(WP_l),\\
a_2 &= -\frac{1}{r}\,\mathrm{tr}(WYWP_l),\\
a_3 &= \frac{1}{r}\{\mathrm{tr}(YWYWP_lW) - \mathrm{tr}(Y^2WP_lWP_lW)\},\\
B_1 &= -YWP_l - P_lWY,\\
B_2 &= YWP_lWY + YWYWP_l - Y^2WP_lWP_l + P_lWYWY - P_lWP_lWY^2\\
&\quad - P_lWY^2WP_l,\\
B_3 &= Y^2WP_lWYWP_l + P_lWYWP_lWY^2 + Y^2WP_lWP_lWY + YWP_lWP_lWY^2\\
&\quad + Y^2WYWP_lWP_l + P_lWP_lWYWY^2 + YWY^2WP_lWP_l + P_lWP_lWY^2WY\\
&\quad - Y^3WP_lWP_lWP_l - P_lWP_lWP_lWY^3 - YWYWP_lWY - YWP_lWYWY\\
&\quad - YWYWYWP_l - P_lWYWYWY + YWP_lWY^2WP_l + P_lWY^2WP_lWY\\
&\quad + P_lWY^2WYWP_l + P_lWYWY^2WP_l - P_lWY^3WP_lWP_l\\
&\quad - P_lWP_lWY^3WP_l
\end{aligned}
$$

7. MAXIMA AND MINIMA

One important application of derivatives involves finding the maxima or minima of a function. A function f has a local maximum at an $n \times 1$ point $\boldsymbol{a}$ if for some $\delta > 0$, $f(\boldsymbol{a}) \geq f(\boldsymbol{a}+\boldsymbol{x})$ whenever $\boldsymbol{x}'\boldsymbol{x} < \delta$. This function has an absolute maximum at $\boldsymbol{a}$ if $f(\boldsymbol{a}) \geq f(\boldsymbol{x})$ for all $\boldsymbol{x}$ for which f is defined. Similar definitions hold for a local minimum and an absolute minimum; in fact, if f has a local minimum at a point $\boldsymbol{a}$, then $-f$ has a local maximum at $\boldsymbol{a}$, and if f has an absolute minimum at $\boldsymbol{a}$, then $-f$ has an absolute maximum at $\boldsymbol{a}$. For this reason, we will at times confine our discussion to only the case of a maximum. In this section and the next section, we state some results that are helpful in finding local maxima and minima. For proofs of these results the reader is referred to Khuri (1993) or Magnus and Neudecker (1988). Our first result gives a necessary condition for a function f to have a local maximum at $\boldsymbol{a}$.

Theorem 8.8. Suppose the function $f(\boldsymbol{x})$ is defined for all $n \times 1$ vectors $\boldsymbol{x} \in S$, where S is some subset of R^n. Let $\boldsymbol{a}$ be an interior point of S; that is, there exists a $\delta > 0$ such that $\boldsymbol{a} + \boldsymbol{u} \in S$ for all $\boldsymbol{u}'\boldsymbol{u} < \delta$. If f has a local maximuim at $\boldsymbol{a}$ and f is differentiable at $\boldsymbol{a}$, then

$$\frac{\partial}{\partial \boldsymbol{a}'} f(\boldsymbol{a}) = \mathbf{0}' \tag{8.29}$$

Any point $\boldsymbol{a}$ satisfying (8.29) is called a stationary point of f. While Theorem 8.8 indicates that any point at which a local maximum or local minimum occurs must be a stationary point, the converse does not hold. A stationary point that does not correspond to a local maximum or a local minimum is called a saddle point. Our next result is helpful in determining whether a particular stationary point is a local maximum or minimum in those situations in which the function f is twice differentiable.

Theorem 8.9. Suppose the function $f(\boldsymbol{x})$ is defined for all $n \times 1$ vectors $\boldsymbol{x} \in S$, where S is some subset of R^n. Suppose also that f is twice differentiable at the interior point $\boldsymbol{a}$ of S. If $\boldsymbol{a}$ is a stationary point of f and H_f is the Hessian matrix of f at $\boldsymbol{a}$, then

(a) f has a local minimum at $\boldsymbol{a}$ if H_f is positive definite,
(b) f has a local maximum at $\boldsymbol{a}$ if H_f is negative definite,
(c) f has a saddle point at $\boldsymbol{a}$ if H_f is nonsingular but not positive definite or negative definite,
(d) f may have a local minimum, a local maximum, or a saddle point at $\boldsymbol{a}$ if H_f is singular.

Example 8.4. On several occasions, we have discussed the problem of finding a least squares solution $\hat{\boldsymbol{\beta}}$ to the inconsistent system of equations

$$\boldsymbol{y} = X\boldsymbol{\beta},$$

where $\boldsymbol{y}$ is an $N \times 1$ vector of constants, X is an $N \times (k+1)$ matrix of constants, and $\boldsymbol{\beta}$ is a $(k+1) \times 1$ vector of variables. A solution was obtained in Chapter 2 by using the geometrical properties of least squares regression, while in Chapter 6 we utilized the results developed on least squares generalized inverses. In this example, we will show how the methods of this section may be used to obtain a solution. We will assume that rank$(X) = k \times 1$; that is, the matrix X has full column rank. Recall that a least squares solution $\hat{\boldsymbol{\beta}}$ is any vector which minimizes the sum of squared errors given by

$$f(\hat{\boldsymbol{\beta}}) = (\boldsymbol{y} - X\hat{\boldsymbol{\beta}})'(\boldsymbol{y} - X\hat{\boldsymbol{\beta}})$$

The differential of $f(\hat{\boldsymbol{\beta}})$ is

$$\begin{aligned} \mathrm{d}f &= \mathrm{d}\{(\boldsymbol{y} - X\hat{\boldsymbol{\beta}})'\}(\boldsymbol{y} - X\hat{\boldsymbol{\beta}}) + (\boldsymbol{y} - X\hat{\boldsymbol{\beta}})'\,\mathrm{d}(\boldsymbol{y} - X\hat{\boldsymbol{\beta}}) \\ &= -(\mathrm{d}\hat{\boldsymbol{\beta}})'X'(\boldsymbol{y} - X\hat{\boldsymbol{\beta}}) - (\boldsymbol{y} - X\hat{\boldsymbol{\beta}})'X\,\mathrm{d}\hat{\boldsymbol{\beta}} = -2(\boldsymbol{y} - X\hat{\boldsymbol{\beta}})'X\,\mathrm{d}\hat{\boldsymbol{\beta}}, \end{aligned}$$

so that

$$\frac{\partial}{\partial \hat{\boldsymbol{\beta}}'} f(\hat{\boldsymbol{\beta}}) = -2(\boldsymbol{y} - X\hat{\boldsymbol{\beta}})'X$$

Thus, upon setting this first derivative equal to $\mathbf{0}'$ and rearranging, we find that the stationary values are given by the solutions $\hat{\boldsymbol{\beta}}$ to the system of equations

$$X'X\hat{\boldsymbol{\beta}} = X'\boldsymbol{y} \tag{8.30}$$

Since X has full column rank, $X'X$ is nonsingular, and so the unique solution to (8.30) is

$$\hat{\boldsymbol{\beta}} = (X'X)^{-1}X'\boldsymbol{y} \tag{8.31}$$

In order to verify that this solution minimizes the sum of squared errors, we need to obtain the Hessian matrix H_f. The second differential of $f(\hat{\boldsymbol{\beta}})$ is given by

$$\begin{aligned} \mathrm{d}^2 f &= \mathrm{d}(\mathrm{d}f) = -\mathrm{d}\{2(\boldsymbol{y} - X\hat{\boldsymbol{\beta}})'X\,\mathrm{d}\hat{\boldsymbol{\beta}}\} = -2\{\mathrm{d}(\boldsymbol{y} - X\hat{\boldsymbol{\beta}})\}'X\,\mathrm{d}\hat{\boldsymbol{\beta}} \\ &= 2(\mathrm{d}\hat{\boldsymbol{\beta}})'X'X\,\mathrm{d}\hat{\boldsymbol{\beta}}, \end{aligned}$$

so that

$$H_f = 2X'X$$

Since this matrix is positive definite, it follows from Theorem 8.9 that the solution given in (8.31) minimizes $f(\hat{\boldsymbol{\beta}})$.

Example 8.5. One of the most popular ways of obtaining estimators of unknown parameters is by a method known as maximum likelihood estimation. If we have a random sample of vectors $\boldsymbol{x}_1, \ldots, \boldsymbol{x}_n$ from a population having density function $f(\boldsymbol{x}; \boldsymbol{\theta})$, where $\boldsymbol{\theta}$ is a vector of parameters, then the likelihood function of $\boldsymbol{\theta}$ is defined to be the joint density function of $\boldsymbol{x}_1, \ldots, \boldsymbol{x}_n$ viewed as a function of $\boldsymbol{\theta}$; that is, this likelihood function is given by

$$L(\boldsymbol{\theta}) = \prod_{i=1}^{n} f(\boldsymbol{x}_i; \boldsymbol{\theta})$$

The method of maximum likelihood estimates $\boldsymbol{\theta}$ by the vector $\hat{\boldsymbol{\theta}}$, which maximizes $L(\boldsymbol{\theta})$. In this example, we will use this method to obtain estimates of $\boldsymbol{\mu}$ and Ω when our sample is coming from the normal distribution, $N_m(\boldsymbol{\mu}, \Omega)$. Thus, $\boldsymbol{\mu}$ is an $m \times 1$ vector, Ω is an $m \times m$ positive definite matrix, and the required density function, $f(\boldsymbol{x}; \boldsymbol{\mu}, \Omega)$ is given in (1.13). In deriving the estimates $\hat{\boldsymbol{\mu}}$ and $\hat{\Omega}$, we will find it a little bit easier to maximize the function $\log(L(\boldsymbol{\mu}, \Omega))$, which is, of course, maximized at the same solution as $L(\boldsymbol{\mu}, \Omega)$. After omitting terms from $\log(L(\boldsymbol{\mu}, \Omega))$ that do not involve $\boldsymbol{\mu}$ or Ω, we find that we must maximize the function

$$g(\boldsymbol{\mu}, \Omega) = -\frac{1}{2} n \log|\Omega| - \frac{1}{2} \operatorname{tr}(\Omega^{-1} U),$$

where

$$U = \sum_{i=1}^{n} (\boldsymbol{x}_i - \boldsymbol{\mu})(\boldsymbol{x}_i - \boldsymbol{\mu})'$$

The first differential of g is given by

$$\begin{aligned}
\mathrm{d}g &= -\frac{1}{2} n\, \mathrm{d}(\log|\Omega|) - \frac{1}{2} \operatorname{tr}\{(\mathrm{d}\Omega^{-1})U\} - \frac{1}{2} \operatorname{tr}(\Omega^{-1} \mathrm{d}U) \\
&= -\frac{1}{2} n \operatorname{tr}(\Omega^{-1} \mathrm{d}\Omega) + \frac{1}{2} \operatorname{tr}\{\Omega^{-1}(\mathrm{d}\Omega)\Omega^{-1} U\} \\
&\quad + \frac{1}{2} \operatorname{tr}\left(\Omega^{-1} \left\{(\mathrm{d}\boldsymbol{\mu}) \sum_{i=1}^{n} (\boldsymbol{x}_i - \boldsymbol{\mu})' + \sum_{i=1}^{n} (\boldsymbol{x}_i - \boldsymbol{\mu})\, \mathrm{d}\boldsymbol{\mu}'\right\}\right) \\
&= \frac{1}{2} \operatorname{tr}\{(\mathrm{d}\Omega)\Omega^{-1}(U - n\Omega)\Omega^{-1}\} \\
&\quad + \frac{1}{2} \operatorname{tr}(\Omega^{-1}\{n(\mathrm{d}\boldsymbol{\mu})(\bar{\boldsymbol{x}} - \boldsymbol{\mu})' + n(\bar{\boldsymbol{x}} - \boldsymbol{\mu})\, \mathrm{d}\boldsymbol{\mu}'\})
\end{aligned}$$

$$= \frac{1}{2}\ \mathrm{tr}\{(\mathrm{d}\Omega)\Omega^{-1}(U - n\Omega)\Omega^{-1}\} + n(\bar{\boldsymbol{x}} - \boldsymbol{\mu})'\Omega^{-1}\,\mathrm{d}\boldsymbol{\mu}$$

$$= \frac{1}{2}\ \mathrm{vec}(\mathrm{d}\Omega)'(\Omega^{-1} \otimes \Omega^{-1})\mathrm{vec}(U - n\Omega) + n(\bar{\boldsymbol{x}} - \boldsymbol{\mu})'\Omega^{-1}\,\mathrm{d}\boldsymbol{\mu},$$

where the second equality used Corollary 8.1.1 and Theorem 8.2, and the fifth used Theorem 7.17. Since Ω is symmetric, $\mathrm{vec}(\mathrm{d}\Omega) = \mathrm{d}\,\mathrm{vec}(\Omega) = D_m\,\mathrm{d}\,\mathrm{v}(\Omega)$, and so the differential may be reexpressed as

$$\mathrm{d}g = \frac{1}{2}\ \{\mathrm{d}\,\mathrm{v}(\Omega)\}'D_m'(\Omega^{-1} \otimes \Omega^{-1})\mathrm{vec}(U - n\Omega) + n(\bar{\boldsymbol{x}} - \boldsymbol{\mu})'\Omega^{-1}\,\mathrm{d}\boldsymbol{\mu}, \qquad (8.32)$$

and thus,

$$\frac{\partial}{\partial\boldsymbol{\mu}'}\ g = n(\bar{\boldsymbol{x}} - \boldsymbol{\mu})'\Omega^{-1}, \qquad \frac{\partial}{\partial\,\mathrm{v}(\Omega)'}\ g = \frac{1}{2}\ \{\mathrm{vec}(U - n\Omega)\}'(\Omega^{-1} \otimes \Omega^{-1})D_m$$

Upon equating these first derivatives to null vectors, we obtain the equations

$$n\Omega^{-1}(\bar{\boldsymbol{x}} - \boldsymbol{\mu}) = \mathbf{0},$$
$$D_m'(\Omega^{-1} \otimes \Omega^{-1})\mathrm{vec}(U - n\Omega) = \mathbf{0}$$

From the first of these two equations, we obtain the solution for $\boldsymbol{\mu}$, $\hat{\boldsymbol{\mu}} = \bar{\boldsymbol{x}}$, while the second can be rewritten as

$$D_m'(\Omega^{-1} \otimes \Omega^{-1})D_m\,\mathrm{v}(U - n\Omega) = \mathbf{0},$$

since the symmetry of $(U - n\Omega)$ implies that $\mathrm{vec}(U - n\Omega) = D_m\,\mathrm{v}(U - n\Omega)$. Premultiplying this equation by $D_m^+(\Omega \otimes \Omega)D_m^{+\prime}$ and using Theorem 7.38, we find that

$$\mathrm{v}(U - n\Omega) = \mathbf{0}$$

Since $(U - n\Omega)$ is symmetric this implies that $(U - n\Omega) = (0)$, and so the solution for Ω is $\hat{\Omega} = n^{-1}U$. All that remains is to show that the solution $(\hat{\boldsymbol{\mu}}, \hat{\Omega})$ yields a maximum. By differentiating (8.32), we find that

$$\begin{aligned} d^2g &= \frac{1}{2}\{d\,v(\Omega)\}'D_m'\{d(\Omega^{-1}\otimes\Omega^{-1})\}\text{vec}(U-n\Omega) \\ &\quad + \frac{1}{2}\{d\,v\{\Omega\}\}'D_m'(\Omega^{-1}\otimes\Omega^{-1})\text{vec}(dU-n\,d\Omega) \\ &\quad - n(d\boldsymbol{\mu})'\Omega^{-1}\,d\boldsymbol{\mu} + n(\bar{\boldsymbol{x}}-\boldsymbol{\mu})'(d\Omega^{-1})\,d\boldsymbol{\mu} \end{aligned}$$

Evaluating this at $\boldsymbol{\mu} = \bar{\boldsymbol{x}}$ and $\Omega = n^{-1}U$, we find that the first and the fourth terms on the right-hand side of the equation above vanish. In addition, note that

$$dU = n(d\boldsymbol{\mu})(\bar{\boldsymbol{x}}-\boldsymbol{\mu})' + n(\bar{\boldsymbol{x}}-\boldsymbol{\mu})\,d\boldsymbol{\mu}'$$

also vanishes when evaluated at $\boldsymbol{\mu} = \bar{\boldsymbol{x}}$. Thus, at $\boldsymbol{\mu} = \bar{\boldsymbol{x}}$ and $\Omega = n^{-1}U$,

$$\begin{aligned} d^2g &= -\frac{n}{2}\{d\,v(\Omega)\}'D_m'(\Omega^{-1}\otimes\Omega^{-1})D_m\,d\,v(\Omega) - n(d\boldsymbol{\mu})'\Omega^{-1}\,d\boldsymbol{\mu}, \\ &= [d\boldsymbol{\mu}' \quad \{d\,v(\Omega)\}']H_g\begin{bmatrix} d\boldsymbol{\mu} \\ d\,v(\Omega) \end{bmatrix}, \end{aligned}$$

where

$$H_g = \begin{bmatrix} -n\Omega^{-1} & (0) \\ (0) & -\frac{n}{2}D_m'(\Omega^{-1}\otimes\Omega^{-1})D_m \end{bmatrix}$$

Clearly, H_g is negative definite since Ω^{-1} and $D_m'(\Omega^{-1}\otimes\Omega^{-1})D_m$ are positive definite matrices. This then establishes that the solution $(\hat{\boldsymbol{\mu}}, \hat{\Omega}) = (\bar{\boldsymbol{x}}, n^{-1}U)$ yields a maximum.

8. CONVEX AND CONCAVE FUNCTIONS

In Section 2.10, we discussed convex sets. Here we will extend the concept of convexity to functions and obtain some special results that apply to this class of functions.

Definition 8.1. Let $f(\boldsymbol{x})$ be a real-valued function defined for all $\boldsymbol{x} \in S$, where S is a convex subset of R^m. Then $f(\boldsymbol{x})$ is a convex function on S, if

$$f(c\boldsymbol{x}_1 + (1-c)\boldsymbol{x}_2) \leq cf(\boldsymbol{x}_1) + (1-c)f(\boldsymbol{x}_2)$$

for all $\boldsymbol{x}_1 \in S$, $\boldsymbol{x}_2 \in S$, and $0 \leq c \leq 1$. If $-f(\boldsymbol{x})$ is a convex function, then $f(\boldsymbol{x})$ is said to be a concave function.

If $f(\boldsymbol{x})$ is a convex function, then it is easily verified that the set defined by

$$T = \{\boldsymbol{z} = (\boldsymbol{x}', y)' : \boldsymbol{x} \in S,\ y \geq f(\boldsymbol{x})\}$$

is a convex subset of R^{m+1}. For instance, if $m = 1$, then T will be a convex subset of R^2. In this case, for any $a \in S$, the point $(a, f(a))$ will be a boundary point of the set T. Now from the supporting hyperplane theorem, Theorem 2.27, we know that there is a line passing through the point $(a, f(a))$ such that the function $f(x)$ is never below this line. Since this line passes through the point $(a, f(a))$, it can be written in the form $g(x) = f(a) + t(x - a)$, where t is the slope of the line, and thus, for all $x \in S$, we have

$$f(x) \geq f(a) + t(x - a) \tag{8.33}$$

The generalization of this result to arbitrary m is given below.

Theorem 8.10. Let $f(\boldsymbol{x})$ be a real-valued convex function defined for all $\boldsymbol{x} \in S$, where S is a convex subset of R^m. Then, corresponding to each interior point $\boldsymbol{a} \in S$, there exists an $m \times 1$ vector $\boldsymbol{t}$ such that

$$f(\boldsymbol{x}) \geq f(\boldsymbol{a}) + \boldsymbol{t}'(\boldsymbol{x} - \boldsymbol{a}) \tag{8.34}$$

for all $\boldsymbol{x} \in S$.

Proof. For any $\boldsymbol{a} \in S$, the point $\boldsymbol{z}_* = (\boldsymbol{a}', f(\boldsymbol{a}))'$ is a boundary point of the convex set T defined above, and so it follows from Theorem 2.27 that there exists an $(m + 1) \times 1$ vector $\boldsymbol{b} = (\boldsymbol{b}_1', b_{m+1})' \neq \boldsymbol{0}$ for which $\boldsymbol{b}'\boldsymbol{z} \geq \boldsymbol{b}'\boldsymbol{z}_*$ for all $\boldsymbol{z} \in T$. Clearly, for any $\boldsymbol{z} = (\boldsymbol{x}', y)' \in T$, we can arbitrarily increase the value of y and get another point in T. For this reason, we see that b_{m+1} cannot be negative since if it were, we would be able to make $\boldsymbol{b}'\boldsymbol{z}$ arbitrarily small and, in particular, less than $\boldsymbol{b}'\boldsymbol{z}_*$. Thus, b_{m+1} is either positive or 0. Now for any $\boldsymbol{x} \in S$, $(\boldsymbol{x}', f(\boldsymbol{x}))' \in T$ and so for this choice of $\boldsymbol{z}$ in the inequality $\boldsymbol{b}'\boldsymbol{z} \geq \boldsymbol{b}'\boldsymbol{z}_*$, we get

$$\boldsymbol{b}_1'\boldsymbol{x} + b_{m+1} f(\boldsymbol{x}) \geq \boldsymbol{b}_1'\boldsymbol{a} + b_{m+1} f(\boldsymbol{a})$$

If b_{m+1} is positive, then the inequality above may be rearranged to the form given in (8.34) with $\boldsymbol{t} = -b_{m+1}^{-1}\boldsymbol{b}_1$. If, on the other hand, $b_{m+1} = 0$, then $\boldsymbol{b}'\boldsymbol{z} \geq \boldsymbol{b}'\boldsymbol{z}_*$ reduces to

$$\boldsymbol{b}_1'\boldsymbol{x} \geq \boldsymbol{b}_1'\boldsymbol{a},$$

which implies that $\boldsymbol{a}$ is a boundary point of S. Thus, the proof is complete. □

If f is a differentiable function, then the hyperplane given on the right-hand side of (8.34) will be given by the tangent hyperplane to $f(\boldsymbol{x})$ at $\boldsymbol{x} = \boldsymbol{a}$.

Theorem 8.11. Let $f(\boldsymbol{x})$ be a real-valued convex function defined for all $\boldsymbol{x} \in S$, where S is an open convex subset of R^m. If $f(\boldsymbol{x})$ is differentiable and $\boldsymbol{a} \in S$, then

$$f(\boldsymbol{x}) \geq f(\boldsymbol{a}) + \left(\frac{\partial}{\partial \boldsymbol{a}'} f(\boldsymbol{a}) \right) (\boldsymbol{x} - \boldsymbol{a})$$

for all $\boldsymbol{x} \in S$.

Proof. Suppose that $\boldsymbol{x} \in S$ and $\boldsymbol{a} \in S$, and let $\boldsymbol{y} = \boldsymbol{a} - \boldsymbol{x}$ so that $\boldsymbol{a} = \boldsymbol{x} + \boldsymbol{y}$. Since S is convex, the point

$$c\boldsymbol{a} + (1 - c)\boldsymbol{x} = c(\boldsymbol{x} + \boldsymbol{y}) + (1 - c)\boldsymbol{x} = \boldsymbol{x} + c\boldsymbol{y}$$

is in S for $0 \leq c \leq 1$. Thus, due to the convexity of f, we have

$$f(\boldsymbol{x} + c\boldsymbol{y}) \leq cf(\boldsymbol{x} + \boldsymbol{y}) + (1 - c)f(\boldsymbol{x}) = f(\boldsymbol{x}) + c\{f(\boldsymbol{x} + \boldsymbol{y}) - f(\boldsymbol{x})\},$$

or, equivalently,

$$f(\boldsymbol{x} + \boldsymbol{y}) \geq f(\boldsymbol{x}) + c^{-1}\{f(\boldsymbol{x} + c\boldsymbol{y}) - f(\boldsymbol{x})\} \tag{8.35}$$

Now since f is differentiable, we also have the Taylor formula

$$f(\boldsymbol{x} + c\boldsymbol{y}) = f(\boldsymbol{x}) + \left(\frac{\partial}{\partial \boldsymbol{x}'} f(\boldsymbol{x}) \right) c\boldsymbol{y} + r_1(c\boldsymbol{y}, \boldsymbol{x}), \tag{8.36}$$

where the remainder satisfies $\lim c^{-1} r_1(c\boldsymbol{y}, \boldsymbol{x}) = 0$ as $c \to 0$. Using (8.36) in (8.35), we get

$$f(\boldsymbol{x} + \boldsymbol{y}) \geq f(\boldsymbol{x}) + \left(\frac{\partial}{\partial \boldsymbol{x}'} f(\boldsymbol{x}) \right) \boldsymbol{y} + c^{-1} r_1(c\boldsymbol{y}, \boldsymbol{x}),$$

and so the result follows by letting $c \to 0$. □

The previous theorem can easily be used to show that a stationary point of a convex function will actually be an absolute minimum. Equivalently, a stationary point of a concave function will be an absolute maximum of that function.

Theorem 8.12. Let $f(\boldsymbol{x})$ be a real-valued convex function defined for all $\boldsymbol{x} \in S$, where S is an open convex subset of R^m. If $f(\boldsymbol{x})$ is differentiable and $\boldsymbol{a} \in S$ is a stationary point of f, then f has an absolute minimum at $\boldsymbol{a}$.

Proof. If $\boldsymbol{a}$ is a stationary point of f, then

$$\frac{\partial}{\partial \boldsymbol{a}'} f(\boldsymbol{a}) = \mathbf{0}'$$

Using this in the inequality of Theorem 8.11, we get $f(\boldsymbol{x}) \geq f(\boldsymbol{a})$ for all $\boldsymbol{x} \in S$, and so the result follows. □

The inequality given in (8.34) can be used to prove a very useful inequality involving the moments of a random vector $\boldsymbol{x}$. This inequality is known as Jensen's inequality. But before we can prove this result, we will need the following.

Theorem 8.13. Suppose that S is a convex subset of R^m and $\boldsymbol{y}$ is an $m \times 1$ random vector with finite first moments. If $\mathrm{P}(\boldsymbol{y} \in S) = 1$, then $\mathrm{E}(\boldsymbol{y}) \in S$.

Proof. We will prove the result by induction. Clearly, the result holds if $m = 1$, since in this case S is an interval, and it is easily demonstrated that a random variable y satisfying $\mathrm{P}(a \leq y \leq b) = 1$ for some constants a and b will have $a \leq \mathrm{E}(y) \leq b$. Now assuming that the result holds for dimension $m-1$, we will show that it must then hold for m. Define the convex set $S_* = \{\boldsymbol{x}\colon \boldsymbol{x} = \boldsymbol{u} - \mathrm{E}(\boldsymbol{y}),\ \boldsymbol{u} \in S\}$ so that the proof will be complete if we show that $\mathbf{0} \in S_*$. Now if $\mathbf{0} \notin S_*$, it follows from Theorem 2.27 that there exists an $m \times 1$ vector $\boldsymbol{a} \neq \mathbf{0}$ such that $\boldsymbol{a}'\boldsymbol{x} \geq 0$ for all $\boldsymbol{x} \in S_*$. Consequently, since $\mathrm{P}(\boldsymbol{y} \in S) = \mathrm{P}(\boldsymbol{w} \in S_*) = 1$, where the random vector $\boldsymbol{w} = \boldsymbol{y} - \mathrm{E}(\boldsymbol{y})$, we have $\boldsymbol{a}'\boldsymbol{w} \geq 0$ with probability 1, yet $\mathrm{E}(\boldsymbol{a}'\boldsymbol{w}) = 0$. This is possible only if $\boldsymbol{a}'\boldsymbol{w} = 0$, in which case $\boldsymbol{w}$ is on the hyperplane defined by $\{\boldsymbol{x}\colon \boldsymbol{a}'\boldsymbol{x} = 0\}$, with probability one. But since $\mathrm{P}(\boldsymbol{w} \in S_*) = 1$ as well, we must have $\mathrm{P}(\boldsymbol{w} \in S_0) = 1$, where $S_0 = S_* \cap \{\boldsymbol{x}\colon \boldsymbol{a}'\boldsymbol{x} = 0\}$. Now it follows from Theorem 2.23 that S_0 is a convex set, and it is contained within an $(m-1)$-dimensional vector space since $\{\boldsymbol{x}\colon \boldsymbol{a}'\boldsymbol{x} = 0\}$ is an $(m-1)$-dimensional vector space. Thus, since our result holds for $m-1$-dimensional spaces, we must have $\mathrm{E}(\boldsymbol{w}) = \mathbf{0} \in S_0$. This leads to the contradiction $\mathbf{0} \in S_*$, since $S_0 \subseteq S_*$, and so the proof is complete. □

We now prove Jensen's inequality.

Theorem 8.14. Let $f(\boldsymbol{x})$ be a real-valued convex function defined for all $\boldsymbol{x} \in S$, where S is a convex subset of R^m. If $\boldsymbol{y}$ is an $m \times 1$ random vector with finite first moments and satisfying $\mathrm{P}(\boldsymbol{y} \in S) = 1$, then

$$E(f(\boldsymbol{y})) \geq f(E(\boldsymbol{y}))$$

Proof. The previous theorem guarantees that $E(\boldsymbol{y}) \in S$. We first prove the result for $m = 1$. If $E(y)$ is an interior point of S, the result follows by taking the expected value of both sides of (8.33) when $x = y$ and $a = E(y)$. Since when $m = 1$, S is an interval, $E(y)$ can be a boundary point of S only if S is closed and $P(y = c) = 1$, where c is an endpoint of the interval. In this case, the result is trivial since the terms on the two sides of the inequality above are equal. We will complete the proof by showing that if the result holds for $m-1$, then it must hold for m. If the $m \times 1$ vector $E(\boldsymbol{y})$ is an interior point of S, the result follows by taking the expected value of both sides of (8.34) with $\boldsymbol{x} = \boldsymbol{y}$ and $\boldsymbol{a} = E(\boldsymbol{y})$. If $E(\boldsymbol{y})$ is a boundary point of S, then we know from the supporting hyperplane theorem that there exists an $m \times 1$ unit vector $\boldsymbol{b}$ such that $w = \boldsymbol{b}'\boldsymbol{y} \geq \boldsymbol{b}'E(\boldsymbol{y}) = \mu$ with probability one. But since we also have $E(w) = \boldsymbol{b}'E(\boldsymbol{y}) = \mu$, it follows that $\boldsymbol{b}'\boldsymbol{y} = \mu$ with probability one. Let P be any $m \times m$ orthogonal matrix with its last column given by $\boldsymbol{b}$, so that the vector $\boldsymbol{u} = P'\boldsymbol{y}$ has the form $\boldsymbol{u} = (\boldsymbol{u}_1', \mu)'$, where $\boldsymbol{u}_1$ is an $(m-1) \times 1$ vector. Define the function $g(\boldsymbol{u}_1)$ as

$$g(\boldsymbol{u}_1) = f\left(P\begin{pmatrix}\boldsymbol{u}_1\\ \mu\end{pmatrix}\right) = f(\boldsymbol{y}),$$

for all $\boldsymbol{u}_1 \in S_* = \{\boldsymbol{x}\colon \boldsymbol{x} = P_1'\boldsymbol{y},\ \boldsymbol{y} \in S\}$, where P_1 is the matrix obtained from P by deleting its last column. The convexity of S_* and g follow from the convexity of S and f, and so, since $\boldsymbol{u}_1$ is $(m-1) \times 1$, our result applies to $g(\boldsymbol{u}_1)$. Thus, we have

$$E(f(\boldsymbol{y})) = E(g(\boldsymbol{u}_1)) \geq g(E(\boldsymbol{u}_1)) = f\left(P\begin{pmatrix}E(\boldsymbol{u}_1)\\ \mu\end{pmatrix}\right) = f(E(\boldsymbol{y})) \qquad \square$$

9. THE METHOD OF LAGRANGE MULTIPLIERS

In some situations we may need to find a local maximum of a function $f(\boldsymbol{x})$, where f is defined for all $\boldsymbol{x} \in S$, while the desired maximum is over all $\boldsymbol{x}$ in T, a subset of S. The method of Lagrange multipliers is useful in those situations in which the set T can be expressed in terms of a number of equality constraints; that is, there exist functions $g_1, \ldots, g_m$ such that

$$T = \{\boldsymbol{x}\colon \boldsymbol{x} \in R^n,\ \boldsymbol{g}(\boldsymbol{x}) = \boldsymbol{0}\},$$

where $\boldsymbol{g}(\boldsymbol{x})$ is the $m \times 1$ function given by $(g_1(\boldsymbol{x}), \ldots, g_m(\boldsymbol{x}))'$.

The method of Lagrange multipliers involves the maximization of the

Lagrange function

$$L(\boldsymbol{x}, \boldsymbol{\lambda}) = f(\boldsymbol{x}) - \boldsymbol{\lambda}'\boldsymbol{g}(\boldsymbol{x}),$$

where the components of the $m \times 1$ vector $\boldsymbol{\lambda}, \lambda_1, \ldots, \lambda_m$, are called the Lagrange multipliers. The stationary values of $L(\boldsymbol{x}, \boldsymbol{\lambda})$ are the solutions $(\boldsymbol{x}, \boldsymbol{\lambda})$ satisfying

$$\frac{\partial}{\partial \boldsymbol{x}'} L(\boldsymbol{x}, \boldsymbol{\lambda}) = \frac{\partial}{\partial \boldsymbol{x}'} f(\boldsymbol{x}) - \boldsymbol{\lambda}' \left(\frac{\partial}{\partial \boldsymbol{x}'} \boldsymbol{g}(\boldsymbol{x}) \right) = \boldsymbol{0}', \tag{8.37}$$

$$\frac{\partial}{\partial \boldsymbol{\lambda}'} L(\boldsymbol{x}, \boldsymbol{\lambda}) = -\boldsymbol{g}(\boldsymbol{x})' = \boldsymbol{0}'$$

The second equation above is simply the equality constraints

$$\boldsymbol{g}(\boldsymbol{x}) = \boldsymbol{0} \tag{8.38}$$

that determine the set T. Under certain conditions, the local maximum of the function $f(\boldsymbol{x})$, subject to $\boldsymbol{x} \in T$, will be given by a vector $\boldsymbol{x}$ that, for some $\boldsymbol{\lambda}$, satisfies equations (8.37) and (8.38). We will present a procedure for determining whether a particular solution vector $\boldsymbol{x}$ is a local maximum. This procedure is based on the following result, a proof of which can be found in Magnus and Neudecker (1988).

Theorem 8.15. Suppose the function $f(\boldsymbol{x})$ is defined for all $n \times 1$ vectors $\boldsymbol{x} \in S$, where S is some subset of R^n and $\boldsymbol{g}(\boldsymbol{x})$ is an $m \times 1$ vector function defined for all $\boldsymbol{x} \in S$, where $m < n$. Let $\boldsymbol{a}$ be an interior point of S and suppose that the following conditions hold.

(a) f and $\boldsymbol{g}$ are twice differentiable at $\boldsymbol{a}$.
(b) The first derivative of $\boldsymbol{g}$ at $\boldsymbol{a}$, $(\partial/\partial \boldsymbol{a}')\boldsymbol{g}(\boldsymbol{a})$, has full rank m.
(c) $\boldsymbol{g}(\boldsymbol{a}) = \boldsymbol{0}$.
(d) $(\partial/\partial \boldsymbol{a}')L(\boldsymbol{a}, \boldsymbol{\lambda}) = \boldsymbol{0}'$, where $L(\boldsymbol{x}, \boldsymbol{\lambda}) = f(\boldsymbol{x}) - \boldsymbol{\lambda}'\boldsymbol{g}(\boldsymbol{x})$ and $\boldsymbol{\lambda}$ is $m \times 1$.

Let H_f and H_{g_i} be the Hessian matrices of the functions $f(\boldsymbol{x})$ and $g_i(\boldsymbol{x})$ evaluated at $\boldsymbol{x} = \boldsymbol{a}$ and define

$$A = H_f - \sum_{i=1}^{m} \lambda_i H_{g_i},$$

$$B = \frac{\partial}{\partial \boldsymbol{a}'} \boldsymbol{g}(\boldsymbol{a})$$

Then $f(\boldsymbol{x})$ has a local maximum at $\boldsymbol{x} = \boldsymbol{a}$, subject to $\boldsymbol{g}(\boldsymbol{x}) = \boldsymbol{0}$, if

$$\boldsymbol{x}'A\boldsymbol{x} < 0$$

for all $\boldsymbol{x} \neq \boldsymbol{0}$ for which $B\boldsymbol{x} = \boldsymbol{0}$.

A similar result holds for a local minimum with the inequality $\boldsymbol{x}'A\boldsymbol{x} > 0$ replacing $\boldsymbol{x}'A\boldsymbol{x} < 0$. Our next result provides a method for determining whether $\boldsymbol{x}'A\boldsymbol{x} < 0$ or $\boldsymbol{x}'A\boldsymbol{x} > 0$ holds for all $\boldsymbol{x} \neq \boldsymbol{0}$ satisfying $B\boldsymbol{x} = \boldsymbol{0}$. Again, a proof can be found in Magnus and Neudecker (1988).

Theorem 8.16. Let A be an $n \times n$ symmetric matrix and B be an $m \times n$ matrix. For $r = 1, \dots, n$, let A_{rr} be the $r \times r$ matrix obtained by deleting the last $n - r$ rows and columns of A, and let B_r be the $m \times r$ matrix obtained by deleting the last $n-r$ columns of B. For $r = 1, \dots, n$, define the $(m+r) \times (m+r)$ matrix Δ_r as

$$\Delta_r = \begin{bmatrix} (0) & B_r \\ B_r' & A_{rr} \end{bmatrix}$$

Then, if B_m is nonsingular, $\boldsymbol{x}'A\boldsymbol{x} > 0$ holds for all $\boldsymbol{x} \neq \boldsymbol{0}$ satisfying $B\boldsymbol{x} = \boldsymbol{0}$ if and only if

$$(-1)^m |\Delta_r| > 0$$

for $r = m + 1, \dots, n$, and $\boldsymbol{x}'A\boldsymbol{x} < 0$ holds for all $\boldsymbol{x} \neq \boldsymbol{0}$ satisfying $B\boldsymbol{x} = \boldsymbol{0}$ if and only if

$$(-1)^r |\Delta_r| > 0,$$

for $r = m + 1, \dots, n$.

Example 8.6. We will find solutions $\boldsymbol{x} = (x_1, x_2, x_3)'$, which maximize and minimize the function

$$f(\boldsymbol{x}) = x_1 + x_2 + x_3,$$

subject to the constraints

$$x_1^2 + x_2^2 = 1, \tag{8.39}$$

$$x_3 - x_1 - x_2 = 1 \tag{8.40}$$

Setting the first derivative of the Lagrange function

$$L(\boldsymbol{x}, \boldsymbol{\lambda}) = x_1 + x_2 + x_3 - \lambda_1(x_1^2 + x_2^2 - 1) - \lambda_2(x_3 - x_1 - x_2 - 1),$$

with respect to $\boldsymbol{x}$, equal to $\mathbf{0}'$, we obtain the equations

$$\begin{aligned} 1 - 2\lambda_1 x_1 + \lambda_2 &= 0, \\ 1 - 2\lambda_1 x_2 + \lambda_2 &= 0, \\ 1 - \lambda_2 &= 0 \end{aligned}$$

The third equation gives $\lambda_2 = 1$, and when this is substituted in the first two equations, we find that we must have

$$x_1 = x_2 = \frac{1}{\lambda_1}$$

Using this in (8.39), we find that $\lambda_1 = \pm\sqrt{2}$, and so we have the stationary points

$$(x_1, x_2, x_3) = \left(\frac{1}{\sqrt{2}}, \frac{1}{\sqrt{2}}, 1 + \sqrt{2}\right) \quad \text{when } \lambda_1 = \sqrt{2},$$

$$(x_1, x_2, x_3) = \left(-\frac{1}{\sqrt{2}}, -\frac{1}{\sqrt{2}}, 1 - \sqrt{2}\right) \quad \text{when } \lambda_1 = -\sqrt{2}$$

To determine whether either of these solutions yields a maximum or minimum we use Theorems 8.15 and 8.16. Thus, since $m = 2$ and $n = 3$, we only need the determinant of the matrix

$$\Delta_3 = \begin{bmatrix} 0 & 0 & 2x_1 & 2x_2 & 0 \\ 0 & 0 & -1 & -1 & 1 \\ 2x_1 & -1 & -2\lambda_1 & 0 & 0 \\ 2x_2 & -1 & 0 & -2\lambda_1 & 0 \\ 0 & 1 & 0 & 0 & 0 \end{bmatrix}$$

By using the cofactor expansion formula for a determinant, it is fairly straightforward to show that

$$|\Delta_3| = -8\lambda_1(x_1^2 + x_2^2)$$

Thus, when $(x_1, x_2, x_3, \lambda_1, \lambda_2) = (1/\sqrt{2}, 1/\sqrt{2}, 1+\sqrt{2}, \sqrt{2}, 1)$, we have

$$(-1)^r|\Delta_r| = (-1)^3|\Delta_3| = 8\sqrt{2} > 0,$$

and so the solution $(x_1, x_2, x_3) = (1/\sqrt{2}, 1/\sqrt{2}, 1+\sqrt{2})$ yields a constrained maximum. On the other hand, when $(x_1, x_2, x_3, \lambda_1, \lambda_2) = (-1/\sqrt{2}, -1/\sqrt{2}, 1-\sqrt{2}, -\sqrt{2}, 1)$,

$$(-1)^m|\Delta_r| = (-1)^2|\Delta_3| = 8\sqrt{2} > 0$$

so the solution $(x_1, x_2, x_3) = (-1/\sqrt{2}, -1/\sqrt{2}, 1-\sqrt{2})$ yields a constrained minimum.

In some situations, in the process of obtaining the stationary values of $L(\boldsymbol{x}, \boldsymbol{\lambda})$, it becomes apparent which solution yields a maximum and which solution yields a minimum. Thus, in this case, there will be no need to compute the Δ_r matrices.

Example 8.7. Let A be an $m \times m$ symmetric matrix and $\boldsymbol{x}$ be an $m \times 1$ vector. We saw in Section 3.6 that

$$\frac{\boldsymbol{x}'A\boldsymbol{x}}{\boldsymbol{x}'\boldsymbol{x}} \tag{8.41}$$

has a maximum value of $\lambda_1(A)$ and a minimum value of $\lambda_m(A)$, where $\lambda_1(A) \geq \cdots \geq \lambda_m(A)$ are the eigenvalues of A. We will prove this result again, this time using Lagrange's method. Note that since $\boldsymbol{z} = (\boldsymbol{x}'\boldsymbol{x})^{-1/2}\boldsymbol{x}$ is a unit vector, it follows that maximizing or minimizing (8.41) over all $\boldsymbol{x} \neq \boldsymbol{0}$ is equivalent to maximizing or minimizing the function

$$f(\boldsymbol{z}) = \boldsymbol{z}'A\boldsymbol{z},$$

subject to the constraint

$$\boldsymbol{z}'\boldsymbol{z} = 1 \tag{8.42}$$

Thus, the Lagrange function is

$$L(\boldsymbol{z}, \lambda) = \boldsymbol{z}'A\boldsymbol{z} - \lambda(\boldsymbol{z}'\boldsymbol{z} - 1)$$

Setting its first derivative, with respect to $\boldsymbol{z}$, equal to $\boldsymbol{0}'$, we obtain the equation

$$2Az - 2\lambda z = \mathbf{0},$$

or, equivalently,

$$Az = \lambda z, \tag{8.43}$$

which is the eigenvalue–eigenvector equation. Thus, the Lagrange multiplier λ is an eigenvalue of A. Further, premultiplying (8.43) by z' and using (8.42), we find that

$$\lambda = z'Az;$$

that is, if z is a stationary point of $L(z, \lambda)$, then $z'Az$ must be an eigenvalue of A. Consequently, the maximum value of $z'Az$, subject to $z'z = 1$, is $\lambda_1(A)$, which is attained when z is equal to any unit eigenvector corresponding to $\lambda_1(A)$. Similarly, the minimum value of $z'Az$, subject to $z'z = 1$, is $\lambda_m(A)$, and this is attained at any unit eigenvector associated with $\lambda_m(A)$.

In our final example, we obtain the best quadratic unbiased estimator of σ^2 in the ordinary least squares regression model.

Example 8.8. Consider the multiple regression model $\boldsymbol{y} = X\boldsymbol{\beta} + \boldsymbol{\epsilon}$, where $\boldsymbol{\epsilon} \sim \text{N}_N(\mathbf{0}, \sigma^2\text{I})$. A quadratic estimator of σ^2 is any estimator, $\hat{\sigma}^2$ that takes the form $\hat{\sigma}^2 = \boldsymbol{y}'A\boldsymbol{y}$, where A is a symmetric matrix of constants. We wish to find the choice of A that minimizes $\text{var}(\hat{\sigma}^2)$ over all choices of A for which $\hat{\sigma}^2$ is unbiased. Now since $\text{E}(\boldsymbol{\epsilon}) = \mathbf{0}$ and $\text{E}(\boldsymbol{\epsilon}\boldsymbol{\epsilon}') = \sigma^2\text{I}$, we have

$$\begin{aligned}\text{E}(\boldsymbol{y}'A\boldsymbol{y}) &= \text{E}\{(X\boldsymbol{\beta} + \boldsymbol{\epsilon})'A(X\boldsymbol{\beta} + \boldsymbol{\epsilon})\} = \text{E}\{\boldsymbol{\beta}'X'AX\boldsymbol{\beta} + 2\boldsymbol{\beta}'X'A\boldsymbol{\epsilon} + \boldsymbol{\epsilon}'A\boldsymbol{\epsilon}\} \\ &= \boldsymbol{\beta}'X'AX\boldsymbol{\beta} + \text{tr}\{A\text{E}(\boldsymbol{\epsilon}\boldsymbol{\epsilon}')\} = \boldsymbol{\beta}'X'AX\boldsymbol{\beta} + \sigma^2\text{tr}(A),\end{aligned}$$

and so $\hat{\sigma}^2 = \boldsymbol{y}'A\boldsymbol{y}$ is unbiased regardless of the value of $\boldsymbol{\beta}$ only if

$$X'AX = (0) \tag{8.44}$$

and

$$\text{tr}(A) = 1 \tag{8.45}$$

Using the fact that the components of $\boldsymbol{\epsilon}$ are independently distributed and the first four moments of each component are 0, 1, 0, 3, it is easily verified that

$$\text{var}(\boldsymbol{y}'A\boldsymbol{y}) = 2\sigma^4\,\text{tr}(A^2) + 4\sigma^2\boldsymbol{\beta}'X'A^2X\boldsymbol{\beta}$$

Thus, the required Lagrange function is

$$L(A, \lambda, \Lambda) = 2\sigma^4 \operatorname{tr}(A^2) + 4\sigma^2 \boldsymbol{\beta}' X' A^2 X \boldsymbol{\beta} - \operatorname{tr}(\Lambda X' A X) - \lambda\{\operatorname{tr}(A) - 1\},$$

where the Lagrange multipliers are given by λ and the components of the matrix Λ, which is symmetric since $X'AX$ is symmetric. Differentiation with respect to A yields

$$\begin{aligned} dL &= 2\sigma^4 \operatorname{tr}\{(dA)A + A\,dA\} + 4\sigma^2 \boldsymbol{\beta}' X' \{(dA)A + A\,dA\} X \boldsymbol{\beta} \\ &\quad - \operatorname{tr}\{\Lambda X'(dA)X\} - \lambda \operatorname{tr}(dA) \\ &= \operatorname{tr}(\{4\sigma^4 A + 4\sigma^2 (AX\boldsymbol{\beta}\boldsymbol{\beta}'X' + X\boldsymbol{\beta}\boldsymbol{\beta}'X'A) - X\Lambda X' - \lambda \mathrm{I}_N\}\, dA) \end{aligned}$$

Thus, we must use

$$4\sigma^4 A + 4\sigma^2 (AX\boldsymbol{\beta}\boldsymbol{\beta}'X' + X\boldsymbol{\beta}\boldsymbol{\beta}'X'A) - X\Lambda X' - \lambda \mathrm{I}_N = (0) \qquad (8.46)$$

along with (8.44) and (8.45) to solve for A. Premultiplying and postmultiplying (8.46) by XX^+ and using (8.44) and the fact that $X^+ = (X'X)^+ X'$, we find that

$$X\Lambda X' = -\lambda X X^+$$

Substituting this back into (8.46), we get

$$A = \frac{1}{4}\sigma^{-4}\lambda(\mathrm{I}_N - XX^+) - \sigma^{-2} H, \qquad (8.47)$$

where $H = A\boldsymbol{\gamma}\boldsymbol{\gamma}' + \boldsymbol{\gamma}\boldsymbol{\gamma}'A$ and $\boldsymbol{\gamma} = X\boldsymbol{\beta}$. Putting (8.47) back into (8.46) and simplifying, we obtain

$$H = -\sigma^{-2}(H\boldsymbol{\gamma}\boldsymbol{\gamma}' + \boldsymbol{\gamma}\boldsymbol{\gamma}'H) \qquad (8.48)$$

By postmultiplying (8.48) by $\boldsymbol{\gamma}$, we find that $\boldsymbol{\gamma}$ must be an eigenvector of H, and in light of equation (8.48), this can be true only if H is of the form $H = c\boldsymbol{\gamma}\boldsymbol{\gamma}'$ for some scalar c. Further, when we put $H = c\boldsymbol{\gamma}\boldsymbol{\gamma}'$ in (8.48), we find that we must have $c = 0$; thus $H = (0)$. In addition, if we take the trace of both sides of (8.47) and use (8.45), we see that

$$\lambda = \frac{4\sigma^4}{\operatorname{tr}(\mathrm{I}_N - XX^+)} = \frac{4\sigma^4}{N - r},$$

where r is the rank of X. Consequently, we have shown that (8.47) simplifies to

$$A = (N - r)^{-1}(\mathrm{I}_N - XX^+), \tag{8.49}$$

so that $\hat{\sigma}^2 = \boldsymbol{y}'A\boldsymbol{y} = \mathrm{SSE}/(N - r)$ is the familiar residual variance estimate. We can easily demonstrate that (8.49) yields an absolute minimum by writing an arbitrary symmetric matrix satisfying (8.44) and (8.45), as $A_* = A + B$, where B must then satisfy $\mathrm{tr}(B) = 0$ and $X'BX = (0)$. Then, since $\mathrm{tr}(AB) = 0$ and $AX = (0)$, we have

$$\begin{aligned}
\mathrm{var}(\boldsymbol{y}'A_*\boldsymbol{y}) &= 2\sigma^4\,\mathrm{tr}(A_*^2) + 4\sigma^2\boldsymbol{\beta}'X'A_*^2X\boldsymbol{\beta} \\
&= 2\sigma^4\{\mathrm{tr}(A^2) + \mathrm{tr}(B^2) + 2\,\mathrm{tr}(AB)\} + 4\sigma^2\boldsymbol{\beta}'X' \\
&\quad \cdot (A^2 + B^2 + AB + BA)X\boldsymbol{\beta} \\
&= 2\sigma^4\{\mathrm{tr}(A^2) + \mathrm{tr}(B^2)\} + 4\sigma^2\boldsymbol{\beta}'X'B^2X\boldsymbol{\beta} \\
&\geq 2\sigma^4\,\mathrm{tr}(A^2) = \mathrm{var}(\boldsymbol{y}'A\boldsymbol{y})
\end{aligned}$$

PROBLEMS

1. Consider the natural log function, $f(x) = \log(x)$.
   (a) Obtain the kth-order Taylor formula for $f(1 + u)$ in powers of u.
   (b) Use the formula in part (a) with $k = 5$ to approximate $\log(1.1)$.

2. Suppose the function f of the 2×1 vector $\boldsymbol{x}$ is given by

$$f(\boldsymbol{x}) = \frac{(x_2 - 1)^2}{(x_1 + 1)^3}$$

   Give the second-order Taylor formula for $f(\boldsymbol{0} + \boldsymbol{u})$ in powers of u_1 and u_2.

3. Suppose the 2×1 function $\boldsymbol{f}$ of the 3×1 vector $\boldsymbol{x}$ is given by

$$\boldsymbol{f}(\boldsymbol{x}) = \begin{bmatrix} x_1^2 + x_2^2 + x_3^2 \\ 2x_1 - x_2 - x_3 \end{bmatrix}$$

   and the 2×1 function $\boldsymbol{g}$ of the 2×1 vector $\boldsymbol{z}$ is given by

$$\boldsymbol{g}(\boldsymbol{z}) = \begin{bmatrix} z_2/z_1 \\ z_1 z_2 \end{bmatrix}$$

Use the chain rule to compute

$$\frac{\partial}{\partial \boldsymbol{x}'} \boldsymbol{y}(\boldsymbol{x}),$$

where $\boldsymbol{y}(\boldsymbol{x})$ is the composite function defined by $\boldsymbol{y}(\boldsymbol{x}) = \boldsymbol{g}(\boldsymbol{f}(\boldsymbol{x}))$.

4. Let A and B be $m \times m$ symmetric matrices of constants and $\boldsymbol{x}$ be an $m \times 1$ vector of variables. Find the differential and first derivative of the function

$$f(\boldsymbol{x}) = \frac{\boldsymbol{x}'A\boldsymbol{x}}{\boldsymbol{x}'B\boldsymbol{x}}$$

5. Let A and B be $m \times n$ matrices of constants and X be an $n \times m$ matrix of variables. Find the differential and derivative of
   (a) $\text{tr}(AX)$,
   (b) $\text{tr}(AXBX)$.

6. Let X be an $m \times m$ nonsingular matrix and A be an $m \times m$ matrix of constants. Find the differential and derivative of
   (a) $|X^2|$,
   (b) $\text{tr}(AX^{-1})$.

7. Let X be an $m \times n$ matrix with $\text{rank}(X) = n$. Show that

$$\frac{\partial}{\partial\, \text{vec}(X)'} |X'X| = 2|X'X|(\text{vec}\{X(X'X)^{-1}\})'$$

8. Let X be an $m \times m$ matrix and n be a positive integer. Show that

$$\frac{\partial}{\partial\, \text{vec}(X)'} \text{vec}(X^n) = \sum_{i=1}^{n} \{(X^{n-i})' \otimes X^{i-1}\}$$

9. Let A and B be $n \times m$ and $m \times n$ matrices of constants, respectively. If X is an $m \times m$ nonsingular matrix find the derivatives of
   (a) $\text{vec}(AXB)$,
   (b) $\text{vec}(AX^{-1}B)$.

10. Show that if X is an $m \times m$ nonsingular matrix and $X_{\#}$ is its adjoint matrix, then

$$\frac{\partial}{\partial \operatorname{vec}(X)'} \operatorname{vec}(X_{\#}) = |X|\{\operatorname{vec}(X^{-1})\operatorname{vec}(X^{-1\prime})' - (X^{-1\prime} \otimes X^{-1})\}$$

11. Prove Corollary 8.1.1.

12. Let X be an $m \times m$ symmetric matrix of variables. For each of the following functions, find the Jacobian matrix

$$\frac{\partial}{\partial \operatorname{v}(X)'} \operatorname{vec}(F)$$

(a) $F(X) = AXA'$, where A is an $m \times m$ matrix of constants.
(b) $F(X) = XBX$, where B is an $m \times m$ symmetric matrix of constants.

13. Let X be an $m \times m$ matrix having correlation structure; that is, X is a symmetric matrix of variables except that each of its diagonal elements is equal to one. Show that, if X is nonsingular, then

$$\frac{\partial}{\partial \tilde{\operatorname{v}}(X)'} \tilde{\operatorname{v}}(X^{-1}) = -2\tilde{L}_m(X^{-1} \otimes X^{-1})\tilde{L}_m'$$

14. Suppose that Y is an $m \times m$ symmetric matrix and ϵ is a scalar such that $(\mathrm{I}_m + \epsilon Y)^{-1}$ exists. Let $(\mathrm{I}_m + \epsilon Y)^{-1/2}$ be the symmetric square root of $(\mathrm{I}_m + \epsilon Y)^{-1}$ so that

$$(\mathrm{I}_m + \epsilon Y)^{-1} = (\mathrm{I}_m + \epsilon Y)^{-1/2}(\mathrm{I}_m + \epsilon Y)^{-1/2}$$

Using perturbation methods, show that

$$(\mathrm{I}_m + \epsilon Y)^{-1/2} = \mathrm{I}_m + \sum_{i=1}^{\infty} \epsilon^i B_i,$$

where

$$B_1 = -\tfrac{1}{2}Y,\ B_2 = \tfrac{3}{8}Y^2,\ B_3 = -\tfrac{5}{16}Y^3, \text{ and } B_4 = \tfrac{35}{128}Y^4.$$

15. Let S be an $m \times m$ sample covariance matrix, and suppose that Ω, the corresponding population covariance matrix, has each of its diagonal elements equal to one. Define A to be the difference between these two matrices; that is, $A = S - \Omega$, so that $S = \Omega + A$. Note that the population correlation matrix is also Ω, while the sample correlation matrix is given by $R = D_S^{-1/2} S D_S^{-1/2}$, where $D_S^{-1/2} = \operatorname{diag}(s_{11}^{-1/2}, \ldots, s_{mm}^{-1/2})$. Show that the

approximation $R = \Omega + C_1 + C_2 + C_3$, accurate up through third-order terms in the elements of A, is given by

$$C_1 = A - \frac{1}{2}(\Omega D_A + D_A\Omega),$$

$$C_2 = \frac{3}{8}(D_A^2\Omega + \Omega D_A^2) + \frac{1}{4}D_A\Omega D_A - \frac{1}{2}(AD_A + D_A A),$$

$$C_3 = \frac{3}{8}(D_A^2 A + AD_A^2) + \frac{1}{4}D_A A D_A - \frac{3}{16}(D_A^2\Omega D_A + D_A\Omega D_A^2)$$
$$- \frac{5}{16}(D_A^3\Omega + \Omega D_A^3),$$

where $D_A = \text{diag}(a_{11}, \ldots, a_{mm})$.

16. Derive the results given in Theorem 8.7. First obtain expressions for B_1, B_2, and B_3 by utilizing the equations $(Z + W)\Phi_l = \Phi_l(Z + W)$, $\Phi_l^2 = \Phi_l$, and $\Phi_l' = \Phi_l$. Then obtain expressions for a_1, a_2, and a_3 by using the fact that $\lambda_{l,l+r-1} = r^{-1}\,\text{tr}\{(Z + W)\Phi_l\}$.

17. Let $X = \text{diag}(x_1, \ldots, x_m)$, where $x_1 \geq \cdots \geq x_m$, and suppose that the lth diagonal element is distinct so that $x_l \neq x_i$ if $i \neq l$. Let $\lambda_1 \geq \cdots \geq \lambda_m$ and $\boldsymbol{\gamma}_1, \ldots, \boldsymbol{\gamma}_m$ be the eigenvalues and corresponding eigenvectors of $(\text{I}_m + V)^{-1}(X + U)$, where U and V are $m \times m$ symmetric matrices; that is, for each i

$$(X + U)\boldsymbol{\gamma}_i = \lambda_i(\text{I}_m + V)\boldsymbol{\gamma}_i$$

The purpose of this exercise is to obtain the first-order approximations $\lambda_l = x_l + a_1$ and $\boldsymbol{\gamma}_l = c\boldsymbol{e}_l + \boldsymbol{b}_1$, where $\boldsymbol{e}_l$ is the lth column of I_m. Higher-order approximations can be found in Sugiura (1976). These approximations can be determined by using the eigenvalue–eigenvector equation just given along with the appropriate scale constraint on $\boldsymbol{\gamma}_l$.

(a) Show that $a_1 = u_{ll} - x_l v_{ll}$.

(b) Show that if $c = 1$ and $\boldsymbol{\gamma}_l'\boldsymbol{\gamma}_l = 1$, then

$$b_{l1} = 0, \qquad b_{i1} = -\frac{u_{li} - x_l v_{li}}{x_i - x_l} \quad \text{for all } i \neq l,$$

where b_{i1} is the ith component of the vector $\boldsymbol{b}_1$.

(c) Show that if $c = 1$ and $\boldsymbol{\gamma}_l'(\text{I}_m + V)\boldsymbol{\gamma}_l = 1$, then

$$b_{l1} = -\frac{1}{2}v_{ll}, \qquad b_{i1} = -\frac{u_{li} - x_l v_{li}}{x_i - x_l} \quad \text{for all } i \neq l$$

(d) Show that if $c = x_l^{1/2}$ and $\boldsymbol{\gamma}_l'\boldsymbol{\gamma}_l = \lambda_l$, then

$$b_{l1} = \frac{u_{ll} - x_l v_{ll}}{2x_l^{1/2}}, \qquad b_{i1} = -\frac{x_l^{1/2}(u_{li} - x_l v_{li})}{x_i - x_l}, \qquad \text{for all } i \neq l$$

18. Consider the function f of the 2×1 vector $\boldsymbol{x}$ given by

$$f(\boldsymbol{x}) = 2x_1^3 + x_2^3 - 6x_1 - 27x_2$$

(a) Determine the stationary points of f.
(b) Identify each of the points in part (a) as a maximum, minimum, or saddle point.

19. For each of the following functions determine any local maxima or minima.
(a) $x_1^2 + \frac{1}{2}x_2^2 - 2x_1x_2 + x_1 - 2x_2 + 1$.
(b) $x_1^3 + \frac{3}{2}x_1^2 + x_2^2 - 6x_1 - 2x_2$.
(c) $x_2^3 + 2x_1^2 + x_3^2 + 2x_1x_3 - 3x_2 - x_3$.

20. Let $\boldsymbol{a}$ be an $m \times 1$ vector and B be an $m \times m$ symmetric matrix, each containing constants. Let $\boldsymbol{x}$ be an $m \times 1$ vector of variables.
(a) Show that the function

$$f(\boldsymbol{x}) = \boldsymbol{x}'B\boldsymbol{x} + \boldsymbol{a}'\boldsymbol{x}$$

has stationary solutions given by

$$\boldsymbol{x} = -\frac{1}{2}B^+\boldsymbol{a} + (\mathrm{I} - B^+B)\boldsymbol{y},$$

where $\boldsymbol{y}$ is an arbitrary $m \times 1$ vector.
(b) Show that if B is nonsingular, then there is only one stationary solution. When will this solution yield a maximum or a minimum?

21. If the Hessian matrix H_f of a function f is singular at a stationary point $\boldsymbol{x}$, then we must take a closer look at the behavior of this function in the neighborhood of the point $\boldsymbol{x}$ to determine whether the point is a maximum, minimum, or a saddle point. For each of the functions below, show that $\mathbf{0}$ is a stationary point and the Hessian matrix is singular at $\mathbf{0}$. In each case, determine whether $\mathbf{0}$ yields a maximum, minimum, or a saddlepoint.
(a) $x_1^4 + x_2^4$.
(b) $x_1^2x_2^2 - x_1^4 - x_2^4$.
(c) $x_1^3 - x_2^3$.

22. Suppose that we have independent random samples from each of k multivariate normal distributions with the ith distribution being $N_m(\boldsymbol{\mu}_i, \Omega)$. Thus, these distributions have possibly different mean vectors but identical covariance matrices. If the ith sample is denoted by $\boldsymbol{x}_{i1}, \dots, \boldsymbol{x}_{in_i}$, show that the maximum likelihood estimators of $\boldsymbol{\mu}_i$ and Ω are given by

$$\hat{\boldsymbol{\mu}}_i = \bar{\boldsymbol{x}}_i = \sum_{j=1}^{n_i} \frac{\boldsymbol{x}_{ij}}{n_i}, \qquad \hat{\Omega} = \sum_{i=1}^{k} \sum_{j=1}^{n_i} \frac{(\boldsymbol{x}_{ij} - \bar{\boldsymbol{x}}_i)(\boldsymbol{x}_{ij} - \bar{\boldsymbol{x}}_i)'}{n},$$

where $n = n_1 + \cdots + n_k$.

23. Consider the multiple regression model,

$$\boldsymbol{y} = X\boldsymbol{\beta} + \boldsymbol{\epsilon},$$

where $\boldsymbol{y}$ is $N \times 1$, X is $N \times m$, β is $m \times 1$, and $\boldsymbol{\epsilon}$ is $N \times 1$. Suppose that $\text{rank}(X) = m$ and $\boldsymbol{\epsilon} \sim N_N(\mathbf{0}, \sigma^2 I_N)$, so that $\boldsymbol{y} \sim N_N(X\boldsymbol{\beta}, \sigma^2 I_N)$. Find the maximum likelihood estimates of $\boldsymbol{\beta}$ and σ^2.

24. Let $f(\boldsymbol{x})$ be a real-valued convex function defined for all $\boldsymbol{x} \in S$, where S is a convex subset of R^m. Show that the set $T = \{z = (\boldsymbol{x}', y)' : \boldsymbol{x} \in S, \boldsymbol{y} \geq f(\boldsymbol{x})\}$ is convex.

25. Suppose that $f(\boldsymbol{x})$ and $g(\boldsymbol{x})$ are convex functions both defined on the convex set $S \subseteq R^m$. Show that the function $af(\boldsymbol{x}) + bg(\boldsymbol{x})$ is convex if a and b are nonnegative scalars.

26. Prove the converse of Theorem 8.11; that is, show that if $f(\boldsymbol{x})$ is defined and differentiable on the open convex set S and

$$f(\boldsymbol{x}) \geq f(\boldsymbol{a}) + \left(\frac{\partial}{\partial \boldsymbol{a}'} f(\boldsymbol{a}) \right) (\boldsymbol{x} - \boldsymbol{a})$$

for all $\boldsymbol{x} \in S$ and $\boldsymbol{a} \in S$, then $f(\boldsymbol{x})$ is a convex function.

27. Let $f(\boldsymbol{x})$ be a real-valued function defined for all $\boldsymbol{x} \in S$, where S is an open convex subset of R^m, and suppose that $f(\boldsymbol{x})$ is a twice differentiable function on S. Show that $f(\boldsymbol{x})$ is a convex function if and only if the Hessian matrix H_f is nonnegative definite at each $\boldsymbol{x} \in S$.

28. Let $\boldsymbol{x}$ be a 2×1 vector and consider the function $f(\boldsymbol{x}) = x_1^{\alpha} x_2^{1-\alpha}$ for all $\boldsymbol{x} \in S$, where $0 < \alpha < 1$ and $S = \{\boldsymbol{x} : x_1 > 0, x_2 > 0\}$.

(a) Use the previous exercise to show that $f(\boldsymbol{x})$ is a concave function.

(b) Show that if $\boldsymbol{y}$ is a 2×1 random vector with finite first moments and satisfying $P(\boldsymbol{y} \in S) = 1$, then

$$E(y_1^{\alpha} y_2^{1-\alpha}) \leq \{E(y_1)\}^{\alpha}\{E(y_2)\}^{1-\alpha}$$

if $0 < \alpha < 1$.

29. Let $\boldsymbol{x}$ be a 3×1 vector and define the function

$$f(\boldsymbol{x}) = x_1 + x_2 - x_3$$

Find the maximum and minimum of $f(\boldsymbol{x})$ subject to the constraint $\boldsymbol{x}'\boldsymbol{x} = 1$.

30. Find the shortest distance from the origin to a point on the surface given by

$$x_1^2 + x_2^2 + x_3^2 + 4x_1 - 6x_3 = 2$$

31. Let A be an $m \times m$ positive definite matrix and $\boldsymbol{x}$ be an $m \times 1$ vector. Find the maximum and minimum of the function

$$f(\boldsymbol{x}) = \boldsymbol{x}'\boldsymbol{x},$$

subject to the constraint $\boldsymbol{x}'A\boldsymbol{x} = 1$.

32. Find the maximum and minimum of the function

$$f(\boldsymbol{x}) = x_1(x_2 + x_3),$$

subject to the constraints $x_1^2 + x_2^2 = 1$ and $x_1 x_3 + x_2 = 2$.

33. For a 3×1 vector $\boldsymbol{x}$, maximize the function

$$f(\boldsymbol{x}) = x_1 x_2 x_3,$$

subject to the constraint $x_1 + x_2 + x_3 = a$, where a is some positive number. Use this to establish the inequality

$$(x_1 x_2 x_3)^{1/3} \leq \frac{1}{3}(x_1 + x_2 + x_3)$$

for all positive real numbers x_1, x_2, and x_3. Generalize this result to m

variables; that is, if $\boldsymbol{x}$ is $m \times 1$, show that

$$(x_1 x_2 \cdots x_m)^{1/m} \leq \frac{1}{m}(x_1 + \cdots + x_m)$$

holds for all positive real numbers $x_1, \ldots, x_m$.

34. Let A and B be $m \times m$ matrices, with A being nonnegative definite and B being positive definite. Following the approach of Example 8.7, use the Lagrange method to find the maximum and minimum values of

$$f(\boldsymbol{x}) = \frac{\boldsymbol{x}'A\boldsymbol{x}}{\boldsymbol{x}'B\boldsymbol{x}},$$

over all $\boldsymbol{x} \neq \boldsymbol{0}$.

35. Let $\boldsymbol{a}$ be an $m \times 1$ vector and B be an $m \times m$ positive definite matrix. Using the results of the previous exercise, show that for $\boldsymbol{x} \neq \boldsymbol{0}$,

$$f(\boldsymbol{x}) = \frac{(\boldsymbol{a}'\boldsymbol{x})^2}{\boldsymbol{x}'B\boldsymbol{x}}$$

has a maximum value of

$$\boldsymbol{a}'B^{-1}\boldsymbol{a}$$

This result can be used to obtain the union–intersection test (see Example 3.14) of the multivariate hypothesis H_0: $\boldsymbol{\mu} = \boldsymbol{\mu}_0$ against H_1: $\boldsymbol{\mu} \neq \boldsymbol{\mu}_0$, where $\boldsymbol{\mu}$ represents the $m \times 1$ mean vector of a population and $\boldsymbol{\mu}_0$ is an $m \times 1$ vector of constants. Let $\overline{\boldsymbol{x}}$ and S denote the sample mean vector and sample covariance matrix computed from a sample of size n from this population. Show that if we base the union–intersection procedure on the univariate t statistic

$$t = \frac{(\overline{x} - \mu_0)}{s/\sqrt{n}}$$

for testing H_0: $\mu = \mu_0$, then the union–intersection test can be based on $T^2 = n(\overline{\boldsymbol{x}} - \boldsymbol{\mu}_0)'S^{-1}(\overline{\boldsymbol{x}} - \boldsymbol{\mu}_0)$.

36. Suppose that $x_1, \ldots, x_n$ are independent and identically distributed random variables with mean μ and variance σ^2. Consider a linear estimator of μ which is any estimator of the form $\hat{\mu} = \Sigma a_i x_i$, where $a_1, \ldots, a_n$ are constants.

(a) For what values of $a_1, \ldots, a_n$ will $\hat{\mu}$ be an unbiased estimator of μ?

(b) Use the method of Lagrange multipliers to show that the sample mean $\bar{x}$ is the best linear unbiased estimator of μ; that is, $\bar{x}$ has the smallest variance among all linear unbiased estimators of μ.

37. A random process involves n independent trials, where each trial can result in one of k distinct outcomes. Let p_i denote the probability that a trial results in outcome i and note that then $p_1+\cdots+p_k = 1$. Define the random variables, $x_1, \ldots, x_k$, where x_i counts the number of times that outcome i occurs in the n trials. Then the random vector $\boldsymbol{x} = (x_1, \ldots, x_k)'$ has the multinomial distribution with probability function given by

$$P(x_1 = n_1, \ldots, x_k = n_k) = \frac{n!}{n_1! \cdots n_k!} p_1^{n_1} \cdots p_k^{n_k},$$

where $n_1, \ldots, n_k$ are nonnegative integers satisfying $n_1 + \cdots + n_k = n$. Find the maximum likelihood estimate of $\boldsymbol{p} = (p_1, \ldots, p_k)'$.

38. Suppose that the $m \times m$ positive definite covariance matrix Ω is partitioned in the form

$$\Omega = \begin{bmatrix} \Omega_{11} & \Omega_{12} \\ \Omega_{12}' & \Omega_{22} \end{bmatrix},$$

where Ω_{11} is $m_1 \times m_1$, Ω_{22} is $m_2 \times m_2$, and $m_1 + m_2 = m$. Suppose also that the $m \times 1$ random vector $\boldsymbol{x}$ has covariance matrix Ω and is partitioned as $\boldsymbol{x} = (\boldsymbol{x}_1', \boldsymbol{x}_2')'$, where $\boldsymbol{x}_1$ is $m_1 \times 1$ and $\boldsymbol{x}_2$ is $m_2 \times 1$. If the $m_1 \times 1$ vector $\boldsymbol{a}$ and $m_2 \times 1$ vector $\boldsymbol{b}$ are vectors of constants, then the square of the correlation between the random variables $u = \boldsymbol{a}'\boldsymbol{x}_1$ and $v = \boldsymbol{b}'\boldsymbol{x}_2$ is given by

$$f(\boldsymbol{a}, \boldsymbol{b}) = \frac{(\boldsymbol{a}'\Omega_{12}\boldsymbol{b})^2}{\boldsymbol{a}'\Omega_{11}\boldsymbol{a}\boldsymbol{b}'\Omega_{22}\boldsymbol{b}}$$

Show that the maximum value of $f(\boldsymbol{x})$, that is, the maximum squared correlation between u and v, subject to the constraints

$$\boldsymbol{a}'\Omega_{11}\boldsymbol{a} = 1, \qquad \boldsymbol{b}'\Omega_{22}\boldsymbol{b} = 1$$

is the largest eigenvalue of $\Omega_{11}^{-1}\Omega_{12}\Omega_{22}^{-1}\Omega_{12}'$ or, equivalently, the largest eigenvalue of $\Omega_{22}^{-1}\Omega_{12}'\Omega_{11}^{-1}\Omega_{12}$. What are the vectors $\boldsymbol{a}$ and $\boldsymbol{b}$ that yield this maximum?

39. Consider the function, $f(P) = \text{tr}(PXP'D)$, where P is an $m \times m$ orthogonal matrix, and both X and D are $m \times m$ positive definite matrices. Further, suppose that D is diagonal with distinct, descending, positive diagonal elements; that is, $D = \text{diag}(d_1, \ldots, d_m)$ with $d_1 > \cdots > d_m > 0$.

(a) By working with the Lagrange function,

$$L(P, \Lambda) = \text{tr}(PXP'D) + \text{tr}\{\Lambda(PP' - \text{I}_m)\},$$

where Λ is a symmetric matrix of Lagrange multipliers, show that the stationary points of $f(P)$ occur when PXP' is diagonal.

(b) Use part (a) to show that

$$\max_{P:\, PP' = \text{I}} f(P) = \sum_{i=1}^{m} d_i \lambda_i(X),$$

and

$$\min_{P:\, PP' = \text{I}} f(P) = \sum_{i=1}^{m} d_{m+1-i} \lambda_i(X),$$

where $\lambda_1(X) \geq \cdots \geq \lambda_m(X) > 0$ are the eigenvalues of X.

CHAPTER NINE

Some Special Topics Related to Quadratic Forms

1. INTRODUCTION

We have seen that if A is an $m \times m$ summetric matrix and $\boldsymbol{x}$ is an $m \times 1$ vector, then the function of $\boldsymbol{x}$, $\boldsymbol{x}'A\boldsymbol{x}$, is called a quadratic form in $\boldsymbol{x}$. In many statistical applications, $\boldsymbol{x}$ is a random vector, while A is a matrix of constants. The most common situation is one in which $\boldsymbol{x}$ has as its distribution, or its asymptotic distribution, the multivariate normal distribution. In this chapter, we investigate some of the distributional properties of $\boldsymbol{x}'A\boldsymbol{x}$ in this setting. In particular, we are most interested in determining conditions under which $\boldsymbol{x}'A\boldsymbol{x}$ will have a chi-squared distribution.

2. SOME RESULTS ON IDEMPOTENT MATRICES

We have noted earlier that an $m \times m$ matrix A is said to be idempotent if $A^2 = A$. We will see in the next section that idempotent matrices play an essential role in the discussion of conditions under which a quadratic form in normal variates has a chi-squared distribution. Consequently, this section is devoted to establishing some of the basic results regarding idempotent matrices.

Theorem 9.1. Let A be an $m \times m$ idempotent matrix. Then

(a) $I_m - A$ is also idempotent,

(b) each eigenvalue of A is 0 or 1,

(c) A is diagonalizable,

(d) $\text{rank}(A) = \text{tr}(A)$.

Proof. Since $A^2 = A$, we have

$$(\mathrm{I}_m - A)^2 = \mathrm{I}_m - 2A + A^2 = \mathrm{I}_m - A,$$

and so (a) holds. Let λ be an eigenvalue of A corresponding to the eigenvector $\boldsymbol{x}$ so that $A\boldsymbol{x} = \lambda\boldsymbol{x}$. Then since $A^2 = A$, we find that

$$\lambda\boldsymbol{x} = A\boldsymbol{x} = A^2\boldsymbol{x} = A(A\boldsymbol{x}) = A(\lambda\boldsymbol{x}) = \lambda A\boldsymbol{x} = \lambda^2\boldsymbol{x},$$

which implies that

$$\lambda(\lambda - 1)\boldsymbol{x} = \mathbf{0}$$

Since eigenvectors are nonnull vectors, we must have $\lambda(\lambda - 1) = 0$, and so (b) follows. Let r be the number of eigenvalues of A equal to one, so that $m - r$ is the number of eigenvalues of A equal to zero. As a result, $A - \mathrm{I}_m$ must have r eigenvalues equal to zero and $m - r$ eigenvalues equal to -1. By Theorem 4.8, (c) will follow if we can show that

$$\mathrm{rank}(A) = r, \qquad \mathrm{rank}(A - \mathrm{I}_m) = m - r \tag{9.1}$$

Now from Theorem 4.10, we know that the rank of any square matrix is at least as large as the number of its nonzero eigenvalues, so we must have

$$\mathrm{rank}(A) \geq r, \qquad \mathrm{rank}(A - \mathrm{I}_m) \geq m - r \tag{9.2}$$

But Corollary 2.12.1 gives

$$\mathrm{rank}(A) + \mathrm{rank}(\mathrm{I}_m - A) \leq \mathrm{rank}\{A(\mathrm{I}_m - A)\} + m = \mathrm{rank}\{(0)\} + m = m,$$

which together with (9.2) implies (9.1), so (c) is proven. Finally, (d) is an immediate consequence of (b) and (c). □

Since any matrix with at least one 0 eigenvalue has to be a singular matrix, a nonsingular idempotent matrix has all of its eigenvalues equal to 1. But the only diagonalizable matrix with all of its eigenvalues equal to 1 is the identity matrix; that is, the only nonsingular $m \times m$ idempotent matrix is I_m.

If A is a diagonal matrix, that is, $A = \mathrm{diag}(a_1, \ldots, a_m)$, then $A^2 = \mathrm{diag}(a_1^2, \ldots, a_m^2)$. Equating A and A^2, we find that a diagonal matrix is idempotent if and only if each diagonal element is 0 or 1.

Example 9.1. Although an idempotent matrix has each of its eigenvalues equal to 1 or 0, the converse is not true; that is, a matrix having only eigenvalues of 1 and 0 need not be an idempotent matrix. For instance, it is easily verified that the matrix

$$A = \begin{bmatrix} 0 & 1 & 0 \\ 0 & 0 & 0 \\ 0 & 0 & 1 \end{bmatrix}$$

has eigenvalues 0 and 1 with multiplicities 2 and 1, respectively. However,

$$A^2 = \begin{bmatrix} 0 & 0 & 0 \\ 0 & 0 & 0 \\ 0 & 0 & 1 \end{bmatrix},$$

so that A is not idempotent.

The matrix A in the example above is not idempotent because it is not diagonalizable. In other words, an $m \times m$ matrix A is idempotent if and only if each of its eigenvalues is 0 or 1 and it is diagonalizable. In fact, we have the following special case.

Theorem 9.2. Let A be an $m \times m$ symmetric matrix. Then A is idempotent if and only if each eigenvalue of A is 0 or 1.

Proof. Let $A = X\Lambda X'$ be the spectral decomposition of A, so that X is an orthogonal matrix and Λ is diagonal. Then

$$A^2 = (X\Lambda X')^2 = X\Lambda X'X\Lambda X' = X\Lambda^2 X'$$

Clearly, this equals A if and only if each diagonal element of Λ, that is, each eigenvalue of A, is 0 or 1. □

Our next result gives some conditions for the sum of two idempotent matrices and the product of two idempotent matrices to be idempotent.

Theorem 9.3. Let A and B be $m \times m$ idempotent matrices. Then

(a) $A + B$ is idempotent if and only if $AB = BA = (0)$,
(b) AB is idempotent if $AB = BA$.

Proof. Since A and B are idempotent, we have

$$(A + B)^2 = A^2 + B^2 + AB + BA = A + B + AB + BA,$$

so that $A + B$ will be idempotent if and only if

$$AB = -BA \tag{9.3}$$

Premultiplication of (9.3) by B and postmultiplication by A yields the identity

$$(BA)^2 = -BA, \tag{9.4}$$

since A and B are idempotent. Similarly, premultiplying (9.3) by A and post-multiplying by B, we also find that

$$(AB)^2 = -AB \tag{9.5}$$

Thus, it follows from (9.4) and (9.5) that both $-BA$ and $-AB$ are idempotent matrices, and due to (9.3), so then is AB. Part (a) now follows since the null matrix is the only idempotent matrix whose negative is also idempotent. To prove (b), note that if A and B commute under multiplication, then

$$(AB)^2 = ABAB = A(BA)B = A(AB)B = A^2B^2 = AB,$$

and so the result follows. □

Example 9.2. The conditions given for $(A+B)$ to be idempotent are necessary and sufficient, while the condition given for AB to be idempotent is only sufficient. We can illustrate that this second condition is not necessary through a simple example. Let A and B be defined as

$$A = \begin{bmatrix} 1 & 1 \\ 0 & 0 \end{bmatrix}, \qquad B = \begin{bmatrix} 0 & 0 \\ 1 & 1 \end{bmatrix},$$

and observe that $A^2 = A$ and $B^2 = B$, so that A and B are idempotent. In addition, $AB = A$, so that AB is also idempotent. However, $AB \neq BA$ since $BA = B$.

Most of the statistical applications involving idempotent matrices deal with symmetric idempotent matrices. For this reason, we end this section with some results for this special class of matrices. The first result gives some restrictions on the elements of a symmetric idempotent matrix.

Theorem 9.4. Suppose A is an $m \times m$ symmetric idempotent matrix. Then

(a) $a_{ii} \geq 0$ for $i = 1, \ldots, m$,
(b) $a_{ii} \leq 1$ for $i = 1, \ldots, m$,
(c) $a_{ij} = a_{ji} = 0$, for all $j \neq i$, if $a_{ii} = 0$ or $a_{ii} = 1$.

Proof. Since A is idempotent and symmetric, it follows that

$$a_{ii} = (A)_{ii} = (A^2)_{ii} = (A'A)_{ii} = (A')_{i\cdot}(A)_{\cdot i} = \sum_{j=1}^{m} a_{ji}^2, \tag{9.6}$$

which clearly must be nonnegative. In addition, from (9.6) we have

$$a_{ii} = a_{ii}^2 + \sum_{j \neq i} a_{ji}^2,$$

so that $a_{ii} \geq a_{ii}^2$ and, thus (b) must hold. If $a_{ii} = 0$ or $a_{ii} = 1$, then $a_{ii} = a_{ii}^2$ and so we must have

$$\sum_{j \neq i} a_{ji}^2 = 0,$$

which, along with the symmetry of A, establishes (c). □

The following theorem is useful in those situations in which it is easier to verify an identity such as $A^3 = A^2$ than the identity $A^2 = A$.

Theorem 9.5. Suppose that for some positive integer i, the $m \times m$ symmetric matrix A satisfies $A^{i+1} = A^i$. Then A is an idempotent matrix.

Proof. If $\lambda_1, \ldots, \lambda_m$ are the eigenvalues of A, then $\lambda_1^{i+1}, \ldots, \lambda_m^{i+1}$ and $\lambda_1^i, \ldots, \lambda_m^i$ are the eigenvalues of A^{i+1} and A^i, respectively. But the identity $A^{i+1} = A^i$ implies that $\lambda_j^{i+1} = \lambda_j^i$, for $j = 1, \ldots, m$, so each λ_j must be either 0 or 1. The result now follows from Theorem 9.2. □

3. COCHRAN'S THEOREM

The following result, sometimes referred to as Cochran's Theorem [Cochran (1934)], will be very useful in establishing the independence of several different quadratic forms in the same normal variables.

Theorem 9.6. Let each of the $m \times m$ matrices $A_1, \ldots, A_k$ be symmetric and idempotent, and suppose that $A_1 + \cdots + A_k = I_m$. Then $A_iA_j = (0)$ whenever $i \neq j$.

Proof. Select any one of the matrices, say A_h, and denote its rank by r. Since A_h is symmetric and idempotent, there exists an orthogonal matrix P such that

$$P'A_hP = \text{diag}(\text{I}_r, (0))$$

For $j \neq h$, define $B_j = P'A_jP$, and note that

$$\text{I}_m = P'\text{I}_mP = P'\left(\sum_{j=1}^{k} A_j\right)P = \left(\sum_{j=1}^{k} P'A_jP\right)$$
$$= \text{diag}(\text{I}_r, (0)) + \sum_{j \neq h} B_j,$$

or, equivalently,

$$\sum_{j \neq h} B_j = \text{diag}((0), \text{I}_{m-r})$$

In particular, for $l = 1, \dots, r$,

$$\sum_{j \neq h} (B_j)_{ll} = 0$$

But clearly, B_j is symmetric and idempotent since A_j is, and so, from Theorem 9.4(a), its diagonal elements are nonnegative. Thus, we must have $(B_j)_{ll} = 0$ for each $l = 1, \dots, r$, and this, along with Theorem 9.4(c), implies that B_j must be of the form

$$B_j = \text{diag}((0), C_j),$$

where C_j is an $(m - r) \times (m - r)$ symmetric idempotent matrix. Now, for any $j \neq h$,

$$P'A_h, A_jP = (P'A_hP)(P'A_jP) = \text{diag}(\text{I}_r, (0))\,\text{diag}((0), C_j) = (0),$$

which can be true only if $A_hA_j = (0)$, since P is nonsingular. Our proof is now complete, since h was arbitrary. □

Our next result is an extension of Cochran's Theorem.

Theorem 9.7. Let $A_1, \dots, A_k$ be $m \times m$ symmetric matrices and define $A = A_1 + \cdots + A_k$. Consider the following statements.

(a) A_i is idempotent for $i = 1, \ldots, k$.
(b) A is idempotent.
(c) $A_iA_j = (0)$, for all $i \neq j$.

Then if any two of these conditions hold, the third condition must also hold.

Proof. First we show that (a) and (b) imply (c). Since A is symmetric and idempotent, there exists an orthogonal matrix P such that

$$P'AP = P'(A_1 + \cdots + A_k)P = \text{diag}(\mathrm{I}_r, (0)), \tag{9.7}$$

where $r = \text{rank}(A)$. Let $B_i = P'A_iP$ for $i = 1, \ldots, k$, and note that B_i is symmetric and idempotent. Thus, it follows from (9.7) and Theorem 9.4 that B_i must be of the form $\text{diag}(C_i, (0))$, where the $r \times r$ matrix C_i also must be symmetric and idempotent. But (9.7) also implies that

$$C_1 + \cdots + C_k = \mathrm{I}_r$$

Consequently, $C_1, \ldots, C_k$ satisfy the conditions of Theorem 9.6 and so $C_iC_j = (0)$ for every $i \neq j$. From this we get $B_iB_j = (0)$ and, hence, $A_iA_j = (0)$ for every $i \neq j$ as is required. That (a) and (c) imply (b) follows immediately, since

$$\begin{aligned} A^2 &= \left(\sum_{i=1}^k A_i\right)^2 = \sum_{i=1}^k \sum_{j=1}^k A_iA_j = \sum_{i=1}^k A_i^2 + \sum_{i \neq j}\sum A_iA_j \\ &= \sum_{i=1}^k A_i = A \end{aligned}$$

Finally, we must prove that (b) and (c) imply (a). If (c) holds, then $A_iA_j = A_jA_i$ for all $i \neq j$, and so by Theorem 4.16, the matrices $A_1, \ldots, A_k$ can be simultaneously diagonalized; that is, there exists an orthogonal matrix Q such that

$$Q'A_iQ = D_i,$$

where each of the matrices $D_1, \ldots, D_k$ is diagonal. Further,

$$D_iD_j = Q'A_iQQ'A_jQ = Q'A_iA_jQ = Q'(0)Q = (0), \tag{9.8}$$

for every $i \neq j$. Now since A is symmetric and idempotent so also is the diagonal matrix

$$Q'AQ = D_1 + \cdots + D_k$$

As a result, each diagonal element of $Q'AQ$ must be either 0 or 1, and due to (9.8) the same can be said of the diagonal elements of $D_1, \ldots, D_k$. Thus, for each i, D_i is symmetric and idempotent, and hence so is $A_i = QD_iQ'$. This completes the proof. □

Suppose that the three conditions given in Theorem 9.7 hold. Then (a) implies that $\text{tr}(A_i) = \text{rank}(A_i)$, and (b) implies that

$$\text{rank}(A) = \text{tr}(A) = \text{tr}\left(\sum_{i=1}^{k} A_i\right) = \sum_{i=1}^{k} \text{tr}(A_i) = \sum_{i=1}^{k} \text{rank}(A_i)$$

Thus, we have shown that the conditions in Theorem 9.7 imply the fourth condition

(d) $\text{rank}(A) = \sum_{i=1}^{k} \text{rank}(A_i)$.

Conversely, suppose that conditions (b) and (d) hold. We will show that these imply (a) and (c). Let $H = \text{diag}(A_1, \ldots, A_k)$ and $F = \mathbf{1}_m \otimes \text{I}_m$ so that $A = F'HF$. Then (d) can be written $\text{rank}(F'HF) = \text{rank}(H)$, and so it follows from Theorem 5.24 that $F(F'HF)^- F'$ is a generalized inverse of H for any generalized inverse $(F'HF)^-$, of $F'HF$. But since A is idempotent, $A\text{I}_m A = A$ and hence I_m is a generalized inverse of $A = F'HF$. Thus, FF' is a generalized inverse of H, yielding the equation

$$HFF'H = H,$$

which in partitioned form is

$$\begin{bmatrix} A_1^2 & A_1A_2 & \cdots & A_1A_k \\ A_2A_1 & A_2^2 & \cdots & A_2A_k \\ \vdots & \vdots & & \vdots \\ A_kA_1 & A_kA_2 & \cdots & A_k^2 \end{bmatrix} = \begin{bmatrix} A_1 & (0) & \cdots & (0) \\ (0) & A_2 & \cdots & (0) \\ \vdots & \vdots & & \vdots \\ (0) & (0) & \cdots & A_k \end{bmatrix}$$

This immediately gives conditions (a) and (c). The following result summarizes the relationship among these four conditions.

Corollary 9.7.1. Let $A_1, \dots, A_k$ be $m \times m$ symmetric matrices and define $A = A_1 + \cdots + A_k$. Consider the following statements.

(a) A_i is idempotent for $i = 1, \dots, k$.
(b) A is idempotent.
(c) $A_iA_j = (0)$, for all $i \neq j$.
(d) $\text{rank}(A) = \sum_{i=1}^{k} \text{rank}(A_i)$.

All four of the conditions hold if any two of (a), (b), and (c) hold, or if (b) and (d) hold.

4. DISTRIBUTION OF QUADRATIC FORMS IN NORMAL VARIATES

The relationship between the normal and chi-squared distributions is fundamental in obtaining the distribution of a quadratic form in normal random variables. Recall that if $z_1, \dots, z_r$ are independent random variables with $z_i \sim \text{N}(0, 1)$ for each i, then

$$\sum_{i=1}^{r} z_i^2 \sim \chi_r^2$$

This is used in our first theorem to determine when the quadratic form $\boldsymbol{x}'A\boldsymbol{x}$ has a chi-squared distribution if the components of $\boldsymbol{x}$ are independently distributed, each having the N(0, 1) distribution.

Theorem 9.8. Let $\boldsymbol{x} \sim \text{N}_m(\boldsymbol{0}, \text{I}_m)$, and suppose that the $m \times m$ matrix A is symmetric, idempotent, and has rank r. Then $\boldsymbol{x}'A\boldsymbol{x} \sim \chi_r^2$.

Proof. Since A is symmetric and idempotent, there exists an orthogonal matrix P such that

$$A = PDP',$$

where $D = \text{diag}(\text{I}_r, (0))$. Let $\boldsymbol{z} = P'\boldsymbol{x}$ and note that since $\boldsymbol{x} \sim \text{N}_m(\boldsymbol{0}, \text{I}_m)$,

$$\begin{aligned} \text{E}(\boldsymbol{z}) &= \text{E}(P'\boldsymbol{x}) = P'\text{E}(\boldsymbol{x}) = P'\boldsymbol{0} = \boldsymbol{0}, \\ \text{var}(\boldsymbol{z}) &= \text{var}(P'\boldsymbol{x}) = P'\{\text{var}(\boldsymbol{x})\}P = P'\text{I}_mP = P'P = \text{I}_m, \end{aligned}$$

and so $\boldsymbol{z} \sim \text{N}_m(\boldsymbol{0}, \text{I}_m)$; that is, the components of $\boldsymbol{z}$ are, like the components of $\boldsymbol{x}$, independent standard normal random variables. Now due to the form of D, we find that

$$x'Ax = x'PDP'x = z'Dz = \sum_{i=1}^{r} z_i^2,$$

and the result then follows. □

The result above is a special case of the next theorem in which the multivariate normal distribution has a general nonsingular covariance matrix.

Theorem 9.9. Let $x \sim N_m(\mathbf{0}, \Omega)$, where Ω is a positive definite matrix, and let A be an $m \times m$ symmetric matrix. If $A\Omega$ is idempotent and $\text{rank}(A\Omega) = r$, then $x'Ax \sim \chi_r^2$.

Proof. Since Ω is positive definite, there exists a nonsingular matrix T satisfying $\Omega = TT'$. If we define $z = T^{-1}x$, then $E(z) = T^{-1}E(x) = \mathbf{0}$, and

$$\text{var}(z) = \text{var}(T^{-1}x) = T^{-1}\text{var}(x)T'^{-1} = T^{-1}(TT')T'^{-1} = I_m,$$

so that $z \sim N_m(\mathbf{0}, I_m)$. The quadratic form $x'Ax$ can be written in terms of z since

$$x'Ax = x'T'^{-1}T'ATT^{-1}x = z'T'ATz$$

All that remains is to show that $T'AT$ satisfies the conditions of the previous theorem. Clearly, $T'AT$ is symmetric, since A is, and idempotent since

$$(T'AT)^2 = T'ATT'AT = T'A\Omega AT = T'AT,$$

where the last equality follows from the identity $A\Omega A = A$, which is a consequence of the fact that $A\Omega$ is idempotent and Ω is nonsingular. Finally, since $T'AT$ and $A\Omega$ are idempotent, we have

$$\text{rank}(T'AT) = \text{tr}(T'AT) = \text{tr}(ATT') = \text{tr}(A\Omega) = \text{rank}(A\Omega) = r,$$

and so the proof is complete. □

It is not uncommon to have a quadratic form in a vector that has a singular multivariate normal distribution. Our next result generalizes the previous theorem to this situation.

Theorem 9.10. Let $\boldsymbol{x} \sim N_m(\mathbf{0}, \Omega)$, where Ω is positive semidefinite, and suppose that A is an $m \times m$ symmetric matrix. If $\Omega A \Omega A \Omega = \Omega A \Omega$ and $\text{tr}(A\Omega) = r$, then $\boldsymbol{x}'A\boldsymbol{x} \sim \chi_r^2$.

Proof. Let $n = \text{rank}(\Omega)$, where $n < m$. Then there exists an $m \times m$ orthogonal matrix $P = [P_1 \quad P_2]$ such that

$$\Omega = [P_1 \quad P_2]\begin{bmatrix} \Lambda & (0) \\ (0) & (0) \end{bmatrix}\begin{bmatrix} P_1' \\ P_2' \end{bmatrix} = P_1 \Lambda P_1'$$

where P_1 is $m \times n$ and Λ is an $n \times n$ nonsingular diagonal matrix. Define

$$\boldsymbol{z} = \begin{bmatrix} \boldsymbol{z}_1 \\ \boldsymbol{z}_2 \end{bmatrix} = \begin{bmatrix} P_1'\boldsymbol{x} \\ P_2'\boldsymbol{x} \end{bmatrix} = P'\boldsymbol{x},$$

and note that since $P'\mathbf{0} = \mathbf{0}$ and $P'\Omega P = \text{diag}(\Lambda, (0))$, $\boldsymbol{z} \sim N_m(\mathbf{0}, \text{diag}(\Lambda, (0)))$. But this means that $\boldsymbol{z} = (\boldsymbol{z}_1', \mathbf{0}')'$, where $\boldsymbol{z}_1$ has the nonsingular distribution $N_n(\mathbf{0}, \Lambda)$. Now

$$\boldsymbol{x}'A\boldsymbol{x} = \boldsymbol{x}'PP'APP'\boldsymbol{x} = \boldsymbol{z}'P'AP\boldsymbol{z} = \boldsymbol{z}_1'P_1'AP_1\boldsymbol{z}_1,$$

and so the proof will be complete if we can show that the symmetric matrix $P_1'AP_1$ satisfies the conditions of the previous theorem, namely, that $P_1'AP_1\Lambda$ is idempotent and $\text{rank}(P_1'AP_1\Lambda) = r$. Since $\Omega A \Omega A \Omega = \Omega A \Omega$, we have

$$\begin{aligned}
(\Lambda^{1/2}P_1'AP_1\Lambda^{1/2})^3 &= \Lambda^{1/2}P_1'A(P_1\Lambda P_1')A(P_1\Lambda P_1')AP_1\Lambda^{1/2} \\
&= \Lambda^{1/2}P_1'A\Omega A\Omega AP_1\Lambda^{1/2} = \Lambda^{1/2}P_1'A\Omega AP_1\Lambda^{1/2} \\
&= \Lambda^{1/2}P_1'A(P_1\Lambda P_1')AP_1\Lambda^{1/2} = (\Lambda^{1/2}P_1'AP_1\Lambda^{1/2})^2,
\end{aligned}$$

and so the idempotency of $\Lambda^{1/2}P_1'AP_1\Lambda^{1/2}$ follows from Theorem 9.5. However, this also establishes the idempotency of $P_1'AP_1\Lambda$ since Λ is nonsingular. Its rank is r since

$$\text{rank}(P_1'AP_1\Lambda) = \text{tr}(P_1'AP_1\Lambda) = \text{tr}(AP_1\Lambda P_1') = \text{tr}(A\Omega) = r \qquad \square$$

To this point, all of our results have dealt with normal distributions having the zero mean vector. In some applications, such as the determination of non-null distributions in hypothesis testing situations, we encounter quadratic forms in normal vectors having nonzero means. The next two theorems are helpful in determining whether such a quadratic form has a chi-squared distribution. The proof of the first of these two theorems, which is very similar to that of Theorem 9.9, is left to the reader. It utilizes the relationship between the normal distribution and the noncentral chi-squared distribution; that is, if $y_1, \dots, y_r$ are independently distributed with $y_i \sim \mathrm{N}(\mu_i, 1)$, then

$$\sum_{i=1}^{r} y_i^2 \sim \chi_r^2(\lambda),$$

where the noncentrality parameter of this noncentral chi-squared distribution is given by

$$\lambda = \frac{1}{2} \sum_{i=1}^{r} \mu_i^2$$

Theorem 9.11. Let $\boldsymbol{x} \sim \mathrm{N}_m(\boldsymbol{\mu}, \Omega)$, where Ω is a positive definite matrix, and let A be an $m \times m$ symmetric matrix. If $A\Omega$ is idempotent and $\mathrm{rank}(A\Omega) = r$, then $\boldsymbol{x}'A\boldsymbol{x} \sim \chi_r^2(\lambda)$, where $\lambda = \frac{1}{2}\boldsymbol{\mu}'A\boldsymbol{\mu}$.

Theorem 9.12. Let $\boldsymbol{x} \sim \mathrm{N}_m(\boldsymbol{\mu}, \Omega)$, where Ω is positive semidefinite of rank n, and suppose that A is an $m \times m$ symmetric matrix. Then $\boldsymbol{x}'A\boldsymbol{x} \sim \chi_r^2(\lambda)$, where $\lambda = \frac{1}{2}\boldsymbol{\mu}'A\boldsymbol{\mu}$ if

(a) $\Omega A \Omega A \Omega = \Omega A \Omega$,
(b) $\boldsymbol{\mu}'A\Omega A\Omega = \boldsymbol{\mu}'A\Omega$,
(c) $\boldsymbol{\mu}'A\Omega A\boldsymbol{\mu} = \boldsymbol{\mu}'A\boldsymbol{\mu}$,
(d) $\mathrm{tr}(A\Omega) = r$.

Proof. Let P_1, P_2, and Λ be defined as in the proof of Theorem 9.10 so that $\Omega = P_1\Lambda P_1'$. Put $C = [P_1\Lambda^{-1/2} \quad P_2]$ and note that

$$\boldsymbol{z} = \begin{bmatrix} \boldsymbol{z}_1 \\ \boldsymbol{z}_2 \end{bmatrix} = \begin{bmatrix} \Lambda^{-1/2}P_1'\boldsymbol{x} \\ P_2'\boldsymbol{x} \end{bmatrix} = C'\boldsymbol{x} \sim \mathrm{N}_m\left(\begin{bmatrix} \Lambda^{-1/2}P_1'\boldsymbol{\mu} \\ P_2'\boldsymbol{\mu} \end{bmatrix}, \begin{bmatrix} \mathrm{I}_n & (0) \\ (0) & (0) \end{bmatrix} \right)$$

In other words,

$$\boldsymbol{z} = \begin{bmatrix} \boldsymbol{z}_1 \\ P_2'\boldsymbol{\mu} \end{bmatrix},$$

where $\boldsymbol{z}_1 \sim \mathrm{N}_n(\Lambda^{-1/2}P_1'\boldsymbol{\mu}, \mathrm{I}_n)$. Now since $C'^{-1} = [P_1\Lambda^{1/2} \quad P_2]$, we find that

$$\begin{aligned}
x'Ax &= x'CC^{-1}AC'^{-1}C'x = z'C^{-1}AC'^{-1}z \\
&= [z_1' \quad \boldsymbol{\mu}'P_2]\begin{bmatrix} \Lambda^{1/2}P_1'AP_1\Lambda^{1/2} & \Lambda^{1/2}P_1'AP_2 \\ P_2'AP_1\Lambda^{1/2} & P_2'AP_2 \end{bmatrix}\begin{bmatrix} z_1 \\ P_2'\boldsymbol{\mu} \end{bmatrix} \\
&= z_1'\Lambda^{1/2}P_1'AP_1\Lambda^{1/2}z_1 + \boldsymbol{\mu}'P_2P_2'AP_2P_2'\boldsymbol{\mu} \\
&\quad + 2\boldsymbol{\mu}'P_2P_2'AP_1\Lambda^{1/2}z_1 \qquad (9.9)
\end{aligned}$$

But conditions (a)–(c) imply the identities

(i) $P_1'A\Omega AP_1 = P_1'AP_1$,
(ii) $\boldsymbol{\mu}'P_2P_2'A\Omega AP_1 = \boldsymbol{\mu}'P_2P_2'AP_1$,
(iii) $\boldsymbol{\mu}'P_2P_2'A\Omega A\Omega AP_2P_2'\boldsymbol{\mu} = \boldsymbol{\mu}'P_2P_2'A\Omega AP_2P_2'\boldsymbol{\mu} = \boldsymbol{\mu}'P_2P_2'AP_2P_2'\boldsymbol{\mu}$;

in particular, (a) implies (i), (b) and (i) imply (ii), while (iii) follows from (c), (i) and (ii). Utilizing these identities in (9.9), we obtiain

$$\begin{aligned}
x'Ax &= z_1'\Lambda^{1/2}P_1'AP_1\Lambda^{1/2}z_1 + \boldsymbol{\mu}'P_2P_2'A\Omega A\Omega AP_2P_2'\boldsymbol{\mu} \\
&\quad + 2\boldsymbol{\mu}'P_2P_2'A\Omega AP_1\Lambda^{1/2}z_1 \\
&= (z_1 + \Lambda^{1/2}P_1'AP_2P_2'\boldsymbol{\mu})'\Lambda^{1/2}P_1'AP_1\Lambda^{1/2}(z_1 + \Lambda^{1/2}P_1'AP_2P_2'\boldsymbol{\mu}) \\
&= w'A_*w.
\end{aligned}$$

Now, $w = (z_1 + \Lambda^{1/2}P_1'AP_2P_2'\boldsymbol{\mu}) \sim \mathrm{N}_n(\boldsymbol{\theta}, \mathrm{I}_n)$, where

$$\boldsymbol{\theta} = \Lambda^{-1/2}P_1'\boldsymbol{\mu} + \Lambda^{1/2}P_1'AP_2P_2'\boldsymbol{\mu},$$

and, since $A_* = \Lambda^{1/2}P_1'AP_1\Lambda^{1/2}$ is idempotent, a consequence of (i), we may apply Theorem 9.11; that is, $w'A_*w \sim \chi_r^2(\lambda)$, where

$$r = \mathrm{tr}(A_*\mathrm{I}_n) = \mathrm{tr}(\Lambda^{1/2}P_1'AP_1\Lambda^{1/2}) = \mathrm{tr}(AP_1\Lambda P_1') = \mathrm{tr}(A\Omega),$$

and

$$\begin{aligned}
\lambda &= \frac{1}{2}\,\boldsymbol{\theta}'A_*\boldsymbol{\theta} = \frac{1}{2}\,(\Lambda^{-1/2}P_1'\boldsymbol{\mu} + \Lambda^{1/2}P_1'AP_2P_2'\boldsymbol{\mu})' \\
&\quad \times \Lambda^{1/2}P_1'AP_1\Lambda^{1/2}(\Lambda^{-1/2}P_1'\boldsymbol{\mu} + \Lambda^{1/2}P_1'AP_2P_2'\boldsymbol{\mu}) \\
&= \frac{1}{2}\,(\boldsymbol{\mu}'P_1P_1'AP_1P_1'\boldsymbol{\mu} + \boldsymbol{\mu}'P_2P_2'A\Omega A\Omega AP_2P_2'\boldsymbol{\mu} + 2\boldsymbol{\mu}'P_1P_1'A\Omega AP_2P_2'\boldsymbol{\mu})
\end{aligned}$$

$$= \frac{1}{2} (\boldsymbol{\mu}' P_1 P_1' A P_1 P_1' \boldsymbol{\mu} + \boldsymbol{\mu}' P_2 P_2' A P_2 P_2' \boldsymbol{\mu} + 2\boldsymbol{\mu}' P_1 P_1' A P_2 P_2' \boldsymbol{\mu})$$

$$= \frac{1}{2} \boldsymbol{\mu}'(P_1 P_1' + P_2 P_2') A (P_1 P_1' + P_2 P_2')\boldsymbol{\mu} = \frac{1}{2} \boldsymbol{\mu}' A \boldsymbol{\mu}$$

This completes the proof. □

A matrix A satisfying conditions (a), (b), and (c) of Theorem 9.12 is Ω^+, the Moore–Penrose inverse of Ω. That is, if $\boldsymbol{x} \sim N_m(\boldsymbol{\mu}, \Omega)$, then $\boldsymbol{x}'\Omega^+\boldsymbol{x}$ will have a chi-squared distribution since the identity $\Omega^+\Omega\,\Omega^+ = \Omega^+$ ensures that conditions (a), (b), and (c) hold. The degrees of freedom $r = \text{rank}(\Omega)$ since $\text{rank}(\Omega^+\Omega) = \text{rank}(\Omega)$.

All of the theorems presented in this section give sufficient conditions for a quadratic form to have a chi-squared distribution. Actually, in each case, the stated conditions are necessary conditions as well. This is most easily proven using moment generating functions. For details on this, the interested reader is referred to Mathai and Provost (1992) or Searle (1971).

Example 9.3. Let $x_1, \dots, x_n$ be a random sample from a normal distribution with mean μ and variance σ^2; that is the x_is are independent random variables, each having the distribution $N(\mu, \sigma^2)$. The sample variance s^2 is given by

$$s^2 = \frac{1}{(n-1)} \sum_{i=1}^{n} (x_i - \overline{x})^2$$

We will use the results of this section to show that

$$t = \frac{(n-1)s^2}{\sigma^2} = \sum_{i=1}^{n} \frac{(x_i - \overline{x})^2}{\sigma^2} \sim \chi^2_{n-1}$$

Define the $n \times 1$ vector $\boldsymbol{x} = (x_1, \dots, x_n)'$ so that $\boldsymbol{x} \sim N_n(\mu \mathbf{1}_n, \sigma^2 \mathrm{I}_n)$. Note that if the $n \times n$ matrix $A = (\mathrm{I}_n - n^{-1}\mathbf{1}_n\mathbf{1}_n')/\sigma^2$, then

$$\boldsymbol{x}'A\boldsymbol{x} = \frac{\{\boldsymbol{x}'\boldsymbol{x} - n^{-1}(\mathbf{1}_n'\boldsymbol{x})^2\}}{\sigma^2} = \left\{ \sum_{i=1}^{n} x_i^2 - n^{-1} \left(\sum_{i=1}^{n} x_i \right)^2 \right\} \Big/ \sigma^2$$

$$= \sum_{i=1}^{n} \frac{(x_i - \overline{x})^2}{\sigma^2} = \frac{(n-1)s^2}{\sigma^2} = t,$$

and so t is a quadratic form in the random vector $\boldsymbol{x}$. The matrix $A(\sigma^2 \mathrm{I}_n) = \sigma^2 A$

is idempotent since

$$\begin{aligned}(\sigma^2 A)^2 &= (\mathrm{I}_n - n^{-1}\mathbf{1}_n\mathbf{1}_n')^2 = \mathrm{I}_n - 2n^{-1}\mathbf{1}_n\mathbf{1}_n' + n^{-2}\mathbf{1}_n\mathbf{1}_n'\mathbf{1}_n\mathbf{1}_n' \\ &= \mathrm{I}_n - n^{-1}\mathbf{1}_n\mathbf{1}_n' = \sigma^2 A,\end{aligned}$$

and so by Theorem 9.11, t has a chi-squared distribution. This chi-squared distribution has $n - 1$ degrees of freedom since

$$\mathrm{tr}(\sigma^2 A) = \mathrm{tr}(\mathrm{I}_n - n^{-1}\mathbf{1}_n\mathbf{1}_n') = \mathrm{tr}(\mathrm{I}_n) - n^{-1}\mathrm{tr}(\mathbf{1}_n\mathbf{1}_n') = n - n^{-1}\mathbf{1}_n'\mathbf{1}_n = n - 1,$$

and the noncentrality parameter is given by

$$\begin{aligned}\lambda &= \frac{1}{2}\,\boldsymbol{\mu}'A\boldsymbol{\mu} = \frac{1}{2}\,\frac{\mu^2}{\sigma^2}\,\mathbf{1}_n'(\mathrm{I}_n - n^{-1}\mathbf{1}_n\mathbf{1}_n')\mathbf{1}_n = \frac{1}{2}\,\frac{\mu^2}{\sigma^2}\,(\mathbf{1}_n'\mathbf{1}_n - n^{-1}\mathbf{1}_n'\mathbf{1}_n\mathbf{1}_n'\mathbf{1}_n) \\ &= \frac{1}{2}\,\frac{\mu^2}{\sigma^2}\,(n - n) = 0\end{aligned}$$

Thus, we have shown that $t \sim \chi^2_{n-1}$.

5. INDEPENDENCE OF QUADRATIC FORMS

We now consider the situation in which we have several different quadratic forms, each a function of the same multivariate normal vector. In some settings, it is important to be able to determine whether or not these quadratic forms are distributed independently of one another. For instance, this is useful in the partitioning of chi-squared random variables as well as in the formation of ratios having an F distribution.

We begin with the following basic result regarding the statistical independence of two quadratic forms in the same normal vector.

Theorem 9.13. Let $\boldsymbol{x} \sim N_m(\boldsymbol{\mu}, \Omega)$, where Ω is positive definite, and suppose that A and B are $m \times m$ symmetric matrices. If $A\Omega B = (0)$, then $\boldsymbol{x}'A\boldsymbol{x}$ and $\boldsymbol{x}'B\boldsymbol{x}$ are independently distributed.

Proof. Since Ω is positive definite, there exists a nonsingular matrix T such that $\Omega = TT'$. Define $G = T'AT$ and $H = T'BT$, and note that if $A\Omega B = (0)$, then

$$GH = (T'AT)(T'BT) = T'A\Omega BT = T'(0)T = (0) \tag{9.10}$$

Consequently, due to the symmetry of G and H, we also have

$$(0) = (0)' = (GH)' = H'G' = HG,$$

and so we have established that $GH = HG$. From Theorem 4.15, we know that there exists an orthogonal matrix P that simultaneously diagonalizes G and H; that is, for some diagonal matrices C and D,

$$P'GP = P'T'ATP = C, \qquad P'HP = P'T'BTP = D \tag{9.11}$$

But using (9.10) and (9.11), we find that

$$(0) = GH = PCP'PDP' = PCDP',$$

which only can be true if $CD = (0)$. Since C and D are diagonal matrices, this means that if the ith diagonal element of one of these matrices is nonzero, the ith diagonal element of the other must be zero. As a result, by choosing P appropriately, we may obtain C and D in the form $C = \text{diag}(c_1, \dots, c_{m_1}, 0, \dots, 0)$ and $D = \text{diag}(0, \dots, 0, d_{m_1+1}, \dots, d_m)$ for some integer m_1. If we let $\boldsymbol{y} = P'T^{-1}\boldsymbol{x}$, then our two quadratic forms simplify as

$$\boldsymbol{x}'A\boldsymbol{x} = \boldsymbol{x}'T'^{-1}PP'T'ATPP'T^{-1}\boldsymbol{x} = \boldsymbol{y}'C\boldsymbol{y} = \sum_{i=1}^{m_1} c_i y_i^2,$$

and

$$\boldsymbol{x}'B\boldsymbol{x} = \boldsymbol{x}'T'^{-1}PP'T'BTPP'T^{-1}\boldsymbol{x} = \boldsymbol{y}'D\boldsymbol{y} = \sum_{i=m_1+1}^{m} d_i y_i^2;$$

that is, the first quadratic form is a function only of $y_1, \dots, y_{m_1}$, while the second quadratic form is a function of $y_{m_1+1}, \dots, y_m$. The result now follows from the independence of $y_1, \dots, y_m$, a consequence of the fact that $\boldsymbol{y}$ is normal and

$$\text{var}(\boldsymbol{y}) = \text{var}(P'T^{-1}\boldsymbol{x}) = P'T^{-1}\Omega T'^{-1}P = \text{I}_m$$

□

Example 9.4. Suppose that $\boldsymbol{x}_1, \dots, \boldsymbol{x}_k$ are independently distributed with $\boldsymbol{x}_i = (x_{i1}, \dots, x_{in})' \sim \text{N}_n(\mu\mathbf{1}_n, \sigma^2\text{I}_n)$ for each i. Let t_1 and t_2 be the random quantities defined by

$$t_1 = n\sum_{i=1}^{k} (\bar{x}_i - \bar{x})^2, \qquad t_2 = \sum_{i=1}^{k}\sum_{j=1}^{n} (x_{ij} - \bar{x}_i)^2,$$

where

$$\bar{x}_i = \sum_{j=1}^{n} \frac{x_{ij}}{n}, \qquad \bar{x} = \sum_{i=1}^{k} \frac{\bar{x}_i}{k}$$

Note that t_1 and t_2 are the formulas for the sum of squares for treatments and the sum of squares for error in a balanced one-way classification model (Example 7.4). Now t_1 can be expressed as

$$t_1 = n\left\{\sum_{i=1}^{k} \bar{x}_i^2 - k^{-1}\left(\sum_{i=1}^{k} \bar{x}_i\right)^2\right\} = n\bar{\boldsymbol{x}}'(\mathrm{I}_k - k^{-1}\mathbf{1}_k\mathbf{1}_k')\bar{\boldsymbol{x}},$$

where $\bar{\boldsymbol{x}} = (\bar{x}_1, \ldots, \bar{x}_k)'$. If we define $\boldsymbol{x}$ as $\boldsymbol{x} = (\boldsymbol{x}_1', \ldots, \boldsymbol{x}_k')'$, then $\boldsymbol{x} \sim \mathrm{N}_{kn}(\boldsymbol{\mu}, \boldsymbol{\Omega})$ with $\boldsymbol{\mu} = \mathbf{1}_k \otimes \mu\mathbf{1}_n = \mu\mathbf{1}_{kn}$ and $\boldsymbol{\Omega} = \mathrm{I}_k \otimes \sigma^2\mathrm{I}_n = \sigma^2\mathrm{I}_{kn}$, and $\bar{\boldsymbol{x}} = n^{-1}(\mathrm{I}_k \otimes \mathbf{1}_n')\boldsymbol{x}$, so

$$\begin{aligned} t_1 &= n^{-1}\boldsymbol{x}'(\mathrm{I}_k \otimes \mathbf{1}_n)(\mathrm{I}_k - k^{-1}\mathbf{1}_k\mathbf{1}_k')(\mathrm{I}_k \otimes \mathbf{1}_n')\boldsymbol{x} \\ &= n^{-1}\boldsymbol{x}'\{(\mathrm{I}_k - k^{-1}\mathbf{1}_k\mathbf{1}_k') \otimes \mathbf{1}_n\mathbf{1}_n'\}\boldsymbol{x} = \boldsymbol{x}'A_1\boldsymbol{x}, \end{aligned}$$

where $A_1 = n^{-1}\{(\mathrm{I}_k - k^{-1}\mathbf{1}_k\mathbf{1}_k') \otimes \mathbf{1}_n\mathbf{1}_n'\}$. Since $(\mathbf{1}_n\mathbf{1}_n')^2 = n\mathbf{1}_n\mathbf{1}_n'$ and $(\mathrm{I}_k - k^{-1}\mathbf{1}_k\mathbf{1}_k')^2 = (\mathrm{I}_k - k^{-1}\mathbf{1}_k\mathbf{1}_k')$, we find that A_1 is idempotent and hence so is $(A_1/\sigma^2)\boldsymbol{\Omega}$. Thus, by Theorem 9.11, $\boldsymbol{x}'(A_1/\sigma^2)\boldsymbol{x} = t_1/\sigma^2$ has a chi-squared distribution. This distribution is central since $\lambda = \frac{1}{2}\boldsymbol{\mu}'A_1\boldsymbol{\mu}/\sigma^2 = 0$, which follows from the fact that

$$\begin{aligned} \{(\mathrm{I}_k - k^{-1}\mathbf{1}_k\mathbf{1}_k') \otimes \mathbf{1}_n\mathbf{1}_n'\}(\mathbf{1}_k \otimes \mu\mathbf{1}_n) &= n\mu\{[(\mathrm{I}_k - k^{-1}\mathbf{1}_k\mathbf{1}_k') \otimes \mathbf{1}_n\} \\ &= n\mu\{(\mathbf{1}_k - \mathbf{1}_k) \otimes \mathbf{1}_n\} = \mathbf{0}, \end{aligned}$$

while its degrees of freedom is given by

$$\begin{aligned} r_1 &= \mathrm{tr}\{(A_1/\sigma^2)\boldsymbol{\Omega}\} = \mathrm{tr}(A_1) = n^{-1}\,\mathrm{tr}\{(\mathrm{I}_k - k^{-1}\mathbf{1}_k\mathbf{1}_k') \otimes \mathbf{1}_n\mathbf{1}_n')\} \\ &= n^{-1}\,\mathrm{tr}(\mathrm{I}_k - k^{-1}\mathbf{1}_k\mathbf{1}_k')\mathrm{tr}(\mathbf{1}_n\mathbf{1}_n') = n^{-1}(k-1)n = k-1 \end{aligned}$$

Turning to t_2, observe that it can be written as

$$\begin{aligned} t_2 &= \sum_{i=1}^{k}\left\{\sum_{j=1}^{n} x_{ij}^2 - n^{-1}\left(\sum_{j=1}^{n} x_{ij}\right)^2\right\} = \sum_{i=1}^{k} \boldsymbol{x}_i'(\mathrm{I}_n - n^{-1}\mathbf{1}_n\mathbf{1}_n')\boldsymbol{x}_i \\ &= \boldsymbol{x}'\{\mathrm{I}_k \otimes (\mathrm{I}_n - n^{-1}\mathbf{1}_n\mathbf{1}_n')\}\boldsymbol{x} = \boldsymbol{x}'A_2\boldsymbol{x}, \end{aligned}$$

where $A_2 = \mathrm{I}_k \otimes (\mathrm{I}_n - n^{-1}\mathbf{1}_n\mathbf{1}_n')$. Clearly, A_2 is idempotent since $(\mathrm{I}_n - n^{-1}\mathbf{1}_n\mathbf{1}_n')$ is idempotent. Thus, $(A_2/\sigma^2)\Omega$ is idempotent, and so $\boldsymbol{x}'(A_2/\sigma^2)\boldsymbol{x} = t_2/\sigma^2$ also has a chi-squared distribution. In particular, $t_2/\sigma^2 \sim \chi^2_{k(n-1)}$ since

$$\begin{aligned}\mathrm{tr}\{(A_2/\sigma^2)\Omega\} &= \mathrm{tr}(A_2) = \mathrm{tr}\{\mathrm{I}_k \otimes (\mathrm{I}_n - n^{-1}\mathbf{1}_n\mathbf{1}_n')\}\\ &= \mathrm{tr}(\mathrm{I}_k)\mathrm{tr}(\mathrm{I}_n - n^{-1}\mathbf{1}_n\mathbf{1}_n')\} = k(n-1),\end{aligned}$$

and

$$\begin{aligned}A_2\boldsymbol{\mu} &= \{\mathrm{I}_k \otimes (\mathrm{I}_n - n^{-1}\mathbf{1}_n\mathbf{1}_n')\}(\mathbf{1}_k \otimes \mu\mathbf{1}_n)\\ &= \mathbf{1}_k \otimes \mu(\mathrm{I}_n - n^{-1}\mathbf{1}_n\mathbf{1}_n')\mathbf{1}_n = \mathbf{1}_k \otimes \mu(\mathbf{1}_n - \mathbf{1}_n) = \mathbf{0},\end{aligned}$$

thereby guaranteeing that $\frac{1}{2}\boldsymbol{\mu}'A_2\boldsymbol{\mu}/\sigma^2 = 0$. Finally, we establish the independence of t_1 and t_2 by using Theorem 9.13. This simply involves verifying that $(A_1/\sigma^2)\Omega(A_2/\sigma^2) = A_1A_2/\sigma^2 = (0)$, which is an immediate consequence of the fact that

$$\mathbf{1}_n\mathbf{1}_n'(\mathrm{I}_n - n^{-1}\mathbf{1}_n\mathbf{1}_n') = (0)$$

Example 9.5. Let us return to the general regression model

$$\boldsymbol{y} = X\boldsymbol{\beta} + \boldsymbol{\epsilon},$$

where $\boldsymbol{y}$ and $\boldsymbol{\epsilon}$ are $N \times 1$, X is $N \times m$, and $\boldsymbol{\beta}$ is $m \times 1$. Suppose that $\boldsymbol{\beta}$ and X are partitioned as $\boldsymbol{\beta} = (\boldsymbol{\beta}_1' \quad \boldsymbol{\beta}_2')'$ and $X = (X_1 \quad X_2)$, where $\boldsymbol{\beta}_1$ is $m_1 \times 1$, $\boldsymbol{\beta}_2$ is $m_2 \times 1$, and we wish to test the hypothesis that $\boldsymbol{\beta}_2 = \mathbf{0}$. We will assume that each component of $\boldsymbol{\beta}_2$ is estimable since this test would not be meaningful otherwise. It is easily shown that this then implies that X_2 has full column rank and $\mathrm{rank}(X_1) = r - m_2$, where $r = \mathrm{rank}(X)$. A test of $\boldsymbol{\beta}_2 = \mathbf{0}$ can be constructed by comparing the sum of squared errors for the reduced model $\boldsymbol{y} = X_1\boldsymbol{\beta}_1 + \boldsymbol{\epsilon}$, which is

$$t_1 = (\boldsymbol{y} - X_1\hat{\boldsymbol{\beta}}_1)'(\boldsymbol{y} - X_1\hat{\boldsymbol{\beta}}_1) = \boldsymbol{y}'(\mathrm{I}_N - X_1(X_1'X_1)^- X_1')\boldsymbol{y},$$

to the sum of squared errors for the complete model, which is given by

$$t_2 = (\boldsymbol{y} - X\hat{\boldsymbol{\beta}})'(\boldsymbol{y} - X\hat{\boldsymbol{\beta}}) = \boldsymbol{y}'(\mathrm{I}_N - X(X'X)^- X')\boldsymbol{y}$$

Now if $\boldsymbol{\epsilon} \sim \mathrm{N}_N(\mathbf{0}, \sigma^2\mathrm{I})$, then $\boldsymbol{y} \sim \mathrm{N}_N(X\boldsymbol{\beta}, \sigma^2\mathrm{I})$. Thus, by applying Theorem 9.11 and using the fact that $X(X'X)^- X'X_1 = X_1$, we find that $(t_1 - t_2)/\sigma^2$ is chi-squared since

$$\left\{\frac{(X(X'X)^{-}X' - X_1(X_1'X_1)^{-}X_1')}{\sigma^2}\right\}\{\sigma^2\mathrm{I}\}\left\{\frac{(X(X'X)^{-}X' - X_1(X_1'X_1)^{-}X_1')}{\sigma^2}\right\}$$

$$= \left\{\frac{(X(X'X)^{-}X' - X_1(X_1'X_1)^{-}X_1')}{\sigma^2}\right\}$$

In particular, if $\boldsymbol{\beta}_2 = \mathbf{0}$, $(t_1 - t_2)/\sigma^2 \sim \chi^2_{m_2}$, since

$$\begin{aligned}\operatorname{tr}\{X(X'X)^{-}X' - X_1(X_1'X_1)^{-}X_1'\} &= \operatorname{tr}\{X(X'X)^{-}X'\} - \operatorname{tr}\{X_1(X_1'X_1)^{-}X_1'\}\\ &= r - (r - m_2) = m_2,\end{aligned}$$

and

$$\boldsymbol{\beta}_1'X_1'\left\{\frac{(X(X'X)^{-}X' - X_1(X_1'X_1)^{-}X_1')}{\sigma^2}\right\}X_1\boldsymbol{\beta}_1 = \frac{(\boldsymbol{\beta}_1'X_1'X_1\boldsymbol{\beta}_1 - \boldsymbol{\beta}_1'X_1'X_1\boldsymbol{\beta}_1)}{\sigma^2} = 0$$

By a similar application of Theorem 9.11, we observe that $t_2/\sigma^2 \sim \chi^2_{N-r}$. In addition, it follows from Theorem 9.13 that $(t_1 - t_2)/\sigma^2$ and t_2/σ^2 are independently distributed since

$$\left\{\frac{(X(X'X)^{-}X' - X_1(X_1'X_1)^{-}X_1')}{\sigma^2}\right\}\{\sigma^2\mathrm{I}\}\left\{\frac{(\mathrm{I}_N - X(X'X)^{-}X')}{\sigma^2}\right\} = 0$$

This then permits the construction of an F statistic for testing that $\boldsymbol{\beta}_2 = \mathbf{0}$; that is, if $\boldsymbol{\beta}_2 = \mathbf{0}$, then the statistic

$$\mathrm{F} = \frac{(t_1 - t_2)/m_2}{t_2/(N - r)}$$

has the F distribution with m_2 and $N - r$ degrees of freedom.

The proof of the next result, which is very similar to the proof of Theorem 9.13, is left to the reader as an exercise.

Theorem 9.14. Let $\boldsymbol{x} \sim \mathrm{N}_m(\boldsymbol{\mu}, \Omega)$, where Ω is positive definite, and suppose that A is an $m \times m$ symmetric matrix while B is an $n \times m$ matrix. If $B\Omega A = (0)$, then $\boldsymbol{x}'A\boldsymbol{x}$ and $B\boldsymbol{x}$ are independently distributed.

Example 9.6. Suppose that we have a random sample $x_1, \ldots, x_n$ from a normal distribution with mean μ and variance σ^2. In Example 9.3, it was shown that $(n-1)s^2/\sigma^2 \sim \chi^2_{n-1}$, where s^2, the sample variance, is given by

$$s^2 = \frac{1}{(n-1)} \sum_{i=1}^{n} (x_i - \overline{x})^2$$

We will now use Theorem 9.14 to show that the sample mean,

$$\overline{x} = \frac{1}{n} \sum_{i=1}^{n} x_i,$$

is independently distributed of s^2. In Example 9.3, we saw that s^2 is a scalar multiple of the quadratic form

$$\boldsymbol{x}'(\mathrm{I}_n - n^{-1}\mathbf{1}_n\mathbf{1}_n')\boldsymbol{x},$$

where $\boldsymbol{x} = (x_1, \dots, x_n)' \sim \mathrm{N}_n(\mu\mathbf{1}_n, \sigma^2\mathrm{I}_n)$. On the other hand, $\overline{x}$ can be expressed as

$$\overline{x} = n^{-1}\mathbf{1}_n'\boldsymbol{x}$$

Consequently, the independence of $\overline{x}$ and s^2 follows from the fact that

$$\mathbf{1}_n'(\sigma^2\mathrm{I}_n)(\mathrm{I}_n - n^{-1}\mathbf{1}_n\mathbf{1}_n') = \sigma^2(\mathbf{1}_n' - n^{-1}\mathbf{1}_n'\mathbf{1}_n\mathbf{1}_n') = \sigma^2(\mathbf{1}_n' - \mathbf{1}_n') = \mathbf{0}'$$

When Ω is positive semidefinite, the condition $A\Omega B = (0)$, given in Theorem 9.13, will still guarantee that the two quadratic forms $\boldsymbol{x}'A\boldsymbol{x}$ and $\boldsymbol{x}'B\boldsymbol{x}$ are independently distributed. Likewise, when Ω is positive semidefinite, the condition $B\Omega A = (0)$, given in Theorem 9.14, will still guarantee that $\boldsymbol{x}'A\boldsymbol{x}$ and $B\boldsymbol{x}$ are independently distributed. However, in these situations a weaker set of conditions will guarantee independence. These conditions are given in the following two theorems. The proofs are left as exercises.

Theorem 9.15. Let $\boldsymbol{x} \sim \mathrm{N}_m(\boldsymbol{\mu}, \Omega)$, where Ω is positive semidefinite, and suppose that A and B are $m \times m$ symmetric matrices. Then $\boldsymbol{x}'A\boldsymbol{x}$ and $\boldsymbol{x}'B\boldsymbol{x}$ are independently distributed if

(a) $\Omega A \Omega B \Omega = (0)$,
(b) $\Omega A \Omega B \boldsymbol{\mu} = \mathbf{0}$,
(c) $\Omega B \Omega A \boldsymbol{\mu} = \mathbf{0}$,
(d) $\boldsymbol{\mu}' A \Omega B \boldsymbol{\mu} = 0$.

Theorem 9.16. Let $\boldsymbol{x} \sim N_m(\boldsymbol{\mu}, \Omega)$, where Ω is positive semidefinite, and suppose that A is an $m \times m$ symmetric matrix while B is an $n \times m$ matrix. If $B\Omega A\Omega = (0)$ and $B\Omega A\boldsymbol{\mu} = \mathbf{0}$, then $\boldsymbol{x}'A\boldsymbol{x}$ and $B\boldsymbol{x}$ are independently distributed.

Our final result can be helpful in establishing that several quadratic forms in the same normal random vector are independently distributed, each having a chi-squared distribution.

Theorem 9.17. Let $\boldsymbol{x} \sim N_m(\boldsymbol{\mu}, \Omega)$, where Ω is positive definite. Suppose that A_i is an $m \times m$ symmetric matrix of rank r_i, for $i = 1, \dots, k$, and $A = A_1 + \cdots + A_k$ is of rank r. Consider the conditions

(a) $A_i\Omega$ is idempotent for each i,
(b) $A\Omega$ is idempotent,
(c) $A_i\Omega A_j = (0)$, for all $i \neq j$,
(d) $r = \sum_{i=1}^{k} r_i$.

If any two of (a), (b), and (c) hold, or if (b) and (d) hold, then

(i) $\boldsymbol{x}'A_i\boldsymbol{x} \sim \chi^2_{r_i}(\frac{1}{2}\boldsymbol{\mu}'A_i\boldsymbol{\mu})$,
(ii) $\boldsymbol{x}'A\boldsymbol{x} \sim \chi^2_r(\frac{1}{2}\boldsymbol{\mu}'A\boldsymbol{\mu})$,
(iii) $\boldsymbol{x}'A_1\boldsymbol{x}, \dots, \boldsymbol{x}'A_k\boldsymbol{x}$ are independently distributed.

Proof. Since Ω is positive definite, there exists a nonsingular matrix T satisfying $\Omega = TT'$, and the conditions (a)–(d) can be equivalently expressed as

(a) $T'A_iT$ is idempotent for each i,
(b) $T'AT$ is idempotent,
(c) $(T'A_iT)(T'A_jT) = (0)$, for all $i \neq j$,
(d) $\text{rank}(T'AT) = \sum_{i=1}^{k} \text{rank}(T'A_iT)$.

Since, $T'A_1T, \dots, T'A_kT$ and $T'AT$ satisfy the conditions of Corollary 9.7.1, we are ensured that if any two of (a), (b), and (c) hold or if (b) and (d) hold, then all four of the conditions (a)–(d) above hold. Now using Theorem 9.11, (a) implies (i) and (b) implies (ii), while Theorem 9.13, along with (c), guarantees that (iii) holds. □

6. EXPECTED VALUES OF QUADRATIC FORMS

When a quadratic form satisfies the conditions given in the theorems of Section 6.4, then its moments can be obtained directly from the appropriate chi-squared distribution. In this section, we derive formulas for means, variances, and covariances of quadratic forms that will be useful when this is not the case. We will start with the most general case in which the random vector $\boldsymbol{x}$ has an arbitrary distribution. The expressions we obtain involve the matrix of second moments of $\boldsymbol{x}$, $\text{E}(\boldsymbol{x}\boldsymbol{x}')$, and the matrix of fourth moments $\text{E}(\boldsymbol{x}\boldsymbol{x}' \otimes \boldsymbol{x}\boldsymbol{x}')$.

Theorem 9.18. Let $\boldsymbol{x}$ be an $m \times 1$ random vector having finite fourth moments, so that both $\text{E}(\boldsymbol{x}\boldsymbol{x}')$ and $\text{E}(\boldsymbol{x}\boldsymbol{x}' \otimes \boldsymbol{x}\boldsymbol{x}')$ exist. Denote the mean vector and covariance matrix of $\boldsymbol{x}$ by $\boldsymbol{\mu}$ and Ω. If A and B are $m \times m$ symmetric matrices, then

(a) $\text{E}(\boldsymbol{x}'A\boldsymbol{x}) = \text{tr}\{A\text{E}(\boldsymbol{x}\boldsymbol{x}')\} = \text{tr}(A\Omega) + \boldsymbol{\mu}'A\boldsymbol{\mu}$,

(b) $\text{var}(\boldsymbol{x}'A\boldsymbol{x}) = \text{tr}\{(A \otimes A)\text{E}(\boldsymbol{x}\boldsymbol{x}' \otimes \boldsymbol{x}\boldsymbol{x}')\} - [\text{tr}(A\Omega) + \boldsymbol{\mu}'A\boldsymbol{\mu}]^2$,

(c) $\text{cov}(\boldsymbol{x}'A\boldsymbol{x}, \boldsymbol{x}'B\boldsymbol{x}) = \text{tr}\{(A \otimes B)\text{E}(\boldsymbol{x}\boldsymbol{x}' \otimes \boldsymbol{x}\boldsymbol{x}')\} - [\text{tr}(A\Omega) + \boldsymbol{\mu}'A\boldsymbol{\mu}][\text{tr}(B\Omega) + \boldsymbol{\mu}'B\boldsymbol{\mu}]$.

Proof. The covariance matrix Ω is defined by

$$\Omega = \text{E}\{(\boldsymbol{x} - \boldsymbol{\mu})(\boldsymbol{x} - \boldsymbol{\mu})'\} = \text{E}(\boldsymbol{x}\boldsymbol{x}') - \boldsymbol{\mu}\boldsymbol{\mu}',$$

so that $\text{E}(\boldsymbol{x}\boldsymbol{x}') = \Omega + \boldsymbol{\mu}\boldsymbol{\mu}'$. Since $\boldsymbol{x}'A\boldsymbol{x}$ is a scalar, we have

$$\begin{aligned}\text{E}(\boldsymbol{x}'A\boldsymbol{x}) &= \text{E}\{\text{tr}(\boldsymbol{x}'A\boldsymbol{x})\} = \text{E}\{\text{tr}(A\boldsymbol{x}\boldsymbol{x}')\} = \text{tr}\{A\text{E}(\boldsymbol{x}\boldsymbol{x}')\} = \text{tr}\{A(\Omega + \boldsymbol{\mu}\boldsymbol{\mu}')\}\\ &= \text{tr}(A\Omega) + \text{tr}(A\boldsymbol{\mu}\boldsymbol{\mu}') = \text{tr}(A\Omega) + \boldsymbol{\mu}'A\boldsymbol{\mu},\end{aligned}$$

and so (a) holds. Part (b) will follow from (c) by taking $B = A$. To prove (c), note that

$$\begin{aligned}\text{E}(\boldsymbol{x}'A\boldsymbol{x}\boldsymbol{x}'B\boldsymbol{x}) &= \text{E}[\text{tr}\{(\boldsymbol{x}' \otimes \boldsymbol{x}')(A \otimes B)(\boldsymbol{x} \otimes \boldsymbol{x})\}]\\ &= \text{E}[\text{tr}\{(A \otimes B)(\boldsymbol{x} \otimes \boldsymbol{x})(\boldsymbol{x}' \otimes \boldsymbol{x}')\}]\\ &= \text{tr}\{(A \otimes B)\text{E}(\boldsymbol{x}\boldsymbol{x}' \otimes \boldsymbol{x}\boldsymbol{x}')\}\end{aligned}$$

Then use this, along with part (a), in the equation

$$\text{cov}(\boldsymbol{x}'A\boldsymbol{x}, \boldsymbol{x}'B\boldsymbol{x}) = \text{E}(\boldsymbol{x}'A\boldsymbol{x}\boldsymbol{x}'B\boldsymbol{x}) - \text{E}(\boldsymbol{x}'A\boldsymbol{x})\text{E}(\boldsymbol{x}'B\boldsymbol{x}) \qquad \square$$

When $\boldsymbol{x}$ has a normal distribution, the expressions for variances and covariances, as well as higher moments, simplify somewhat. This is a consequence of the special structure of the moments of the multivariate normal distribution. The commutation matrix K_{mm}, discussed in Chapter 7, plays a crucial role in obtaining some of these matrix expressions. We will also make use of the $m \times m$ matrix T_{ij} defined by

$$T_{ij} = E_{ij} + E_{ji} = \boldsymbol{e}_i\boldsymbol{e}_j' + \boldsymbol{e}_j\boldsymbol{e}_i';$$

that is, all of the elements of T_{ij} are equal to 0 except for the (i, j)th and (j, i)th elements, which equal 1, unless $i = j$, in which case the only nonzero element is a 2 in the (i, i)th position. Before obtaining expressions for the variance and covariance of quadratic forms in normal variates, we will need the following result.

Theorem 9.19. If $\boldsymbol{z} \sim \mathrm{N}_m(\boldsymbol{0}, \mathrm{I}_m)$ and $\boldsymbol{c}$ is a vector of constants, then

(a) $\mathrm{E}(\boldsymbol{z} \otimes \boldsymbol{z}) = \mathrm{vec}(\mathrm{I}_m)$,
(b) $\mathrm{E}(\boldsymbol{c}\boldsymbol{z}' \otimes \boldsymbol{z}\boldsymbol{z}') = (0)$, $\mathrm{E}(\boldsymbol{z}\boldsymbol{c}' \otimes \boldsymbol{z}\boldsymbol{z}') = (0)$, $\mathrm{E}(\boldsymbol{z}\boldsymbol{z}' \otimes \boldsymbol{c}\boldsymbol{z}') = (0)$, $\mathrm{E}(\boldsymbol{z}\boldsymbol{z}' \otimes \boldsymbol{z}\boldsymbol{c}') = (0)$,
(c) $\mathrm{E}(\boldsymbol{z}\boldsymbol{z}' \otimes \boldsymbol{z}\boldsymbol{z}') = 2N_m + \mathrm{vec}(\mathrm{I}_m)\{\mathrm{vec}(\mathrm{I}_m)\}'$,
(d) $\mathrm{var}(\boldsymbol{z} \otimes \boldsymbol{z}) = 2N_m$.

Proof. Since $\mathrm{E}(\boldsymbol{z}) = \boldsymbol{0}$, $\mathrm{I}_m = \mathrm{var}(\boldsymbol{z}) = \mathrm{E}(\boldsymbol{z}\boldsymbol{z}')$ and so

$$\mathrm{E}(\boldsymbol{z} \otimes \boldsymbol{z}) = \mathrm{E}\{\mathrm{vec}(\boldsymbol{z}\boldsymbol{z}')\} = \mathrm{vec}\{\mathrm{E}(\boldsymbol{z}\boldsymbol{z}')\} = \mathrm{vec}(\mathrm{I}_m)$$

It is easily verified using the standard normal moment generating function that

$$\mathrm{E}(z_i^3) = 0, \qquad \mathrm{E}(z_i^4) = 3 \tag{9.12}$$

Each element of the matrices of expected values in (b) will be of the form $c_i\mathrm{E}(z_jz_kz_l)$. Since the components of $\boldsymbol{z}$ are independent, we get

$$\mathrm{E}(z_jz_kz_l) = \mathrm{E}(z_j)\mathrm{E}(z_k)\mathrm{E}(z_l) = 0$$

when the three subscripts are distinct,

$$\mathrm{E}(z_jz_kz_l) = \mathrm{E}(z_j^2)\mathrm{E}(z_l) = 1 \cdot 0 = 0$$

when $j = k \neq l$, and similarly for $j = l \neq k$ and $l = k \neq j$, and

$$\mathrm{E}(z_jz_kz_l) = \mathrm{E}(z_j^3) = 0,$$

when $j = k = l$. This proves (b). Next we consider terms of the form $\mathrm{E}(z_iz_jz_kz_l)$. These equal 1 if $i = j \neq l = k$, $i = k \neq j = l$, or $i = l \neq j = k$, equal 3 if $i = j = k = l$, and equal zero otherwise. This leads to

$$\mathrm{E}(z_iz_j\boldsymbol{z}\boldsymbol{z}') = T_{ij} + \delta_{ij}\mathrm{I}_m,$$

where δ_{ij} is the (i,j)th element of I_m. Thus,

$$\mathrm{E}(zz' \otimes zz') = \mathrm{E}\left\{\left(\sum_{i=1}^{m}\sum_{j=1}^{m} E_{ij}z_iz_j\right) \otimes zz'\right\} = \sum_{i=1}^{m}\sum_{j=1}^{m} \{E_{ij} \otimes \mathrm{E}(z_iz_jzz')\}$$

$$= \sum_{i=1}^{m}\sum_{j=1}^{m} \{E_{ij} \otimes (T_{ij} + \delta_{ij}\mathrm{I}_m)\}$$

$$= \sum_{i=1}^{m}\sum_{j=1}^{m} (E_{ij} \otimes T_{ij}) + \sum_{i=1}^{m}\sum_{j=1}^{m} (\delta_{ij}E_{ij} \otimes \mathrm{I}_m)$$

The third result now follows since

$$\sum_{i=1}^{m}\sum_{j=1}^{m} (E_{ij} \otimes T_{ij}) = \sum_{i=1}^{m}\sum_{j=1}^{m} (E_{ij} \otimes E_{ji}) + \sum_{i=1}^{m}\sum_{j=1}^{m} (E_{ij} \otimes E_{ij})$$

$$= K_{mm} + \left\{\sum_{i=1}^{m} (\boldsymbol{e}_i \otimes \boldsymbol{e}_i)\right\}\left\{\sum_{j=1}^{m} (\boldsymbol{e}_j' \otimes \boldsymbol{e}_j')\right\}$$

$$= K_{mm} + \left\{\sum_{i=1}^{m} \mathrm{vec}(\boldsymbol{e}_i\boldsymbol{e}_i')\right\}\left\{\sum_{j=1}^{m} \{\mathrm{vec}(\boldsymbol{e}_j\boldsymbol{e}_j')\}'\right\}$$

$$= K_{mm} + \mathrm{vec}(\mathrm{I}_m)\{\mathrm{vec}(\mathrm{I}_m)\}',$$

$$\sum_{i=1}^{m}\sum_{j=1}^{m} (\delta_{ij}E_{ij} \otimes \mathrm{I}_m) = \left(\sum_{i=1}^{m} E_{ii}\right) \otimes \mathrm{I}_m = \mathrm{I}_m \otimes \mathrm{I}_m = \mathrm{I}_{m^2},$$

and $\mathrm{I}_{m^2} + K_{mm} = 2N_m$. Finally, (d) is an immediate consequence of (a) and (c). □

The next result generalizes the results of Theorem 9.19 to a multivariate normal distribution having a general positive definite covariance matrix.

Theorem 9.20. Let $\boldsymbol{x} \sim \mathrm{N}_m(\boldsymbol{0}, \Omega)$, where Ω is positive definite, and let $\boldsymbol{c}$ be an $m \times 1$ vector of constants. Then

(a) $\mathrm{E}(\boldsymbol{x} \otimes \boldsymbol{x}) = \mathrm{vec}(\Omega)$,
(b) $\mathrm{E}(\boldsymbol{cx}' \otimes \boldsymbol{xx}') = (0)$, $\mathrm{E}(\boldsymbol{xc}' \otimes \boldsymbol{xx}') = (0)$, $\mathrm{E}(\boldsymbol{xx}' \otimes \boldsymbol{cx}') = (0)$, $\mathrm{E}(\boldsymbol{xx}' \otimes \boldsymbol{xc}') = (0)$,

(c) $\mathrm{E}(\boldsymbol{x}\boldsymbol{x}' \otimes \boldsymbol{x}\boldsymbol{x}') = 2N_m(\Omega \otimes \Omega) + \mathrm{vec}(\Omega)\{\mathrm{vec}(\Omega)\}'$,

(d) $\mathrm{var}(\boldsymbol{x} \otimes \boldsymbol{x}) = 2N_m(\Omega \otimes \Omega)$.

Proof. Let T be any nonsingular matrix satisfying $\Omega = TT'$, so that $\boldsymbol{z} = T^{-1}\boldsymbol{x}$ and $\boldsymbol{x} = T\boldsymbol{z}$, where $\boldsymbol{z} \sim \mathrm{N}_m(\boldsymbol{0}, \mathrm{I}_m)$. Then the results above are consequences of Theorem 9.19 since

$$\mathrm{E}(\boldsymbol{x} \otimes \boldsymbol{x}) = (T \otimes T)\mathrm{E}(\boldsymbol{z} \otimes \boldsymbol{z}) = (T \otimes T)\mathrm{vec}(\mathrm{I}_m) = \mathrm{vec}(TT') = \mathrm{vec}(\Omega),$$

$$\mathrm{E}(\boldsymbol{c}\boldsymbol{x}' \otimes \boldsymbol{x}\boldsymbol{x}') = (\mathrm{I}_m \otimes T)\mathrm{E}(\boldsymbol{c}\boldsymbol{z}' \otimes \boldsymbol{z}\boldsymbol{z}')(T' \otimes T') = (0),$$

and

$$\begin{aligned}
\mathrm{E}(\boldsymbol{x}\boldsymbol{x}' \otimes \boldsymbol{x}\boldsymbol{x}') &= (T \otimes T)\mathrm{E}(\boldsymbol{z}\boldsymbol{z}' \otimes \boldsymbol{z}\boldsymbol{z}')(T' \otimes T') \\
&= (T \otimes T)(2N_m + \mathrm{vec}(\mathrm{I}_m)\{\mathrm{vec}(\mathrm{I}_m)\}')(T' \otimes T') \\
&= 2(T \otimes T)N_m(T' \otimes T') + (T \otimes T)\mathrm{vec}(\mathrm{I}_m)\{\mathrm{vec}(\mathrm{I}_m)\}'(T' \otimes T') \\
&= 2N_m(T \otimes T)(T' \otimes T') + \mathrm{vec}(TT')\{\mathrm{vec}(TT')\}' \\
&= 2N_m(\Omega \otimes \Omega) + \mathrm{vec}(\Omega)\{\mathrm{vec}(\Omega)\}'
\end{aligned}$$

□

We are now ready to obtain simplified expressions for the variance and covariance of quadratic forms in normal variates.

Theorem 9.21. Let A and B be $m \times m$ symmetric matrices and suppose that $\boldsymbol{x} \sim \mathrm{N}_m(\boldsymbol{0}, \Omega)$, where Ω is positive definite. Then

(a) $\mathrm{E}\{\boldsymbol{x}'A\boldsymbol{x}\,\boldsymbol{x}'B\boldsymbol{x}\} = \mathrm{tr}(A\Omega)\mathrm{tr}(B\Omega) + 2\,\mathrm{tr}(A\Omega B\Omega)$,

(b) $\mathrm{cov}(\boldsymbol{x}'A\boldsymbol{x}, \boldsymbol{x}'B\boldsymbol{x}) = 2\,\mathrm{tr}(A\Omega B\Omega)$,

(c) $\mathrm{var}(\boldsymbol{x}'A\boldsymbol{x}) = 2\,\mathrm{tr}\{(A\Omega)^2\}$.

Proof. Since (c) is the special case of (b) in which $B = A$, we only need to prove (a) and (b). Note that by making use of Theorem 9.20, we find that

$$\begin{aligned}
\mathrm{E}\{\boldsymbol{x}'A\boldsymbol{x}\,\boldsymbol{x}'B\boldsymbol{x}\} &= \mathrm{E}\{(\boldsymbol{x}' \otimes \boldsymbol{x}')(A \otimes B)(\boldsymbol{x} \otimes \boldsymbol{x})\} \\
&= \mathrm{E}[\mathrm{tr}\{(A \otimes B)(\boldsymbol{x}\boldsymbol{x}' \otimes \boldsymbol{x}\boldsymbol{x}')\}] = \mathrm{tr}\{(A \otimes B)\mathrm{E}(\boldsymbol{x}\boldsymbol{x}' \otimes \boldsymbol{x}\boldsymbol{x}')\} \\
&= \mathrm{tr}\{(A \otimes B)(2N_m(\Omega \otimes \Omega) + \mathrm{vec}(\Omega)\{\mathrm{vec}(\Omega)\}')\} \\
&= \mathrm{tr}\{(A \otimes B)((\mathrm{I}_{m^2} + K_{mm})(\Omega \otimes \Omega) + \mathrm{vec}(\Omega)\{\mathrm{vec}(\Omega)\}')\} \\
&= \mathrm{tr}\{(A \otimes B)(\Omega \otimes \Omega)\} + \mathrm{tr}\{(A \otimes B)K_{mm}(\Omega \otimes \Omega)\} \\
&\quad + \mathrm{tr}((A \otimes B)\mathrm{vec}(\Omega)\{\mathrm{vec}(\Omega)\}')
\end{aligned}$$

Now

$$\text{tr}\{(A \otimes B)(\Omega \otimes \Omega)\} = \text{tr}(A\Omega \otimes B\Omega) = \text{tr}(A\Omega)\text{tr}(B\Omega)$$

follows directly from Theorem 7.8, while

$$\text{tr}\{(A \otimes B)K_{mm}(\Omega \otimes \Omega)\} = \text{tr}\{(A\Omega \otimes B\Omega)K_{mm}\} = \text{tr}(A\Omega B\Omega)$$

follows from Theorem 7.31. Using the symmetry of A and Ω along with Theorems 7.15 and 7.16, the last term in $\text{E}\{\boldsymbol{x}'A\boldsymbol{x}\boldsymbol{x}'B\boldsymbol{x}\}$ simplifies as

$$\begin{aligned}\text{tr}((A \otimes B)\text{vec}(\Omega)\{\text{vec}(\Omega)\}') &= \{\text{vec}(\Omega)\}'(A \otimes B)\text{vec}(\Omega)\\ &= \{\text{vec}(\Omega)\}' \text{vec}(B\Omega A) = \text{tr}(A\Omega B\Omega)\end{aligned}$$

This then proves (a). To prove (b) we use the definition of covariance and Theorem 9.18(a) to get

$$\text{cov}(\boldsymbol{x}'A\boldsymbol{x}, \boldsymbol{x}'B\boldsymbol{x}) = \text{E}\{\boldsymbol{x}'A\boldsymbol{x}\boldsymbol{x}'B\boldsymbol{x}\} - \text{E}(\boldsymbol{x}'A\boldsymbol{x})\text{E}(\boldsymbol{x}'B\boldsymbol{x}) = 2\,\text{tr}(A\Omega B\Omega) \qquad \square$$

The formulas given in the previous theorem become somewhat more complicated when the normal distribution has a nonnull mean vector. These formulas are given in the following theorem.

Theorem 9.22. Let A and B be symmetric $m \times m$ matrices and suppose that $\boldsymbol{x} \sim \text{N}_m(\boldsymbol{\mu}, \Omega)$, where Ω is positive definite. Then

(a) $\text{E}\{\boldsymbol{x}'A\boldsymbol{x}\boldsymbol{x}'B\boldsymbol{x}\} = \text{tr}(A\Omega)\,\text{tr}(B\Omega) + 2\,\text{tr}(A\Omega B\Omega) + \text{tr}(A\Omega)\boldsymbol{\mu}'B\boldsymbol{\mu} + 4\boldsymbol{\mu}'A\Omega B\boldsymbol{\mu}$ $+ \boldsymbol{\mu}'A\boldsymbol{\mu}\,\text{tr}(B\Omega) + \boldsymbol{\mu}'A\boldsymbol{\mu}\boldsymbol{\mu}'B\boldsymbol{\mu}$,

(b) $\text{cov}(\boldsymbol{x}'A\boldsymbol{x}, \boldsymbol{x}'B\boldsymbol{x}) = 2\,\text{tr}(A\Omega B\Omega) + 4\boldsymbol{\mu}'A\Omega B\boldsymbol{\mu}$,

(c) $\text{var}(\boldsymbol{x}'A\boldsymbol{x}) = 2\,\text{tr}\{(A\Omega)^2\} + 4\boldsymbol{\mu}'A\Omega A\boldsymbol{\mu}$.

Proof. Again (c) is a special case of (b), so we only need to prove (a) and (b). We can write $\boldsymbol{x} = \boldsymbol{y} + \boldsymbol{\mu}$, where $\boldsymbol{y} \sim \text{N}_m(\boldsymbol{0}, \Omega)$ and, consequently,

$$\begin{aligned}\text{E}\{\boldsymbol{x}'A\boldsymbol{x}\boldsymbol{x}'B\boldsymbol{x}\} &= \text{E}\{(\boldsymbol{y}+\boldsymbol{\mu})'A(\boldsymbol{y}+\boldsymbol{\mu})(\boldsymbol{y}+\boldsymbol{\mu})'B(\boldsymbol{y}+\boldsymbol{\mu})\}\\ &= \text{E}\{(\boldsymbol{y}'A\boldsymbol{y} + 2\boldsymbol{\mu}'A\boldsymbol{y} + \boldsymbol{\mu}'A\boldsymbol{\mu})(\boldsymbol{y}'B\boldsymbol{y} + 2\boldsymbol{\mu}'B\boldsymbol{y} + \boldsymbol{\mu}'B\boldsymbol{\mu})\}\\ &= \text{E}\{\boldsymbol{y}'A\boldsymbol{y}\,\boldsymbol{y}'B\boldsymbol{y}\} + 2\text{E}\{\boldsymbol{y}'A\boldsymbol{y}\,\boldsymbol{\mu}'B\boldsymbol{y}\} + \text{E}(\boldsymbol{y}'A\boldsymbol{y})\boldsymbol{\mu}'B\boldsymbol{\mu}\\ &\quad + 2\text{E}\{\boldsymbol{\mu}'A\boldsymbol{y}\,\boldsymbol{y}'B\boldsymbol{y}\} + 4\text{E}(\boldsymbol{\mu}'A\boldsymbol{y}\,\boldsymbol{\mu}'B\boldsymbol{y}) + 2\text{E}(\boldsymbol{\mu}'A\boldsymbol{y})\boldsymbol{\mu}'B\boldsymbol{\mu}\\ &\quad + \boldsymbol{\mu}'A\boldsymbol{\mu}\,\text{E}(\boldsymbol{y}'B\boldsymbol{y}) + 2\boldsymbol{\mu}'A\boldsymbol{\mu}\,\text{E}(\boldsymbol{\mu}'B\boldsymbol{y}) + \boldsymbol{\mu}'A\boldsymbol{\mu}\,\boldsymbol{\mu}'B\boldsymbol{\mu}\end{aligned}$$

The sixth and eighth terms in this last expression are zero since $\text{E}(\boldsymbol{y}) = \boldsymbol{0}$, while it follows from Theorem 9.20(b) that the second and fourth terms are zero. To simplify the fifth term note that

$$\begin{aligned}
E(\boldsymbol{\mu}'Ay\,\boldsymbol{\mu}'By) &= E\{(\boldsymbol{\mu}'A \otimes \boldsymbol{\mu}'B)(y \otimes y)\} = (A\boldsymbol{\mu} \otimes B\boldsymbol{\mu})'E\{(y \otimes y)\} \\
&= \{\text{vec}(B\boldsymbol{\mu}\boldsymbol{\mu}'A)\}' \text{vec}(\Omega) = \text{tr}\{(B\boldsymbol{\mu}\boldsymbol{\mu}'A)'\Omega\} \\
&= \text{tr}(A\boldsymbol{\mu}\boldsymbol{\mu}'B\Omega) = \boldsymbol{\mu}'A\Omega B\boldsymbol{\mu}
\end{aligned}$$

Thus, using this and Theorems 9.18(a) and 9.21(a), we find that

$$\begin{aligned}
E\{\boldsymbol{x}'A\boldsymbol{x}\,\boldsymbol{x}'B\boldsymbol{x}\} &= \text{tr}(A\Omega)\,\text{tr}(B\Omega) + 2\,\text{tr}(A\Omega B\Omega) + \text{tr}(A\Omega)\boldsymbol{\mu}'B\boldsymbol{\mu} + 4\boldsymbol{\mu}'A\Omega B\boldsymbol{\mu} \\
&\quad + \boldsymbol{\mu}'A\boldsymbol{\mu}\,\text{tr}(B\Omega) + \boldsymbol{\mu}'A\boldsymbol{\mu}\,\boldsymbol{\mu}'B\boldsymbol{\mu},
\end{aligned}$$

thereby proving (a); (b) then follows immediately from the definition of covariance and Theorem 9.18(a). □

Example 9.7. Let us return to the subject of Example 9.4, where we defined

$$A_1 = n^{-1}\{(\text{I}_k - k^{-1}\mathbf{1}_k\mathbf{1}_k') \otimes \mathbf{1}_n\mathbf{1}_n'\}$$

and

$$A_2 = \text{I}_k \otimes (\text{I}_n - n^{-1}\mathbf{1}_n\mathbf{1}_n')$$

It was shown that if $\boldsymbol{x} = (\boldsymbol{x}_1', \ldots, \boldsymbol{x}_k')' \sim N_{kn}(\boldsymbol{\mu}, \Omega)$ with $\boldsymbol{\mu} = \mathbf{1}_k \otimes \mu\mathbf{1}_n$ and $\Omega = \text{I}_k \otimes \sigma^2\text{I}_n$, then $t_1/\sigma^2 = \boldsymbol{x}'(A_1/\sigma^2)\boldsymbol{x} \sim \chi^2_{k-1}$, $t_2/\sigma^2 = \boldsymbol{x}'(A_2/\sigma^2)\boldsymbol{x} \sim \chi^2_{k(n-1)}$, independently. Since the mean of a chi-squared random variable equals its degrees of freedom, while the variance is two times the degrees of freedom, we can easily calculate the mean and variance of t_1 and t_2 without employing the results of this section; in particular, we have

$$\begin{aligned}
E(t_1) &= \sigma^2(k-1), & \text{var}(t_1) &= 2\sigma^4(k-1), \\
E(t_2) &= \sigma^2 k(n-1), & \text{var}(t_2) &= 2\sigma^4 k(n-1)
\end{aligned}$$

Suppose now that $\boldsymbol{x}_i \sim N_n(\mu\mathbf{1}_n, \sigma_i^2\text{I}_n)$ so that $\Omega = \text{var}(\boldsymbol{x}) = D \otimes \text{I}_n$, where $D = \text{diag}(\sigma_1^2, \ldots, \sigma_k^2)$. It can be easily verified that, in this case, t_1/σ^2 and t_2/σ^2 no longer satisfy the conditions of Theorem 9.11 for chi-squaredness, but are still independently distributed. The mean and variance of t_1 and t_2 can be computed by using Theorems 9.18 and 9.22. For instance, the mean of t_2 is given by

$$\begin{aligned}
E(t_2) &= E(\boldsymbol{x}'A_2\boldsymbol{x}) = \text{tr}(A_2\Omega) + \boldsymbol{\mu}'A_2\boldsymbol{\mu} \\
&= \text{tr}(\{\text{I}_k \otimes (\text{I}_n - n^{-1}\mathbf{1}_n\mathbf{1}_n')\}(D \otimes \text{I}_n)) \\
&\quad + \mu^2(\mathbf{1}_k' \otimes \mathbf{1}_n')\{\text{I}_k \otimes (\text{I}_n - n^{-1}\mathbf{1}_n\mathbf{1}_n')\}(\mathbf{1}_k \otimes \mathbf{1}_n)
\end{aligned}$$

$$= \text{tr}(D)\text{tr}(\text{I}_n - n^{-1}\mathbf{1}_n\mathbf{1}_n') + \mu^2(\mathbf{1}_k'\mathbf{1}_k)\{\mathbf{1}_n'(\text{I}_n - n^{-1}\mathbf{1}_n\mathbf{1}_n')\mathbf{1}_n\}$$

$$= (n-1)\sum_{i=1}^{k} \sigma_i^2,$$

while its variance is

$$\begin{aligned}
\text{var}(t_2) &= \text{var}(\boldsymbol{x}'A_2\boldsymbol{x}) = 2\,\text{tr}\{(A_2\Omega)^2\} + 4\boldsymbol{\mu}'A_2\Omega A_2\boldsymbol{\mu} \\
&= 2\,\text{tr}\{D^2 \otimes (\text{I}_n - n^{-1}\mathbf{1}_n\mathbf{1}_n')\} \\
&\quad + 4\mu^2(\mathbf{1}_k' \otimes \mathbf{1}_n')\{D \otimes (\text{I}_n - n^{-1}\mathbf{1}_n\mathbf{1}_n')\}(\mathbf{1}_k \otimes \mathbf{1}_n) \\
&= 2\,\text{tr}(D^2)\text{tr}\{(\text{I}_n - n^{-1}\mathbf{1}_n\mathbf{1}_n')\} + 4\mu^2(\mathbf{1}_k'D\mathbf{1}_k)\{\mathbf{1}_n'(\text{I}_n - n^{-1}\mathbf{1}_n\mathbf{1}_n')\mathbf{1}_n\} \\
&= 2(n-1)\sum_{i=1}^{k} \sigma_i^4
\end{aligned}$$

We will leave it to the reader to verify that

$$\text{E}(t_1) = (1 - k^{-1})\sum_{i=1}^{k} \sigma_i^2,$$

$$\text{var}(t_1) = 2\left\{(1 - 2k^{-1})\sum_{i=1}^{k} \sigma_i^4 + k^{-2}\left(\sum_{i=1}^{k} \sigma_i^2\right)^2\right\}$$

So far we have considered the expectation of a quadratic form as well as the expectation of a product of two quadratic forms. A more general situation is one in which we need the expected value of the product of n quadratic forms. This expectation becomes more tedious to compute as n increases. For example, if A, B, and C are $m \times m$ symmetric matrices and $\boldsymbol{x} \sim \text{N}_m(\mathbf{0}, \Omega)$, the expected value $\text{E}(\boldsymbol{x}'A\boldsymbol{x}\boldsymbol{x}'B\boldsymbol{x}\boldsymbol{x}'C\boldsymbol{x})$ can be obtained by first computing $\text{E}(\boldsymbol{x}\boldsymbol{x}' \otimes \boldsymbol{x}\boldsymbol{x}' \otimes \boldsymbol{x}\boldsymbol{x}')$, and then using this in the identity

$$\text{E}(\boldsymbol{x}'A\boldsymbol{x}\boldsymbol{x}'B\boldsymbol{x}\boldsymbol{x}'C\boldsymbol{x}) = \text{tr}\{(A \otimes B \otimes C)\text{E}(\boldsymbol{x}\boldsymbol{x}' \otimes \boldsymbol{x}\boldsymbol{x}' \otimes \boldsymbol{x}\boldsymbol{x}')\}$$

The details of this derivation are left as an exercise. Magnus (1978) used an alternative method, utilizing the cumulants of a distribution and their relationship to the moments of a distribution, to obtain the expectation of the product of an arbitrary number of quadratic forms. The results for a product of three and four quadratic forms are summarized below.

Theorem 9.23. Let A, B, C, and D be symmetric $m \times m$ matrices and suppose that $\boldsymbol{x} \sim \mathrm{N}_m(\mathbf{0}, \mathrm{I}_m)$. Then

(a) $$\mathrm{E}(\boldsymbol{x}'A\boldsymbol{x}\,\boldsymbol{x}'B\boldsymbol{x}\boldsymbol{x}'C\boldsymbol{x}) = \mathrm{tr}(A)\,\mathrm{tr}(B)\,\mathrm{tr}(C) + 2\{\mathrm{tr}(A)\,\mathrm{tr}(BC) + \mathrm{tr}(B)\,\mathrm{tr}(AC) + \mathrm{tr}(C)\,\mathrm{tr}(AB)\} + 8\,\mathrm{tr}(ABC),$$

(b) $$\begin{aligned}&\mathrm{E}(\boldsymbol{x}'A\boldsymbol{x}\,\boldsymbol{x}'B\boldsymbol{x}\boldsymbol{x}'C\boldsymbol{x}\boldsymbol{x}'D\boldsymbol{x})\\ &\quad= \mathrm{tr}(A)\,\mathrm{tr}(B)\,\mathrm{tr}(C)\,\mathrm{tr}(D) + 8\{\mathrm{tr}(A)\,\mathrm{tr}(BCD)\\ &\qquad+ \mathrm{tr}(B)\,\mathrm{tr}(ACD) + \mathrm{tr}(C)\,\mathrm{tr}(ABD) + \mathrm{tr}(D)\mathrm{tr}(ABC)\} + 4\{\mathrm{tr}(AB)\,\mathrm{tr}(CD)\\ &\qquad+ \mathrm{tr}(AC)\,\mathrm{tr}(BD) + \mathrm{tr}(AD)\,\mathrm{tr}(BC)\} + 2\{\mathrm{tr}(A)\,\mathrm{tr}(B)\,\mathrm{tr}(CD)\\ &\qquad+ \mathrm{tr}(A)\,\mathrm{tr}(C)\,\mathrm{tr}(BD) + \mathrm{tr}(A)\,\mathrm{tr}(D)\,\mathrm{tr}(BC) + \mathrm{tr}(\mathrm{B})\,\mathrm{tr}(C)\,\mathrm{tr}(AD)\\ &\qquad+ \mathrm{tr}(B)\,\mathrm{tr}(D)\,\mathrm{tr}(AC) + \mathrm{tr}(C)\,\mathrm{tr}(D)\,\mathrm{tr}(AB)\}\\ &\qquad+ 16\{\mathrm{tr}(ABCD) + \mathrm{tr}(ABDC) + \mathrm{tr}(ACBD)\}\end{aligned}$$

If $\boldsymbol{x} \sim \mathrm{N}_m(\mathbf{0}, \Omega)$, where Ω is positive definite, then A, B, C, and D appearing in the right-hand side of the equations in Theorem 9.23 are replaced by $A\Omega$, $B\Omega$, $C\Omega$, and $D\Omega$.

An alternative approach to the calculation of moments of quadratic forms utilizes tensor methods. This approach may be particularly appealing in those situations in which higher ordered moments are needed or the random vector $\boldsymbol{x}$ does not have a multivariate normal distribution. A detailed discussion of these tensor methods can be found in McCullagh (1987).

7. THE WISHART DISTRIBUTION

When $x_1, \ldots, x_n$ are independently distributed, with $x_i \sim \mathrm{N}(0, \sigma^2)$ for every i, then

$$\boldsymbol{x}'\boldsymbol{x} = \sum_{i=1}^{n} x_i^2 \sim \sigma^2\chi_n^2,$$

where $\boldsymbol{x}' = (x_1, \ldots, x_n)$; that is, $\boldsymbol{x}'\boldsymbol{x}/\sigma^2$ has a chi-squared distribution with n degrees of freedom. A natural matrix generalization of this situation, one which has important applications in multivariate analysis, involves the distribution of

$$X'X = \sum_{i=1}^{n} \boldsymbol{x}_i\boldsymbol{x}_i',$$

where $X' = (\boldsymbol{x}_1, \ldots, \boldsymbol{x}_n)$ is an $m \times n$ matrix such that $\boldsymbol{x}_1, \ldots, \boldsymbol{x}_n$ are independent and $\boldsymbol{x}_i \sim \mathrm{N}_m(\mathbf{0}, \Omega)$ for each i. Thus, the components of the jth column of

X are independently distributed each as $\mathrm{N}(0, \sigma_{jj})$, where σ_{jj} is the jth diagonal element of Ω, so that the jth diagonal element of $X'X$ has the distribution $\sigma_{jj}\chi_n^2$. The joint distribution of all of the elements of the $m \times m$ matrix $X'X$ is called the Wishart distribution with scale matrix Ω and degrees of freedom n, and will be denoted by $\mathrm{W}_m(\Omega, n)$. This Wishart distribution, like the chi-squared distribution χ_n^2, is said to be central. More generally, if $\boldsymbol{x}_1, \dots, \boldsymbol{x}_n$ are independent and $\boldsymbol{x}_i \sim \mathrm{N}_m(\boldsymbol{\mu}_i, \Omega)$, then $X'X$ has the noncentral Wishart distribution with noncentrality matrix $\Phi = \frac{1}{2}M'M$, where M' is the $m \times n$ matrix given by $M' = (\boldsymbol{\mu}_1, \dots, \boldsymbol{\mu}_n)$. We will denote this noncentral Wishart distribution as $W_m(\Omega, n, \Phi)$. Additional information regarding the Wishart distribution, such as the form of its density function, can be found in texts on multivariate analysis such as Srivastava and Khatri (1979) and Muirhead (1982).

If A is an $n \times n$ symmetric matrix and X' is an $m \times n$ matrix, then the matrix $X'AX$ is sometimes called a generalized quadratic form. The following theorem gives some generalizations of the results obtained in Sections 9.4 and 9.5 regarding quadratic forms to these generalized quadratic forms.

Theorem 9.24. Let X' be an $m \times n$ matrix whose columns are independently distributed, with the ith column having the $\mathrm{N}_m(\boldsymbol{\mu}_i, \Omega)$ distribution, where Ω is positive definite. Suppose that A and B are $n \times n$ symmetric matrices while C is $k \times n$. Let $M' = (\boldsymbol{\mu}_1, \dots, \boldsymbol{\mu}_n)$, $\Phi = \frac{1}{2}M'AM$, and $r = \mathrm{rank}(A)$. Then

(a) $X'AX \sim \mathrm{W}_m(\Omega, r, \Phi)$, if A is idempotent,

(b) $X'AX$ and $X'BX$ are independently distributed if $AB = (0)$,

(c) $X'AX$ and CX are independently distributed if $CA = (0)$.

Proof. The proof of (a) will be complete if we can show that there exists an $m \times r$ matrix Y' such that $X'AX = Y'Y$, where the columns of Y' are independently distributed each having a normal distribution with the same covariance matrix Ω, and $\frac{1}{2}\mathrm{E}(Y')\mathrm{E}(Y) = \Phi$. Since the columns of X' are independently distributed, it follows that

$$\mathrm{vec}(\mathrm{X}') \sim \mathrm{N}_{nm}(\mathrm{vec}(M'), \mathrm{I}_n \otimes \Omega)$$

Since A is symmetric, idempotent, and has rank r, there must exist an $n \times r$ matrix P satisfying $A = PP'$ and $P'P = \mathrm{I}_r$. Consequently, $X'AX = Y'Y$, where the $m \times r$ matrix $Y' = X'P$ so that

$$\begin{aligned} \mathrm{vec}(Y') &= \mathrm{vec}(X'P) = (P' \otimes \mathrm{I}_m)\mathrm{vec}(X') \\ &\sim \mathrm{N}_{mr}((P' \otimes \mathrm{I}_m)\mathrm{vec}(M'), (P' \otimes \mathrm{I}_m)(\mathrm{I}_n \otimes \Omega)(P \otimes \mathrm{I}_m)) \\ &\sim \mathrm{N}_{mr}(\mathrm{vec}(M'P), (\mathrm{I}_r \otimes \Omega)) \end{aligned}$$

But this means that the columns of Y' are independently and normally distributed, each with covariance matrix Ω. Further,

$$\frac{1}{2}\,\mathrm{E}(Y')\mathrm{E}(Y) = \frac{1}{2}\,M'PP'M = \frac{1}{2}\,M'AM = \boldsymbol{\Phi},$$

and so (a) follows. To prove (b), note that since A and B are symmetric, $AB = (0)$ implies that $AB = BA$, so A and B are diagonalized by the same orthogonal matrix; that is, there exist diagonal matrices C and D and an orthogonal matrix Q such that $Q'AQ = C$ and $Q'BQ = D$. Further, $AB = (0)$ implies that $CD = (0)$, so that by appropriately choosing Q we will have $C = \mathrm{diag}(c_1, \dots, c_h, 0, \dots, 0)$ and $D = \mathrm{diag}(0, \dots, 0, d_{h+1}, \dots, d_n)$ for some h. Thus, if we let $U = QX$, we find that

$$X'AX = U'CU = \sum_{i=1}^{h} c_i \boldsymbol{u}_i \boldsymbol{u}_i', \qquad X'BX = U'DU = \sum_{i=h+1}^{n} d_i \boldsymbol{u}_i \boldsymbol{u}_i',$$

where $\boldsymbol{u}_i$ is the ith column of U'. Since $\mathrm{vec}(U') \sim N_{nm}(\mathrm{vec}(M'Q'), (I_n \otimes \Omega))$, these columns are independently distributed and so (b) follows. The proof of (c) is similar to that of (b). □

If the columns of the $m \times n$ matrix X' are independent and identically distributed as $N_m(\mathbf{0}, \Omega)$ and M' is an $m \times n$ matrix of constants, then $V = (X + M)'(X + M)$ has the Wishart distribution $W_m(\Omega, n, \frac{1}{2}M'M)$. A more general situation is one in which the columns of X' are independent and identically distributed having zero mean vector and some nonnormal multivariate distribution. In this case, the distribution of $V = (X + M)'(X + M)$, which may be very complicated, will depend on the specific nonnormal distribution. In particular, the moments of V are directly related to the moments of the columns of X'. Our next result gives expressions for the first two moments of V when $M = (0)$. Since V is a matrix and joint distributions are more conveniently handled in the form of vectors, we will vectorize V; that is, for instance, variances and covariances of the elements of V can be obtained from the matrix $\mathrm{var}\{\mathrm{vec}(V)\}$.

Theorem 9.25. Let the columns of the $m \times n$ matrix $X' = (\boldsymbol{x}_1, \dots, \boldsymbol{x}_n)$ be independently and identically distributed with $\mathrm{E}(\boldsymbol{x}_i) = \mathbf{0}$, $\mathrm{var}(\boldsymbol{x}_i) = \Omega$, and $\mathrm{E}(\boldsymbol{x}_i\boldsymbol{x}_i' \otimes \boldsymbol{x}_i\boldsymbol{x}_i') = \Psi$. If $V = X'X$, then

(a) $\mathrm{E}(V) = n\Omega$,

(b) $\mathrm{var}\{\mathrm{vec}(V)\} = n\{\Psi - \mathrm{vec}(\Omega)\mathrm{vec}(\Omega)'\}$.

Proof. Since $\mathrm{E}(\boldsymbol{x}_i) = \mathbf{0}$, we have $\Omega = \mathrm{E}(\boldsymbol{x}_i\boldsymbol{x}_i')$ and so

$$\mathrm{E}(V) = \mathrm{E}(X'X) = \sum_{i=1}^{n} \mathrm{E}(\boldsymbol{x}_i\boldsymbol{x}_i') = \sum_{i=1}^{n} \Omega = n\Omega$$

In addition, since $\boldsymbol{x}_1, \ldots, \boldsymbol{x}_n$ are independent, we have

$$\begin{aligned}\mathrm{var}\{\mathrm{vec}(V)\} &= \mathrm{var}\left\{\mathrm{vec}\left(\sum_{i=1}^{n} \boldsymbol{x}_i\boldsymbol{x}_i'\right)\right\} = \mathrm{var}\left\{\sum_{i=1}^{n} \mathrm{vec}(\boldsymbol{x}_i\boldsymbol{x}_i')\right\} \\ &= \sum_{i=1}^{n} \mathrm{var}\{\mathrm{vec}(\boldsymbol{x}_i\boldsymbol{x}_i')\} \\ &= \sum_{i=1}^{n} \mathrm{var}(\boldsymbol{x}_i \otimes \boldsymbol{x}_i) = \sum_{i=1}^{n} \{\mathrm{E}(\boldsymbol{x}_i\boldsymbol{x}_i' \otimes \boldsymbol{x}_i\boldsymbol{x}_i') - \mathrm{E}(\boldsymbol{x}_i \otimes \boldsymbol{x}_i)\mathrm{E}(\boldsymbol{x}_i' \otimes \boldsymbol{x}_i')\} \\ &= \sum_{i=1}^{n} \{\Psi - \mathrm{vec}(\Omega)\mathrm{vec}(\Omega)'\} = n\{\Psi - \mathrm{vec}(\Omega)\mathrm{vec}(\Omega)'\}. \qquad \square\end{aligned}$$

The expression for $\mathrm{var}\{\mathrm{vec}(V)\}$ simplifies when V has a Wishart distribution due to the special structure of the fourth moments of the normal distribution. This simplified expression is given in our next theorem. Note that although this theorem is stated for normally distributed columns, the first result given applies to the general case as well.

Theorem 9.26. Let the columns of the $m \times n$ matrix X' be independently and identically distributed as $\mathrm{N}_m(\mathbf{0}, \Omega)$. Define $V = (X+M)'(X+M)$, where $M' = (\boldsymbol{\mu}_1, \ldots, \boldsymbol{\mu}_n)$ is an $m \times n$ matrix of constants, so that $V \sim \mathrm{W}_m(\Omega, n, \frac{1}{2}M'M)$. Then

(a) $\mathrm{E}(V) = n\Omega + M'M$,

(b) $\mathrm{var}\{\mathrm{vec}(V)\} = 2N_m\{n(\Omega \otimes \Omega) + \Omega \otimes M'M + M'M \otimes \Omega\}$.

Proof. Since $\mathrm{E}(X) = (0)$ and $\mathrm{E}(X'X) = n\Omega$ from the previous theorem, it follows that

$$\mathrm{E}(V) = \mathrm{E}(X'X + X'M + M'X + M'M) = \mathrm{E}(X'X) + M'M = n\Omega + M'M$$

Proceeding as in the proof of Theorem 9.25, we obtain

$$\mathrm{var}\{\mathrm{vec}(V)\} = \sum_{i=1}^{n} \mathrm{var}\{(\boldsymbol{x}_i + \boldsymbol{\mu}_i) \otimes (\boldsymbol{x}_i + \boldsymbol{\mu}_i)\} \tag{9.13}$$

But

$$\begin{aligned}(\boldsymbol{x}_i + \boldsymbol{\mu}_i) \otimes (\boldsymbol{x}_i + \boldsymbol{\mu}_i) &= \boldsymbol{x}_i \otimes \boldsymbol{x}_i + \boldsymbol{x}_i \otimes \boldsymbol{\mu}_i + \boldsymbol{\mu}_i \otimes \boldsymbol{x}_i + \boldsymbol{\mu}_i \otimes \boldsymbol{\mu}_i \\ &= \boldsymbol{x}_i \otimes \boldsymbol{x}_i + (\mathrm{I}_m + K_{mm})(\boldsymbol{x}_i \otimes \boldsymbol{\mu}_i) + \boldsymbol{\mu}_i \otimes \boldsymbol{\mu}_i \\ &= \boldsymbol{x}_i \otimes \boldsymbol{x}_i + 2N_m(\mathrm{I}_m \otimes \boldsymbol{\mu}_i)\boldsymbol{x}_i + \boldsymbol{\mu}_i \otimes \boldsymbol{\mu}_i\end{aligned}$$

Since all first and third order moments of $\boldsymbol{x}_i$ are equal to 0, $\boldsymbol{x}_i \otimes \boldsymbol{x}_i$ and $\boldsymbol{x}_i$ are uncorrelated, and so using Theorem 9.20 and Problem 7.52, we find that

$$\begin{aligned}\mathrm{var}\{(\boldsymbol{x}_i + \boldsymbol{\mu}_i) \otimes (\boldsymbol{x}_i + \boldsymbol{\mu}_i)\} &= \mathrm{var}(\boldsymbol{x}_i \otimes \boldsymbol{x}_i) + \mathrm{var}\{2N_m(\mathrm{I}_m \otimes \boldsymbol{\mu}_i)\boldsymbol{x}_i\} \\ &= 2N_m(\Omega \otimes \Omega) + 4N_m(\mathrm{I}_m \otimes \boldsymbol{\mu}_i)\Omega(\mathrm{I}_m \otimes \boldsymbol{\mu}_i')N_m \\ &= 2N_m(\Omega \otimes \Omega) + 4N_m(\Omega \otimes \boldsymbol{\mu}_i\boldsymbol{\mu}_i')N_m \\ &= 2N_m(\Omega \otimes \Omega + \Omega \otimes \boldsymbol{\mu}_i\boldsymbol{\mu}_i' + \boldsymbol{\mu}_i\boldsymbol{\mu}_i' \otimes \Omega)\end{aligned} \tag{9.14}$$

Now substituting (9.14) in (9.13) and simplifying, we obtain (b). □

Example 9.8. In Examples 9.3 and 9.6 it was shown that, when sampling from a normal distribution, a constant multiple of the sample variance s^2 has a chi-squared distribution, and it is independently distributed of the sample mean $\bar{x}$. In this example, we consider the multivariate version of this problem involving $\bar{\boldsymbol{x}}$ and S; that is, suppose that $\boldsymbol{x}_1, \ldots, \boldsymbol{x}_n$ are independently distributed with $\boldsymbol{x}_i \sim \mathrm{N}_m(\boldsymbol{\mu}, \Omega)$ for each i, and define X' to be the $m \times n$ matrix $(\boldsymbol{x}_1, \ldots, \boldsymbol{x}_n)$. Then the sample mean vector and sample covariance matrix can be expressed as

$$\bar{\boldsymbol{x}} = \frac{1}{n} \sum_{i=1}^{n} \boldsymbol{x}_i = \frac{1}{n} X'\mathbf{1}_n,$$

and

$$\begin{aligned}S &= \frac{1}{n-1} \sum_{i=1}^{n} (\boldsymbol{x}_i - \bar{\boldsymbol{x}})(\boldsymbol{x}_i - \bar{\boldsymbol{x}})' = \frac{1}{n-1} \left(\sum_{i=1}^{n} \boldsymbol{x}_i\boldsymbol{x}_i' - n\bar{\boldsymbol{x}}\bar{\boldsymbol{x}}' \right) \\ &= \frac{1}{n-1} (X'X - n^{-1}X'\mathbf{1}_n\mathbf{1}_n'X) = \frac{1}{n-1} X'(\mathrm{I}_n - n^{-1}\mathbf{1}_n\mathbf{1}_n')X\end{aligned}$$

Since $A = (\mathrm{I}_n - n^{-1}\mathbf{1}_n\mathbf{1}_n')$ is idempotent and $\mathrm{rank}(A) = \mathrm{tr}(A) = n - 1$, it follows from Theorem 9.24(a) that $(n-1)S$ has a Wishart distribution. To determine its noncentrality matrix, note that $M' = (\boldsymbol{\mu}, \dots, \boldsymbol{\mu}) = \boldsymbol{\mu}\mathbf{1}_n'$, so that

$$M'AM = \boldsymbol{\mu}\mathbf{1}_n'(\mathrm{I}_n - n^{-1}\mathbf{1}_n\mathbf{1}_n')\mathbf{1}_n\boldsymbol{\mu}' = \boldsymbol{\mu}(n-n)\boldsymbol{\mu}' = (0)$$

Thus, $(n-1)S$ has the central Wishart distribution $\mathrm{W}_m(\Omega, n-1)$. Further, using Theorem 9.24(c), we see that S and $\bar{\boldsymbol{x}}$ are independently distributed since

$$\mathbf{1}_n'(\mathrm{I}_n - n^{-1}\mathbf{1}_n\mathbf{1}_n') = (\mathbf{1}_n' - \mathbf{1}_n') = \mathbf{0}'$$

In addition, it follows from Theorem 9.26 that

$$\mathrm{E}(S) = \Omega, \qquad \mathrm{var}\{\mathrm{vec}(S)\} = \frac{2}{n-1} N_m(\Omega \otimes \Omega) = \frac{2}{n-1} N_m(\Omega \otimes \Omega)N_m$$

The redundant elements in $\mathrm{vec}(S)$ can be eliminated by utilizing $\mathrm{v}(S)$. Since $\mathrm{v}(S) = D_m^+ \,\mathrm{vec}(S)$, where D_m is the duplication matrix discussed in Section 7.8, we have

$$\mathrm{var}\{\mathrm{v}(S)\} = \frac{2}{n-1} D_m^+ N_m(\Omega \otimes \Omega)N_m D_m^{+\prime}$$

In some situations, we may be interested only in the sample variances and not the sample covariances; that is, the random vector of interest here is the $m \times 1$ vector $\boldsymbol{s} = (s_{11}, \dots, s_{mm})'$. Expressions for the mean vector and covariance matrix of $\boldsymbol{s}$ are easily obtained from the formulas in this example since $\boldsymbol{s} = \mathrm{w}(S) = \Psi_m \,\mathrm{vec}(S)$ as seen in Problem 7.45, where

$$\Psi_m = \sum_{i=1}^{m} \boldsymbol{e}_{i,m}(\boldsymbol{e}_{i,m} \otimes \boldsymbol{e}_{i,m})'$$

Thus, using the properties of Ψ_m obtained in Problem 7.45, we find that

$$\begin{aligned}
\mathrm{E}(\boldsymbol{s}) &= \Psi_m \,\mathrm{vec}\{\mathrm{E}(S)\} = \Psi_m \,\mathrm{vec}(\Omega) = \mathrm{w}(\Omega), \\
\mathrm{var}(\boldsymbol{s}) &= \Psi_m \,\mathrm{var}\{\mathrm{vec}(S)\}\Psi_m' = \Psi_m \left\{ \frac{2}{n-1} N_m(\Omega \otimes \Omega)N_m \right\} \Psi_m' \\
&= \frac{2}{n-1} \Psi_m(\Omega \otimes \Omega)\Psi_m' = \frac{2}{n-1} (\Omega \odot \Omega),
\end{aligned}$$

where $\odot$ is the Hadamard product.

Example 9.9. The perturbation formulas for eigenvalues and eigenvectors of a symmetric matrix obtained in Section 8.6 can be used to approximate the distributions of an eigenvalue or an eigenvector of a matrix having a Wishart distribution. One important application in statistics that utilizes these asymptotic distributions is principal components analysis, an analysis involving the eigenvalues and eigenvectors of the $m \times m$ sample covariance matrix S. The exact distributions of an eigenvalue and an eigenvector of S are rather complicated, while their asymptotic distributions follow in a fairly straightforward manner from the asymptotic distribution of S. Now it can be shown by using the central limit theorem [see Muirhead, (1982)] that $\sqrt{n-1}\,\text{vec}(S)$ has an asymptotic normal distribution. In particular, using results from Example 9.8, we have, asymptotically,

$$\sqrt{n-1}\{\text{vec}(S) - \text{vec}(\Omega)\} \sim \text{N}_{m^2}(\mathbf{0}, 2N_m(\Omega \otimes \Omega)),$$

where Ω is the population covariance matrix. Let $W = S - \Omega$ and $W_* = \sqrt{n-1}W$, so that $\text{vec}(W_*)$ has the asymptotic normal distribution indicated above. Suppose that $\boldsymbol{\gamma}_i$ is a normalized eigenvector of $S = \Omega + W$ corresponding to the ith largest eigenvalue λ_i, while $\boldsymbol{q}_i$ is a normalized eigenvector of Ω corresponding to its ith largest eigenvalue x_i. Now if x_i is a distinct eigenvalue of Ω, then we have the first-order approximations from Section 8.6

$$\begin{aligned} \lambda_i &= x_i + \boldsymbol{q}_i'W\boldsymbol{q}_i = x_i + (\boldsymbol{q}_i' \otimes \boldsymbol{q}_i')\text{vec}(W), \\ \boldsymbol{\gamma}_i &= \boldsymbol{q}_i - (\Omega - x_i\text{I}_m)^+W\boldsymbol{q}_i = \boldsymbol{q}_i - \{\boldsymbol{q}_i' \otimes (\Omega - x_i\text{I}_m)^+\}\text{vec}(W) \end{aligned} \tag{9.15}$$

Thus, the asymptotic normality of $a_i = \sqrt{n-1}(\lambda_i - x_i)$ follows from the asymptotic normality of $\text{vec}(W_*)$. Further, we have, asymptotically,

$$\begin{aligned} \text{E}(a_i) &= (\boldsymbol{q}_i' \otimes \boldsymbol{q}_i')\text{E}\{\text{vec}(W_*)\} = (\boldsymbol{q}_i' \otimes \boldsymbol{q}_i')\mathbf{0} = \mathbf{0}, \\ \text{var}(a_i) &= (\boldsymbol{q}_i' \otimes \boldsymbol{q}_i')(\text{var}\{\text{vec}(W_*)\})(\boldsymbol{q}_i \otimes \boldsymbol{q}_i) \\ &= (\boldsymbol{q}_i' \otimes \boldsymbol{q}_i')(2N_m(\Omega \otimes \Omega))(\boldsymbol{q}_i \otimes \boldsymbol{q}_i) = 2(\boldsymbol{q}_i'\Omega\boldsymbol{q}_i \otimes \boldsymbol{q}_i'\Omega\boldsymbol{q}_i) = 2x_i^2; \end{aligned}$$

that is, for large n, $\lambda_i \sim \text{N}(x_i, 2x_i^2/(n-1))$, approximately. Similarly, $\boldsymbol{b}_i = \sqrt{n-1}(\boldsymbol{\gamma}_i - \boldsymbol{q}_i)$ is asymptotically normal with

$$\text{E}(\boldsymbol{b}_i) = -\{\boldsymbol{q}_i' \otimes (\Omega - x_i\text{I}_m)^+\}\text{E}\{\text{vec}(W_*)\} = -\{\boldsymbol{q}_i' \otimes (\Omega - x_i\text{I}_m)^+\}\mathbf{0} = \mathbf{0},$$

$$\begin{aligned} \Phi = \text{var}(\boldsymbol{b}_1) &= \{\boldsymbol{q}_i' \otimes (\Omega - x_i\text{I}_m)^+\}\{\text{var}\{\text{vec}(W_*)\}\}\{\boldsymbol{q}_i' \otimes (\Omega - x_i\text{I}_m)^+\}' \\ &= \{\boldsymbol{q}_i' \otimes (\Omega - x_i\text{I}_m)^+\}\{2N_m(\Omega \otimes \Omega)\}\{\boldsymbol{q}_i \otimes (\Omega - x_i\text{I}_m)^+\} \end{aligned}$$

$$= \{(\Omega - x_i \mathrm{I}_m)^+ \otimes \boldsymbol{q}_i' + \boldsymbol{q}_i' \otimes (\Omega - x_i \mathrm{I}_m)^+\}\{(\Omega \otimes \Omega)\}\{\boldsymbol{q}_i \otimes (\Omega - x_i \mathrm{I}_m)^+\}$$

$$= \boldsymbol{q}_i'\Omega \boldsymbol{q}_i \otimes (\Omega - x_i \mathrm{I}_m)^+\Omega(\Omega - x_i \mathrm{I}_m)^+ = \lambda_i \left\{ \sum_{j \neq i} \frac{x_j}{(x_j - x_i)^2} \boldsymbol{q}_j \boldsymbol{q}_j' \right\},$$

and so for large n, $\boldsymbol{\gamma}_i \sim \mathrm{N}_m(\boldsymbol{q}_i, (n-1)^{-1}\boldsymbol{\Phi})$, approximately. While the first-order approximations in (9.15) can be used to obtain the asymptotic distributions, higher-order approximations, such as those given in Theorem 8.5, can be used to further improve the performance of these asymptotic distributions. The most common application of this process involves asymptotic chi-squared distributions, so we will illustrate the basic idea with the statistic

$$t = \frac{(n-1)(\lambda_i - x_i)^2}{2x_i^2},$$

which, due to the asymptotic normality of a_i, is asymptotically chi-squared with one degree of freedom. The mean of this chi-squared distribution is 1, while the exact mean of t is of the form

$$\mathrm{E}(t) = 1 + \sum_{j=1}^{\infty} \frac{c_j}{(n-1)^{(j+1)/2}},$$

where the c_js are constants. The higher-order approximations of λ_i can be used to determine the first constant c_1, and then this may be used to compute an adjusted statistic

$$t_* = \left\{1 - \frac{c_1}{(n-1)}\right\} t$$

The mean of this adjusted statistic is

$$\begin{aligned}\mathrm{E}(t_*) &= \left\{1 - \frac{c_1}{(n-1)}\right\} \mathrm{E}(t) \\ &= \left\{1 - \frac{c_1}{(n-1)}\right\}\left(1 + \sum_{j=1}^{\infty} \frac{c_j}{(n-1)^{(j+1)/2}}\right) \\ &= 1 + \sum_{j=2}^{\infty} \frac{d_j}{(n-1)^{(j+1)/2}},\end{aligned}$$

where the d_js are constants that are functions of the c_js. Note that the mean of t_* converges to 1 at a faster rate than does $\mathrm{E}(t)$. For this reason, the chi-squared distribution with one degree of freedom should approximate the distribution of this adjusted statistic better than it would approximate the distribution of t. This type of adjustment of asymptotically chi-squared statistics is commonly referred to as a Bartlett adjustment [Bartlett (1937, 1947)]. Some discussion of Bartlett adjustments can be found in Barndorff-Nielsen and Cox (1994).

Some of the inequalities for eigenvalues developed in Chapter 3 have important applications regarding the distributions of eigenvalues of certain functions of Wishart matrices. One such application is illustrated in our next example.

Example 9.10. A multivariate analysis of variance, such as the multivariate one-way classification model discussed in Example 3.14, utilizes the eigenvalues of BW^{-1}, where the $m \times m$ matrices B and W are independently distributed with $B \sim W_m(I_m, b, \Phi)$ and $W \sim W_m(I_m, w)$ (Problem 9.30). We will show that if the rank of the noncentrality matrix Φ is $r < m$ and V_1 and V_2 are independently distributed with $V_1 \sim W_{m-r}(I_{m-r}, b-r)$ and $V_2 \sim W_{m-r}(I_{m-r}, w)$, then

$$P\{\lambda_{r+i}(BW^{-1}) > c\} \leq P\{\lambda_i(V_1 V_2^{-1}) > c\},$$

for $i = 1, \dots, m-r$ and any constant c. This result is useful in determining the the dimensionality in a canonical variate analysis [see Schott (1984)]. Since $\text{rank}(\Phi) = r$, there exists an $r \times m$ matrix T such that $\frac{1}{2}T'T = \Phi$. If we define the $m \times b$ matrix $M' = (T' \quad (0))$, then since $\frac{1}{2}M'M = \Phi$ and $B \sim W_m(I_m, b, \Phi)$, it follows that B can be expressed as $B = X'X$, where X' is a $m \times b$ matrix for which $\text{vec}(X') \sim N_{bm}(\text{vec}(M'), I_b \otimes I_m)$. Partitioning X' as $X' = (X_1' \quad X_2')$, where X_1' is $m \times r$, we find that

$$B = X_1'X_1 + X_2'X_2 = B_1 + B_2,$$

where $B_1 \sim W_m(I_m, r, \Phi)$ and $B_2 \sim W_m(I_m, b-r)$ since $\text{vec}(X_1') \sim N_{rm}(\text{vec}(T'), I_r \otimes I_m)$ and $\text{vec}(X_2') \sim N_{(b-r)m}(\text{vec}\{(0)\}, I_{b-r} \otimes I_m)$. Now for fixed B_1, let F be any $m \times (m-r)$ matrix satisfying $F'B_1F = (0)$ and $F'F = I_{m-r}$, and define the sets

$$S_1(B_1) = \{B_2,\ W\colon \lambda_{r+i}(BW^{-1}) > c\},$$
$$S_2(B_1) = \{B_2,\ W\colon \lambda_i\{(F'B_2F)(F'WF)^{-1}\}) > c\}$$

It follows from Problem 3.32(a) that

$$\lambda_i\{(F'BF)(F'WF)^{-1}\} = \lambda_i\{(F'B_2F)(F'WF)^{-1}\} \geq \lambda_{r+i}(BW^{-1}),$$

so for each fixed $B_1, S_1(B_1) \subseteq S_2(B_1)$, and it can be easily verified that $V_1 = F'B_2F \sim W_{m-r}(\mathrm{I}_{m-r}, b-r)$ and $V_2 = F'WF \sim W_{m-r}(\mathrm{I}_{m-r}, w)$. Consequently, if $g(W)$,$f_1(B_1)$, and $f_2(B_2)$ are the density functions for W, B_1, and B_2, respectively, then

$$\int_{S_1(B_1)} g(W)f_2(B_2)\,\mathrm{d}W\,\mathrm{d}B_2 \leq \int_{S_2(B_1)} g(W)f_2(B_2)\,\mathrm{d}W\,\mathrm{d}B_2 = \mathrm{P}\{\lambda_i(V_1V_2^{-1}) > c\}$$

If we also define the sets

$$C_1 = \{B_1,\ B_2,\ W\colon \lambda_{r+i}(BW^{-1}) > c\}$$
$$C_2 = \{B_1 : B_1 \text{ positive definite}\},$$

then the desired result follows since

$$\begin{aligned}
\mathrm{P}\{\lambda_{r+i}(BW^{-1}) > c\} &= \int_{C_1} g(W)f_1(B_1)f_2(B_2)\,\mathrm{d}W\,\mathrm{d}B_1\,\mathrm{d}B_2 \\
&= \int_{C_2} \left\{\int_{S_1(B_1)} g(W)f_2(B_2)\,\mathrm{d}W\,\mathrm{d}B_2\right\} f_1(B_1)\,\mathrm{d}B_1 \\
&\leq \int_{C_2} \mathrm{P}\{\lambda_i(V_1V_2^{-1}) > c\}f_1(B_1)\,\mathrm{d}B_1 \\
&= \mathrm{P}\{\lambda_i(V_1V_2^{-1}) > c\}
\end{aligned}$$

The relationship between the sample correlation and covariance matrices and the expression for var{vec(S)} given in Example 9.8 can be used to obtain an expression for the asymptotic covariance matrix of vec(R). This is the subject of our final example.

Example 9.11. As in Example 9.8, let $\boldsymbol{x}_1, \dots, \boldsymbol{x}_n$ be independently distributed with $\boldsymbol{x}_i \sim \mathrm{N}_m(\boldsymbol{\mu}, \Omega)$, for each i, and let S and R be the sample covariance and correlation matrices computed from this sample. Thus, if we use the notation $D_X^a = \mathrm{diag}(x_{11}^a, \dots, x_{mm}^a)$, where X is an $m \times m$ matrix, then the sample correlation matrix can be expressed as

$$R = D_S^{-1/2} S D_S^{-1/2},$$

while the population correlation matrix is given by

$$P = D_\Omega^{-1/2} \Omega D_\Omega^{-1/2}$$

Note that if we define $\boldsymbol{y}_i = D_\Omega^{-1/2} \boldsymbol{x}_i$, then $\boldsymbol{y}_1, \ldots, \boldsymbol{y}_n$ are independently distributed with $\boldsymbol{y}_i \sim N_m(D_\Omega^{-1/2} \boldsymbol{\mu}, P)$. If S_* is the sample covariance matrix computed from the $\boldsymbol{y}_i$s, then $S_* = D_\Omega^{-1/2} S D_\Omega^{-1/2}$, $D_{S*}^{-1/2} = D_S^{-1/2} D_\Omega^{1/2} = D_\Omega^{1/2} D_S^{-1/2}$, and so

$$\begin{aligned} D_{S*}^{1/2} S_* D_{S*}^{1/2} &= D_S^{-1/2} D_\Omega^{1/2} (D_\Omega^{-1/2} S D_\Omega^{-1/2}) D_\Omega^{1/2} D_S^{-1/2} \\ &= D_S^{-1/2} S D_S^{-1/2} = R; \end{aligned}$$

that is, the sample correlation matrix computed from the $\boldsymbol{y}_i$s is the same as that computed from the $\boldsymbol{x}_i$s. If $A = S_* - P$, then the first-order approximation for R is given by (see Problem 8.15)

$$R = P + A - \frac{1}{2} (PD_A + D_A P),$$

and so

$$\begin{aligned} \text{vec}(R) &= \text{vec}(P) + \text{vec}(A) - \frac{1}{2} \{\text{vec}(PD_A) + \text{vec}(D_A P)\} \\ &= \text{vec}(P) + \text{vec}(A) - \frac{1}{2} \{(\text{I}_m \otimes P) + (P \otimes \text{I}_m)\} \text{vec}(D_A) \\ &= \text{vec}(P) + \left(\text{I}_{m^2} - \frac{1}{2} \{(\text{I}_m \otimes P) + (P \otimes \text{I}_m)\} \Lambda_m \right) \text{vec}(A), \quad (9.16) \end{aligned}$$

where

$$\Lambda_m = \sum_{i=1}^{m} (E_{ii} \otimes E_{ii})$$

Thus, since

$$\text{var}\{\text{vec}(A)\} = \text{var}\{\text{vec}(S_*)\} = \frac{2}{n-1} N_m (P \otimes P) N_m,$$

we get the first-order approximation

$$\text{var}\{\text{vec}(R)\} = \frac{2}{n-1} H N_m (P \otimes P) N_m H',$$

where the matrix H is the premultiplier on vec(A) in the last expression given in (9.16). Simplification (see Problem 9.33) leads to

$$\text{var}\{\text{vec}(R)\} = \frac{2}{n-1} N_m \Phi N_m, \tag{9.17}$$

where

$$\Phi = \{\mathrm{I}_{m^2} - (\mathrm{I}_m \otimes P)\Lambda_m\}(P \otimes P)\{\mathrm{I}_{m^2} - \Lambda_m(\mathrm{I}_m \otimes P)\}$$

Since R is symmetric and has each diagonal element equal to one, its redundant and nonrandom elements can be eliminated by utilizing $\tilde{\text{v}}(R)$. Since $\tilde{\text{v}}(R) = \tilde{L}_m \,\text{vec}(R)$, where $\tilde{L}_m$ is the matrix discussed in Section 7.8, we find that the asymptotic covariance matrix of $\tilde{\text{v}}(R)$ is given by

$$\text{var}\{\tilde{\text{v}}(R)\} = \frac{2}{n-1} \tilde{L}_m N_m \Phi N_m \tilde{L}_m'$$

The Hadamard product and its associated properties can be useful in analyses involving the manipulation of Φ since

$$\begin{aligned}\Phi &= P \otimes P - (\mathrm{I}_m \otimes P)\Lambda_m(P \otimes P) - (P \otimes P)\Lambda_m(\mathrm{I}_m \otimes P)\\ &\quad +(\mathrm{I}_m \otimes P)\Lambda_m(P \otimes P)\Lambda_m(\mathrm{I}_m \otimes P),\end{aligned}$$

and the last term on the right-hand side of this equation can be expressed as

$$(\mathrm{I}_m \otimes P)\Lambda_m(P \otimes P)\Lambda_m(\mathrm{I}_m \otimes P) = (\mathrm{I}_m \otimes P)\Psi_m'(P \odot P)\Psi_m(\mathrm{I}_m \otimes P)$$

PROBLEMS

1. We saw in the proof of Theorem 9.1 that if A is an $m \times m$ idempotent matrix, then rank(A) + rank($\mathrm{I}_m - A$) = m. Prove the converse; that is, show that if A is an $m \times m$ matrix satisfying rank(A) + rank($\mathrm{I}_m - A$) = m, then A is idempotent.

2. Suppose that A is an $m \times m$ idempotent matrix. Show that each of the following matrices is also idempotent.
   (a) A'.
   (b) BAB^{-1}, where B is any $m \times m$ nonsingular matrix.
   (c) A^n, where n is a positive integer.

3. Let A be an $m \times n$ matrix. Show that each of the following matrices is idempotent.

(a) AA^-.
(b) A^-A.
(c) $A(A'A)^-A'$.

4. Determine the class of $m \times 1$ vectors $\{\boldsymbol{x}\}$, for which $\boldsymbol{x}\boldsymbol{x}'$ is idempotent.

5. Determine constants a, b, and c so that each of the following is an idempotent matrix.
(a) $a\mathbf{1}_m\mathbf{1}_m'$.
(b) $b\mathrm{I}_m + c\mathbf{1}_m\mathbf{1}_m'$.

6. Let A be an $m \times n$ matrix with $\text{rank}(A) = m$. Show that $A'(AA')^{-1}A$ is symmetric and idempotent and find its rank.

7. Let A and B be $m \times m$ matrices. Show that if B is nonsingular and AB is idempotent, then BA is also idempotent.

8. Show that if A is an $m \times m$ matrix and $A^2 = mA$ for some scalar m, then

$$\text{tr}(A) = m\,\text{rank}(A)$$

9. Give an example of a collection of matrices $A_1, \ldots, A_k$ that satisfies conditions (a) and (d) of Corollary 9.7.1, but does not satisfy conditions (b) and (c). Similarly, find a collection of matrices that satisfies conditions (c) and (d) but does not satisfy conditions (a) and (b).

10. Prove Theorem 9.11.

11. Let A be an $m \times m$ symmetric matrix with $r = \text{rank}(A)$ and suppose that $\boldsymbol{x} \sim N_m(\mathbf{0}, \mathrm{I}_m)$. Show that the distribution of $\boldsymbol{x}'A\boldsymbol{x}$ can be expressed as a linear combination of r independent chi-squared random variables, each with 1 degree of freedom. What are the coefficients in this linear combination when A is idempotent?

12. Extend the result of Problem 11 to the situation in which $\boldsymbol{x} \sim N_m(\mathbf{0}, \Omega)$, where Ω is nonnegative definite; that is, show that if A is a symmetric matrix, then $\boldsymbol{x}'A\boldsymbol{x}$ can be expressed as a linear combination of independent chi-squared random variables each having one degree of freedom. How many chi-squared random variables are in this linear combination?

13. Let $x_1, \ldots, x_n$ be a random sample from a normal distribution with mean μ and variance σ^2, and let $\bar{x}$ be the sample mean. Write

$$t = \frac{n(\bar{x} - \mu)^2}{\sigma^2}$$

as a quadratic form in the vector $(\boldsymbol{x} - \mu \mathbf{1}_n)$, where $\boldsymbol{x} = (x_1, \dots, x_n)'$. What is the distribution of t?

14. Suppose that $\boldsymbol{x} \sim N_n(\boldsymbol{\mu}, \Omega)$, where Ω is positive definite. Partition $\boldsymbol{x}$, $\boldsymbol{\mu}$, and Ω as

$$\boldsymbol{x} = \begin{bmatrix} \boldsymbol{x}_1 \\ \boldsymbol{x}_2 \end{bmatrix}, \qquad \boldsymbol{\mu} = \begin{bmatrix} \boldsymbol{\mu}_1 \\ \boldsymbol{\mu}_2 \end{bmatrix}, \qquad \Omega = \begin{bmatrix} \Omega_{11} & \Omega_{12} \\ \Omega'_{12} & \Omega_{22} \end{bmatrix},$$

where $\boldsymbol{x}_1$ is $r \times 1$ and $\boldsymbol{x}_2$ is $(n - r) \times 1$. Show that
(a) $t_1 = (\boldsymbol{x}_1 - \boldsymbol{\mu}_1)'\Omega_{11}^{-1}(\boldsymbol{x}_1 - \boldsymbol{\mu}_1) \sim \chi_r^2$,
(b) $t_2 = (\boldsymbol{x} - \boldsymbol{\mu})'\Omega^{-1}(\boldsymbol{x} - \boldsymbol{\mu}) - (\boldsymbol{x}_1 - \boldsymbol{\mu}_1)'\Omega_{11}^{-1}(\boldsymbol{x}_1 - \boldsymbol{\mu}_1) \sim \chi_{n-r}^2$,
(c) t_1 and t_2 are independently distributed.

15. Prove Theorem 9.14.

16. Pearson's chi-squared statistic is given by

$$t = \sum_{i=1}^{m} \frac{(nx_i - n\mu_i)^2}{n\mu_i},$$

where n is a positive integer, the x_is are random variables, and the μ_is are nonnegative constants satisfying $\mu_1 + \cdots + \mu_m = 1$. Let $\boldsymbol{x} = (x_1, \dots, x_m)'$, $\boldsymbol{\mu} = (\mu_1, \dots, \mu_m)'$, and $\Omega = D - \boldsymbol{\mu}\boldsymbol{\mu}'$, where $D = \text{diag}(\mu_1, \dots, \mu_m)$.
(a) Show that Ω is a singular matrix.
(b) Show that if $\sqrt{n}(\boldsymbol{x} - \boldsymbol{\mu}) \sim N_m(\mathbf{0}, \Omega)$, then $t \sim \chi_{m-1}^2$.

17. Suppose that $\boldsymbol{x} \sim N_4(\mathbf{0}, I_4)$ and consider the three functions of the components of $\boldsymbol{x}$ given by

$$t_1 = \frac{1}{4}(x_1 + x_2 + x_3 + x_4)^2 + \frac{1}{2}(x_1 - x_2)^2,$$

$$t_2 = \frac{1}{12}(x_1 + x_2 + x_3 - 3x_4)^2,$$

$$t_3 = (x_1 + x_2 - 2x_3)^2 + (x_3 - x_4)^2$$

(a) Write t_1, t_2, and t_3 as quadratic forms in $\boldsymbol{x}$.
(b) Which of these statistics have chi-squared distributions?
(c) Which of the pairs t_1 and t_2, t_1 and t_3, and t_2 and t_3 are independently distributed?

18. Suppose that $\boldsymbol{x} \sim N_4(\boldsymbol{\mu}, \Omega)$, where $\boldsymbol{\mu} = (1, -1, 1, -1)'$ and $\Omega = I_4 + \mathbf{1}_4\mathbf{1}_4'$. Define

$$t_1 = \frac{1}{2}(x_1 - x_2)^2 + \frac{1}{2}(x_3 - x_4)^2,$$

$$t_2 = \frac{1}{2}(x_1 + x_2)^2 + \frac{1}{2}(x_3 + x_4)^2$$

(a) Does t_1 or t_2 have a chi-squared distribution? If so, identify the parameters of the distribution.
(b) Are t_1 and t_2 independently distributed?

19. Prove Theorem 9.15.

20. Prove Theorem 9.16.

21. The purpose of this exercise is to generalize the results of Example 9.5 to a test of the hypothesis that $H\boldsymbol{\beta} = \boldsymbol{c}$, where H is an $m_2 \times m$ matrix having rank m_2 and $\boldsymbol{c}$ is an $m_2 \times 1$ vector; Example 9.5 dealt with the special case in which $H = ((0) \quad I_{m_2})$ and $\boldsymbol{c} = \mathbf{0}$. Let G be an $(m - m_2) \times m$ matrix having rank $m - m_2$ and satisfying $HG' = (0)$. Show that the reduced model may be written as

$$\boldsymbol{y}_* = X_*\boldsymbol{\beta}_* + \boldsymbol{\epsilon},$$

where $\boldsymbol{y}_* = \boldsymbol{y} - XH'(HH')^{-1}\boldsymbol{c}$, $X_* = XG'(GG')^{-1}$, and $\boldsymbol{\beta}_* = G\boldsymbol{\beta}$. Use the sum of squared errors for this reduced model and the sum of squared errors for the complete model to construct the appropriate F statistic.

22. Suppose that $\boldsymbol{x} \sim N_m(\mathbf{0}, \Omega)$, where $r = \text{rank}(\Omega) < m$. If T is any $m \times r$ matrix satisfying $TT' = \Omega$, and $\boldsymbol{z} \sim N_r(\mathbf{0}, I_r)$, then $\boldsymbol{x}$ is distributed the same as $T\boldsymbol{z}$. Use this to show that the formulas given in Theorem 9.21 for positive definite Ω also hold when Ω is positive semidefinite.

23. Let $\boldsymbol{x} \sim N_m(\mathbf{0}, I_m)$. Use the fact that the first six moments of the standard normal distribution are 0, 1, 0, 3, 0, and 15 to show that

$$E(xx' \otimes xx' \otimes xx') = I_{m^3} + \frac{1}{2} \sum_{i=1}^{m} \sum_{j=1}^{m} (I_m \otimes T_{ij} \otimes T_{ij} + T_{ij} \otimes I_m \otimes T_{ij}$$
$$+ T_{ij} \otimes T_{ij} \otimes I_m)$$
$$+ \sum_{i=1}^{m} \sum_{j=1}^{m} \sum_{k=1}^{m} (T_{ij} \otimes T_{ik} \otimes T_{jk}),$$

where $T_{ij} = E_{ij} + E_{ji}$.

24. Let A, B, and C be $m \times m$ symmetric matrices and suppose that $x \sim N_m(\mathbf{0}, I_m)$.
   (a) Show that
   $$E(x'Axx'Bxx'Cx) = \text{tr}\{(A \otimes B \otimes C)E(xx' \otimes xx' \otimes xx')\}$$
   (b) Use part (a) and the result of the previous exercise to derive the formula for $E(x'Axx'Bxx'Cx)$ given in Theorem 9.23.

25. Let $x \sim N_m(\mu, \Omega)$, where Ω is positive definite.
   (a) Using Theorem 9.20, show that
   $$\text{var}(x \otimes x) = 2N_m(\Omega \otimes \Omega + \Omega \otimes \mu\mu' + \mu\mu' \otimes \Omega)$$
   (b) Show that the matrix $(\Omega \otimes \Omega + \Omega \otimes \mu\mu' + \mu\mu' \otimes \Omega)$ is nonsingular.
   (c) Determine the eigenvalues of N_m. Use these along with part (b) to show that $\text{rank}\{\text{var}(x \otimes x)\} = m(m+1)/2$.

26. Suppose that the $m \times 1$ vector x and the $n \times 1$ vector y are independently distributed with $E(x) = \mu_1$, $E(y) = \mu_2$, $E(xx') = V_1$, and $E(yy') = V_2$. Show that
   (a) $E(xy' \otimes xy') = \text{vec}(V_1)\{\text{vec}(V_2)\}'$,
   (b) $E(xy' \otimes yx') = (V_1 \otimes V_2)K_{mn} = K_{mn}(V_2 \otimes V_1)$,
   (c) $E(x \otimes x \otimes y \otimes y) = \text{vec}(V_1) \otimes \text{vec}(V_2)$,
   (d) $E(x \otimes y \otimes x \otimes y) = (I_m \otimes K_{nm} \otimes I_n)\{\text{vec}(V_1) \otimes \text{vec}(V_2)\}$,
   (e) $\text{var}(x \otimes y) = V_1 \otimes V_2 - \mu_1\mu_1' \otimes \mu_2\mu_2'$.

27. Let A, B, and C be $m \times m$ symmetric matrices, and let a and b be $m \times 1$ vectors of constants. If $x \sim N_m(\mathbf{0}, \Omega)$, show that
   (a) $E(x'Aax'Bb) = a'A\Omega Bb$,
   (b) $E(x'Aax'Bbx'Cx) = a'A\Omega Bb\,\text{tr}(\Omega C) + 2a'A\Omega C\Omega Bb$.

28. Suppose that $\boldsymbol{x} \sim N_4(\boldsymbol{\mu}, \Omega)$, where $\boldsymbol{\mu} = \mathbf{1}_4$ and $\Omega = 4I_4 + \mathbf{1}_4\mathbf{1}_4'$. Let the random variables t_1 and t_2 be defined by

$$t_1 = (x_1 + x_2 - 2x_3)^2 + (x_3 - x_4)^2,$$
$$t_2 = (x_1 - x_2 - x_3)^2 + (x_1 + x_2 - x_4)^2$$

Use Theorem 9.22 to find
(a) $\text{var}(t_1)$,
(b) $\text{var}(t_2)$,
(c) $\text{cov}(t_1, t_2)$.

29. Verify the formulas given at the end of Example 9.7 for $E(t_1)$ and $\text{var}(t_1)$.

30. Suppose that the $m \times 1$ vectors $\{\boldsymbol{y}_{ij}, 1 \leq i \leq k, 1 \leq j \leq n_i\}$ are independently distributed with $\boldsymbol{y}_{ij} \sim N_m(\boldsymbol{\mu}_i, \Omega)$. A multivariate analysis of variance utilizes the matrices (Example 3.14)

$$B = \sum_{i=1}^{k} n_i(\overline{\boldsymbol{y}}_i - \overline{\boldsymbol{y}})(\overline{\boldsymbol{y}}_i - \overline{\boldsymbol{y}})', \qquad W = \sum_{i=1}^{k} \sum_{j=1}^{n_i} (\boldsymbol{y}_{ij} - \overline{\boldsymbol{y}}_i)(\boldsymbol{y}_{ij} - \overline{\boldsymbol{y}}_i)',$$

where

$$\overline{\boldsymbol{y}}_i = \sum_{j=1}^{n_i} \frac{\boldsymbol{y}_{ij}}{n_i}, \qquad \overline{\boldsymbol{y}} = \sum_{i=1}^{k} \frac{n_i \overline{\boldsymbol{y}}_i}{n}, \qquad n = \sum_{i=1}^{k} n_i$$

Use Theorem 9.24 to show that W and B are independently distributed, $W \sim W_m(\Omega, w)$, and $B \sim W_m(\Omega, b, \Phi)$, where $w = n - k, b = k - 1$, and

$$\Phi = \frac{1}{2} \sum_{i=1}^{k} n_i(\boldsymbol{\mu}_i - \overline{\boldsymbol{\mu}})(\boldsymbol{\mu}_i - \overline{\boldsymbol{\mu}})', \qquad \overline{\boldsymbol{\mu}} = \sum_{i=1}^{k} \frac{n_i \boldsymbol{\mu}_i}{n}$$

31. Let $X' = (\boldsymbol{x}_1, \ldots, \boldsymbol{x}_n)$ be an $m \times n$ matrix, where $\boldsymbol{x}_1, \ldots, \boldsymbol{x}_n$ are independent and $\boldsymbol{x}_i \sim N_m(\mathbf{0}, \Omega)$ for each i. Show that

$$\begin{aligned} E(X \otimes X \otimes X \otimes X) = {} & \{\text{vec}(I_n) \otimes \text{vec}(I_n)\}\{\text{vec}(\Omega) \otimes \text{vec}(\Omega)\}' \\ & + \text{vec}(I_n \otimes I_n)\{\text{vec}(\Omega \otimes \Omega)\}' + \text{vec}(K_{nn}) \\ & \cdot [\text{vec}\{K_{mm}(\Omega \otimes \Omega)\}]' \end{aligned}$$

32. Suppose that the columns of $X' = (\boldsymbol{x}_1, \dots, \boldsymbol{x}_n)$ are independently distributed with $\boldsymbol{x}_i \sim N_m(\boldsymbol{\mu}_i, \Omega)$. Let A be an $m \times m$ symmetric matrix, and let $M' = (\boldsymbol{\mu}_1, \dots, \boldsymbol{\mu}_n)$. Use the spectral decomposition of A to show that
(a) $E(X'AX) = \text{tr}(A)\Omega + M'AM$,
(b) $\text{var}\{\text{vec}(X'AX)\} = 2N_m\{\text{tr}(A^2)(\Omega \otimes \Omega) + \Omega \otimes M'A^2M + M'A^2M \otimes \Omega\}$

33. Use the results of Problems 7.45(e) and 7.52 to show that

$$\left(I_{m^2} - \frac{1}{2} \{(I_m \otimes P) + (P \otimes I_m)\}\Lambda_m \right) N_m = N_m\{I_{m^2} - (I_m \otimes P)\Lambda_m\}$$

thereby verifying the simplified formula for $\text{var}\{\text{vec}(R)\}$ given in (9.17).

References

Agaian, S. S. (1985). *Hadamard Matrices and Their Applications*. Springer-Verlag, Berlin.

Anderson, T. W. (1955). The integral of a symmetric unimodal function over a symmetric convex set and some probability inequalities. *Proceedings of the American Mathematical Society*, **6,** 170–176.

Barndorff-Nielsen, O. E. and Cox, D. R. (1994). *Inference and Asymptotics*. Chapman and Hall, London.

Bartlett, M. S. (1937). Properties of sufficiency and statistical tests. *Proceedings of the Royal Society of London, Ser. A*, **160,** 268–282.

Bartlett, M. S. (1947). Multivariate analysis. *Journal of the Royal Statistical Society Supplement, Ser. B*, **9,** 176–197.

Basilevsky, A. (1983). *Applied Matrix Algebra in the Statistical Sciences*. North-Holland, New York.

Bellman, R. (1970). *Introduction to Matrix Analysis*. McGraw-Hill, New York.

Ben-Israel, A. (1966). A note on an iterative method for generalized inversion of matrices. *Mathematics of Computation*, **20,** 439–440.

Ben-Israel, A. and Greville, T. N. E. (1974). *Generalized Inverses: Theory and Applications*. John Wiley, New York.

Berman, A. and Plemmons, R. J. (1994). *Nonnegative Matrices in the Mathematical Sciences*. Society for Industrial and Applied Mathematics, Philadelphia.

Bhattacharya, R. N. and Waymire, E. C. (1990). *Stochastic Processes with Applications*. John Wiley, New York.

Boullion, T. L. and Odell, P. L. (1971). *Generalized Inverse Matrices*. John Wiley, New York.

Campbell, S. L. and Meyer, C. D. (1979). *Generalized Inverses of Linear Transformations*. Pitman, London.

Casella, G. and Berger, R. L. (1990). *Statistical Inference*. Wadsworth & Brooks/Cole, Pacific Grove, CA.

Cline, R. E. (1964a). Note on the generalized inverse of the product of matrices. *SIAM Review*, **6,** 57–58.

Cline, R. E. (1964b). Representations for the generalized inverse of a partitioned matrix. *SIAM Journal of Applied Mathematics*, **12,** 588–600.

Cline, R. E. (1965). Representations for the generalized inverse of sums of matrices. *SIAM Journal of Numerical Analysis*, **2,** 99–114.

Cochran, W. G. (1934). The distribution of quadratic forms in a normal system with applications to the analysis of variance. *Proceedings of the Cambridge Philosophical Society*, **30,** 178–191.

Davis, P. J. (1979). *Circulant Matrices*. John Wiley, New York.

Duff, I. S., Erisman, A. M., and Reid, J. K. (1986). *Direct Methods for Sparse Matrices*. Oxford University Press.

Elsner, L. (1982). On the variation of the spectra of matrices. *Linear Algebra and Its Applications*, **47,** 127–138.

Eubank, R. L. and Webster, J. T. (1985). The singular-value decomposition as a tool for solving estimability problems. *American Statistician*, **39,** 64–66.

Fan, K. (1949). On a theorem of Weyl concerning eigenvalues of linear transformations, I. *Proceedings of the National Academy of Sciences of the USA*, **35,** 652–655.

Ferguson, T. S. (1967). *Mathematical Statistics: A Decision Theoretic Approach*. Academic Press, New York.

Gantmacher, F. R. (1959). *The Theory of Matrices*, Volumes I and II. Chelsea, New York.

Golub, G. H. and Van Loan, C. F. (1989). *Matrix Computations*. Johns Hopkins University Press, Baltimore.

Graybill, F. A. (1983). *Matrices With Applications in Statistics*, 2nd ed. Wadsworth, Belmont, CA.

Grenander, U. and Szego, G. (1984). *Toeplitz Forms and Their Applications*. Chelsea, New York.

Greville, T. N. E. (1960). Some applications of the pseudoinverse of a matrix. *SIAM Review*, **2,** 15–22.

Greville, T. N. E. (1966). Note on the generalized inverse of a matrix product. *SIAM Review*, **8,** 518–521.

Hageman, L. A. and Young, D. M. (1981). *Applied Iterative Methods*. Academic Press, New York.

Hammarling, S. J. (1970). *Latent Roots and Latent Vectors*. University of Toronto Press.

Healy, M. J. R. (1986). *Matrices for Statistics*. Clarendon Press, Oxford.

Hedayat, A. and Wallis, W. D. (1978). Hadamard matrices and their applications. *Annals of Statistics*, **6,** 1184–1238.

Heinig, G. and Rost, K. (1984). *Algebraic Methods for Toeplitz-like Matrices and Operators*. Birkhäuser, Basel.

Henderson, H. V. and Searle, S. R. (1979). Vec and vech operators for matrices, with some uses in Jacobians and multivariate statistics. *Canadian Journal of Statistics*, **7,** 65–81.

Hinch, E. J. (1991). *Perturbation Methods*. Cambridge University Press.

Horn, R. A. and Johnson, C. R. (1985). *Matrix Analysis*. Cambridge University Press.

Horn, R. A. and Johnson, C. R. (1991). *Topics in Matrix Analysis*. Cambridge University Press.

Hotelling, H. (1933). Analysis of a complex of statistical variables into principal components. *Journal of Educational Psychology*, **24,** 417–441, 498–520.

Huberty, C. J. (1994). *Applied Discriminant Analysis*. John Wiley, New York.

Jackson, J. E. (1991). *A User's Guide to Principal Components*. John Wiley, New York.

Jolliffe, I. T. (1986). *Principal Component Analysis*. Springer-Verlag, New York.

Kato, T. (1982). *A Short Introduction to Perturbation Theory for Linear Operators*. Springer-Verlag, New York.

Kelly, P. J. and Weiss, M. L. (1979). *Geometry and Convexity*. John Wiley, New York.

Khuri, A. (1993). *Advanced Calculus with Applications in Statistics*. John Wiley, New York.

Krzanowski, W. J. (1988). *Principles of Multivariate Analysis: A User's Perspective*. Clarendon Press, Oxford.

Lanczos, C. (1950). An iteration method for the solution of the eigenvalue problem of linear differential and integral operators. *Journal of Research of the National Bureau of Standards*, **45,** 255–282.

Lay, S. R. (1982). *Convex Sets and Their Applications*. John Wiley, New York.

Lindgren, B. W. (1993). *Statistical Theory*, 4th ed. Chapman and Hall, New York.

Magnus, J. R. (1978). The moments of products of quadratic forms in normal variables. *Statistica Neerlandica* **32,** 201–210.

Magnus, J. R. (1988). *Linear Structures*. Charles Griffin, London.

Magnus, J. R. and Neudecker, H. (1979). The commutation matrix: some properties and applications. *Annals of Statistics*, **7,** 381–394.

Magnus, J. R. and Neudecker, H. (1988). *Matrix Differential Calculus with Applications in Statistics and Econometrics*. John Wiley, New York.

Mandel, J. (1982). Use of the singular value decomposition in regression analysis. *American Statistician*, **36,** 15–24.

Mardia, K. V., Kent, J. T., and Bibby, J. M. (1979). *Multivariate Analysis*. Academic Press, New York.

Mathai, A. M. and Provost, S. B. (1992). *Quadratic Forms in Random Variables*. Marcel Dekker, New York.

McCullagh, P. (1987). *Tensor Methods in Statistics*. Chapman and Hall, London.

McLachlan, G. J. (1992). *Discriminant Analysis and Statistical Pattern Recognition*. John Wiley, New York.

Medhi, J. (1994). *Stochastic Processes*. John Wiley, New York.

Miller, R. G., Jr. (1981). *Simultaneous Statistical Inference*, 2nd ed. Springer-Verlag, New York.

Minc, H. (1988). *Nonnegative Matrices*. John Wiley, New York.

Moore, E. H. (1920). On the reciprocal of the general algebraic matrix (Abstract). *Bulletin of the American Mathematical Society*, **26,** 394–395.

Moore, E. H. (1935). General analysis. *Memoirs of the American Philosophical Society*, **1,** 147–209.

Morrison, D. F. (1990). *Multivariate Statistical Methods*. McGraw-Hill, New York.

Muirhead, R. J. (1982). *Aspects of Multivariate Statistical Theory*. John Wiley, New York.

Nayfeh, A. H. (1981). *Introduction to Perturbation Techniques*. John Wiley, New York.

Nel, D. G. (1980). On matrix differentiation in statistics. *South African Statistical Journal*, **14,** 137–193.

Nelder, J. A. (1985). An alternative interpretation of the singular-value decomposition in regression. *American Statistician*, **39,** 63–64.

Neter, J, Wasserman, W., and Kutner, M. H. (1985). *Applied Linear Statistical Models: Regression, Analysis of Variance, and Experimental Design*. Irwin, Homewood, IL.

Olkin, I. and Tomsky, J. L. (1981). A new class of multivariate tests based on the union–intersection principle. *Annals of Statistics*, **9,** 792–802.

Ostrowski, A. M. (1973). *Solution of Equations in Euclidean and Banach Spaces*. Academic Press, New York.

Penrose, R. (1955). A generalized inverse for matrices. *Proceedings of the Cambridge Philosophical Society*, **51,** 406–413.

Penrose, R. (1956). On best approximate solutions of linear matrix equations. *Proceedings of the Cambridge Philosophical Society*, **52,** 17–19.

Poincaré, H. (1890). Sur les équations aux dérivées partielles de la physique mathématique. *American Journal of Mathematics*, **12,** 211–294.

Press, W. H., Flannery, B. P., Teukolsky, S. A., and Vetterline, W. T. (1992). *Numerical Recipes in FORTRAN: The Art of Scientific Computing*. Cambridge University Press.

Pringle, R. M. and Rayner, A. A. (1971). *Generalized Inverse Matrices with Applications to Statistics*. Charles Griffin, London.

Rao, C. R. (1973). *Linear Statistical Inference and Its Applications*, 2nd ed. John Wiley, New York.

Rao, C. R. and Mitra, S. K. (1971). *Generalized Inverse of Matrices and Its Applications*, John Wiley, New York.

Rockafellar, R. T. (1970). *Convex Analysis*. Princeton University Press.

Scheffé, H. (1953). A method for judging all contrasts in the analysis of variance. *Biometrika*, **40,** 87–104.

Schott, J. R. (1984). Optimal bounds for the distribution of some test criteria for tests of dimensionality. *Biometrika*, **71,** 561–567.

Searle, S. R. (1971). *Linear Models*. John Wiley, New York.

Searle, S. R. (1982). *Matrix Algebra Useful for Statistics*. John Wiley, New York.

Sen, A. K. and Srivastava, M. S. (1990). *Regression Analysis: Theory, Methods, and Applications*. Springer-Verlag, New York.

Seneta, E. (1973). *Non-negative Matrices: An Introduction to Theory and Applications*. John Wiley, New York.

Srivastava, M. S. and Khatri, C. G. (1979). *An Introduction to Multivariate Analysis*. North-Holland, New York.

Styan, G. P. H. (1973). Hadamard products and multivariate statistical analysis. *Linear Algebra and Its Applications*, **6,** 217–240.

Sugiura, N. (1976). Asymptotic expansions of the distributions of the latent roots and the latent vector of the Wishart and multivariate F matrices. *Journal of Multivariate Analysis*, **6,** 500–525.

Taylor, H. M. and Karlin, S. (1984). *An Introduction to Stochastic Modeling*. Academic Press, Orlando.

Young, D. M. (1971). *Iterative Solution of Large Linear Systems*. Academic Press, New York.

Index

WILEY SERIES IN PROBABILITY AND STATISTICS

ESTABLISHED BY WALTER A. SHEWHART AND SAMUEL S. WILKS

Editors

Vic Barnett, Ralph A. Bradley, Nicholas I. Fisher, J. Stuart Hunter, J. B. Kadane, David G. Kendall, David W. Scott, Adrian F. M. Smith, Jozef L. Teugels, Geoffrey S. Watson

Probability and Statistics

ANDERSON · An Introduction to Multivariate Statistical Analysis, *Second Edition*
*ANDERSON · The Statistical Analysis of Time Series
ARNOLD, BALAKRISHNAN, and NAGARAJA · A First Course in Order Statistics
BACCELLI, COHEN, OLSDER, and QUADRAT · Synchronization and Linearity: An Algebra for Discrete Event Systems
BARTOSZYNSKI and NIEWIADOMSKA-BUGAJ · Probability and Statistical Inference
BERNARDO and SMITH · Bayesian Statistical Concepts and Theory
BHATTACHARYYA and JOHNSON · Statistical Concepts and Methods
BILLINGSLEY · Convergence of Probability Measures
BILLINGSLEY · Probability and Measure, *Second Edition*
BOROVKOV · Asymptotic Methods in Queuing Theory
BRANDT, FRANKEN, and LISEK · Stationary Stochastic Models
CAINES · Linear Stochastic Systems
CAIROLI and DALANG · Sequential Stochastic Optimization
CHEN · Recursive Estimation and Control for Stochastic Systems
CONSTANTINE · Combinatorial Theory and Statistical Design
COOK and WEISBERG · An Introduction to Regression Graphics
COVER and THOMAS · Elements of Information Theory
CSÖRGŐ and HORVÁTH · Weighted Approximations in Probability Statistics
*DOOB · Stochastic Processes
DUDEWICZ and MISHRA · Modern Mathematical Statistics
DUPUIS · A Weak Convergence Approach to the Theory of Large Deviations
ETHIER and KURTZ · Markov Processes: Characterization and Convergence
FELLER · An Introduction to Probability Theory and Its Applications, Volume 1, *Third Edition,* Revised; Volume II, *Second Edition*
FREEMAN and SMITH · Aspects of Uncertainty: A Tribute to D. V. Lindley
FULLER · Introduction to Statistical Time Series, *Second Edition*
FULLER · Measurement Error Models
GHOSH · Sequential Estimation
GIFI · Nonlinear Multivariate Analysis
GUTTORP · Statistical Inference for Branching Processes
HALD · A History of Probability and Statistics and Their Applications before 1750
HALL · Introduction to the Theory of Coverage Processes
HANNAN and DEISTLER · The Statistical Theory of Linear Systems
HEDAYAT and SINHA · Design and Inference in Finite Population Sampling
HOEL · Introduction to Mathematical Statistics, *Fifth Edition*
HUBER · Robust Statistics
IMAN and CONOVER · A Modern Approach to Statistics
JUREK and MASON · Operator-Limit Distributions in Probability Theory
KAUFMAN and ROUSSEEUW · Finding Groups in Data: An Introduction to Cluster Analysis
KOTZ · Leading Personalities in Statistical Sciences from the Seventeenth Century to the Present
LAMPERTI · Probability: A Survey of the Mathematical Theory, *Second Edition*

*Now available in a lower priced paperback edition in the Wiley Classics Library.

Probability and Statistics (Continued)

LARSON · Introduction to Probability Theory and Statistical Inference, *Third Edition*
LESSLER and KALSBEEK · Nonsampling Error in Surveys
LINDVALL · Lectures on the Coupling Method
MANTON, WOODBURY, and TOLLEY · Statistical Applications Using Fuzzy Sets
MARDIA · The Art of Statistical Science: A Tribute to G. S. Watson
MORGENTHALER and TUKEY · Configural Polysampling: A Route to Practical Robustness
MUIRHEAD · Aspects of Multivariate Statistical Theory
OLIVER and SMITH · Influence Diagrams, Belief Nets and Decision Analysis
*PARZEN · Modern Probability Theory and Its Applications
PRESS · Bayesian Statistics: Principles, Models, and Applications
PUKELSHEIM · Optimal Experimental Design
PURI and SEN · Nonparametric Methods in General Linear Models
PURI, VILAPLANA, and WERTZ · New Perspectives in Theoretical and Applied Statistics
RAO · Asymptotic Theory of Statistical Inference
RAO · Linear Statistical Inference and Its Applications, *Second Edition*
*RAO and SHANBHAG · Choquet-Deny Type Functional Equations with Applications to Stochastic Models
RENCHER · Methods of Multivariate Analysis
ROBERTSON, WRIGHT, and DYKSTRA · Order Restricted Statistical Inference
ROGERS and WILLIAMS · Diffusions, Markov Processes, and Martingales, Volume I: Foundations, *Second Edition;* Volume II: Îto Calculus
ROHATGI · An Introduction to Probability Theory and Mathematical Statistics
ROSS · Stochastic Processes
RUBINSTEIN · Simulation and the Monte Carlo Method
RUBINSTEIN and SHAPIRO · Discrete Event Systems: Sensitivity Analysis and Stochastic Optimization by the Score Function Method
RUZSA and SZEKELY · Algebraic Probability Theory
SCHEFFE · The Analysis of Variance
SEBER · Linear Regression Analysis
SEBER · Multivariate Observations
SEBER and WILD · Nonlinear Regression
SERFLING · Approximation Theorems of Mathematical Statistics
SHORACK and WELLNER · Empirical Processes with Applications to Statistics
SMALL and McLEISH · Hilbert Space Methods in Probability and Statistical Inference
STAPLETON · Linear Statistical Models
STAUDTE and SHEATHER · Robust Estimation and Testing
STOYANOV · Counterexamples in Probability
STYAN · The Collected Papers of T. W. Anderson: 1943–1985
TANAKA · Time Series Analysis: Nonstationary and Noninvertible Distribution Theory
THOMPSON and SEBER · Adaptive Sampling
WELSH · Aspects of Statistical Inference
WHITTAKER · Graphical Models in Applied Multivariate Statistics
YANG · The Construction Theory of Denumerable Markov Processes

Applied Probability and Statistics

ABRAHAM and LEDOLTER · Statistical Methods for Forecasting
AGRESTI · Analysis of Ordinal Categorical Data
AGRESTI · Categorical Data Analysis
AGRESTI · An Introduction to Categorical Data Analysis
ANDERSON and LOYNES · The Teaching of Practical Statistics

*Now available in a lower priced paperback edition in the Wiley Classics Library.

Applied Probability and Statistics (Continued)

ANDERSON, AUQUIER, HAUCK, OAKES, VANDAELE, and WEISBERG · Statistical Methods for Comparative Studies
ARMITAGE and DAVID (editors) · Advances in Biometry
*ARTHANARI and DODGE · Mathematical Programming in Statistics
ASMUSSEN · Applied Probability and Queues
*BAILEY · The Elements of Stochastic Processes with Applications to the Natural Sciences
BARNETT and LEWIS · Outliers in Statistical Data, *Second Edition*
BARTHOLOMEW, FORBES, and McLEAN · Statistical Techniques for Manpower Planning, *Second Edition*
BATES and WATTS · Nonlinear Regression Analysis and Its Applications
BECHHOFER, SANTNER, and GOLDSMAN · Design and Analysis of Experiments for Statistical Selection, Screening, and Multiple Comparisons
BELSLEY · Conditioning Diagnostics: Collinearity and Weak Data in Regression
BELSLEY, KUH, and WELSCH · Regression Diagnostics: Identifying Influential Data and Sources of Collinearity
BERRY · Bayesian Analysis in Statistics and Econometrics: Essays in Honor of Arnold Zellner
BERRY, CHALONER, and GEWEKE · Bayesian Analysis in Statistics and Econometrics: Essays in Honor of Arnold Zellner
BHAT · Elements of Applied Stochastic Processes, *Second Edition*
BHATTACHARYA and WAYMIRE · Stochastic Processes with Applications
BIEMER, GROVES, LYBERG, MATHIOWETZ, and SUDMAN · Measurement Errors in Surveys
BIRKES and DODGE · Alternative Methods of Regression
BLOOMFIELD · Fourier Analysis of Time Series: An Introduction
BOLLEN · Structural Equations with Latent Variables
BOULEAU · Numerical Methods for Stochastic Processes
BOX · R. A. Fisher, the Life of a Scientist
BOX and DRAPER · Empirical Model-Building and Response Surfaces
BOX and DRAPER · Evolutionary Operation: A Statistical Method for Process Improvement
BOX, HUNTER, and HUNTER · Statistics for Experimenters: An Introduction to Design, Data Analysis, and Model Building
BROWN and HOLLANDER · Statistics: A Biomedical Introduction
BUCKLEW · Large Deviation Techniques in Decision, Simulation, and Estimation
BUNKE and BUNKE · Nonlinear Regression, Functional Relations and Robust Methods: Statistical Methods of Model Building
CHATTERJEE and HADI · Sensitivity Analysis in Linear Regression
CHATTERJEE and PRICE · Regression Analysis by Example, *Second Edition*
CLARKE and DISNEY · Probability and Random Processes: A First Course with Applications, *Second Edition*
COCHRAN · Sampling Techniques, *Third Edition*
*COCHRAN and COX · Experimental Designs, *Second Edition*
CONOVER · Practical Nonparametric Statistics, *Second Edition*
CONOVER and IMAN · Introduction to Modern Business Statistics
CORNELL · Experiments with Mixtures, Designs, Models, and the Analysis of Mixture Data, *Second Edition*
COX · A Handbook of Introductory Statistical Methods
*COX · Planning of Experiments
COX, BINDER, CHINNAPPA, CHRISTIANSON, COLLEDGE, and KOTT · Business Survey Methods
CRESSIE · Statistics for Spatial Data, *Revised Edition*
DANIEL · Applications of Statistics to Industrial Experimentation

*Now available in a lower priced paperback edition in the Wiley Classics Library.

Applied Probability and Statistics (Continued)

DANIEL · Biostatistics: A Foundation for Analysis in the Health Sciences, *Sixth Edition*
DAVID · Order Statistics, *Second Edition*
*DEGROOT, FIENBERG, and KADANE · Statistics and the Law
*DEMING · Sample Design in Business Research
DILLON and GOLDSTEIN · Multivariate Analysis: Methods and Applications
DODGE and ROMIG · Sampling Inspection Tables, *Second Edition*
DOWDY and WEARDEN · Statistics for Research, *Second Edition*
DRAPER and SMITH · Applied Regression Analysis, *Second Edition*
DUNN · Basic Statistics: A Primer for the Biomedical Sciences, *Second Edition*
DUNN and CLARK · Applied Statistics: Analysis of Variance and Regression, *Second Edition*
ELANDT-JOHNSON and JOHNSON · Survival Models and Data Analysis
EVANS, PEACOCK, and HASTINGS · Statistical Distributions, *Second Edition*
FISHER and VAN BELLE · Biostatistics: A Methodology for the Health Sciences
FLEISS · The Design and Analysis of Clinical Experiments
FLEISS · Statistical Methods for Rates and Proportions, *Second Edition*
FLEMING and HARRINGTON · Counting Processes and Survival Analysis
FLURY · Common Principal Components and Related Multivariate Models
GALLANT · Nonlinear Statistical Models
GLASSERMAN and YAO · Monotone Structure in Discrete-Event Systems
GNANADESIKAN · Methods for Statistical Data Analysis of Multivariate Observations, *Second Edition*
GREENWOOD and NIKULIN · A Guide to Chi-Squared Testing
GROSS and HARRIS · Fundamentals of Queueing Theory, *Second Edition*
GROVES · Survey Errors and Survey Costs
GROVES, BIEMER, LYBERG, MASSEY, NICHOLLS, and WAKSBERG · Telephone Survey Methodology
HAHN and MEEKER · Statistical Intervals: A Guide for Practitioners
HAND · Discrimination and Classification
*HANSEN, HURWITZ, and MADOW · Sample Survey Methods and Theory, Volume I: Methods and Applications
*HANSEN, HURWITZ, and MADOW · Sample Survey Methods and Theory, Volume II: Theory
HEIBERGER · Computation for the Analysis of Designed Experiments
HELLER · MACSYMA for Statisticians
HINKELMAN and KEMPTHORNE: · Design and Analysis of Experiments, Volume 1: Introduction to Experimental Design
HOAGLIN, MOSTELLER, and TUKEY · Exploratory Approach to Analysis of Variance
HOAGLIN, MOSTELLER, and TUKEY · Exploring Data Tables, Trends and Shapes
HOAGLIN, MOSTELLER, and TUKEY · Understanding Robust and Exploratory Data Analysis
HOCHBERG and TAMHANE · Multiple Comparison Procedures
HOCKING · Methods and Applications of Linear Models: Regression and the Analysis of Variables
HOEL · Elementary Statistics, *Fifth Edition*
HOGG and KLUGMAN · Loss Distributions
HOLLANDER and WOLFE · Nonparametric Statistical Methods
HOSMER and LEMESHOW · Applied Logistic Regression
HØYLAND and RAUSAND · System Reliability Theory: Models and Statistical Methods
HUBERTY · Applied Discriminant Analysis
IMAN and CONOVER · Modern Business Statistics
JACKSON · A User's Guide to Principle Components
JOHN · Statistical Methods in Engineering and Quality Assurance

*Now available in a lower priced paperback edition in the Wiley Classics Library.

Applied Probability and Statistics (Continued)

JOHNSON · Multivariate Statistical Simulation

JOHNSON and KOTZ · Distributions in Statistics
Continuous Univariate Distributions—2
Continuous Multivariate Distributions

JOHNSON, KOTZ, and BALAKRISHNAN · Continuous Univariate Distributions, Volume 1, *Second Edition*

JOHNSON, KOTZ, and BALAKRISHNAN · Discrete Multivariate Distributions

JOHNSON, KOTZ, and KEMP · Univariate Discrete Distributions, *Second Edition*

JUDGE, GRIFFITHS, HILL, LÜTKEPOHL, and LEE · The Theory and Practice of Econometrics, *Second Edition*

JUDGE, HILL, GRIFFITHS, LÜTKEPOHL, and LEE · Introduction to the Theory and Practice of Econometrics, *Second Edition*

JUREČKOVÁ and SEN · Robust Statistical Procedures: Aymptotics and Interrelations

KADANE · Bayesian Methods and Ethics in a Clinical Trial Design

KADANE AND SCHUM · A Probabilistic Analysis of the Sacco and Vanzetti Evidence

KALBFLEISCH and PRENTICE · The Statistical Analysis of Failure Time Data

KASPRZYK, DUNCAN, KALTON, and SINGH · Panel Surveys

KISH · Statistical Design for Research

*KISH · Survey Sampling

LAD · Operational Subjective Statistical Methods: A Mathematical, Philosophical, and Historical Introduction

LANGE, RYAN, BILLARD, BRILLINGER, CONQUEST, and GREENHOUSE · Case Studies in Biometry

LAWLESS · Statistical Models and Methods for Lifetime Data

LEBART, MORINEAU., and WARWICK · Multivariate Descriptive Statistical Analysis: Correspondence Analysis and Related Techniques for Large Matrices

LEE · Statistical Methods for Survival Data Analysis, *Second Edition*

LePAGE and BILLARD · Exploring the Limits of Bootstrap

LEVY and LEMESHOW · Sampling of Populations: Methods and Applications

LINHART and ZUCCHINI · Model Selection

LITTLE and RUBIN · Statistical Analysis with Missing Data

LYBERG · Survey Measurement

MAGNUS and NEUDECKER · Matrix Differential Calculus with Applications in Statistics and Econometrics

MAINDONALD · Statistical Computation

MALLOWS · Design, Data, and Analysis by Some Friends of Cuthbert Daniel

MANN, SCHAFER, and SINGPURWALLA · Methods for Statistical Analysis of Reliability and Life Data

MASON, GUNST, and HESS · Statistical Design and Analysis of Experiments with Applications to Engineering and Science

McLACHLAN and KRISHNAN · The EM Algorithm and Extensions

McLACHLAN · Discriminant Analysis and Statistical Pattern Recognition

McNEIL · Epidemiological Research Methods

MILLER · Survival Analysis

MONTGOMERY and MYERS · Response Surface Methodology: Process and Product in Optimization Using Designed Experiments

MONTGOMERY and PECK · Introduction to Linear Regression Analysis, *Second Edition*

NELSON · Accelerated Testing, Statistical Models, Test Plans, and Data Analyses

NELSON · Applied Life Data Analysis

OCHI · Applied Probability and Stochastic Processes in Engineering and Physical Sciences

OKABE, BOOTS, and SUGIHARA · Spatial Tesselations: Concepts and Applications of Voronoi Diagrams

*Now available in a lower priced paperback edition in the Wiley Classics Library.

Applied Probability and Statistics (Continued)

OSBORNE · Finite Algorithms in Optimization and Data Analysis
PANKRATZ · Forecasting with Dynamic Regression Models
PANKRATZ · Forecasting with Univariate Box-Jenkins Models: Concepts and Cases
PORT · Theoretical Probability for Applications
PUTERMAN · Markov Decision Processes: Discrete Stochastic Dynamic Programming
RACHEV · Probability Metrics and the Stability of Stochastic Models
RÉNYI · A Diary on Information Theory
RIPLEY · Spatial Statistics
RIPLEY · Stochastic Simulation
ROSS · Introduction to Probability and Statistics for Engineers and Scientists
ROUSSEEUW and LEROY · Robust Regression and Outlier Detection
RUBIN · Multiple Imputation for Nonresponse in Surveys
RYAN · Modern Regression Methods
RYAN · Statistical Methods for Quality Improvement
SCHOTT · Matrix Analysis for Statistics
SCHUSS · Theory and Applications of Stochastic Differential Equations
SCOTT · Multivariate Density Estimation: Theory, Practice, and Visualization
SEARLE · Linear Models
SEARLE · Linear Models for Unbalanced Data
SEARLE · Matrix Algebra Useful for Statistics
SEARLE, CASELLA, and McCULLOCH · Variance Components
SKINNER, HOLT, and SMITH · Analysis of Complex Surveys
STOYAN, KENDALL, and MECKE · Stochastic Geometry and Its Applications, *Second Edition*
STOYAN and STOYAN · Fractals, Random Shapes and Point Fields: Methods of Geometrical Statistics
THOMPSON · Empirical Model Building
THOMPSON · Sampling
TIERNEY · LISP-STAT: An Object-Oriented Environment for Statistical Computing and Dynamic Graphics
TIJMS · Stochastic Modeling and Analysis: A Computational Approach
TITTERINGTON, SMITH, and MAKOV · Statistical Analysis of Finite Mixture Distributions
UPTON and FINGLETON · Spatial Data Analysis by Example, Volume 1: Point Pattern and Quantitative Data
UPTON and FINGLETON · Spatial Data Analysis by Example, Volume II: Categorical and Directional Data
VAN RIJCKEVORSEL and DE LEEUW · Component and Correspondence Analysis
WEISBERG · Applied Linear Regression, *Second Edition*
WESTFALL and YOUNG · Resampling-Based Multiple Testing: Examples and Methods for p-Value Adjustment
WHITTLE · Optimization Over Time: Dynamic Programming and Stochastic Control, Volume I and Volume II
WHITTLE · Systems in Stochastic Equilibrium
WONNACOTT and WONNACOTT · Econometrics, *Second Edition*
WONNACOTT and WONNACOTT · Introductory Statistics, *Fifth Edition*
WONNACOTT and WONNACOTT · Introductory Statistics for Business and Economics, *Fourth Edition*
WOODING · Planning Pharmaceutical Clinical Trials: Basic Statistical Principles
WOOLSON · Statistical Methods for the Analysis of Biomedical Data
*ZELLNER · An Introduction to Bayesian Inference in Econometrics

Tracts on Probability and Statistics

BILLINGSLEY · Convergence of Probability Measures
KELLY · Reversability and Stochastic Networks
TOUTENBURG · Prior Information in Linear Models